Power Electronics

Power Electronics

VOLUME 2 :

DRIVE TECHNOLOGY and MOTION CONTROL

JEAN POLLEFLIET

ELSEVIER

ACADEMIC PRESS
An imprint of Elsevier

Academic Press is an imprint of Elsevier
125 London Wall, London EC2Y 5AS, United Kingdom
525 B Street, Suite 1800, San Diego, CA 92101-4495, United States
50 Hampshire Street, 5th Floor, Cambridge, MA 02139, United States
The Boulevard, Langford Lane, Kidlington, Oxford OX5 1GB, United Kingdom

Notices

Knowledge and best practice in this field are constantly changing. As new research and experience broaden our understanding, changes in research methods, professional practices, or medical treatment may become necessary.

Practitioners and researchers must always rely on their own experience and knowledge in evaluating and using any information, methods, compounds, or experiments described herein. In using such information or methods they should be mindful of their own safety and the safety of others, including parties for whom they have a professional responsibility.

To the fullest extent of the law, neither the Publisher nor the authors, contributors, or editors, assume any liability for any injury and/or damage to persons or property as a matter of products liability, negligence or otherwise, or from any use or operation of any methods, products, instructions, or ideas contained in the material herein.

British Library Cataloguing in Publication Data
A catalogue record for this book is available from the British Library

Library of Congress Cataloging-in-Publication Data
A catalog record for this book is available from the Library of Congress

ISBN: 978-0-12-814641-5

For information on all Academic Press Publications
visit our website at https://www.elsevier.com/

Publisher: Katie Hammon
Acquisition Editor: Lisa Reading
Editorial Project Manager: Natasha Welford
Production Project Manager: Mohana Priyan Rajendran
Designer: Jean Polleflfiet (lay-out, illustrations, cover) and Mark Rogers

Typeset by SPi Global, India

To my wife Gilberte

PREFACE

Our Dutch textbook “Electronic Power Control” first appeared in 1986 and has reached the eighth edition. The present book “Power Electronics” is the translated version of this eighth edition.

Every edition saw continuous updating rearranging as well as addition of material and chapters. At the same time attention was also paid to the didactic aspects. This is not just important for students but also for the large group of people who use the book for self study.

New in the eighth edition was a brief study of standing waves in transmission lines, of importance for a longue line between frequency converter and three phase motor. Also new was an introduction to the principles of 3-level inverters and the study of the synchronous reluctance motor.

In this edition we continue to use the tradition of white and green pages. The green pages contain the mathematical derivations which in the first case are not necessary for studying the electronics. Once a sufficiently high level and the desire for specialist knowledge the reader can choose to make use of the green pages without disturbing the continuity of the study.

To mention a few numerical details, this book contains more than seven hundred figures, a hundred photos and more than fifty fully worked problems.

The purpose of the book is to explain the principles and applications of power electronics. Electronic switches and converters are studied in volume 1 and drive technology and motion control are dealt with in volume 2.

The largest part of this book is distilled from more than 40 years of lessons, talks and projects. The most important source of information is my students, especially the few hundred of whom I was the mentor I guided during their thesis for Master of Applied Engineering Sciences.

These I quided in wich I remain thankful and indebted to them.

To Lisa Reading and Nathasa Welford of Elsevier I wish to express my thanks for the pleasant cooperation.

I would also like to thank Prof. dr. ir. Bernard Baeyens of the Ibague University (Colombia) for correcting and improving the Spanish technical vocabulary.

Last but not least, we have to thank a number of manufacturers for their support in providing photos and data of their products.

In conclusion we wish the readers of this book a fruitful study.

Oostende, Belgium, September 2017

Jean.Pollefliet@telenet.be

With thanks for the cover photos:

TABLE OF CONTENTS

VOLUME 1: SWITCHES and CONVERTERS

- **Semiconductor switches:** diodes, transistors, thyristors
- **Electronic power converters:** DC and AC controllers , choppers, SMPS, inverters
- **Applications of power electronics**
- **Computer simulations**

PRINCIPAL SYMBOLS

α	*transistor current gain*
α	*firing angle thyristor* (rad)
β	*conduction angle thyristor* (rad)
B	*magnetic flux density* (T = Wb/m²)
AC	*alternating current*
DC	*direct current*
δ	*duty ratio (%)*
e	*instantaneous e.m.f.* (V)
E	*RMS-value elektromotive force (e.m.f.)* (V) / *DC-e.m.f.* (V)
E	*electric field intensity* (V/m)
E_{on} ,E_{off}	*energy dissipation during transistor switching "on" and "off" respectively* (J)
f	*frequency* (Hz)
Φ	*flux per pole DC-machine / rotating air gap flux induction motor* (Wb)
Φ_{S1}	*flux one stator winding of an induction motor* (Wb)
g_{fs}	*transconductance* (Siemens / mho)
g_m	*transconductance coefficient* (Siemens / mho)
H	*magnetic field intensity* (A/m)
h_{FE}	*current gain common emitter connection*
i / $\hat{\imath}$	*instantaneous current* (A) / *peak value of sinusoidal current* (A)
i_0 / I_0	*output current of a circuit* (A)
i_μ	*magnetizing current* (A)
$\hat{\imath}_\mu$	*peak value of magnetizing current* (A)
I_{AV}	*average value of a semiconductor current* (A)
I_{RMS} / I	*r.m.s. value of current* (A) / *DC-current* (A)
J	*(polar) moment of inertia* (kgm²)
L_b	*load self inductance*
L_o	*magnetizing inductance (transformer / induction motor)* (H)
$\mathscr{L}$	*Laplace transform*
μ_0	*permeability of free space* ($4.\pi.10^{-7}$ H/m)
μ_r	*relative permeability*
M	*momentum (of torque)* (Nm)
M_{em}	*electromagnetic momentum(of torque)* (Nm)
M_J	*accelerating or decelerating momentum (of torque) due to inertia* (Nm)
$M_{max} = M_{po}$	*peak value momentum (of torque) induction motor* (Nm)
M_t	*total momentum of load torque (mechanical load* M_L *+ static friction torque* M_F *+ windage torques* M_W *...)* (Nm)

R_μ	*reluctance* (A/Wb)
F_μ	*magnetomotive force (m.m.f.)* (Aw)
N_{Se}	*equivalent sinusoidal (stator) winding induction motor*
n	*motor speed* (r.p.m. or rad/s)
n_S	*synchronous speed (rotating stator field) induction motor* (r.p.m.)
n_R	*speed rotating rotor field induction motor* (r.p.m.)
η	*efficiency of operation (%)*
P	*DC-power (W) / average power* (W)
P_e	*eddy current loss density* (W/m³)
P_h	*hysteresis loss density* (W/m³)
p	*number of pole pairs DC-machine*
p	*number of pole pairs stator winding induction motor*
$\sigma_R.L_0$	*leakage inductance rotor induction motor*
$\sigma_S.L_0$	*leakage inductance stator induction motor*
R_b	*load resistance*
s	*Laplace operator*
T	*time period* (s)
T	*temperature* (°C ; K)
t_{on}	*time to switch on a power semiconductor (switch)* (µs; ns)
t_{off}	*time to switch off a power semiconductor (switch)* (µs; ns)
t_{ON}	*time that the power semiconductor is conducting (ON-state)* (µs ; ms)
t_{OFF}	*time that the power semiconductor is blocking (OFF-state)* (µs ; ms)
t_d	*delay time (to switch a transistor on)* (µs ; ns)
t_f	*fall time during switching off transistor* (µs ; ns)
t_r	*rise time during switching on transistor* (µs ; ns)
t_s	*storage time (to build off the space charge in a BJT)* (µs ; ns)
τ	*time constant* (s)
v	*instantaneous voltage* (V)
v_0 / V_0	*output voltage of a circuit* (V)
v_s / V_s	*supply voltage* (V)
$\hat{v}$	*peak value sinusoidal voltage* (V)
V	*voltage (DC, average, ...)* (V)
V_L / V_F	*line voltage / phase voltage in a three-phase system* (V)
V_{RMS}	*root mean square voltage* (V)
V_{di}	*(dc-) average voltage for ideal rectifier* (V)
$V_{di\alpha}$	*(dc-) average voltage for ideal controlled rectifier with firing angle α* (V)
v	*speed* (m/s)
W	*energy* (J)
ω	*angular frequency* (rad/s)

16 ELECTRIC MACHINES

16.1 TRANSFORMERS

CONTENTS

1.1 Transformer at no load

The simplest form of a single phase static transformer consists of a ferromagnetic circuit of Si-steel plates upon which two separate windings have been placed (fig. 16-1). The primary coil p has N_p windings and the secondary s has N_s windings. As long as no load is connected to the secondary we refer to the unloaded transformer or transformer at no-load. We now connect the voltage $v = \hat{v}_p \, . \, sin \, \omega t$ to the primary. A primary sinusoidal current flows, resulting in a sinusoidal flux in the core. The self inductance with respect to this flux Φ_0 is L_0. Primary current: $\approx \frac{V_p}{\sqrt{R_p^2 + \omega^2 \, . \, L_0^2}}$

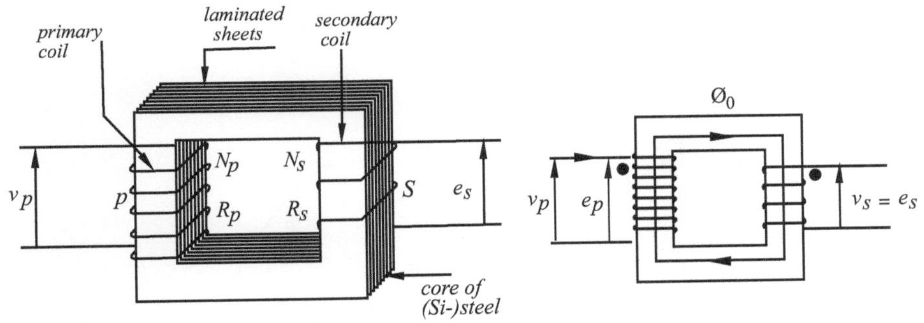

Fig. 16-1: Single phase transformer, no load

Power Electronics: DRIVE TECHNOLOGY and MOTION CONTROL
http://dx.doi.org/10.1016/B978-0-12-814641-5.50001-9,

The primary current cannot yet be exactly determined since we first need to take the leakage reactance in the transformer into account. If we neglect the resistance R_p of the coil, then $I_\mu \approx \frac{V_p}{\omega . L_0}$ = magnetizing current. This wattless current lags 90° behind V_p (fig. 16-2) and creates the flux $\Phi_0 = \frac{N_p . I_\mu}{R_\mu}$. Here $R_\mu = \frac{l}{\mu_0 \mu_r A}$ is the reluctance of the magnetic circuit, with l being the average length of the field lines and A being the cross-sectional area of the core. I_μ is calculated later when we take R_p and the leakage flux into account.

Flux Φ_0 →

- is in phase with I_μ and $\pi/2$ behind $v_p = \hat{v}_p . \sin \omega.t \rightarrow \Phi_0 = \hat{\Phi}_0 . \sin(\omega t - \pi/2)$
- produces iron losses (see number 1.7.2): $P_{Fe} \approx V_p . I_v$; I_v is in phase with V_p
- $\vec{I}_\mu + \vec{I}_v = \vec{I}_n$ = no-load current, lags almost $\pi/2$ behind V_p
- induces an emf $e = N . \frac{d\Phi_0}{dt}$ in each coil

Primary:

- $e_p = N_p . \frac{\hat{\Phi}_0 . d \sin(\omega.t - \frac{\pi}{2})}{dt} = N_p . \omega . \hat{\Phi}_0 . \sin \omega.t$
- effective value: $$E_p = \frac{N_p . \omega . \hat{\Phi}_0}{\sqrt{2}} \quad (16\text{-}1)$$
- E_p is in phase with V_p and is the self induced emf of the primary.

Secondary

- $e_s = N_s \frac{\hat{\Phi}_0 . d(\sin(\omega.t - \frac{\pi}{2}))}{dt} = N_s . \omega . \hat{\Phi}_0 . \sin \omega.t$
- effective value: $$E_s = \frac{N_S . \omega . \hat{\Phi}_0}{\sqrt{2}} \quad (16\text{-}2)$$

A secondary emf is produced with the same frequency as the applied primary voltage. The primary emf E_p is eliminated by the applied voltage V_p . If we ignore the voltage losses: $E_p = V_p$ = the applied primary voltage.

The magnetizing current can be written as: $$I_\mu = \frac{E_p}{\omega.L_0} \quad (16\text{-}3)$$

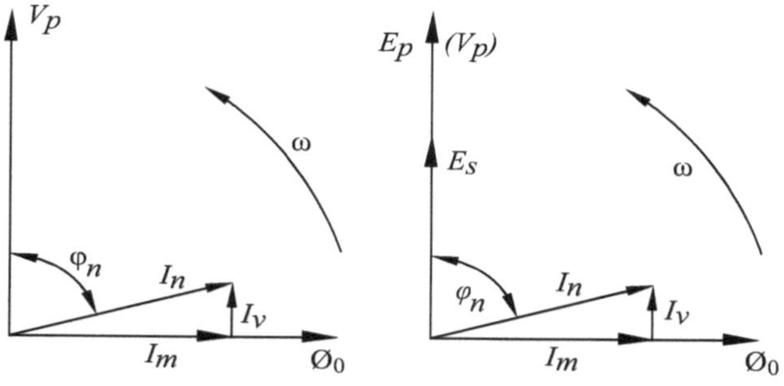

Fig. 16-2: Vector diagram of no-load transformer

From (16-1) and (16-2) follows: $\frac{E_p}{E_s}$ = transformer ratio = $\frac{N_p}{N_s}$

At no-load $E_p \approx V_p$ and $E_s = V_s$ so that $\frac{V_p}{V_s} = \frac{N_p}{N_s} = k$ = turns ratio (16-4)

In addition: $E_p = \frac{N_p \,.\, \omega \,.\, \hat{\Phi}_0}{\sqrt{2}} = N_p \,.\, \frac{2 \,.\, \pi \,.\, f}{\sqrt{2}} \,.\, \hat{\Phi}_0 = 4.44 \,.\, N_p \,.\, f \,.\, \hat{\Phi}_0$ (16-5)

By neglecting the losses, the flux is directly proportional to the primary voltage (assuming that the frequency is constant).

1.2 Transformer with load

1.2.1 Secondary and primary currents

When a load is connected to the secondary, then a current I_s flows. The power consumed by the secondary load is drawn from the net by the primary, which means that the current I_p is larger than the no-load current I_n.

With V_p constant, $E_p \approx V_p$ = constant. From (16-3) and (16-5), it follows that I_μ and Φ_0 are practically unchanged. Constant flux means unchanged iron losses (I_v) so that the current I_n also does not change. In other words, the secondary current I_s and the primary current I_p produce the same flux Φ_0 as I_n at no-load (see fig. 16-3a).

$$N_p \,.\, \vec{I}_p - N_s \,.\, \vec{I}_s = N_p \,.\, \vec{I}_n \rightarrow \vec{I}_p - \frac{\vec{I}_s}{k} = \vec{I}_n \rightarrow \quad \vec{I}_p = \vec{I}_n + \frac{\vec{I}_s}{k} \qquad (16\text{-}6)$$

From (16-6) the vectorial construction of $\vec{I}_p$ in fig. 16-3b follows.

Since with a good transformer I_n is quite small with respect to I_p, we find from (16-6) that

$$\vec{I}_p \approx \frac{\vec{I}_s}{k} \quad \text{or:} \quad \frac{I_p}{I_s} \approx \frac{1}{k} = \frac{N_s}{N_p} \qquad (16\text{-}7)$$

From (16-7) and (16-4) it follows, by approximation: $\frac{V_p}{V_s} = \frac{I_s}{I_p} = \frac{N_p}{N_s} = k$ (16-8)

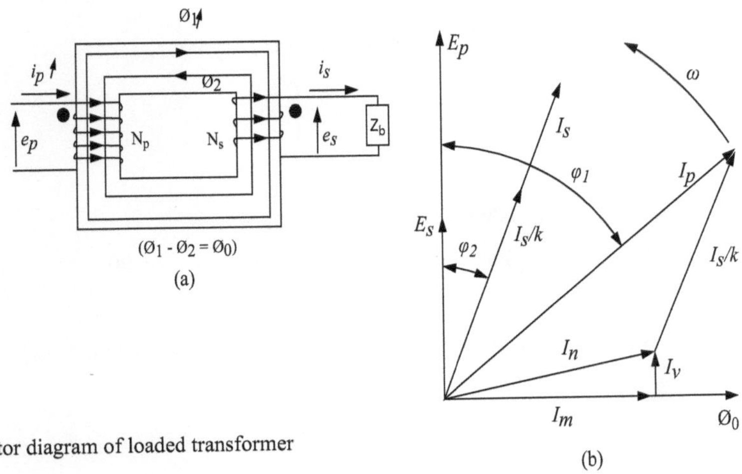

Fig. 16-3: Vector diagram of loaded transformer

1.2.2 Leakage flux

The current I_p produces a flux Φ_P which is mostly enclosed (Φ_1) within the core. A small part Φ_{pl} is not coupled with the secondary coil, so that $\Phi_P = \Phi_1 + \Phi_{pl}$. We refer to Φ_{pl} as the primary leakage flux. On the secondary side the current I_s produces a flux Φ_S which for the most part (Φ_2) flows through the primary and a small component Φ_{sl} (leakage flux) that is not linked to the primary: $\Phi_S = \Phi_2 + \Phi_{sl}$.

We therefore have: $\Phi_P = \dfrac{N_p \cdot I_p}{R_\mu}$ in phase with I_p and $\Phi_S = \dfrac{N_s \cdot I_s}{R_\mu}$ in phase with I_s.

In fig. 16-4, the instantaneous currents and voltages are drawn.

Here we see that Φ_1 and Φ_2 oppose each other, so that the resulting flux $\Phi_0 = \Phi_1 - \Phi_2$ = flux which was considered at no-load (= no-load flux !). This resulting flux Φ_0 is practically constant for every load and includes the emf's E_P and E_S as already shown.

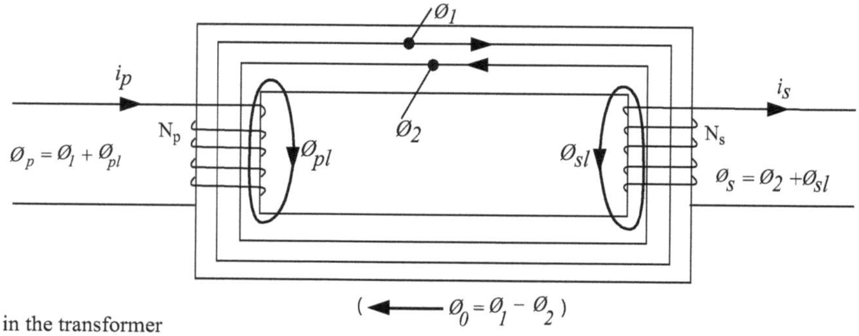

Fig. 16-4: Fluxes in the transformer

1.3 Transformer vector diagram

Primary:

- Leakage flux Φ_{pl} it's practically proportional to I_p : $\dfrac{N_p \cdot (\hat{\Phi}_{pl})}{\sqrt{2}} = s_p \cdot I_p$

  s_p = coefficient of primary self inductance with respect to leakage flux Φ_{pl}
- The leakage flux produces an emf in the primary coil: $e_{pl} = N_p \cdot \dfrac{d(\Phi_{pl})}{dt} = s_p \cdot \dfrac{d\, i_p}{dt}$
- If $i_p = \hat{i}_p \cdot \sin \omega.t$, then $e_{pl} = s_p \cdot \hat{i}_p \cdot \dfrac{d(\sin \omega.t)}{dt} = s_p \cdot \omega \cdot \hat{i}_p \cdot \sin(\omega.t + \frac{\pi}{2})$

  e_{pl}
  - is a sinusoidal emf , which leads I_p by 90°
  - effective value: $E_{pl} = \omega \cdot s_p \cdot I_p$

Summary:

Primary:

1. Induced counter-EMF E_P that leads Φ_0 by 90°
2. Induced counter-EMF $E_{pl} = \omega \cdot s_p \cdot I_P$ which leads I_P by 90°
3. Voltage drop $I_P \cdot R_P$ in phase with I_P
4. If we include the voltage losses $I_P \cdot R_P$ and $\omega \cdot s_P \cdot I_P$ we find:

$$\vec{V}_P = \vec{E}_P + \vec{E}_{pl} + \vec{I}_P \cdot R_P \qquad (16\text{-}9)$$

Secondary: 1. Flux Φ_0 produces an emf E_S which leads Φ_0 by 90°
2. Flux Φ_{sl} produces an emf $E_{sl} = \omega \cdot s_s \cdot I_S$ which leads I_S by 90°
3. Voltage drop $I_S \cdot R_S$ in phase with I_S
4. Terminal voltage V_S is formed by: $\vec{V}_S = \vec{E}_S - \omega \cdot s_s \cdot \vec{I}_S - \vec{I}_S \cdot R_S$ (16-10)

With what we have considered up to now, we can create a diagram in fig. 16-5 of a loaded transformer.

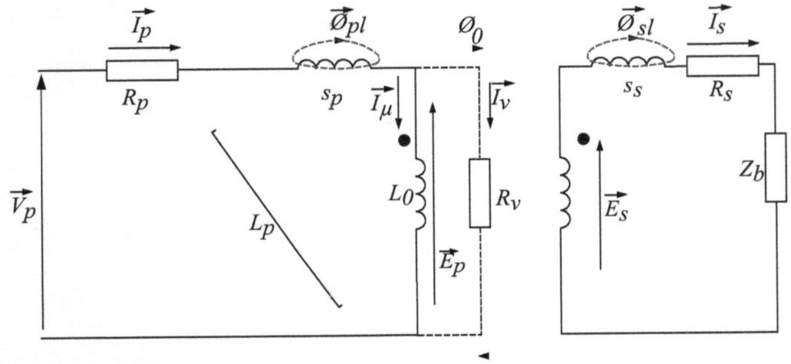

Fig. 16-5: Transformer with losses and secondary load

With the help of (16-9) and (16-10) we now construct fig. 16-6.

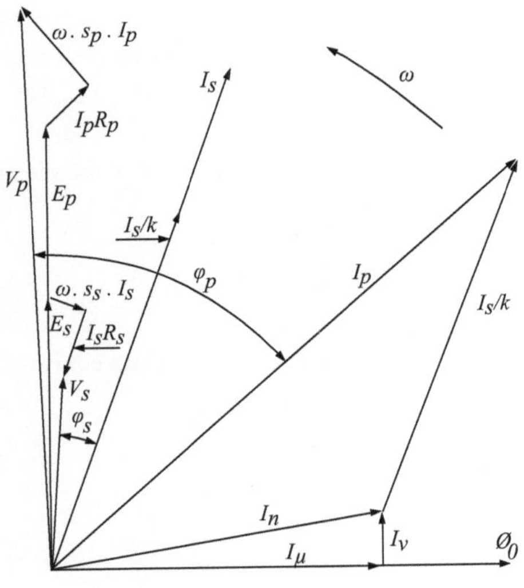

Fig. 16-6: Vector diagram of transformer with inductive load

Remark

From the figures 16-2 and 16-6 we see that the displacement angle between primary current and voltage decreases from almost 90° (at no-load) to a value determined by $\vec{I}_n$ and $\vec{I}_S$.

1.4 Impedance transformation

1.4.1 Transformation formula

Fig. 16-7a shows an ideal transformer, loaded with a series *R-L-C* circuit. An ideal transformer is a transformer without losses.

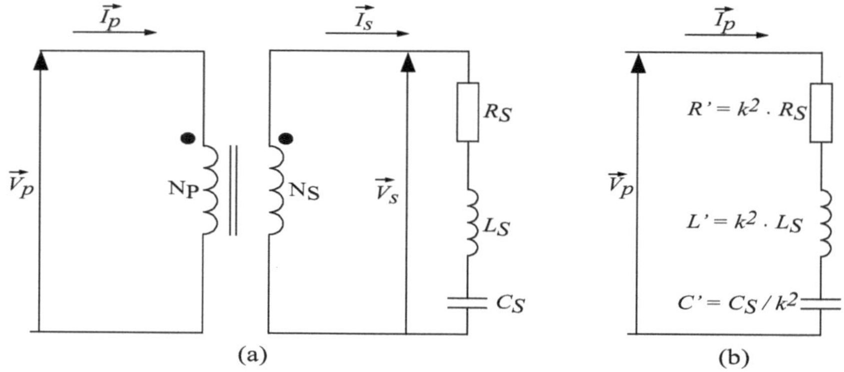

Fig. 16-7: Ideal transformer, loaded with a series *R-L-C* circuit: impedance transformation

Secondary: $v_s = R_s . i_s + L_s . \frac{di_s}{dt} + \frac{1}{C_s}\int_0^t i_p . dt$

Application of (16-8) gives: $v_p = k^2 . R_s . i_p + k^2 . L_s . \frac{di_p}{dt} + k^2 . \frac{1}{C_s}\int_0^t i_p . dt$

For an ideal transformer, the secondary load can be represented as an equivalent circuit seen from the primary side (fig. 16-7b), as long as: $R' = k^2 . R_s$; $L' = k^2 . L_s$; $C' = C_S / k^2$.

More generally: an ideal transformer with secondary impedance $Z_{sec.}$ may be seen as a

primary impedance:

$$Z'_{prim.} = \left(\frac{N_p}{N_s}\right)^2 . Z_{sec.} \quad (\Omega) \qquad (16\text{-}11)$$

From (16-11) it follows that we can transform a primary impedance to an equivalent secondary

impedance:

$$Z'_{sec.} = \left(\frac{N_s}{N_p}\right)^2 . Z_{prim.} \quad (\Omega) \qquad (16\text{-}12)$$

1.4.2 Numeric example 16-1:

1. An electrical oven is supplied with 46 volt and has a power of 4 kW. The supply network is 230V-50 Hz. If we had an ideal transformer available, what is then:
   a) the transformation ratio
   b) the primary and secondary current
   c) impedance seen from the 230 V-50Hz net?

Solution:

a) $k = \frac{N_P}{N_S} = \frac{230}{46} = 5$

b) secondary current: $I_S = \frac{P_S}{V_S} = \frac{4000}{46} = 86.95\text{A}$

primary current: $I_P = \frac{I_S}{k} = \frac{86.95}{5} = 17.39\text{A}$

c) load impedance: $Z_S = \frac{V_S}{I_S} = \frac{46}{86.95} = 0.529 + j.0\ \Omega$

Transformed impedance seen from the source: $Z_{prim.} = k^2 \cdot Z_{sec.} = 5^2 \times 0.529 = 13.23\Omega$

Proof: $Z_{prim.} \times I_P = 13.23 \times 17.39 = 230\text{V}$!!

2. If maximum power transfer is required from the generator to the consumer, then the consumers impedance should be the complex conjugate value of the generator impedance.
We have a power amplifier with an output resistance of 48Ω and wish to connect a loudspeaker with the following characteristics: 30 W - 4Ω .
Maximum power transfer is possible by placing an impedance transformer between amplifier and loudspeaker. The turns ratio should be $k = \frac{N_P}{N_S} = \sqrt{\frac{Z_{prim.}}{Z_{sec.}}} = \sqrt{\frac{48}{4}} = 3.46$.

1.5 Magnetizing inductance

With a magnetising current I_μ the magnetic field strength in the core is $H = \frac{N_P \cdot I_\mu}{l_k}$ and the magnetic induction is $B = \mu_0 \cdot \mu_r \cdot H$ so that the flux in the core of the transformer is:

$$\Phi_0 = B \cdot A_k = \frac{\mu_0 \cdot \mu_r \cdot A_k}{l_k} \cdot N_p \cdot I_\mu$$

Whereby:
- Φ_0 (Wb): no-load flux ≈ resulting flux $(\Phi_1 - \Phi_2)$ with load
- B (Wb/m²): magnetic induction in the core
- μ_r : relative permeability of core material
- μ_0 : $= 4 \cdot \pi \cdot 10^{-7}$ H/m
- l_k (m): average length of field line in the core
- A_k (m²): cross-sectional area of core
- I_μ (A): magnetising current of the transformer.

If we call L_0 the self inductance of the primary with respect to the flux Φ_0 in the core, then we may write: $N_p \cdot \Phi_0 = L_0 \cdot I_\mu$ so that: $N_p \cdot \Phi_0 = \frac{\mu_r \cdot \mu_0 \cdot A_k}{l_k} \cdot N_p^2 \cdot I_\mu = L_0 \cdot I_\mu$

from which follows:

$$L_0 = N_p^2 \cdot \frac{\mu_r \cdot \mu_0 \cdot A_k}{l_k} \quad \text{(H)} \qquad (16\text{-}13)$$

Numeric example 16-2:

1. A ring core transformer (fig. 16-8) consists of:

Core:	average radius 60 mm; cross-section of torus 45 mm; μ_r =1600.
Insulation layer:	1 mm thick
Primary:	3 layers: respectively 201, 189 and 140 windings AWG 18.
	Each layer is separated by 1 mm thick insulation.
Insulation layer:	4 mm thick
Secondary:	two layers: 50 and 22 windings AWG 10, separated by 1 mm of insulation

2. Extract from winding wire table (AWG = American wire gauge)

AWG	diameter (with insulation) in mm min.	max.	resistance (per 100 m) Ω	admitted current (on base of 2A/mm²) A
10	2.64	2.69	0.3276	10.38
18	1.08	1.11	2.095	1.624

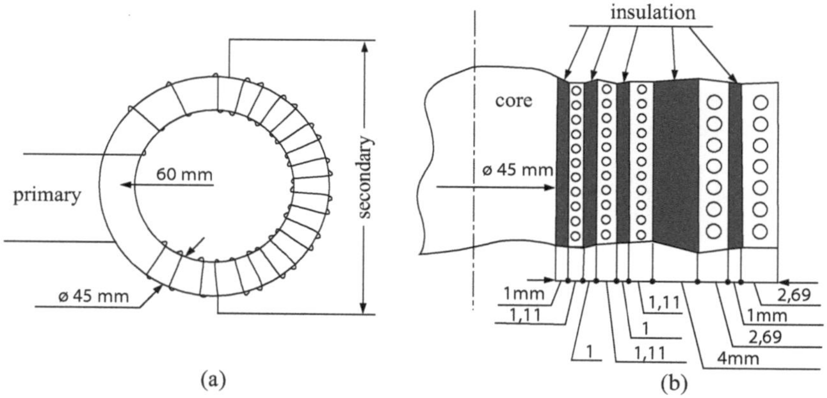

Fig. 16-8: Ring core transformer (a) cross-section (b) core and windings

Question:

1. Resistance of primary and secondary coil
2. Magnetising inductance

Solution:

1.1 Primary resistance

LAYER 1: length of one winding: π x 0.04811 = 0.15114 m

total length: 201 x 0.15114 = 30.38 m

LAYER 2:	length of one winding:	$\pi \cdot 0.05233 = 0.1644$ m
	total length:	$189 \cdot 0.1644 = 31.07$ m
LAYER 3:	length of one winding:	$\pi \cdot 0.05655 = 0.1776$ m
	total length:	$140 \cdot 0.1776 = 24.87$ m

Total length primary winding: 86.32 m

Resistance: $R_p = \frac{86.32}{100} \cdot 2.095 = 1.8\ \Omega$

1.2 Secondary resistance

LAYER 1:	length of one winding:	$\pi \cdot 0.06835 = 0.2147$ m
	total length:	$50 \cdot 0.2147 = 10.73$ m
LAYER 2:	length of one winding:	$\pi \cdot 0.07537 = 0.238$ m
	total length:	$22 \cdot 0.238 = 5.234$ m

Total length of secondary winding: 15.96 m

Resistance: $R_s = \frac{15.96}{100} \cdot 0.3276 = 0.0523\ \Omega$

2. Magnetizing inductance

$$L_0 = N_p^2 \cdot \frac{\mu_0 \cdot \mu_r \cdot A_k}{l_k} = 530^2 \cdot \frac{4 \cdot \pi \cdot 10^{-7} \cdot 1600 \cdot \pi \cdot 0.0225^2}{2 \cdot \pi \cdot 0.06} = 2.38\ \text{H}$$

1.6 Leakage inductance

From the viewpoint of voltage loss leakage inductance is undesirable. Transformers are therefore constructed to minimise the leakage fluxes. Fig. 16-9 shows, for example, how a coaxial implementation of primary and secondary coils minimises the leakage reactance by minimising the distance between consecutive coils. On the other hand, possible short circuit currents are limited by the leakage reactance, which can form a protection for the transformer. In practice distribution transformers are constructed with sufficient leakage reactance, so that short-circuit current is limited to 8 or 10 times the full load current.

In electronic power supplies ring core transformers are frequently used. Due to the construction method they have a minimum leakage reactance. Electronic technicians talk about "hard" transformers since large variations in the load coupled with low leakage inductance can produce large current spikes. These varying load conditions occur for example during commutation of one rectifier element to another on the secondary side of three-phase transformers.

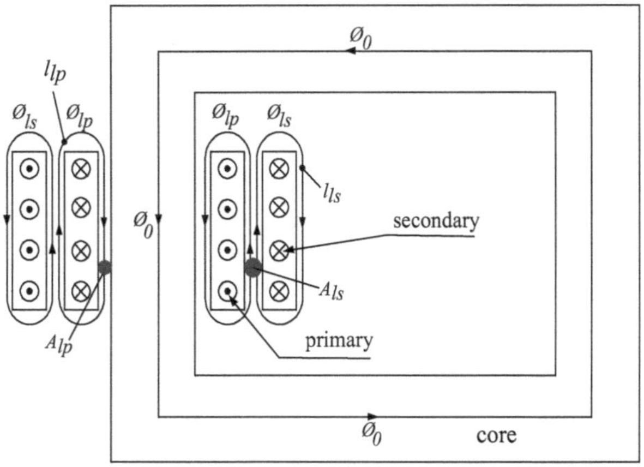

Fig. 16-9: Leakage fluxes by a coaxially wound transformer

To determine the leakage inductance, we consider the primary leakage flux (the same reasoning is valid for the secondary side). We can not make an accurate calculation since the cross-sectional area through which the flux flows can not be accurately determined. It is possible to make an approximate calculation. If the cross-sectional area where in the leakage flux flows is A_{lp} and the average length of the field line is l_{lp} then similar to expression (16-13), it may be written as:

$$L_{lp} = s_P = N_P^2 \cdot \mu_0 \cdot \frac{A_{lp}}{l_{lp}} \qquad (16\text{-}14)$$

The field lines of the leakage flux complete their circuit through the air ($\mu_r = 1$) instead of through the ferromagnetic core (μ_r), which explains the difference with expression (16-13).

Numeric example 16-3:

We reuse the data of numeric example 16-2 ensure the possible parts of the leakage fluxes in fig. 16-10a and fig. 16-10b.

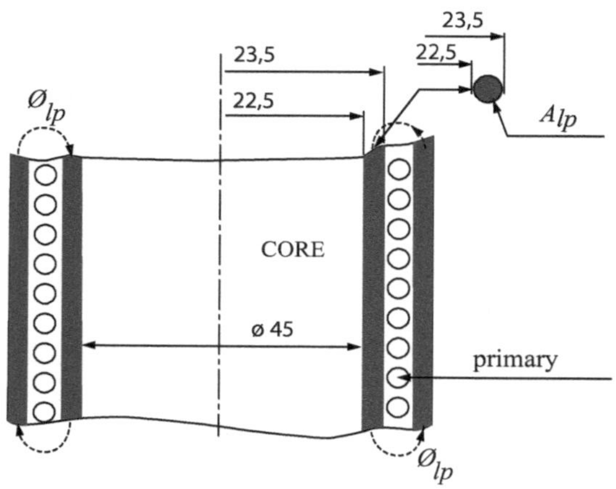

Fig. 16-10a: Primary leakage flux of transformer in fig. 16-8a

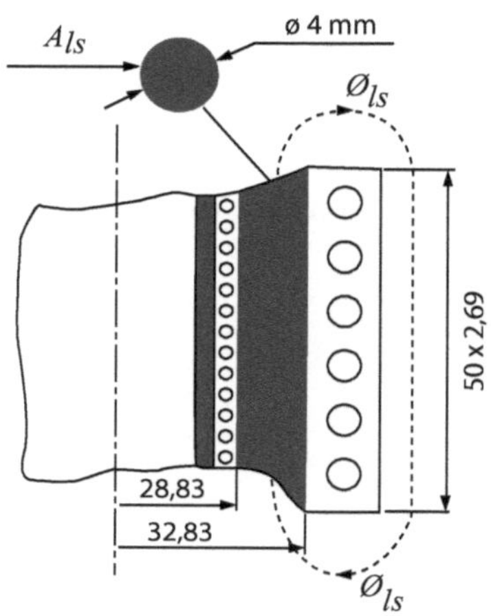

Fig. 16-10b: Secondary leakage flux fig. 16-8b

Primary leakage inductance

$$s_p = N_p^2 \cdot \frac{\mu_0 \cdot A_{lp}}{l_{lp}} = 530^2 \cdot \frac{4 \cdot \pi \cdot 10^{-7} \cdot \pi \cdot (23.5^2 - 22.5^2) \cdot 10^{-6}}{201 \cdot 1.11 \cdot 10^{-3}} = 228\ \mu H$$

Secondary leakage inductance

$$s_s = N_s^2 \cdot \frac{\mu_0 \cdot A_{ls}}{l_{ls}} = 72^2 \cdot \frac{4 \cdot \pi \cdot 10^{-7} \cdot \pi \cdot (32.83^2 - 28.83^2) \cdot 10^{-6}}{50 \cdot 2.69 \cdot 10^{-3}} = 37.42\ \mu H$$

If we realise that the magnetising inductance for this transformer is 2.38 H then we see that the leakage inductance is indeed minimal.
It is clear that the path of the leakage fluxes depends upon the practical implementation of the transformer windings. The present numeric example gives us a rough idea of the relative magnitude of the leakage inductance.

1.7 Energy losses

1.7.1. Copper losses

In the primary and secondary windings energy losses occur. If R_P and R_S are the respective resistances of the windings then the losses may be written as $R_P \cdot I_P^2$ and $R_S \cdot I_S^2$. The sum of both is the total energy loss. This is referred to as the copper losses of the transformer.

1.7.2 Iron losses

In ferromagnetic materials, subjected to a varying magnetic field, hysteresis losses occur:

$$P_h = k_h \cdot f \cdot \hat{B}^n \quad \text{W/kg} \qquad (16\text{-}15)$$

whereby:
- k_h = material constant of the ferromagnetic material used in relation to hysteresis losses.
- f = frequency (Hz)
- $\hat{B}$ = amplitude of the magnetic induction (T = Wb/m²)
- n = empirical constant for the magnetic material ($1 < n < 3$).

Since the magnetic circuit of a transformer is constructed from metal plates, hysteresis losses occur. To limit these losses, it is desirable that the material constant be as small as possible.
A possibility in this case is an iron alloy using silicon (e.g. 3% silicon). If the core was made from solid iron, then considerable eddy currents would occur. These can be dramatically limited by making the magnetic circuit from plates which are insulated from each other and the surface of which is in the direction of the flux. As result of this the path of the eddy currents is limited.
The eddy current losses P_w can be determined with a formula in the following form:

$$P_w = k_w \cdot \delta^2 \cdot f^2 \cdot \hat{B}^2 \quad \text{W/kg} \qquad (16\text{-}16)$$

Here in :
- k_w = material constant with respect to the eddy current losses
- δ = plate thickness in mm.

By adding silicon the electrical resistance is also increased as a result of which the eddy current losses are reduced. According to the last formula, it is advantageous to have the plates as thin as possible. Typical plate thickness lies between 0.3 and 1mm for 50 Hz operation. The plates can be 0.02 mm for high frequencies. For band wound cores thicknesses of 0.003 to 0.3mm are possible.

In many applications it is possible that non-sinusoidal waveforms occur. The eddy current losses are proportional to the square of the form factor $a = \frac{V_{RMS}}{V_{AV}}$ so that:

$$P_w = \frac{k_w}{1.11^2} \cdot a^2 \cdot \delta^2 \cdot f^2 \cdot \hat{B}^2 \tag{16-17}$$

1.11 = form factor of sinusoidal voltage
a = form factor of the actual voltage.

Hysteresis and eddy current losses form the iron losses. They are sometimes called the constant losses of the transformer since they do not depend upon the load but only the magnetic induction. The magnetic induction only depends upon the applied voltage. The following table provides an idea of the iron losses with plates between 0.2 and 0.5 mm thick, and a frequency of 50 Hz with an induction of 1 Tesla.

Material	Losses in W/kg
commercial iron	5 ... 10
Si-Fe, warm rolled	1 ... 3
Si-Fe, cold rolled and crystal orientated	0.3 ... 0.6
50% Ni-Fe	0.2
approximately 65% Ni-Fe	0.06

Fig. 16-11 shows the total iron losses at 50 Hz for toroidal band wound cores of 0.3mm (data for cold rolled 3% Si-Fe cores). In fig. 16-12, we see the influence of the frequency on the total iron losses for the same material. Such cores are used for power transformers, impulse transformers, welding transformers, line transformers, etc... .

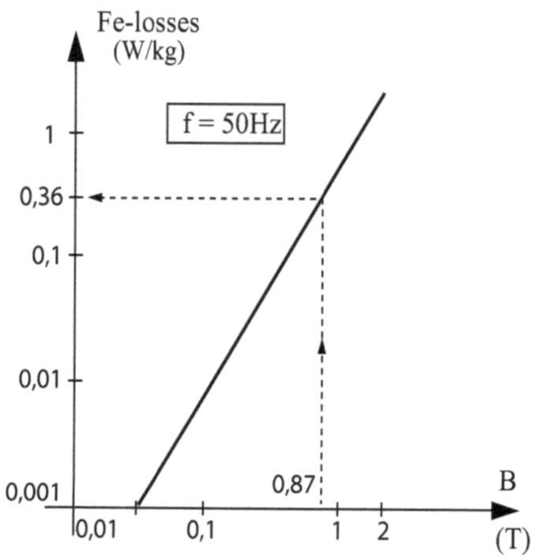

Fig. 16-11: Iron losses as a function of induction

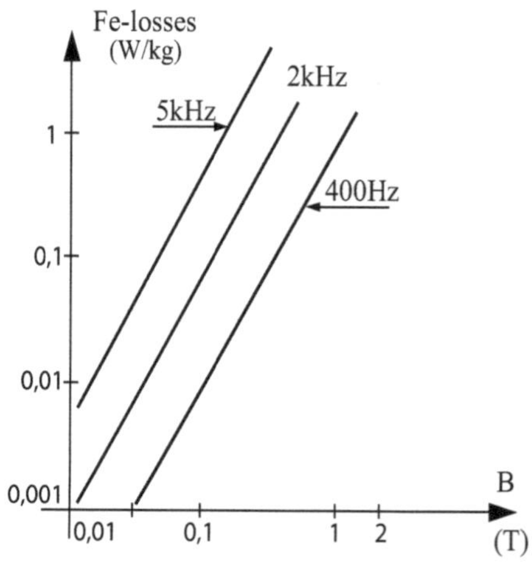

Fig. 16-12: Iron losses with the frequency as a parameter

In order to limit the eddy current losses at higher frequencies, ferromagnetic cores are used. These ferrites consist of an alloy of iron oxide with modern materials such as manganese, nickel,... Oxides have a low electrical conductivity. Ferrites can be made with no losses into the MHz-range. The maximum self induction of ferrites (0.3 to 0.5T) is less than that of Si-steel plates (1 to 1.5T) since a large part of the volume is composed of oxygen atoms that are obviously non-magnetic. Fig. 16-13 shows the core losses for ferrite (Siemens) as a function of the inductance, while fig. 16-14 shows the losses as a function of the frequency for the same material.

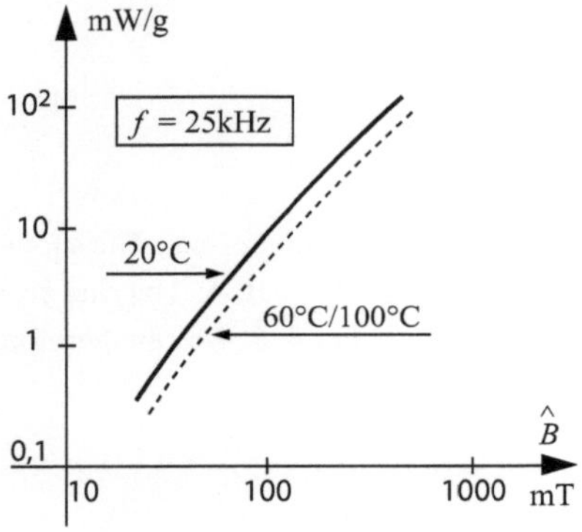

Fig. 16-13: Core losses N27 (for R16 ring cores)

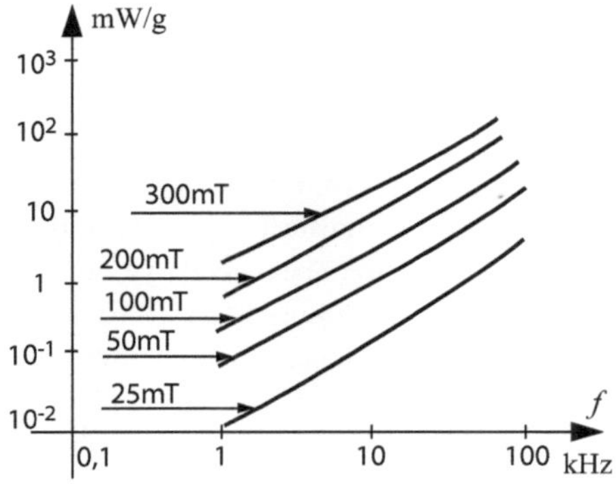

Fig.16-14: Losses N27 (for R16 ring cores)

1.8 Equivalent diagram

1.8.1 T-equivalent of a transformer

Making use of the vector diagram in fig. 16-6 enables us to draw an equivalent diagram of unloaded transformer in fig. 16-15.

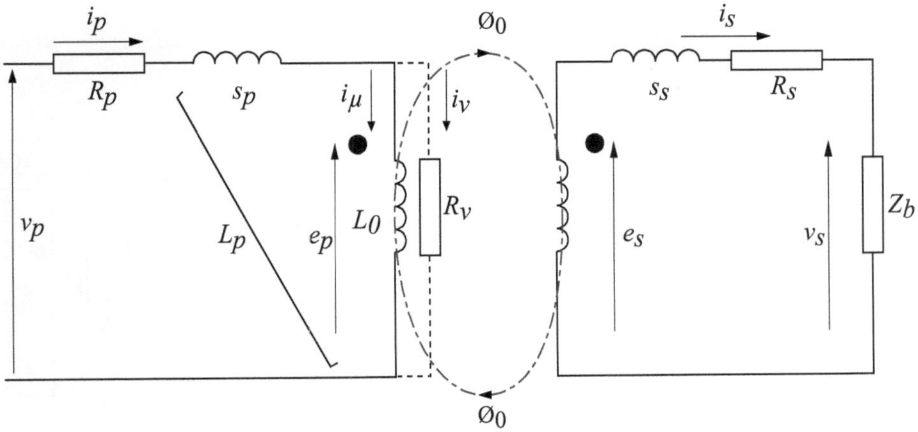

Fig. 16-15: Equivalent diagram of unloaded transformer

Reflecting the secondary impedances to the primary side gives us the T-equivalent. We have placed all losses in the primary side so that on the right-hand side of fig. 16-16 an ideal transformer remains.

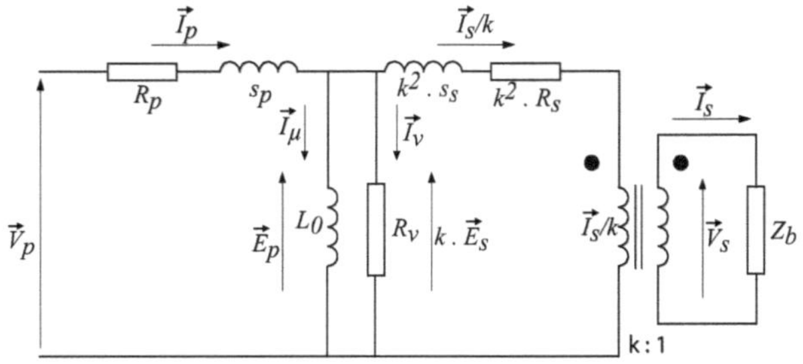

Fig. 16-16: T-equivalent of a loaded transformer

Numeric example 16-4:

We want to draw a T-equivalent and a vector diagram of a real transformer. The supply voltage is 230V - 50 Hz. The secondary of the transformer can supply 30 V - 10 A. The ring core transformer in numeric example 16-2 will be designed with these details in mind. We can therefore use the previously calculated values of this transformer.

Transformer ratio: $k = \frac{N_p}{N_s} = \frac{530}{72} = 7.361$; $L_0 = 2.38\text{H}$; $R_p = 1.8\ \Omega$

$R_s = 0.0523\ \Omega$; $s_p = 228\ \mu\text{H}$; $s_s = 37.42\ \mu\text{H}$

If we ignore the primary voltage drop, we can calculate the flux in the core using (16-5):

$$\hat{\Phi}_0 = \frac{230}{4.44 \cdot 530 \cdot 50} = 0.0019547 \text{ Wb so that:}$$

$$B = \frac{\hat{\Phi}_0}{A_k \cdot \sqrt{2}} = \frac{0.0019547}{\pi \cdot 22.5^2 \cdot 10^{-6} \cdot \sqrt{2}} = 0.869 \text{ T.}$$

If the core is composed of a band wound strip of 0.3 mm then the power loss for this induction and a mass of 4.67 kg is $0.36 \cdot 4.67 = 1.68\text{W}$ (fig. 16-11)

From $N_p . \Phi_0 = L_0 \cdot I_\mu$ it follows that the magnetizing current is:

$$I_\mu = \frac{N_p \cdot \Phi_0}{L_0} = \frac{530 \cdot 0.0019547}{2.38 \cdot \sqrt{2}} = 0.307 \text{ A.}$$

We are now able to draw the T-equivalent diagram for the fully loaded ring core transformer in our numeric example.

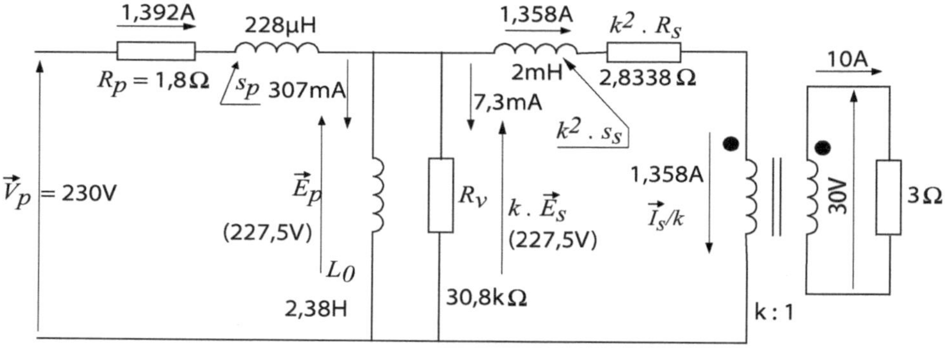

Fig. 16-17: T-equivalent for a ring core transformer 230 V/30 V-10 A

The current that corresponds to the iron losses has been determined from the energy loss of 1.68 W at 230V.

In fig. 16-18, the vector diagram of the fully loaded transformer is drawn with the assumption that the secondary load is purely resistive. Take note that the primary phase displacement of almost 90° (at no-load) is reduced to about 13°. If the magnetising induction is even larger, and therefore the magnetising current smaller then the primary phase displacement would be extremely small.

A transformer which is very low loss would, in the case of a resistive load, behave almost as a resistor for the supply network.

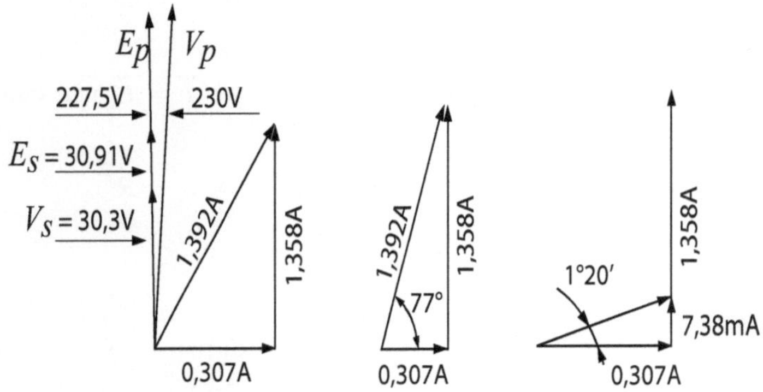

Fig. 16-18: Vector diagram for a fully loaded ring core transformer 230V /30 V-10A (vector E_p calculated from fig. 16-6!!)

1.9 No-load and short circuit test

Fig. 16-16 can be further simplified by neglecting the reasonably small voltage loss $I_p \cdot R_p$ and $I_p \cdot \omega \cdot s_p$ and then placing R_p and s_p to the secondary side (fig. 16-19).

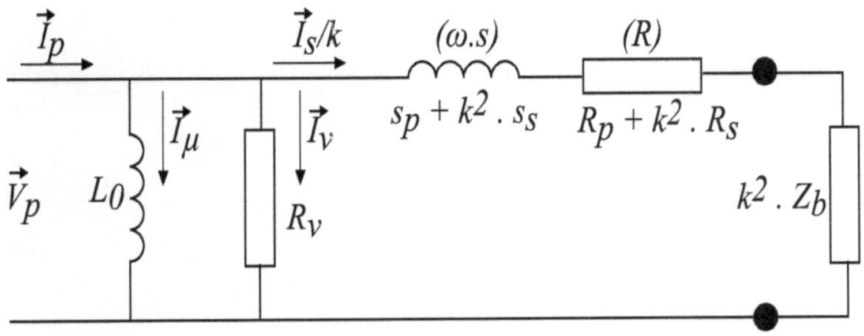

Fig. 16-19: Simplified T-equivalent of a transformer

1.9.1 . No-load test

If we connect the primary side to the nominal voltage and leave the secondary side open the transformer only dissipates the no-load losses. These are composed almost entirely of iron losses. Indeed, the no-load current is small compared to the nominal current so that the copper losses $I_n^2 \cdot R_p$ are extremely small. Note that the no-load test is usually carried out on the low voltage side of the transformer since on the high voltage side the current is small and the voltage is high.

1.9.2. Short circuit test

If short-circuit the secondary side (e.g. with an amp meter), and apply a reduced voltage to the primary side in such a way that the nominal primary (and also secondary) current flow. The reduced voltage at which this occurs is referred to as to short-circuit voltage V_K. Since this voltage is only a few percent of the nominal voltage the iron losses (proportional to the square of the inductance and therefore the voltage) is negligibly small. With a wattmeter in the primary, we can measure the copper losses at full load.

The impedance $Z_k = \frac{V_k}{I_{p(nom)}}$ = transformer short-circuit impedance.
In fig. 16-19 with a short-circuit of $k^2 . Z_b$ and by neglecting $\vec{I}_n (= \vec{I}_\mu + \vec{I}_v)$ the short circuit impedance becomes: $Z_k = \sqrt{(\omega.s)^2 + R^2}$.
The resistive component follows from $R = \frac{P_j}{I_p^2}$. Where: $(\omega.s) = \sqrt{Z_k^2 - R^2}$.

1.9.3 Numeric example 16-5:

A 600VA (230 V/ 400V) transformer has the following data:

No-load test:	$V_P = 230\text{V}$	Short-circuit test:	$V_k = 13\text{V}$
	$I_n = 0.3\text{A}$		$I_P = 2.6\text{A}$
	$P_{Fe} = 5\text{W}$		$I_S = 1.5\text{A}$
			$P_J = 12\text{W}$

Question:
Determine the equivalent elements as shown in fig. 16-19.

Solution:
At no-load (fig. 16-2) $P_n = V_p . I_n . \cos\varphi_n = V_p . I_v = P_{Fe} + I_n^2 . R_p \approx P_{Fe}$

From which: $I_v \approx \frac{P_{Fe}}{V_p} = \frac{5}{230} = 0.02174\text{ A}$; $R_v = \frac{V_p}{I_v} = \frac{230}{0.02174} = 10.58\text{ k}\Omega$

$$I_\mu = \sqrt{I_n^2 - I_v^2} = \sqrt{0.3^2 - 0.02174^2} = 0.299\text{ A}$$

$$\omega.L_0 \approx \frac{230}{0.299} = 769.23\ \Omega \rightarrow L_0 = 2.448\text{ H}$$

short circuit impedance: $Z_k = \frac{V_k}{I_p} = \frac{13}{2.6} = 5\ \Omega$; $R = \frac{P_j}{I_p^2} = \frac{12}{2.6^2} = 1.775\ \Omega$

$$\omega.s = \sqrt{Z_k^2 - R^2} = 4.67\ \Omega \rightarrow \left(s_p + k^2 . s_s = \frac{4.67}{\omega} = 14.87\text{ mH}\right)$$

1.10 Nominal values of a transformer

We refer to a transformer with:

- apparent power (VA or kVA)
- primary and secondary voltages
- secondary and primary currents
- the working frequency

1.11 Three-phase transformers

In three-phase networks we use either three single phase transformers or a three-phase transformer with three cores (with a primary and secondary coil on each core). Normally the secondary coil is composed of two halves, so that in addition to star or delta it may be connected in zigzag. In zigzag one half of the secondary of one leg is connected in series with a half winding of another leg. The result is a more even power distribution across the three phases of the transformer. The connection (star, delta, zigzag) of the primary and secondary do not have to be the same but when the secondary is connected in parallel the secondary voltages need to be in phase.
The most common connections are divided into four categories. Only transformers of the same category can be connected in parallel, since within this same group the secondary voltages are in phase. This phase displacement is indicated with a number called the clock number.

Numeric example 16-6:

A three-phase transformer of 70kVA has a transformation ratio of 10kV/230V. The resistance of one primary phase winding is 1.2Ω and of one secondary 0.015Ω . The total iron losses are 700 W. The transformer primary is connected in delta and the secondary in star. The primary line voltage is 3x10kV.

Question:

a) calculatethe efficiency when the transformer provides 10kW of the secondary side with a power factor = 0.8. Determine the primary and secondary line and phase currents.

b) calculate the efficiency when the secondary load is 40kW at a $\cos\varphi = 0.6$.

Solution:

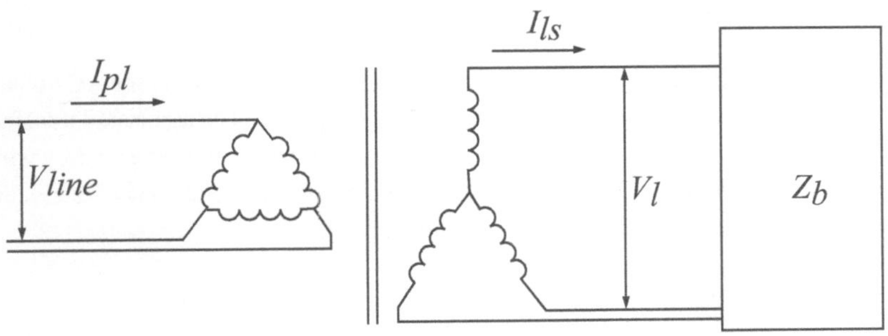

a)

Secondary phase voltage: 10kV/230V → $V_f = 230V$

Secondary line voltage: $V_l = \sqrt{3} \cdot V_f = 400V$

Secondary line current: $10^4 = \sqrt{3} \cdot V_l \cdot I_l \cdot \cos\varphi = \sqrt{3} \cdot 400 \cdot I_l \cdot 0.8 \rightarrow I_{ls} = 18A$

Primary phase current: $I_{pf} = \frac{230}{10{,}000} \cdot 18 = 0.415A$ (no load current : is neglected!)

Primary line current: $I_{pl} = \sqrt{3} \cdot 0.415 = 0.7187A$

Joule losses of all phase windings together: $3 \cdot [(0.415)^2 \cdot 1.2 + (18)^2 \cdot 0.015] = 15.2$ W

Total losses: $P_{Fe} + P_j = 715.2$W

Primary power = 10,000 + 715.2 = 10,715.2W

Efficiency: $\eta = \frac{10{,}000}{10{,}715.2} = 93.32\%$

b)

Secondary current $= \frac{40{,}000}{\sqrt{3} \cdot 400 \cdot 0.6} = 96.225$A

Primary phase current $= 96.225 \cdot \frac{230}{10{,}000} = 2.2132$A

Joule losses $= 3 . [(2.2132)^2 \cdot 1.2 + (96.225)^2 \cdot 0.015] = 434.3$W

Total losses $= P_{Fe} + P_j = 1134.3$W

Efficiency $= \frac{40{,}000}{41{,}134.3} = 97.24\%$

1.12. Transformer types

The construction of a transformer depends partially upon the specific application for which the transformer will be used.

1. Distribution or net transformer

Used for the distribution of electrical energy between the central generator and the consumer. These net transformers are three-phase.

2. Supply transformers

In electronic equipment, a rectifier circuit is often present to transform the AC voltage to an useful DC voltage. We normally talk about the supply in the device. The associated transformer therefore also acquires the name supplied transformer. Often multiple secondary windings are present, so that the possibility exists for series and parallel combinations of the secondary allowing for different secondary voltages and currents.

3. Isolation transformer

To isolate and separate any appliance from the distribution net. We use a transformer with transformation ratio of $k = 1$. For example: a 230/230 V transformer of 250 VA.

4. Welding transformer

To create an arc 60 to 80 V is necessary. To maintain the arc 20 to 40 V is sufficient with a current between 80 and 1000A. The transformer needs a voltage-current characteristic such that the no-load voltage of 80V drops to 30 V when loaded.

5. Auto-transformer

This contains only one winding of which a part is common to both primary and secondary coil (fig. 16-20). The secondary voltage can obviously be higher than the primary voltage. With this configuration, copper can be saved.
Take note that there is no galvanic isolation between the consumer and the net.

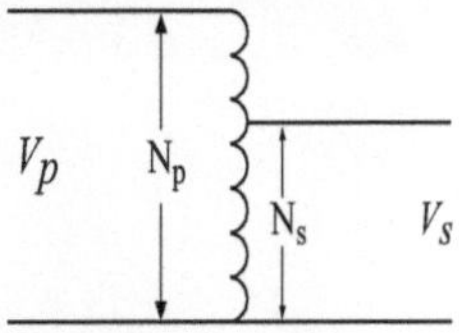

Fig. 16-20: Auto-transformer

6. Safety transformer

With an isolated primary and secondary and a safe (e.g. 24V) secondary voltage.

7. Measurement transformers

Measurement transformers are special transformers of low-power and high accuracy. They are used to transform high voltages and large currents to lower values so that:

1. Ordinary measurement instruments can be used
2. A galvanic separation is achieved between measuring instrument and supplying net (possibly high voltage). We distinguish between voltage and current transformers.

8. Impulse transformers

For triggering thyristors often special impulse transformers are used. The purpose is to implement a galvanic separation between the electronic control circuit and the power circuit of a controlled rectifier. These transformers with little distortion are capable of transferring impulses with an amplitude of for example 7V for a time duration of 100μs.

9. Impedance transformer

In order to optimally load an electronic circuit a specific impedance is required. If the load to be connected (for example a speaker) does not have the desired impedance an impedance transformer can be used with a correct winding ratio (see numeric example 2 under number 1.4.2).

10. SMPS transformer

In switch mode power supplies, which mostly operates up to 20 kHz (or higher), ferromagnetic cored transformers are used, for amongst other reasons as galvanic separation between the AC supply network and the consumers side of the supply.

16.2 DC COMMUTATOR MACHINES

CONTENTS

2.1 DC GENERATOR

2.1.1 Operating principle

DC generators (dynamos) are a direct application of the laws of induction. We consider fig. 16-21. A conductor is moved in a homogenous magnetic field. The conductor is moved in such a way that it cuts the magnetic field lines at 90°. In the conductor an induced voltage is created the value of which is given by:

$$e = B \,.\, l \,.\, v \qquad (16\text{-}18)$$

e = induced emf (V)
B = magnetic induction (Wb/m²)
l = length of conductor perpendicular to the magnetic field (m)
v = relative velocity of conductor with respect to the magnetic field (m/s)

The direction of the induced voltage is indicated by the rule of the three fingers of the right-hand.

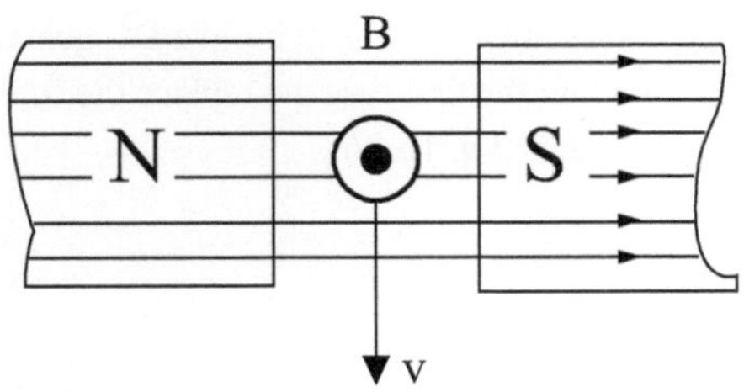

Fig. 16-21: Moving conductor with respect to pole

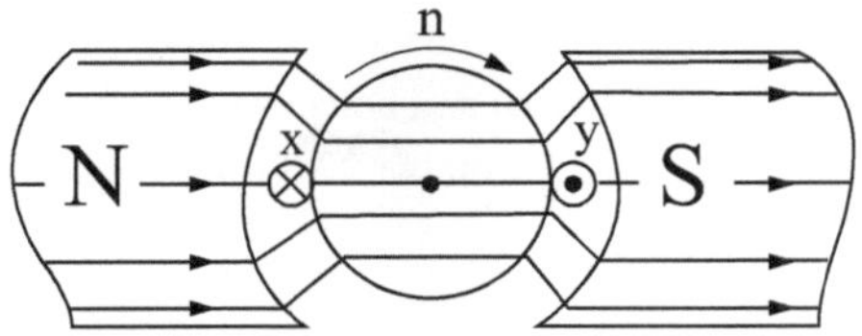

Fig. 16-22: Drum armature

To obtain a continuous voltage the wire of fig. 16-21 is wound on a drum and to minimise the size of the air gap the poles have been cylindrically engineered (fig. 16-22).

Next a wire frame is created (fig. 16-23). The ends of the frame are in contact with two rings, insulated on the shaft of the drum.

The external electrical contact is achieved using two brushes (A and B) which press on the rings.

The speed v_1 with which the conductors perpendicularly cut the field lines (fig. 16-25) is indicated by : $v_1 = v \cdot \sin\alpha$. From (16-18) it follows that : $e = B \cdot l \cdot v \cdot \sin\alpha$.

For $\alpha = 90°$ then $\hat{e} = B \cdot l \cdot v$ = maximal value of the induced emf. We find therefore: $e = \hat{e} \cdot \sin\alpha$.

If the wire frame is at an angle α then the enclosed flux is: (16-19)

$$\Phi = \hat{\Phi} \cdot \cos\alpha$$

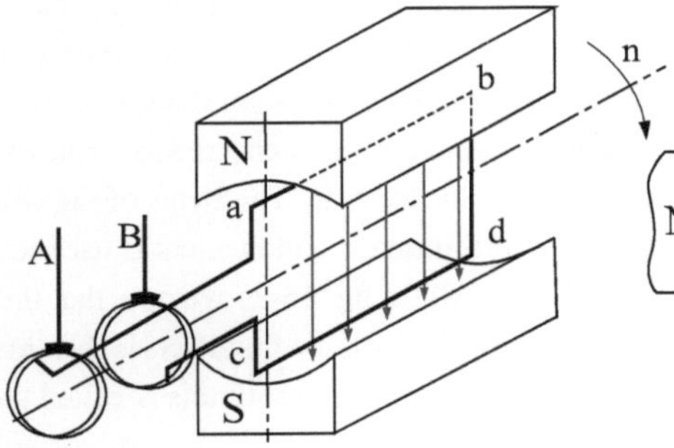

Fig. 16-23: Inducing a voltage

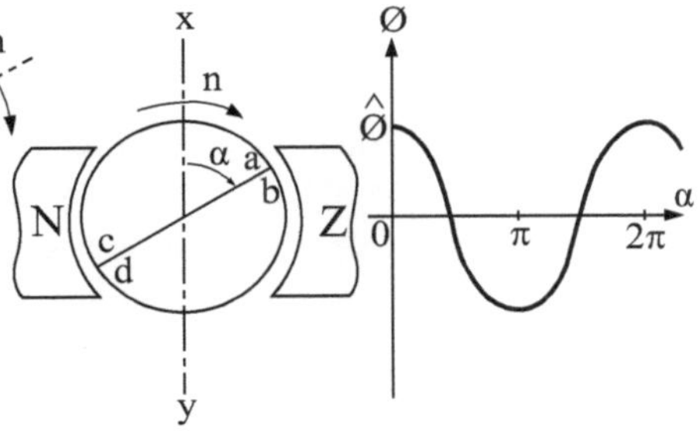

Fig. 16-24: The enclosed flux

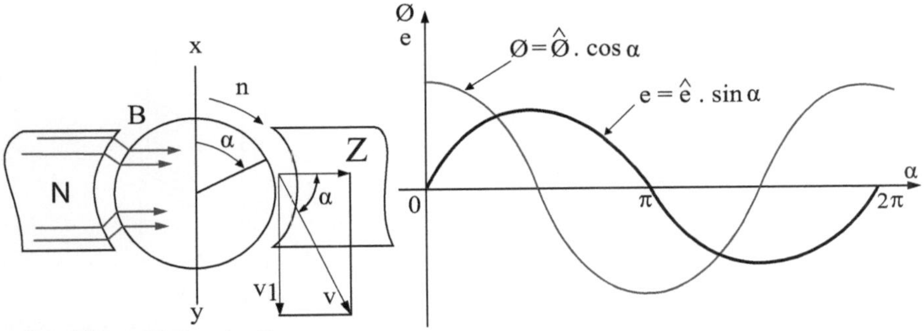

Fig. 16-25: The enclosed flux and induced voltage

2.1.2 Commutator

The form of the induced voltage is AC. The intention however was to build a DC generator (= dynamo) and not an AC generator (= alternator). In the first case we replace the full rings of fig. 16-23 with two half rings insulated from each other (fig. 16-26).

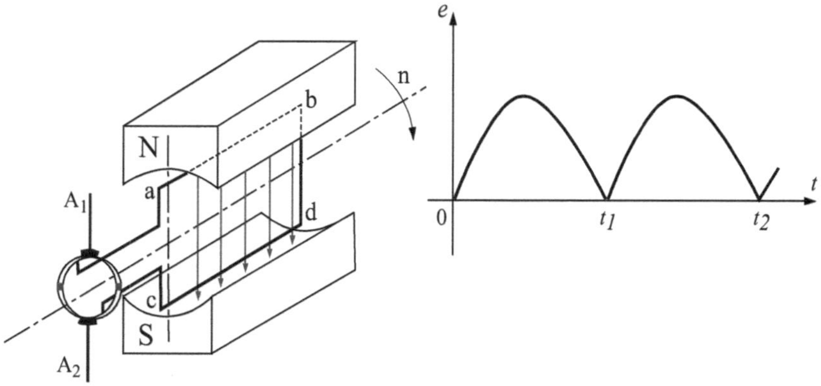

Fig. 16-26: Reverse the polarity of the induced voltage

By replacing the rings with half cylinders we create a pulsing DC voltage between the brushes A_1 and A_2 . A consumer connected to these brushes would receive pulsing DC voltage. The two half cylinders play the role of commutator. Since, in addition, they draw the current from the generator they are also called the collector. To make maximum use of the space on the drum, conductors (in slots) are placed over the entire surface and are connected to a part of the hollow cylinder (fig. 16-27). Each conductor has its own collector strip. The strips of the collector are insulated from each other. Rather than using the wireframe, a coil element is used to produce a larger emf between the two collector strips (fig. 16-28). In fig. 16-25 we note that the induced emf in the coil is zero if the flux **change** through the coil is zero, the flux itself is then maximum. In fig. 16-28, the voltage is zero at the horizontal position of the coil: this is called the **neutral line** of the machine.

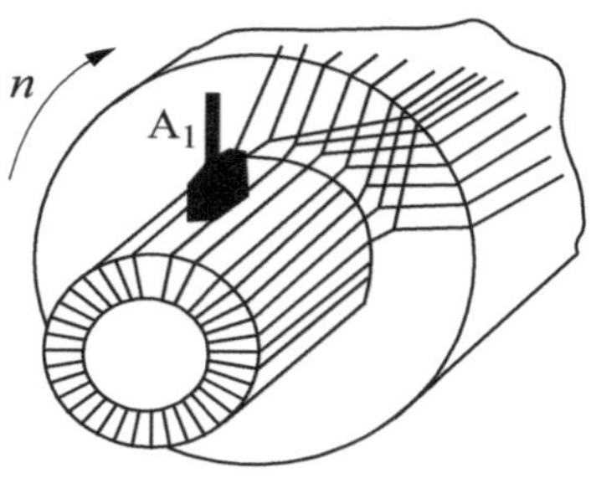

Fig. 16-27: Collector

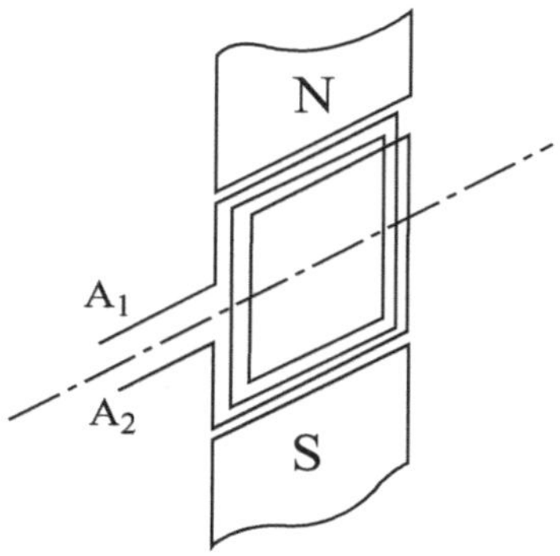

Fig. 16-28: Neutral line of machine

2.1.3 Main components

a. Poles and magnetic circuit

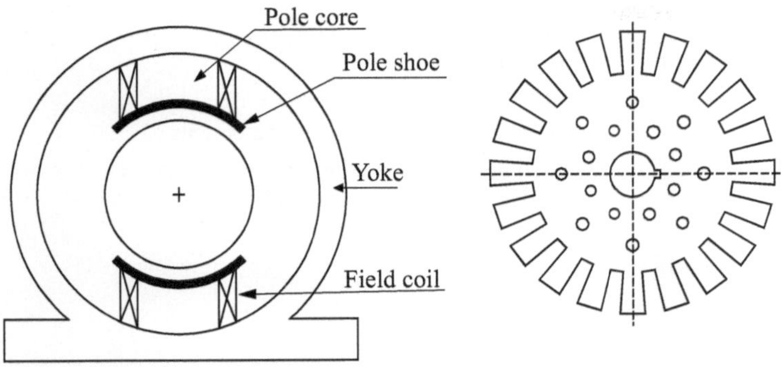

(a) Magnetic system of a two-pole machine (b) Si-steel sheet for slotted armature

Fig. 16-29: Parts of a dynamo

To achieve high levels of induction (B) electromagnets are used. That is why field windings are placed around the (protruding) poles (fig. 16-29a). These are referred to as the inductor winding, excitation winding or field coil. For small machines permanent magnets may be used.
The induction varies between 0.6 and 0.9Wb/m^2 depending on the size of the machine.

b. The drum armature with the armature winding

The coils within which the emf is induced lie in the grooves found around the circumference of the rotor. For DC machines we don't refer to the rotor but rather to the armature. The conductors or the armature winding are wound according to certain rules (partly in series and partly in parallel). The rotating amature is subject to a changing magnetic field. To limit the iron losses the armature is built up from Si-steel plates (0.35 to 0.5mm thick with 0.5 to 5% Si). These plates are fabricated with sufficient slots and holes (fig. 16-29b). Thereafter, these plates are insulated from each other and then pressed together to form a solid structure. Iron losses were discussed during the study of the transformer. The losses caused by the plates at B = 1 Tesla, f = 50Hz and T = 30° C are called the power loss. Similar to the transformer the losses associated with Si-steel plates can be less than 1W/kg.

c. The collector or commutator

The collector is formed by a large amount of copper strips insulated from each other and from the shaft. The brushes which rest in brush holders are in contact with the collector. The pressure from springs ($\approx$ 1.5 N/cm^2) see to it that the brush makes contact with the collector. The maximum current for the brushes (e.g. 10A/cm^2) depends upon the material used in the brushes.

d. The mechanical parts

Bearings, bearing covers, ventilators and pulley complete the mechanical configuration of the dynamo.

2.1.4 Value of the induced emf

Assuming:

Φ	=	flux per pole (Wb)
$2.p$	=	number of poles
d	=	drum diameter (m)
l	=	useful drum length (m)
n	=	armature speed (rpm)
$2.a$	=	number of parallel armature elements
N	=	total number of windings at the circumference
E	=	induced emf of dynamo (V)

Total flux in machine : $\Phi_l = 2 \,.\, p \,.\, \Phi$

Reference surface armature drum: $A = \pi \,.\, d \,.\, l$

Induction in the air gap: $B = \dfrac{\Phi_l}{A} = \dfrac{2 \,.\, p \,.\, \Phi}{\pi \,.\, d \,.\, l}$

Circumferential velocity: $v = \dfrac{\pi \,.\, d \,.\, n}{60}$ (m/s)

Induced emf in one armature winding: $E_l = B \,.\, l \,.\, v = \dfrac{2 \,.\, p \,.\, \Phi}{\pi \,.\, d \,.\, l} \,.\, l \,.\, \dfrac{\pi \,.\, d \,.\, n}{60} = 2 \,.\, p \,.\, \Phi \,.\, \dfrac{n}{60}$ (V)

Where N = total number of armature windings and $2.a$ = number of parallel armature elements, the total emf is then: $E = \dfrac{N}{2 \,.\, a} \,.\, E_l = \dfrac{N}{2 \,.\, a} \,.\, 2 \,.\, p \,.\, \Phi \,.\, \dfrac{n}{60} = \dfrac{N}{a} \,.\, \dfrac{p}{60} \,.\, n \,.\, \Phi$ (V)

For a particular type of construction p, a and N are constants, then: $E = k_l \cdot n \cdot \Phi$ (16-20)

E	=	induced emf dynamo (V)
k_l	=	machine constant ($= \dfrac{p}{a} \,.\, \dfrac{N}{60}$)
Φ	=	flux of single pole (Wb)

If we use the angular velocity ω of the armature instead of the speed in revolutions, then we find:

$$E = k_l \cdot n \cdot \Phi \cdot \frac{\pi}{30} \cdot \frac{30}{\pi} = k_l \cdot \Phi \cdot \frac{30}{\pi} \cdot \omega = K_G \cdot \omega \quad \rightarrow \quad E = K_G \cdot \omega \qquad (16\text{-}21)$$

$$K_G = k_l \cdot \Phi \cdot \frac{30}{\pi} = \frac{p}{a} \cdot \frac{N}{60} \cdot \frac{30}{\pi} \cdot \Phi = \text{generator constant of machine} \qquad (16\text{-}22)$$

2.1.5 Armature reaction

2.1.5.1 The armature reaction

When the dynamo is connected to a load, current flows in the armature windings. This results in an armature flux. The main flux (fig. 16-30) is increased as it exits the poles and weakens where it enters the poles. The mixture of armature flux and main flux results in the neutral line shifting in the direction of rotation.

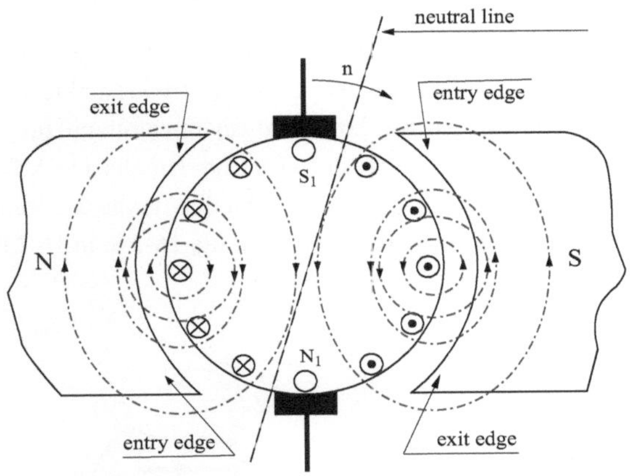

Fig. 16-30: Armature reaction

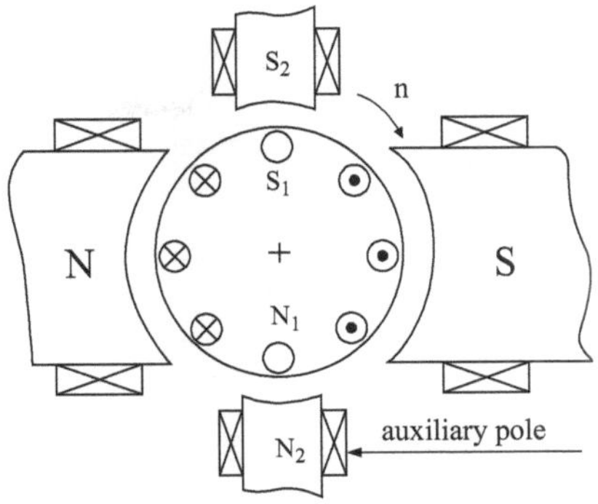

Fig. 16-31: Auxiliary poles

2 . 1 . 5 . 2. Disadvantages of the armature reaction

1. **voltage drop:** since we operate close to saturation of the magnetic circuit in such a machine, the flux reduction (pole entry) will have more effect than the flux increase (pole exit) so that the total flux is reduced as a result of the armature reaction. This results in a voltage drop at the terminals of the dynamo.

2. **displacement of the neutral line** in the direction of rotation. Displacement angle depends upon the load so that the brushes (which need to be on the neutral line) are usually in the wrong position.

3. **measures to counteract the armature reaction:**

a. use of auxiliary poles

In the armature a N_1 and S_1 pole are created as a result of the current through the armature windings. If we now place two poles on the neutral line (N_2 and S_2 in fig. 16-31) which are just as strong as N_1 and S_1 , then only the original *N-S* will remain. The auxiliary pole should have the same name as the following main pole. If the auxiliary pole winding B_1B_2 is connected in series with the armature, then the strength of the auxiliary pole will automatically increase and decrease with the load current. Auxiliary poles are used with machines with a power from 1kW.

b. use of compensation winding

Due to do their local operation, auxiliary poles are not sufficient to counteract the armature reaction in large machines (from 100 kW). A compensation winding C_1C_2 is placed in the opening of the pole shoes. These compensation windings are connected in series with the armature and thus are part of the resistance R_i of the armature winding.

c. by making **the flux of the main poles large**

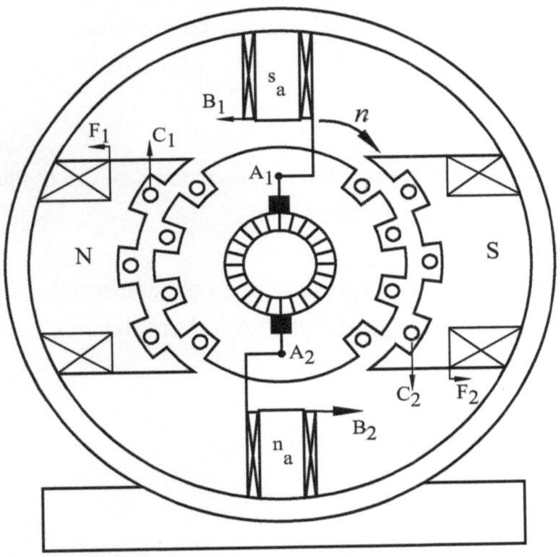

Fig. 16-32: DC machine with compensation winding

2.1.6 Commutation

Commutation is the reversal of the current through a coil of the armature winding. In simple terms, we can understand the phenomena by considering the coil PQ (in fig. 16-33) and following it during its movement together with the collector. The brush is obviously remaining in place. Note that the direction of the current in PQ reverses when the brush slides from lamella a to b. Due to the self induced emf, which occurs in PQ a relatively large current can flow in the temporarily short-circuited coil (fig. 16-33b). It is obvious that in the transition from fig. 16-33b to 16-33c an arc exists between the brush and collector.

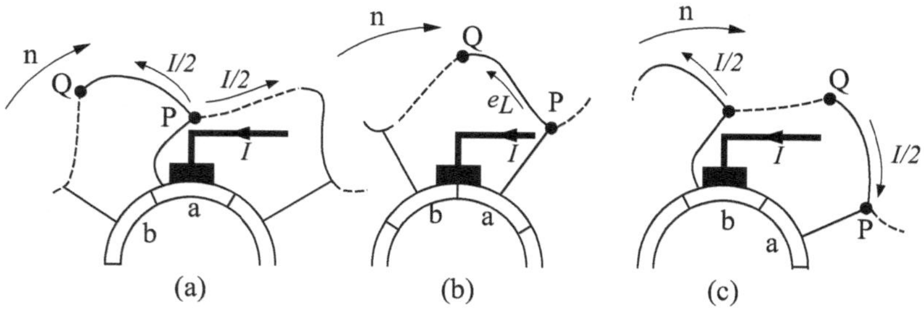

Fig. 16-33: Commutation of the armature current in one of the armature coils

To prevent the self induced voltage becoming too large in the coil which is commutating we attempt to produce an emf that opposes the self induced voltage. The required field can be produced by the auxiliary pole (already in use to counteract the armature reaction). This is the reason that auxiliary poles are called internationally "commutation poles". As an alternative to auxiliary poles commutation can be improved by arranging for the brushes to be offset from the neutral line (in the direction of rotation of the armature). Commutation is one of the factors that limit the maximum emf of the dynamo to about 3500V and in most cases much less than that.

2.1.7 Methods for exciting the dynamo

The excitation results in the production of a magnetic flux in the machine. This almost always occurs by sending an electrical current through a field winding. Special machines of very small power level can use permanent magnets.

a. Independent excitation
The inductor current is provided by a DC current source that is independent of the dynamo. This DC current source can be a battery or a small mains rectifier.

b. Self excitation
The dynamo is excited by a current that it produces. Depending upon the construction of the machine and the electrical connections we can distinguish between a shunt dynamo, series dynamo and the compound dynamo.

c. Separate form of excitation
The dynamo is excited by a field dynamo, mechanically connected to it.

2.1.8 Terminal terminology

To distinguish between the machine windings we use the following symbols: armature: *A*; auxiliary poles: *B*; compensation winding: *C*; series winding: *D*; shunt winding: *E*; separate and independent excitation: *F*.
Both terminals bear the index "1" and "2". For example, armature winding: A_1-A_2.

2.2 DC MOTOR

2.2.1 Operating principle

In two slots of a drum we place a current carrying conductor in a homogenous magnetic field (fig. 16-34). The Lorentz force will operate on the current carrying conductors:

- **magnitude:** $F_l = B \,.\, I \,.\, l$ (N)
- **direction:** the three fingers of the left-hand (for result see fig. 16-34).

On the conductors (and therefore also the drum) a torque will exist resulting in **rotation** of the drum. We now take a DC machine, as shown in figure 16-32 and send a DC current through the field winding and through the armature winding. As shown in fig. 16-34 the armature will begin to rotate. The machine is now operating as a **motor**. Due to the construction of the collector (fig. 16-27) the current in the armature winding will always have the same direction when in the vicinity of each pole: **the motor remains rotating**. The same machine can operate as a motor or a generator.

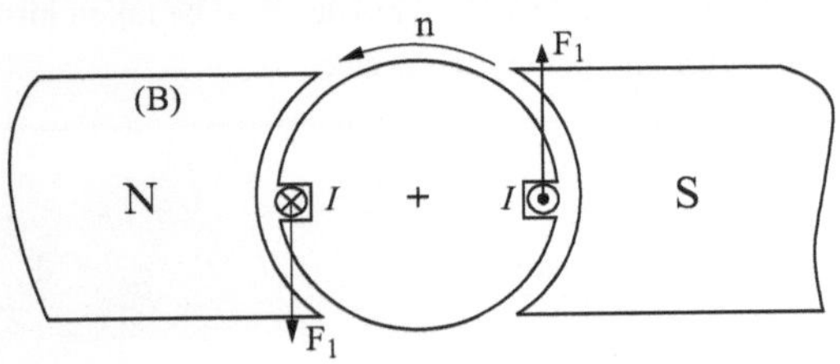

Fig. 16-34: Principle of DC motor

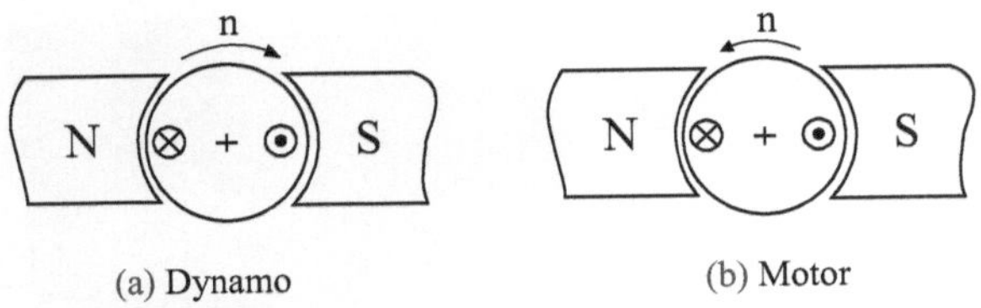

Fig. 16-35: Rotation direction DC machine

The direction of rotation of the motor changes if the current direction through the armature or through the field winding changes.
In fig. 16-35 we can see that with the same direction of armature current in a dynamo and a motor, the rotation direction of both machines is opposite.

2.2.2 Value of the counter emf

The armature winding of a rotating motor is subject to the same conditions as the armature of a rotating generator. An induced voltage is also produced in the armature winding of the motor ($E = k_1 \, . \, n \, . \, \Phi$). If we determine the direction of this emf in fig. 16-35b, then we see that it opposes the armature current: we have a **counter emf** in the armature.

If V_a is the applied armature voltage and E the counter emf, then we have:

$V_a = E + I_a \, . \, R_i$ with: $R_i = R_a + R_m + R_h$ for a series motor

$R_i = R_a + R_h$ with an independently excited motor.

R_i = total internal resistance of the armature circuit;

R_a = armature resistance (including the resistance of the brushes and any current sensors,...);

R_h = resistance of the auxiliary pole winding;

R_m = resistance of the magnetizing winding.

2.2.3 Motor speed

From $E = k_1 \, . \, n \, . \, \Phi$ it follows: $n = \frac{E}{k_1 \, . \, \Phi}$ so that:

$$n_{motor} = \frac{V_a - I_a \, . \, R_i}{k_1 \, . \, \Phi} \qquad (16\text{-}23)$$

For a DC generator:

$$n_{generator} = \frac{E}{k_1 \, . \, \Phi} = \frac{V_a + I_a \, . \, R_i}{k_1 \, . \, \Phi} \qquad (16\text{-}24)$$

For the same V_a, I_a and Φ, a motor will have a lower speed than will be necessary for a generator to operate. The difference can be for example 10%. This needs to be taken into account if a single machine is to be used for both motor and generator service.

Remark

With $I_a \, . \, R_i$ for example 5% of V_a, then we can write:

$$n = \frac{V_a - I_a \, . \, R_i}{k_1 \, . \, \Phi} \approx \frac{V_a}{k_1 \, . \, \Phi}$$

The speed of a DC motor can be controlled by controlling V_a or Φ.

Photo Leroy-Somer: View of a DC motor from the LSK series. Compact square implementation with visible lamella structure, forced air cooling (ICOG), insulation class H.

2.2.4 Momentum (of torque)

If: I_a = total armature current (A)
I = current in a single armature winding (A)
l = useful length of armature winding (m)
$2.a$ = number of parallel armature windings
B = magnetic induction (Wb/m^2)
N = total number of armature windings
r = radius of armature (m)
b = pole length (pole length measured in meters, according to circumference)
$2.p$ = number of poles per machine
Φ = flux (Wb) of one pole
M_{em} = electromechanical momentum (of torque) in the machine (Nm)
M = momentum (of torque) on the shaft of the machine (Nm)

We find in fig. 16-34: $F_1 = B \,.\, l \,.\, I$ (Newton)

$I = \frac{I_a}{2\,.\,a}$ gives: $F_1 = B \,.\, \frac{I_a}{2\,.\,a} \,.\, I$

Armature windings per meter of circumference: $N_1 = \frac{N}{2\,.\,\pi\,.\,r}$

Total pole length of machine: $2\,.\,p\,.\,b$

Number of windings in front of the poles in the magnetic field:

$$N_2 = N_1 \,.\, 2\,.\,p\,.\,b = \frac{N}{2\,.\,\pi\,.\,r} \,.\, 2\,.\,p\,.\,b = \frac{N\,.\,b}{\pi\,.\,r} \,.\, p$$

Total force on circumference: $F = F_1 \,.\, N_2 = B \,.\, \frac{I_a}{2\,.\,a} \,.\, l \,.\, \frac{N\,.\,b}{\pi\,.\,r} \,.\, p$

But: $B\,.\,l\,.\,b = \Phi$ therefore: $F = \frac{p}{a} \,.\, \Phi \,.\, I_a \,.\, N \,.\, \frac{1}{2\,.\,\pi\,.\,r}$

Momentum of electromagnetic torque: $M_{em} = F\,.\,r = \frac{p}{a} \,.\, \Phi \,.\, I_a \,.\, N \,.\, \frac{1}{2\,.\,\pi}$ (Nm)

For a specific machine p, a and N are constants and: $$M_{em} = k_2 \,.\, I_a \,.\, \Phi \tag{16-25}$$

M_{em} = momentum of electromechanical torque

$k_2 = \frac{p}{a} \,.\, \frac{N}{2\,.\,\pi}$ = constant ; I_a = armature current (A); Φ = flux per pole (Wb)

Remark

If $k_2 \,.\, \Phi = K_M$, then: $$M_{em} = K_M \,.\, I_a \tag{16-26}$$

K_M = motor constant

$K_M = k_2 \,.\, \Phi = \frac{p}{a} \,.\, \frac{N}{2\,.\,\pi} \,.\, \Phi = \frac{p}{a} \,.\, \frac{N}{60} \,.\, \frac{30}{\pi} \,.\, \Phi$. From (16-22) it follows: $$K_M = K_G \tag{16-27}$$

K_M has the same value +as K_G, but K_M is expressed in Nm/A and K_G in $\frac{V}{rad/s}$!

2.2.5 Power

If we call $P_{sup.}$ the total supplied electrical power, then:

$P_{sup.} = V_a \cdot I_a + V_m \cdot I_m$ for an independently excited motor (fig. 16-36a).

$P_{sup.} = V_a \cdot I_a$ for a permanent magnet motor and for a series motor (fig. 16-40).

We call P the mechanical power on the shaft of the motor, this is the power stated on the name plate of the motor.

The efficiency of the motor can be written as: $\eta = P/P_{sup.}$.

We now consider the practical example of an **independently excited motor.**

Supplied electrical energy in the armature circuit: $P_{sup.} = V_a \cdot I_a$

Internal Joule loss in armature circuit: $P_j = I_a^2 \cdot R_i$

Remainder: $P_{em} = P_a - P_j = V_a \cdot I_a - I_a^2 \cdot R_i = (V_a - I_a \cdot R_i) . I_a = E \cdot I_a$ = electromechanical power.

P_{em} is the electrical power converted into mechanical power. This is sometimes referred to as the internal electrical power.

On the shaft of the motor we have the following power: $P = P_{em} - P_{Fe} - P_{fric.}$

P_{Fe} = iron losses in armature = sum of hysteresis and eddy current losses

$P_{fric.}$ = friction losses of
- the shaft in its bearings
- brushes on the commutator
- ventilator losses of armature with respect to the air.

Fig.16-36b shows these different powers once again in a didactic manner.

A certain M_{em} corresponds with P_{em} : $P_{em} = M_{em} . \omega$. Here in ω is the angular velocity of the armature (rads/s). For every P on the shaft of the motor there is a corresponding moment of torque M:

$$P = \omega \cdot M \qquad (16\text{-}28)$$

If we make use of the data on the name plate of the motor then there is a practical formula to determine the output torque of the motor:

$M = \frac{P}{\omega} = \frac{P \cdot 60}{2 \cdot \pi \cdot n} = \frac{9.55 \cdot P}{n}$ With P in kW we find:

$$M = \frac{9550 \cdot P}{n} \qquad (16\text{-}29)$$

P : power in kW

n : speed in rpm

M: moment of torque in Nm

On the shaft of the motor we can distinguish four mechanical variables each electronically controllable:

Variable	Unit	Regulation	Example
Velocity n	rpm	Speed control	Drive system
Momentum of torque M	Nm	Torque control	Lifting equipment
Angular position of shaft θ	radian	Motion control	CNC-machine
Power P	W	Power control	Wind up equipment

The speed and torque together determine the power (expression 16-28!) so that in fact we could create a table with three variables: n, M, θ .

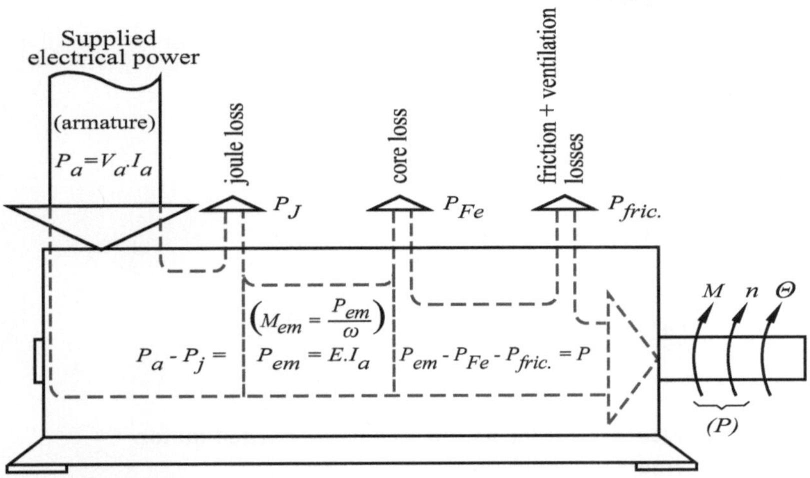

Fig. 16-36b: Relationship between the powers in an independently excited motor

Numeric example 16-7:

Independently excited motor

440 V - 13 A - 1540 rpm - 4.7kW-

R_i = 0.68 Ω - Excitation: 300 V - 1A

Question:

- Efficiency
- Machine constants K_M and K_G
- Powers and torques which are indicated in figure 16-36b.

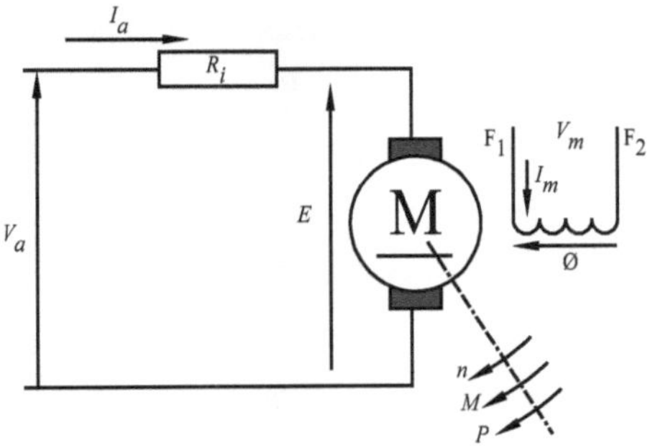

Fig. 16-36a: Configuration of independently excited motor

Solution:

$P_{sup.}$ = 400 x 13 + 300 x 1 = 6020W

Efficiency of machine: $\eta = \frac{4700}{6020} = 78\%$

Losses in armature circuit: $P_j = I_a^2 \cdot R_i = 13^2 \cdot 0.68 = 114.92\text{W}$

Emf at full load and 1540 rpm: $E = V_a - I_a \cdot R_i = 440 - 13 \cdot 0.68 = 431.16\text{V}$

Normalized emf: $e_N = \frac{431.16}{1540} = 0.28$ V/ rpm

(16-21): $E = K_G \cdot \omega = 0.28 = K_G \cdot \frac{2 \cdot \pi \cdot 1}{60} \rightarrow K_G = 2.6738 \frac{\text{V}}{\text{rad/s}}$

$K_M = [\ K_G\ ] = 2.6738$ Nm/A

$P_{em} = P_a - P_j = 440 \cdot 13 - 114.92 = 5605\text{W}$

$M_{em} = \frac{P_{em}}{\omega} = \frac{5605 \cdot 60}{2 \cdot \pi \cdot 1540} = 34.75\text{Nm}$

$P_{Fe} + P_{fric.} = P_{em} - P = 5605 - 4700 = 905\text{W}$

$M = \frac{9550 \cdot 4.7}{1540} = 29.14\text{Nm}$

Losses in percent:

$P_{j\,(rotor+stator)} = 114.92 + 300 = 414.92\text{W}$

$P_j : \frac{414.92}{6020} \cdot 100\% = 6.89\%$

$P_{Fe} + P_{fric.} : \frac{905}{6020} \cdot 100\% = 15\%$

2.2.6 Equivalent circuit for an independently excited motor. Mathematical model.

The ideal DC motor can be described with two equations:

$E = k_1 \cdot n \cdot \Phi = K_G \cdot \omega$ (V)

$M \approx M_{em} = k_2 \cdot I_a\ \Phi = K_M \cdot I_a$ (Nm)

In reality, the armature of the DC motor has not only resistance R_a but also self inductance L_a. On the other hand, the motor is driving a mechanical load and so the properties of this load (moment of inertia J_m, load torque M_l) need to be also taken into account. Fig. 16-37 shows the equivalent circuit for a loaded DC commutating motor.

The electrical equations associated with fig. 16-37 are:

$$v_a = e + i_a \cdot R_i + L_a \cdot \frac{di_a}{dt} = i_a \cdot R_i + L_a \cdot \frac{di_a}{dt} + k_1 \cdot n \cdot \Phi \quad (16\text{-}30)$$

In nominal service: $V_a = I_a \cdot R_i + E$ (16-31)

$$\Phi = k_3 \cdot i_m \quad (16\text{-}32)$$

$$v_m = i_m \cdot R_m + \frac{d\Phi}{dt} = i_m \cdot R_m + L_M \cdot \frac{di_m}{dt} \quad (16\text{-}33)$$

Take note that the term $L_a \cdot \frac{di_a}{dt}$ is only relevant when the current changes quickly. In nominal service : $I_a = C^{te}$ and $L_a \cdot \frac{di_a}{dt} = 0$.

The torque M on the shaft of the motor serves to counteract the load torque M_t.
Any eventual difference $M - M_t = M_v$ will accelerate or reduce the speed of the total moment of inertia (armature + load), depending on the polarity of M_v:

$$M_v = J_m \cdot \frac{d\omega}{dt} \tag{16-34}$$

$M \approx k_2 \cdot I_a \cdot \Phi = M_t + M_v$
In reality $M < M_{em} = k_2 \cdot I_a \cdot \Phi$, but for low speeds the difference is minimal.
We can also incorporate the friction and iron losses in M_t, in this case $k_2 \cdot I_a \cdot \Phi = M_t + M_v$.
Table 16-1 shows the mathematical model for an independently excited motor.

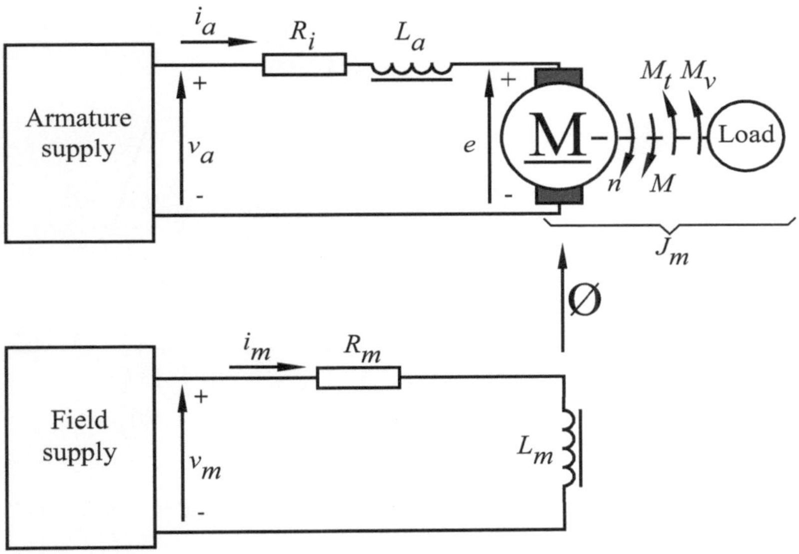

Fig. 16-37: Independently excited motor

Table 16-1

$v_a = e + i_a \cdot R_i + L_a \cdot \frac{di_a}{dt}$	(16-30)	$M_v = J_m \cdot d\omega/dt$ (Nm)	(16-34)
nominal: $V_a = I_a \cdot R_i + E$	(16-31)	$M \approx M_{em} = k_2 \cdot I_a \cdot \Phi = M_t + M_v$	(16-35)
$\Phi = k_3 \cdot I_m$ (Wb)	(16-32)	$P = \omega . M = (V_a . I_a - I_a^2 . R_i) - P_{Fe} - P_{fric.}$	(16-36)
$E = k_1 \cdot n \cdot \Phi$	(16-20)	$M = \frac{9550 \cdot P}{n}$ (Nm), with P in kW!	(16-29)

2.2.7 *M-n* curves

Assume the flux Φ is constant and equal to its nominal value. The armature is connected to a variable voltage V_a. The motor is loaded with a reverse operating torque M_t.
The relationship between M_t and n is provided. For example, a ventilator or a centrifugal pump with an almost quadratic characteristic (16-38a). From the equations (16-20), 16-31) and (16-35) we find, after rearrangement (and in nominal service):

$$n = \frac{V_a - I_a \cdot R_i}{k_1 \cdot \Phi} = \frac{V_a}{k_1 \cdot \Phi} - \frac{R_i \cdot M}{k_5 \cdot \Phi^2} \qquad (16\text{-}37).$$

This relationship is graphically demonstrated in fig. 16-38b for three values of V_a and with Φ = constant.

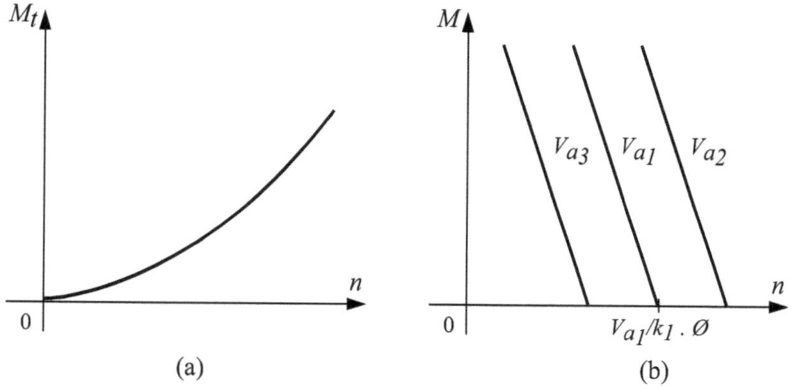

Fig. 16-38: *M-n* load curves and motor

To know where the operating point of the motor is, we draw both curves (fig. 16-38a and b) together in one *M-n* diagram in fig. 16-39.

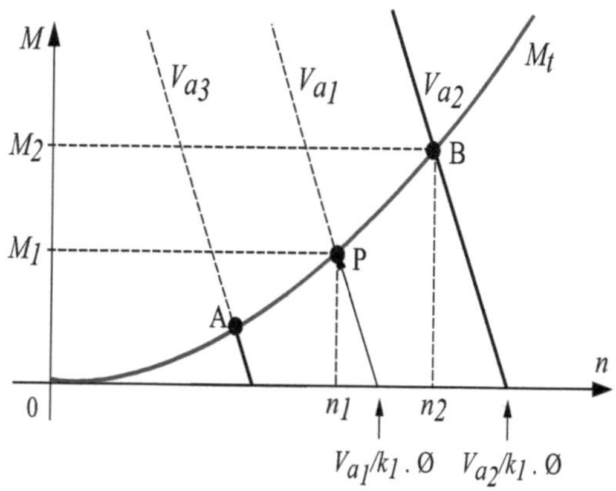

With an armature voltage V_{a1} the operating point is P (speed n_1, working torque M_1)
Assume a (fast) change in the voltage V_{a1} to the value V_{a2}.
In the first instance, this produces a momentary increasing drive torque, so that the motor accelerates, according to:

$$M_2 - M_1 = M_v = J_m \cdot \frac{d\omega}{dt}$$

At $M = M_2 = M_t$ equilibrium has returned (operating point B!). Now the motor rotates with a speed n_2 and an armature voltage V_{a2}.

Fig. 16-39: Determining the operating point of a motor.

2.2.8 Series motor

Fig. 16-14 shows the electrical connections. This is a self exciting machine. The field winding is in series with the armature. Since the (large) armature current flows through this coil, we only need a few windings (thick wire). With an independently excited motor the field coil is connected to a high voltage and many windings of thin wire with a small field current (3 to 5% of armature current) create a sufficiently large flux.

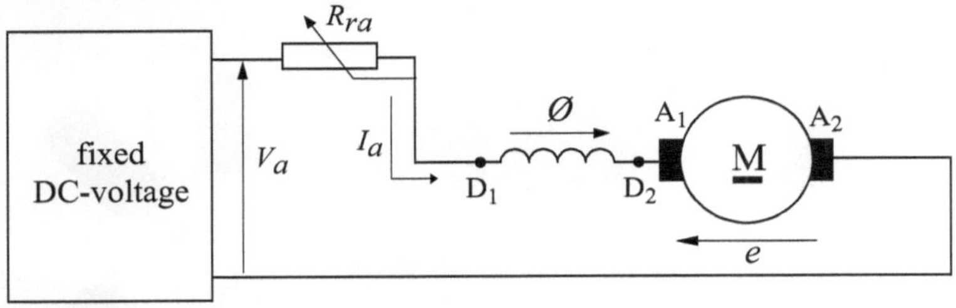

Fig. 16-40: Series motor

Formulae:

$\Phi = k \cdot I_a$; $E = k_1 \cdot n \cdot \Phi$; $V_a = E + I_a \cdot R_i$; $M = k_2 \cdot I_a \cdot \Phi = k_3 \cdot I_a^2$ (valid when poles not saturated!). From $M = k_3 \cdot I_a^2$ and with I_{start} = 2 to 3 times $I_{a\,(nom.)}$ we see that the motor has a large starting torque. From $n = \frac{E}{k_1\,\Phi} \approx \frac{V_a}{k_4 \cdot I_a} \approx \frac{1}{k_5 \cdot I_a}$ we derive fig. 16-41a.

The expression $M = k_3 \cdot I_a^2$ leads to fig. 16-41b. From fig. 16-41a and b follows fig.16-41c.

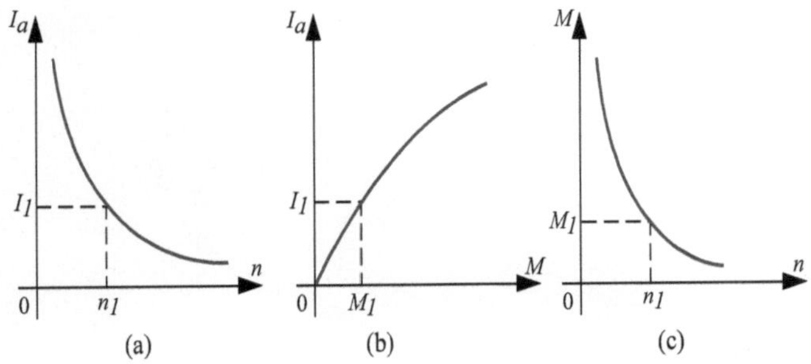

Fig. 16-41: Curves associated with a series motor

In the past, a series motor was predominantly used for traction applications because of its extremely interesting property of a large starting torque.

In fig. 16-41c, we note that for a small mechanical load the speed is extremely high. At no load for a series motor we lose control and the machine can be damaged by the resulting centrifugal forces. Since at high speeds the armature current is very small (fig. 16-41a) the motor will not burn out as a result of this loss of control. For safety reasons a series motor is connected directly or via gears to the load.

A much used derivative of the series motor is the AC version. In this case we're talking about a universal motor since it can operate on AC or DC (see chapter 21).

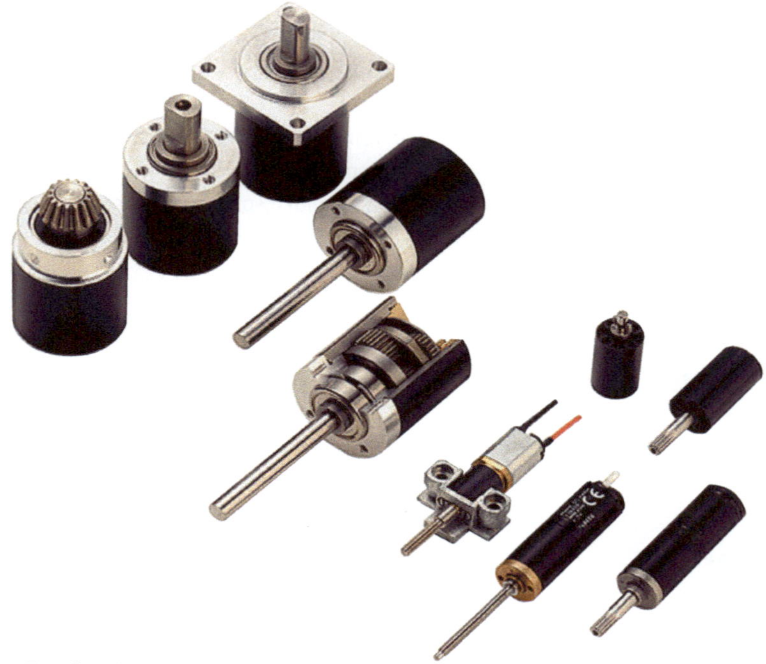

Photo Maxon motor Benelux: Customer specific gear heads

An ideal solution when performance of the motor is sufficient, but the motor speed is too high and torque too low. Example: special planetary gear heads developed for customer specifications. Gear wheels and planetary gear heads with reduction ratios of 6:1 to 5721:1 and 4:1 to 6285:1. Output torques up to 120 Nm.

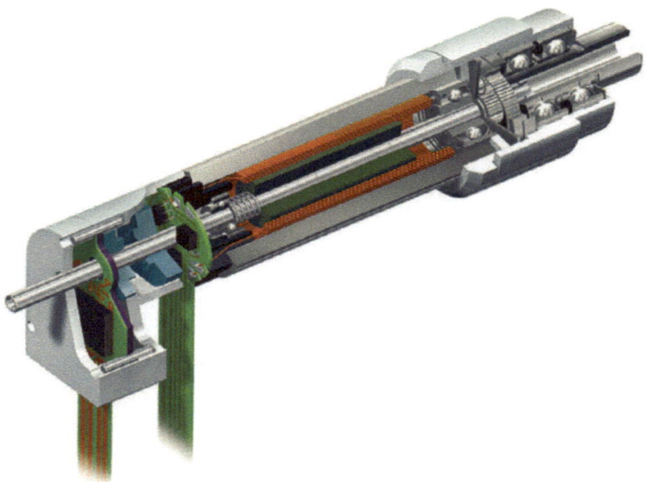

Photo Maxon motor Benelux: miniature servo system

Miniature servo comprised of a brushless EC-motor $\Phi = 6$mm with a magnetic encoder and a low tolerance $\Phi = 8$mm harmonic drive gearbox ®. Continuous 1.2 W despite the small dimensions.

16.3 Three-phase asynchronous motor

CONTENTS

3.1 Rotating stator field

3.1.1 Qualitative study

Assume a stator with 12 slots as shown in fig. 16-42. In the first case we place a single insulated conductor in both the upper and lower slot. We connect these two conductors so that we have a coil as shown in fig. 16-43.

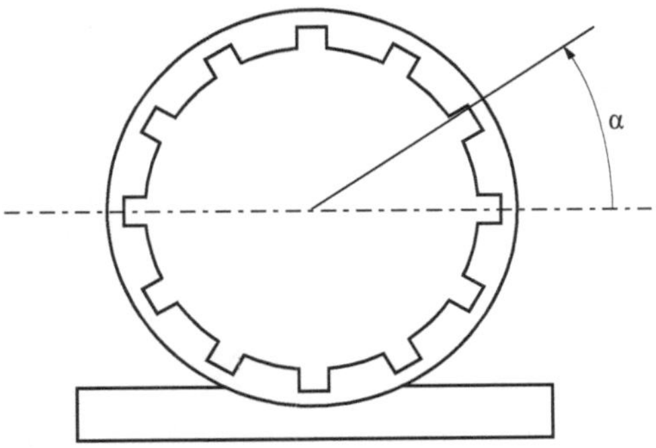

Fig. 16-42: Stator of an asynchronous motor

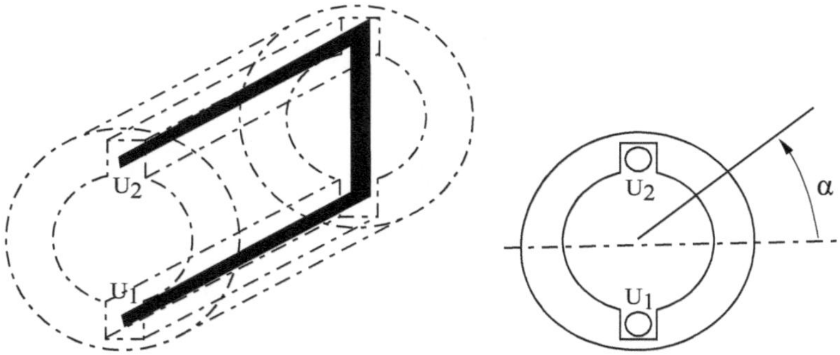

Fig. 16-43: U_1 - U_2 coil with a single winding

We now place, not one, but a total of three coils in specific slots of the stator shown in fig. 16-42 and we name these coils: U_1-U_2, V_1-V_2, W_1-W_2 (see fig. 16-44).

The three coils may be connected to a three phase supply in either star or in delta. Fig. 16-45 shows star connection. A three phase current flows through the stator coils.

The three phase currents i_{S1}, i_{S2}, i_{S3} are drawn (fig. 16-46) and we consider the points in time 1 to 7.

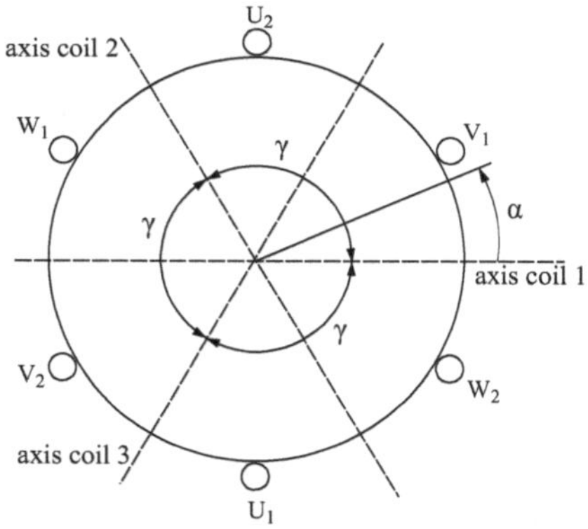

Fig. 16-44: Simple stator winding of three phase motor

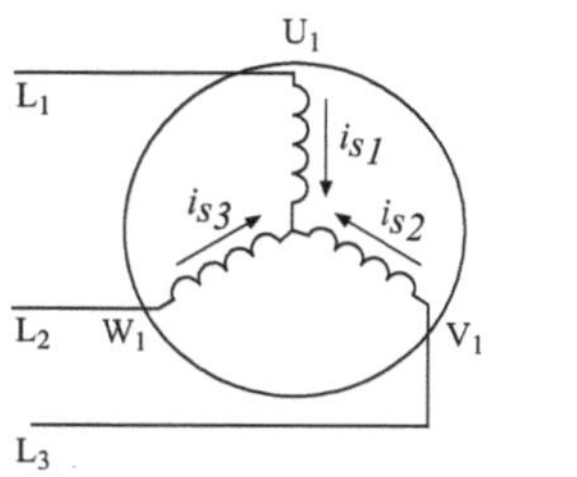

Fig. 16-45: Stator connected to the grid

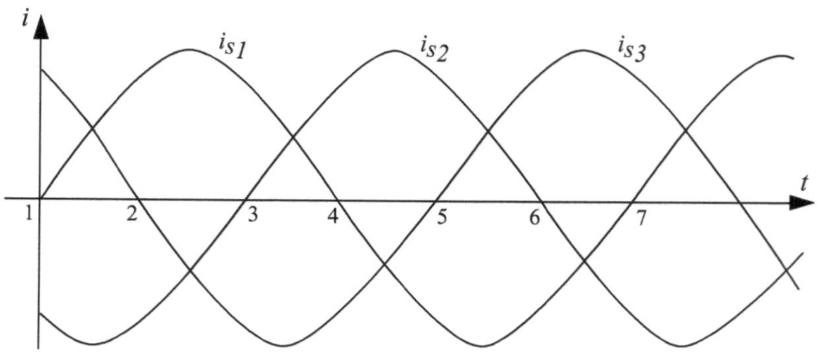

Fig. 16-46: Three phase currents in the stator windings of fig. 16-44

Every current in its respective stator coil produces a varying magnetic field. We now draw the stator coils in fig. 16-47 and determine (for the seven points in time) the resultant magnetic field in the stator. We refer to the phase currents as positive if they flow respectively from U_1 to U_2, from V_1 to V_2 and from W_1 to W_2.

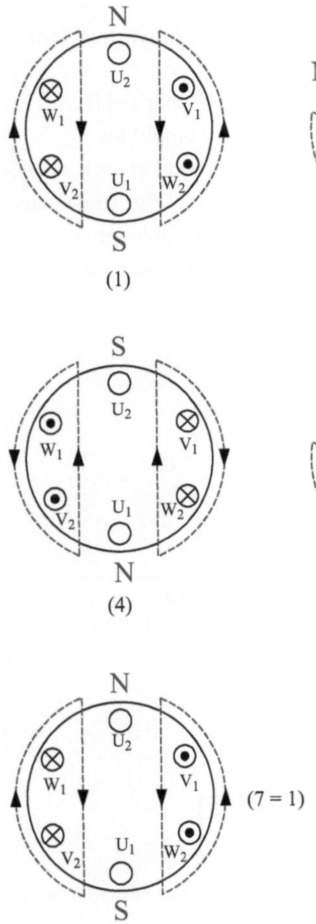

Fig. 16-47: Magnetic field in the stator at carefully selected moments

We note in fig. 16-47 that a rotating magnetic field is produced in the stator iron of the machine. In the future we simply talk about the rotating stator field. The stator field rotates counter clockwise (left rotating field).

The configuration in fig. 16-44 is for a two pole machine. In one period of the three-phase current the rotating field has covered two pole pitches. With p pole pairs (or 2.p poles) and a supply frequency of f_S the speed of the rotating field (rpm) will be :

$$n_s = \frac{60 \cdot f_s}{p} \qquad (16\text{-}38)$$

If we had wound the forward going and returning conductor of every coil with a displacement of 90° instead of 180°, then the rotating field would have been composed of two north and two south poles. We would then have a four pole machine (p = 2).

3.1.2 . Sinusoidal distribution of windings

Fig. 16-44 shows the principle of a three-phase winding, however in practical motors we would have:

1. more than one winding per phase coil (to create a strong stator field)
2. attempt to create a sinusoidal rotating field with constant amplitude.

To achieve this last point the coils of every phase winding are wound in such a way that the air gap flux is sinusoidally distributed around the stator circumference. Fig. 16-48 shows a theoretical example of phase winding S_1 (this is coil U_1- U_2). For clarity in the drawing we have for now not yet included the other coils. In the future we take coil S_1 as reference and the shaft of this coil is used as the **stator reference axis** in our further study.

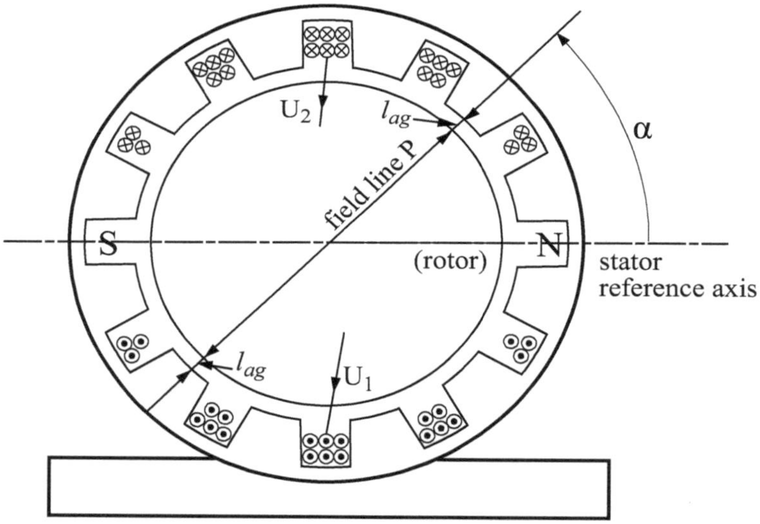

Fig. 16-48: Three-phase motor with one phase shown which produces (by approximation) a sinusoidal flux distribution around the stator circumference

If the phase winding S_1 has N_S windings in the slots, then the ideal winding density is given by:

$$n_{S1} = \frac{N_S}{2} . \sin \alpha \quad (16\text{-}39)$$

If the current through S_1 is i_{S1} , then the magnetomotive force (mmf) in fig.16-48 according to field line P is:

$$F_{\mu,P} = \int_{\alpha}^{\alpha+\pi} n_{S1} . i_{S1} . d\alpha \quad \rightarrow \quad F_{\mu,P} = N_S . i_{S1} .\cos \alpha \quad (16\text{-}40)$$

As we shall see in 3.2 a rotor is placed in the machine. The length of the airgap between rotor and stator is indicated by l_{ag} . This is the effective length, taking the rotor and stator positions into account.

Since there are now two airgaps in series, the magnetomotive force in one airgap is:

$$F_{\mu,ag\,S_1(\alpha)} = \frac{N_{S1} \cdot i_{S1}}{2} \cdot \cos\alpha \tag{16-41}$$

The induction in the airgap is:

$$B_{ag\,S_1(\alpha)} = \frac{\mu_0 \cdot F_{\mu,ag\,S1(\alpha)}}{l_{ag}} = \frac{\mu_0 \cdot N_S \cdot i_{S1}}{2 \cdot l_{ag}} \cdot \cos\alpha \tag{16-42}$$

For the other coils V_1-V_2 and W_1-W_2 the winding densities are:

$$n_{S2} = \frac{N_S}{2} \cdot \sin\left(\alpha - \frac{2 \cdot \pi}{3}\right) \quad \text{and} \quad n_{S3} = \frac{N_S}{2} \cdot \sin\left(\alpha - \frac{4 \cdot \pi}{3}\right)$$

Photo Siemens: High Efficiency motors with efficiency class of IE 4.
The Simotics IE4 motors are designed according to DIN EN 50347 standard, which means that existing IE1, IE2 or IE3 motors can be replaced easily.

Based on the 1LE1 platform, highly-efficient IE4 motors with a choice of aluminum housing (Simotics General Purpose) in the range from 2.2 to 18 kW, or cast-iron housing (Simotics Severe Duty) from 2.2 to 200 kW in two and four-pole versions are available.

Numeric example 16-8:

We reconsider a stator with 12 slots (fig.16-49). The coils N_1 to N_6 are in series and form one phase. Assume that one such complete phase winding has 1200 windings. We wish to distribute these windings as sinusoidal as possible. Determine the number of windings for these six coils, as well as the mmf in the airgap if the current is 1A.

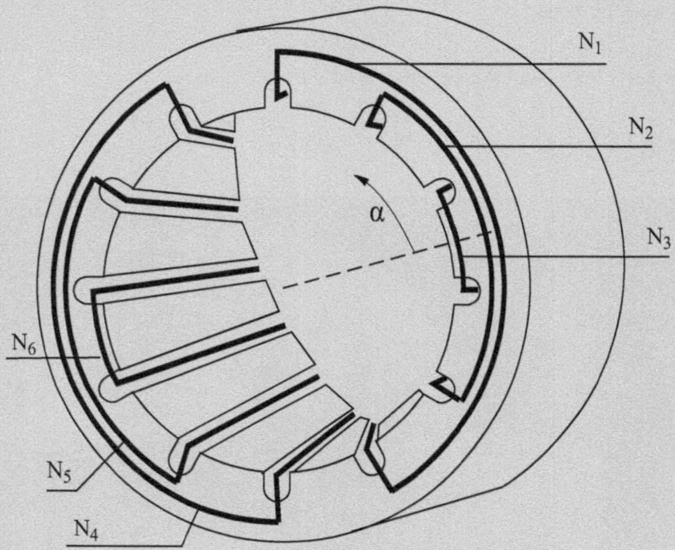

Fig.16-49: Stator iron wound with one phase coil S_1 (= U_1U_2)

Solution:

In a stator with m slots we are able to determine the number of windings in slot n by taking the ideal sinusoidal distribution over an angle $\frac{2.\pi}{m}$ around slot n:

$$N_k = \int_{2\pi(n-1)/m}^{2\pi n/m} n_{S1} \,.\, d\alpha = \frac{N_S}{2} \int_{2\pi(n-1)/m}^{2\pi n/m} \sin\alpha \,.\, d\alpha \qquad \text{with } n = 1, 2, ..., 6 \qquad (16\text{-}43)$$

With m = 12 and N_S = 1200 we find:

$$N_3 = N_6 = \frac{1200}{2} \int_0^{\pi/6} \sin\alpha \,.\, d\alpha = 80 \;;\; N_2 = N_5 = \frac{1200}{2} \int_{\pi/6}^{\pi/3} \sin\alpha \,.\, d\alpha = 220\,;$$

$$N_1 = N_4 = \frac{1200}{2} \int_{\pi/3}^{\pi/2} \sin\alpha \,.\, d\alpha = 300$$

With i_{S1} =1A we find as a result of N_1 and N_4 in each air gap an mmf of 300Aw across a width of $5.\pi/6$. The coils N_2 and N_5 produce a field of 220 Aw across $\pi/2$ radians. N_6 and N_3 are responsible for an extra 80 Aw across $\pi/6$ radians. This has been drawn in fig.16-50.

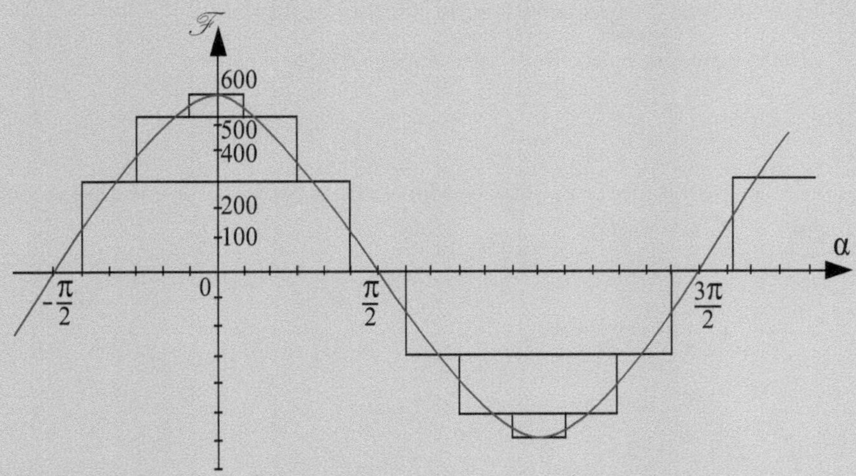

Fig.16-50: Magnetomotive force in the airgap of the configuration of fig. 16-49

Application of the Fourier series shows that only odd cosine terms can exist.

We calculate these Fourier components: $F_{\mu,S1,\alpha} = \frac{4}{\pi}\int_0^{\pi/2} F_\mu \cdot \cos k\alpha \cdot d\alpha$ with k = 1, 3, 5, ...

The fundamental harmonic (k = 1) has an amplitude:

$$F_{\mu,1\,(S1)} = \frac{4}{\pi}\cdot\left[300\cdot\sin\alpha\Big|_0^{5.\pi/12} + 220\cdot\sin\alpha\Big|_0^{\pi/4} + 80\cdot\sin\alpha\Big|_0^{\pi/12}\right] = 593.39\text{ Aw}$$

The amplitude of the third harmonic (k = 3) is given by:

$$F_{\mu,3\,(S1)} = \frac{4}{3\cdot\pi}\cdot\left[300\cdot\sin\alpha\Big|_0^{15.\pi/12} + 220\cdot\sin\alpha\Big|_0^{3.\pi/4} + 80\cdot\sin\alpha\Big|_0^{\pi/4}\right] = 0\text{ Aw}$$

Fifth, seventh and ninth harmonics prove to be zero.

$$F_{\mu,11\,(S1)} = \frac{4}{11\cdot\pi}\cdot\left[300\cdot\sin\alpha\Big|_0^{55.\pi/12} + 220\cdot\sin\alpha\Big|_0^{11.\pi/4} + 80\cdot\sin\alpha\Big|_0^{11.\pi/12}\right] = 53.94\text{ Aw}$$

$$F_{\mu,13\,(S1)} = 35.84\text{ Aw}$$

The eleventh harmonic is $\frac{1}{11} \approx 9\%$ of the fundamental.

According to fig. 16-50 the maximum mmf is 600Aw. The amplitude of the fundamental according to our calculation is 593.39 Aw.

Usually we define an equivalent sinusoidal winding as :

$$N_{Se} = 1200\cdot\frac{593.39}{600} = 1184.78\text{ Aw}$$

3.1.3 Analytical expression for the rotating stator field

Assume three sinusoidal distributed windings S_1, S_2, S_3 whereby:

$$S_1\,(=U_1U_2)\;,\quad S_2\,(=V_1V_2)\;,\quad S_3\,(=W_1W_2)$$

So, as to avoid cluttering the drawing with details, in fig. 16-51, only one central conductor per phase winding is shown.

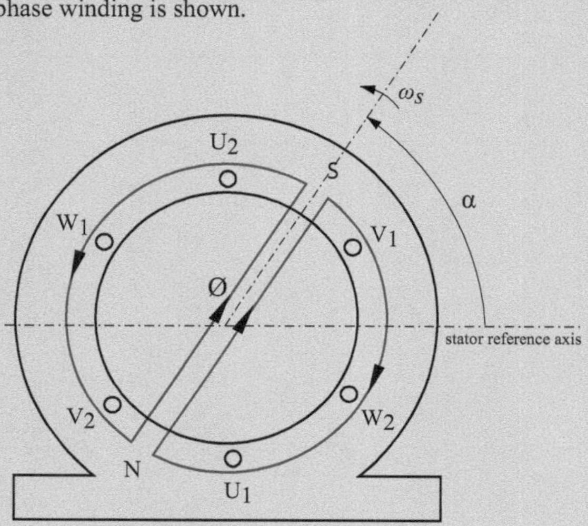

Fig. 16-51: Stator with three-phase winding

The phase current can be written as:

$$i_{S1} = \hat{i}_\mu \cdot \cos(\omega_S \cdot t + \varphi_\mu)$$

$$i_{S2} = \hat{i}_\mu \cdot \cos(\omega_S \cdot t + \varphi_\mu - \frac{2.\pi}{3})$$

$$i_{S3} = \hat{i}_\mu \cdot \cos(\omega_S \cdot t + \varphi_\mu - \frac{4.\pi}{3})$$

$\hat{i}_\mu$ = peak value of magnetising current

$\omega_S = 2 \cdot \pi \cdot f_S$ = pulsation of the three-phase stator currents.

Each of these stator currents produces an mmf so that at an angle α of the stator the total magnetising mmf (in the airgap) can be written as : $F_{\mu,\alpha} = F_{agS1\,(\alpha)} + F_{agS2\,(\alpha)} + F_{agS3\,(\alpha)}$

Application of 16-41 gives:

$$F_{\mu,\alpha} = \frac{N_{Se}}{2} \cdot \hat{i}_\mu \cdot \Big[\cos(\omega_S \cdot t + \varphi_\mu) \cdot \cos\alpha + \cos(\omega_S \cdot t + \varphi_\mu - \frac{2.\pi}{3}) \cdot \cos(\alpha - \frac{2.\pi}{3})$$

$$+ \cos(\omega_S \cdot t + \varphi_\mu - \frac{4.\pi}{3}) \cdot \cos(\alpha - \frac{4.\pi}{3})\Big]$$

$$F_{\mu,\alpha} = \frac{N_{Se} \cdot \hat{i}_\mu}{2} \cdot \left[\frac{3}{2} \cdot \cos(\omega_S \cdot t + \varphi_\mu - \alpha) \right]$$

$$F_{\mu,\alpha} = \frac{3 \cdot N_{Se} \cdot \hat{i}_\mu}{4} \cdot \cos(\omega_S \cdot t + \varphi_\mu - \alpha) \qquad (16\text{-}44)$$

Expression 16-44 indicates that the magneto motive force is sinusoidally distributed in space and rotates with an angular velocity of ω_S. The amplitude is determined by $N_{Se} \cdot \hat{i}_\mu$.
The maximum of the magnetic field at each instant occurs at an angle α, given by:
$\alpha = \omega_S \cdot t + \varphi_\mu$ (fig.16-52).

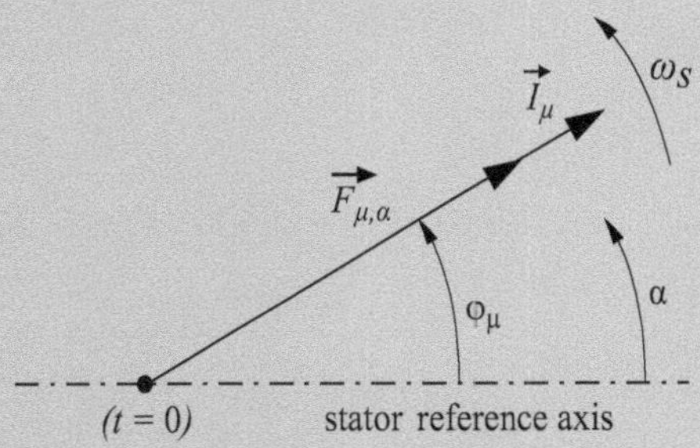

Fig. 16-52: Space vector of the mmf in the air gap of an asynchronous motor

The angle φ_μ is the displacement angle between the applied phase voltage $v_{S1} = \hat{v}_S \cdot \cos \omega_S . t$ and the magnetizing current i_μ.
With the choice of phase order U-V-W we obtain a counter clockwise rotating stator field. If the connection to any two windings are reversed (e.g. U-W-V) then the stator field rotates clockwise.

3.1.4 Magnetising inductance

The induction in the air gap is calculated with: $\vec{B}_{ag} = \frac{\mu_0 \cdot \vec{F}_{\mu,\alpha}}{l_{ag}}$

From (16-44) we can calculate the effective value of $F_{\mu,\alpha}$ so that we can write for B_{ag}:

$$\vec{B}_{ag} = \frac{3 \cdot N_{Se} \cdot \mu_0}{4 \cdot l_{ag}} \cdot \vec{I}_\mu \quad \text{(Tesla)} \qquad (16\text{-}45)$$

We consider fig. 16-51.
We are looking for the instant at which $i_{S1} = \hat{i}_\mu \cdot \cos(\omega_S \cdot t + \varphi_\mu)$, this occurs at $\omega_S \cdot t = -\varphi_\mu$.
At that instant the magnetic induction (vector) is directed according to the reference shaft of the stator (this is coil U_1U_2). The value of the mmf follows from (16-44):

$$F_{\mu,\alpha} = \frac{3 \cdot N_{Se} \cdot \hat{i}_\mu}{4} \cdot \cos\alpha \quad \text{so that: } B_{ag,\alpha} = \frac{\mu_0 \cdot F_{\mu,\alpha}}{l_{ag}} = \frac{3 \cdot N_{Se} \cdot \mu_0}{4 \cdot l_{ag}} \cdot \hat{i}_\mu \cdot \cos\alpha = \hat{B}_{ag} \cdot \cos\alpha$$

The magnetic induction is sinusoidally distributed in space just like the winding.

We will now determine the flux in the coil S_1 and that leads us to the magnetizing inductance. Assume in the first case a winding of coil U_1-U_2 at an angle α (and α + π) to the stator. For a stator with internal radius *r* and axial length *l* the flux of one winding is:

$$\Phi_{winding} = 2 \,.\, \hat{B}_{ag} \,.\, l \,.\, r \,.\, \sin\alpha$$

To find the maximum of all the linked flux of a winding U_1-U_2 we have to integrate $\Phi_{winding}$ across the winding distribution:

$$\hat{\Phi}_{S1} = \int_{-\pi/2}^{+\pi/2} n_{S1} \,.\, \Phi_{winding} \,.\, d\alpha$$

Applying (16-39): $\hat{\Phi}_{S1} = \int_{-\pi/2}^{+\pi/2} \frac{N_{Se}}{2} \,.\, \sin\alpha \,.\, (2 \,.\, \hat{B}_{ag} \,.\, l \,.\, r \,.\, \sin\alpha) \,.\, d\alpha$

$$\hat{\Phi}_{S1} = \frac{\pi}{2} \,.\, N_{Se} \,.\, \hat{B}_{ag} \,.\, l \,.\, r \quad \text{(Wb)} \qquad (16\text{-}46)$$

If we replace (16-45) in expression (16-46), then we find the effective value of the stator flux, generated in one coil (S_1):

$$\vec{\Phi}_{S1} = \frac{3 \,.\, \pi}{8} \,.\, N_{Se}^2 \,.\, \frac{\mu_0 \,.\, l \,.\, r}{l_{ag}} \,.\, \vec{I}_\mu \qquad (16\text{-}47)$$

The linked flux $\vec{\Phi}_{S1}$ is proportional to the magnetizing current. Flux and current are in phase. We refer to $\frac{\vec{\Phi}_{S1}}{\vec{I}_\mu} = L_0$ as the magnetizing inductance:

$$L_0 = \frac{3 \,.\, \pi}{8} \,.\, N_{Se}^2 \,.\, \frac{\mu_0 \,.\, l \,.\, r}{l_{ag}} \qquad (16\text{-}48)$$

Fig. 16-53 shows the magnetizing curve. The induced emf in one phase winding is given by:

$$\vec{E}_{S1} = j \,.\, \omega_S \,.\, \vec{\Phi}_{S1} = j \,.\, \omega_S \,.\, L_0 \,.\, \vec{I}_\mu \qquad (16\text{-}49)$$

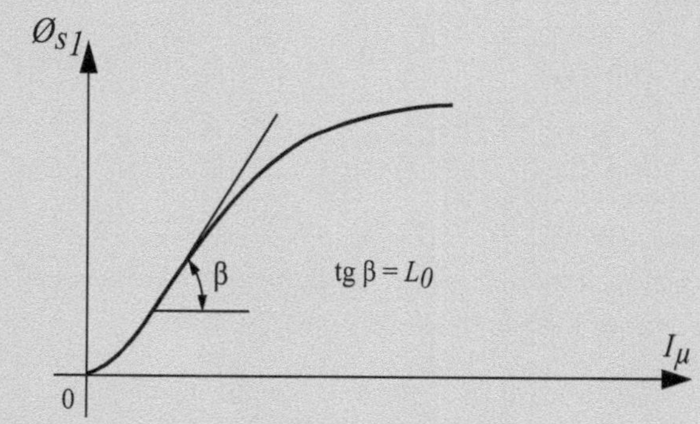

Fig. 16-53: Magnetising curve of an asynchronous machine

Numeric example 16-9:

A two pole three-phase machine has a rotor with a length of 100mm. The radius of the rotor is 50 mm while the air gap has an effective length of 0.6 mm. The effective number of sinusoidally distributed windings is 160. The motor is connected in star to a 400V-50Hz power grid.

Question:

1. The nominal induction in the air gap
2. The magnetising inductance and the magnetising current.

Solution:

If we neglect the resistance R_S of the stator coil, then: $E_{S1} = V_{S1} = \frac{400}{\sqrt{3}} = 230\text{V}$

(16-49): $\Phi_{S1} = \frac{V_{S1}}{\omega_S} = \frac{230}{2 \cdot \pi \cdot 50} = 0.732\text{Wb}$

(16-46): $\hat{B}_l = \frac{2 \,.\, \sqrt{2} \,.\, \Phi_{S1}}{\pi \,.\, N_{Se} \,.\, l \,.\, r} = \frac{2 \cdot \sqrt{2} \cdot 0.732}{\pi \cdot 160 \cdot 0.1 \cdot 0.05} = 0.824\text{Tesla}$

A value of 0.8 to 1 T is normal. The maximum induction in the stator teeth is usually 1.6 to 1.8T. If the stator teeth are as wide as the slots, then the average induction is 0.8 to 0.9T.

Magnetising inductance (16-48): $L_0 = \frac{3 \cdot \pi}{8} \cdot 160^2 \cdot \frac{4 \cdot \pi \cdot 10^{-7} \cdot 0.1 \cdot 0.05}{0.0006} = 315\text{mH}$

Magnetising current (16-49): $I_\mu = \frac{V_{S1}}{\omega_S \cdot L_0} = \frac{230}{2 \cdot \pi \cdot 50 \cdot 315 \cdot 10^{-3}} = 2.324\text{A}$

1LE1 Standard induction motor
0.55 to 200 kW
9.9 to 1546 Nm
50/60 Hz
2-4-6 poles
USA (also 8 poles)
General purpose in aluminium and "Severe Duty" in cast iron for more demanding industry.
Rotor in cast copper for higher efficiency (IE1 to IE3)
Modular design: a number of options are easly to add (see p.20.15).

Photo Siemens: Cross section of a 1LE1 asynchronous motor

3.2 Three-phase induction motor

3.2.1 Configuration of a three-phase induction motor

Up to now we have considered the rotating field in the stator iron of a three-phase machine. To limit the iron losses the motor is built up out of Si-steel sheets just as was the case in the transformer.

The induction motor is composed of two important parts:

1 stator:

Three-phase windings in slots on the inside of laminated stator iron. These three-phase windings are connected to a power grid.

A rotating field is created with speed: $n_S = \frac{60 \cdot f_S}{p}$ (16-38)

2. Rotor

A) can be **wound** (in slots around the circumference of the rotor plates). Needs to be multi phase and have the same number of poles as the stator winding. The rotor is connected in star and the ends of the windings are brought out of the motor via slip rings and brushes. The rotor is not connected to the power grid. This type of motor is seldom used these days.

B) Almost exclusively used these days is the **cage rotor** (photos p. 16.47 and p. 16.54). The rotor bars are placed in slots and at both ends of the rotor these bars are short circuited with a ring. This is called a cage induction motor or a squirrel cage motor. The cage is usually made from aluminium. The cage does not need to be insulated from the rotor iron.

3.2.2 Influence of rotating field on the rotor

To determine the influence of the rotating stator field on the rotor we assume in the first case a wound rotor. Assume for now that the rotor winding is identical to the stator winding. Fig. 16-54 shows the setup.

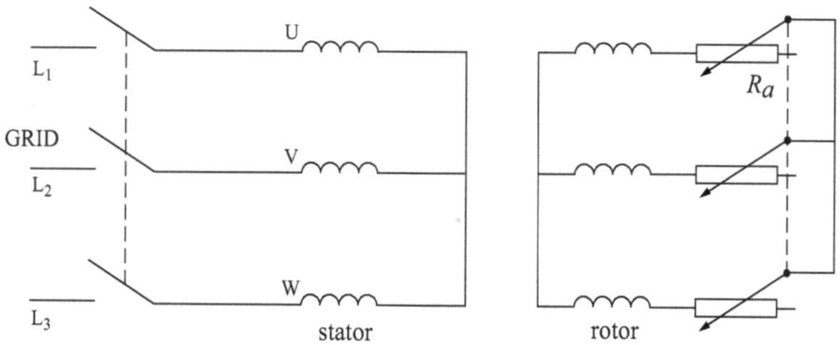

Fig. 16-54: Three-phase induction motor with wound rotor

We make three steps to explain the influence of the stator field on the rotor.

1. We leave **the rotor windings open** (e.g. the resistor wiper makes no contact with the resistor). V_{S1} is the phase voltage of one stator phase. Just as with the transformer the magnetic field (in this case the rotating field!) will produce a reverse emf E_{S1} equal to V_{S1} . In addition the rotating field will produce an emf $E_{R1standstill} = E_{R1st.st}$ in rotor coil nr. 1. The frequency f_R of this rotor emf is equal to the power grid frequency while the rotor is stationary. **In this situation the motor is**

comparable to a transformer of which the secondary operates in no load mode. The stator plays the role of primary and the rotor is the secondary. Both windings are only linked via the magnetic flux which is rotating. The ratio $\frac{E_{S1}}{E_{R1st.st}} = k =$ transformation ratio of the motor. There is a big difference with a transformer though since the no load current of the motor is large. This is because the reluctance of the motor is much larger (air gap) than that of the transformer. For a motor the no load current is between 20 to 50% of the full load current. For a transformer the no load current is only a few percent of the full load current.

2. We now connect the current paths of the rotor windings but keep the rotor securely locked. In fig. 16-55 rotor conductor (A) is shown. Assume a two pole stator field. An induced voltage is produced in A ($e = B . l . v$). Since the rotor circuit is electrically closed an induced current flows in A. The direction of the induced voltage (and current) is indicated by the right hand rule. On the rotor conductor a force $F = B . l . I$ will exist. The direction of this force is indicated by the left hand rule.

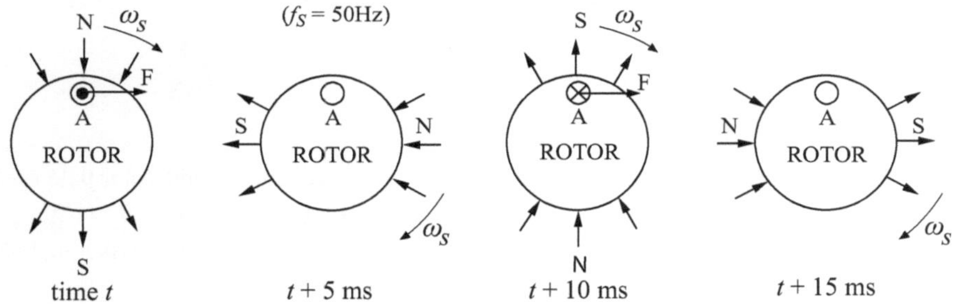

Fig. 16-55: Induced current in the rotor conductor and the Lorentz force on this conductor

A torque exists on the rotor in the same direction as the rotating stator field. The torque is maximum when the rotor current is in phase with the stator flux. Since the rotor conductors are in a medium with a large permeability the rotor windings have a high self inductance which causes the rotor current to heavily lag the rotor emf ($tg\ \varphi = \frac{\omega . L_R}{R_R}$).

3. We unlock the rotor and it rotates under the influence of the torque on the current carrying rotor conductors. It appears that the rotor rotates in the same direction as the stator rotating field. As the rotor increases in speed the relative speed difference between the stator field and rotor is reduced. As a result the emf E_{R1} decreases as does the rotor frequency f_R . If the motor reaches the (synchronous) speed of the stator rotating field then the rotor conductors would experience no changing field. The induced voltage and current would cease to exist and also the torque. With this type of motor we can never reach synchronous speed, it is an **asynchronous motor**.
Since the operation depends on the induction of currents in the rotor conductors it is also called an **induction motor**. As previously mentioned the induced emf decreases with increasing speed. There must be a certain speed that produces a maximum torque. In paragraph 3.3.1 we will calculate this torque.Once the motor is up to speed the external resistors are switched out of circuit and the rotor is short-circuited. The wound rotor then behaves as a short-circuited cage. In the following study we only deal with the squirrel cage motor.

3.2.3 Slip

The slip (g) of an induction motor is the relative difference of the rotor speed (n) with respect to the synchronous (stator rotating field) speed n_S :

$$g = \frac{n_S - n}{n_S} \quad \text{from which} \quad n = n_S \cdot (1 - g) \tag{16-50}$$

Slip is a fraction but is usually expressed in %. The slip scale is a linear scale (inverted with respect to the rotor speed): at standstill ($n = 0$) is $g = 1$ and at synchronous speed ($n = n_S$) is $g = 0$. The slip is negative if the rotor speed (e.g. with the aid of other machines) is increased beyond synchronous speed. If the rotor is externally driven and in the opposite direction to the rotating field then g is larger than 1.

Rotor frequency

The rotor frequency is dependent upon the speed difference between stator field and rotor.

$$f_R = \frac{p \cdot (n_S - n)}{60} = \frac{p \cdot g \cdot n_S}{60} = g \cdot \frac{p \cdot n_S}{60} = g \cdot f_S \rightarrow f_R = g \cdot f_S \tag{16-51}$$

Rotor-emf

At rotor stand still the speed difference between rotor conductors and stator field is equal to n_S.
The rotor emf at stand still ($E_{Rst.st}$) is determined with the formula $e = B\,.\,l\,.\,v_{stator}$.
The rotor emf at speed n depends on the difference in speed between the rotating field and the rotor:

$e = B \cdot l \cdot (v_{stator} - v_{rotor}) = B \cdot l \cdot g \cdot v_{stator} = g \cdot (B \cdot l \cdot v_{stator})$.

For a slip g the rotor emf can be determined from: $E_R = g \cdot E_{Rstandstill}$ (16-52)

3.2.4 Speed of rotor field

The rotor current with frequency $f_R = g \cdot f_S$ produces a magneto motive force which rotates with a speed ω_R ($= 2 \cdot \pi \cdot f_R$) with respect to the rotating rotor: $\omega_R = g \cdot \omega_S$.
The rotor itself rotates with $\omega = \frac{2 \cdot \pi \cdot n}{60}$ rad/s. Here by $\omega = (1 - g) \cdot \omega_S$. For an observer on the stator the rotor field rotates at $(\omega + \omega_R)$ rad/s or $(\omega + \omega_R) = (1 - g) \cdot \omega_S + g \cdot \omega_S = \omega_S$ .The rotor field rotates with respect to the stator reference axis with the same synchronous speed as the stator rotating field.

3.2.5 Multipolar machine

The previous theory was developed for a two pole machine. In practice this is seldom used for two reasons:

1. The synchronous speed of 3000 rpm is too high for most driven loads.
2. In a two pole machine the iron frame is poorly used. The magnetic field lines of four pole and six pole machines are significantly shorter than in the case of the two polar machine.

In a 2p-poled machine the maximum air gap induction is the same as for a two pole machine since $\vec{B}_{ag}$ is limited by the saturation of the stator and rotor teeth.
The current densities in stator and rotor windings will also not change since the determining factor (the heating!) is the same. For the same number of windings N_{Se} (sinusoidally distributed) the winding density in expression (16-39) is:

$$n_{S1} = \frac{N_{Se}}{2 \, . \, p} \, . \, \sin \alpha \qquad (16\text{-}53)$$

With 2p poles the total contained flux is now:

$$\vec{\Phi}_{S1} = \frac{\pi}{2 \, . \, p} \, . \, N_{Se} \, . \, \vec{B}_{ag} \, . \, l \, . \, r \qquad (16\text{-}54)$$

The magnetising inductance $L_0 = \frac{\vec{\Phi}_{S1}}{\vec{I}_{\mu}}$, taking (16-45) into account is:

$$L_0 = \frac{3 \, . \, \pi}{8 \, . \, p} \, . \, N_{Se}^2 \, . \, \frac{\mu_0 \, . \, l \, . \, r}{l_{ag}} \qquad (16\text{-}55)$$

3.2.6 T-equivalent of an induction motor

1. Leakage inductance - equivalent transformer

As was the case with the transformer we have field lines from the stator field which do not reach the rotor coil and field lines from the rotor that do not reach the stator. We therefore have a stator and rotor leakage inductance.

In fig. 16-56 we draw an equivalent transformer for one phase of a three-phase motor.

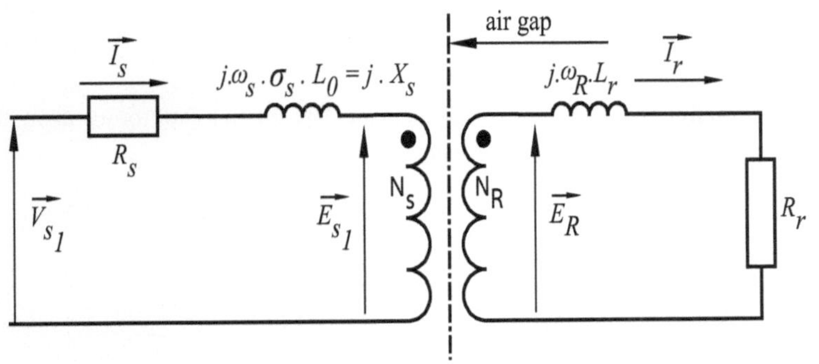

Fig. 16-56: Equivalent transformer for one phase of the induction motor

2. Induction motor T-equivalent

With the actual rotor resistance of one phase (R_r) and the actual rotor leakage inductance of one phase (L_r) we can write for the rotor current: $\vec{I}_r = \dfrac{\vec{E}_R}{R_r + j \,.\, \omega_R \,.\, L_r}$

In order to be able to draw a T-equivalent diagram as shown in fig. 16-57 (with the associated vector diagram) then obviously we need to work with the same frequency on the rotor side as on the stator side. On the rotor side we have the same frequency at stand still:

$$\vec{I}_r = \frac{\vec{E}_R}{R_r + j \,.\, \omega_R \,.\, L_r} = g \,.\, \frac{\vec{E}_{Rstandstill}}{R_r + j \,.\, \omega_R \,.\, L_r} = \frac{\vec{E}_{Rstandstill}}{\frac{R_r}{g} + j \,.\, \frac{\omega_R}{g} \,.\, L_r} = \frac{\vec{E}_{Rstandstill}}{\frac{R_r}{g} + j \,.\, \omega_S \,.\, L_r}$$

In addition we also need to take into account the winding ratio of stator and rotor coils:

$$R_R = \left(\frac{N_S}{N_R}\right)^2 \text{ x actual resistance } (R_r) \rightarrow \frac{R_R}{g} = \frac{R_r}{g} \,.\left(\frac{N_S}{N_R}\right)^2$$

$$\sigma_R \,.\, L_0 = \left(\frac{N_S}{N_R}\right)^2 \text{ x actual leakage inductance } (L_r) \rightarrow X_R = \omega_S \,.\, \sigma_R \,.\, L_0$$

In the T-equivalent of fig. 16-57 we have connected the rotor to $E_{S1} = E_{Rstandstill} \cdot \left(\frac{N_S}{N_R}\right)$ and the rotor current is given by $I_R = I_r \,.\left(\frac{N_R}{N_S}\right)$

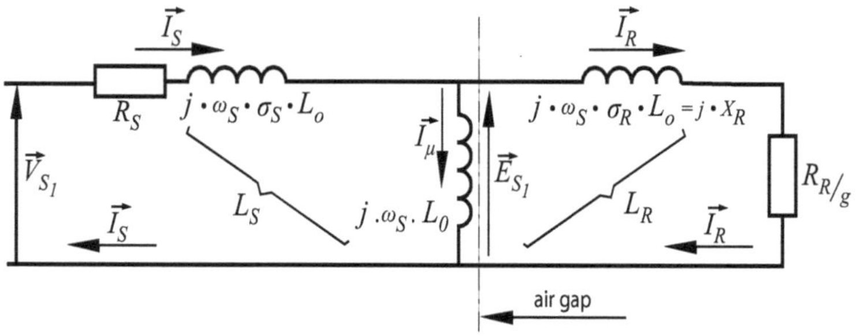

Fig. 16-57: T-equivalent of one phase of a three-phase asynchronous motor

3.3 *M-n* curve of an induction motor

3.3.1 Torque

Continuing from fig. 16-57 we are now able to calculate the (air gap) torque using the power P_1 of one phase, together with the angular velocity: ω_S :

$$M = \frac{3 \,.\, P_1}{\frac{\omega_S}{p}} = \frac{3 \,.p}{\omega_S} \,.\left(\frac{R_R}{g}\right) .\, I_R^2 = \frac{3 \,.\, p}{\omega_S} \,.\left(\frac{\omega_S \,.R_R}{\omega_R}\right) .\, I_R^2$$

$$= \frac{3 \,.\, p \,.\, R_R}{\omega_R} \cdot \frac{{E_{S1}}^2}{\left(\frac{R_R \,.\omega_S}{\omega_R}\right)^2 + (\omega_S \,.\, \sigma_R \,.\, L_0)^2}$$

With $V_{S1} \approx E_{S1} = \Phi_{S1} \cdot \omega_S$ the expression for the moment of torque is:

$$M = \frac{3 \cdot p \cdot R_R}{\omega_R} \cdot \frac{\Phi_{S1}^2}{\left[\left(\frac{R_R}{\omega_R}\right)^2 + (\sigma_R \cdot L_0)^2\right]} \tag{16-56}$$

The maximum torque occurs when $\frac{dM}{d\omega_R} = 0$, this is when $\omega_R = \frac{R_R}{\sigma_R \cdot L_0}$

If we replace ω_R for this value, then we find:

$$M_{max.} = \frac{3 \cdot p \cdot \Phi_{S1}^2}{2 \cdot \sigma_R \cdot L_0} \quad \text{(Nm)} \tag{16-57}$$

The *M-n* curve looks as shown in fig. 16-58. We have also drawn in a *g*- and ω_R-scale. The maximum torque is also called the pull out torque. $M_{po} = M_{max.}$.

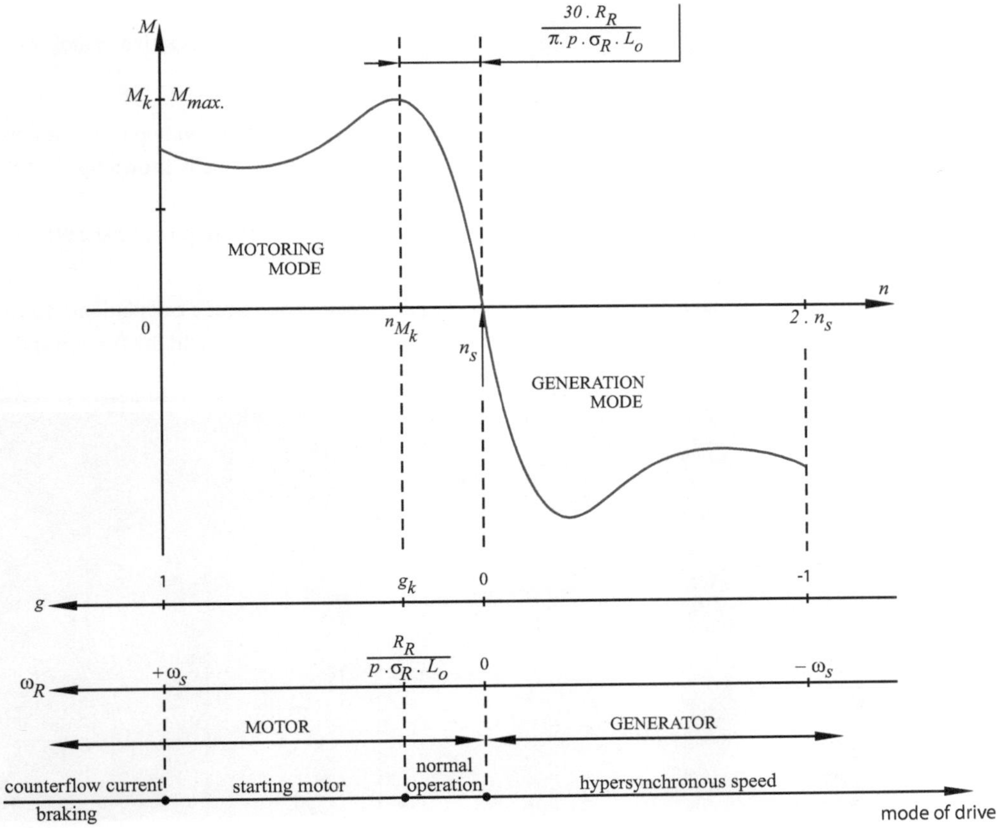

Fig. 16-58: *M-n* curves of a three-phase asynchronous motor

Remarks

1. Speed at pull out torque:

Maximum torque will occur with a pulsation of the rotor current given by:

$\omega_R = \frac{R_R}{\sigma_R \cdot L_0}$. The maximum torque lies in fact at a distance $\Delta\omega = \omega_R = \frac{R_R}{\sigma_R \cdot L_0}$ from the synchronous speed (see fig. 16-58). If 2.p poles are implemented, then this distance becomes

$\Delta\omega = \frac{1}{p} \cdot \frac{R_R}{\sigma_R \cdot L_0}$. The pull out torque therefore appears at: $\omega = \omega_S - \frac{1}{p} \cdot \frac{R_R}{\sigma_R \cdot L_0}$

If we are using the speed we find:

$$n_{M_{po}} = n_S - \frac{30}{\pi \cdot p} \cdot \frac{R_R}{\sigma_R \cdot L_0} \qquad (16\text{-}58)$$

The speed at pull out torque is influenced by the value of the rotor resistance.

2. From $M_{max.} = \frac{3 \cdot p \cdot \Phi_{S1}^2}{2 \cdot \sigma_R \cdot L_0}$ it follows that the pull out torque is independent of the rotor resistance.

3. Generator service:

In

$$M = \frac{3 \cdot p \cdot R_R}{\omega_R} \cdot \frac{\Phi_{S1}^2}{\left[\left(\frac{R_R}{\omega_R}\right)^2 + (\sigma_R \cdot L_0)^2\right]}$$

we see that for negative values of ω_R , the value of the torque is negative. The path of the torque curve will be symmetrical with respect to the point $\omega_R = 0$, this is with respect to the synchronous speed. This is shown fig. 16-58.

For $0 < n < n_S$ the machine operates as a motor (see fig. 16-58).

For $n > n_S$ the machine operates as a generator (negative torque with positive speed).

4. Air gap torque:

Up to now we have always calculated the air gap torque in our torque calculations. To calculate the torque on the shaft of the motor we need to subtract the rotor losses and the friction and ventilator losses from the air gap torque (see further).

Photo Siemens: Squirrel cage of a 2 to 16 pole H-modyn three-phase asynchronous motor.
Standard versions for 6/10 kV - 50Hz and 13.2 kV - 60Hz.
Power: 2-38 MW (dependent from number of poles!). Examples: 6 poles (7-38 MW);16 poles: 2-9 MW.

3.3.2 Stable operating range

For driving a load M_t we distinguish two operating areas in the *M-n* curves of fig. 16-59:
- an unstable area: left of M_{po} : reverse operating torque M_{t1}
- a stable area: to the right of M_{po} : reverse operating torque M_{t2}

In the stable area the following is true $M_t < M_{po}$

For example the point P is a stable operating point for the load M_{t2} .

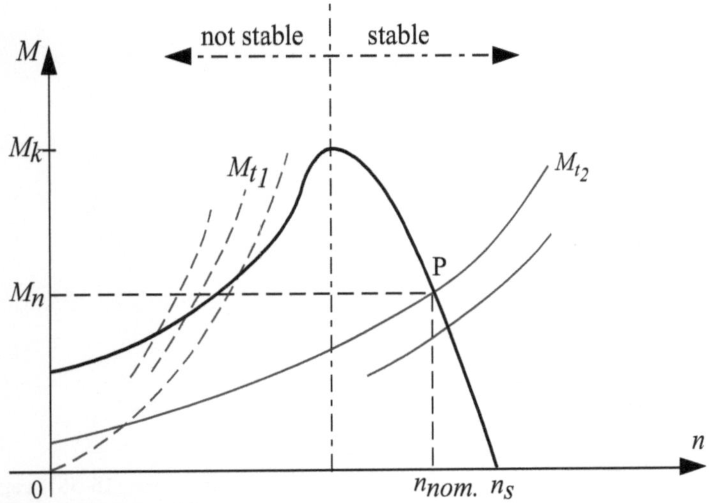

Fig. 16-59: Stable operating point on an M-n curve

3.3.3 Nominal operating torque

Assume a constant supply voltage and frequency, then $\Phi_S = \frac{V_S}{\omega_S} = C^{te}$. In this case according to expression (16-57) the pull out torque is inversely proportional to the leakage inductance of the motor. Standard induction motors are designed so that M_{po} = 1.75 to 3. M_n . Here in M_n is the nominal torque of the motor. Mostly $M_{po} \approx 2 \, . \, M_n$.

The nominal torque is determined using the data on the motor tag plate.

Example:

Three-phase motor : 3x230/400V - delta/star - 15.6/9A
n = 1428 rpm
P = 4 kW

As was the case with the DC motor: $M_n = \frac{9550 \cdot P}{n} = \frac{9550 \cdot 4}{1428} = 26.75$ Nm

For this motor it is probably the case that $M_{po} \approx 2 \cdot M_n = 53.5$ Nm.

If we include a torque sensor on the shaft of this motor we would need to take this M_{po} into account in determining the maximum allowable torque applied to the sensor, since at start up we always pass the point M_{po} and with a very small M_t practically the maximum torque is applied to the sensor.

3.3.4 Starting torque

We reconsider a part of fig. 16-58 in fig. 16-60. With a reverse operating torque M_{t1} the motor will accelerate from stand still to the operating point $P\,(n_1\,;\,M_{t1})$.

The difference $M - M_t = M_v = J_m \cdot \frac{d\omega}{dt}$ is indicated at certain times with an arrow as shown in fig. 16-60.

With a reverse operating torque M_{t2} the motor cannot start.

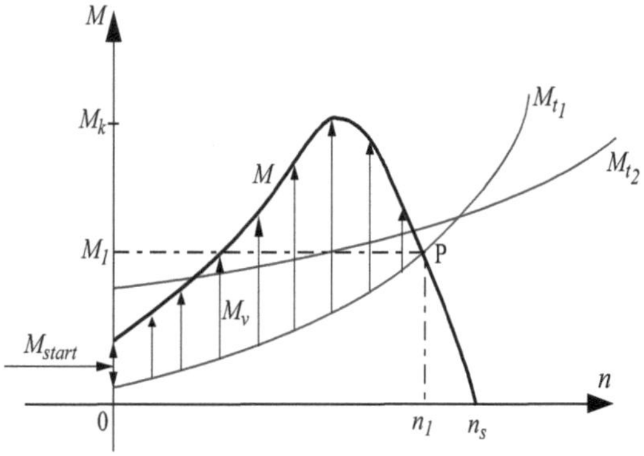

Fig. 16-60: Starting or stalling of a motor

To determine the start torque set $n = 0$.

In addition $\omega_R >> \frac{R_R}{\sigma_R \cdot L_0}$ or $\sigma_R \cdot L_0 >> \frac{R_R}{\omega_R}$, so that expression (16-56) is approximated by:

$$M_{start} \approx \frac{3 \cdot p \cdot R_R}{\omega_R} \cdot \left(\frac{\Phi_{S1}}{\sigma_R \cdot L_0} \right)^2 \qquad (16\text{-}59)$$

To obtain a sufficiently large starting torque for a motor connected to a constant supply voltage there are in practice two methods used:

1. Motor with slip ring rotor
2. Motor with double cage rotor

1. Slip ring rotor:

Instead of the squirrel cage motor considered up to now the rotor may also be wound with a three-phase winding. This is then connected in star and the free ends are brought outside the motor via slip rings. Externally a set of (start) resistors are connected (fig. 16-61).

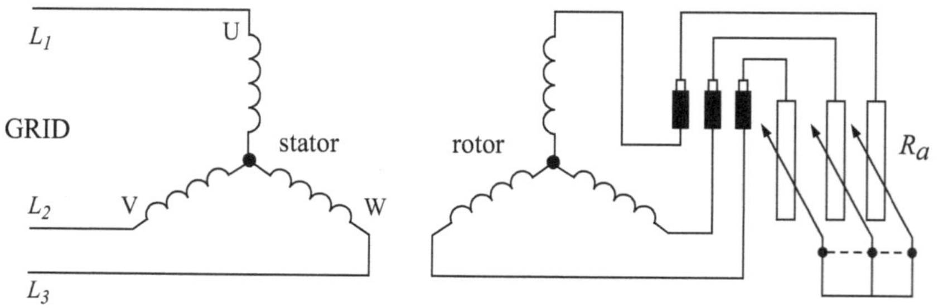

Fig. 16-61: Electrical connections of a slip ring motor

Photo Siemens: 1 MW 3600 rpm pump drive, used in the production of atomized metal powder by the firm Coldstream in Ath Belgium. A Masterdrive Vector Control controls this heavy motor.

Photo Siemens: Integrated Drive System (IDS). Integration in three ways: down and across, life time integration

The rotor resistance increases because of the series connected resistors R_a .
We consider the influence of the rotor resistance on three important areas of the *M-n* curve:

1. Maximum torque (pull out torque):

$$M_{max.} = \frac{3 .p . \Phi_{S1}^2}{2 . \sigma_R . L_0}$$: the value of pull out torque is independent of the value of R_R

2. Start or run in torque:

(16-59): a larger resistance increases the start torque

3. The path of the torque between $n_{M_{po}}$ and n_S (fig. 16-62). In this area $\frac{R_R}{\omega_R} > \sigma_R . L_0$ so that from (16-56) it follows that with an increasing R_R the value of torque will decrease.

Fig. 16-62 shows the influence of the rotor resistance on the *M-n* curve.
The choice of $R_{R2} = R_R + R_{start\ resistance}$ can facilitate that the motor starting torque be maximum torque.

Such motors with wound rotors were used extensively in the past from power levels from about 4kW, since with specially produced resistors the speed of the motor could also be varied. This is totally outdated now since we nowadays use variable frequency drives that are ideally suited to control the speed of an asynchronous (squirrel cage) motor with high efficiency.

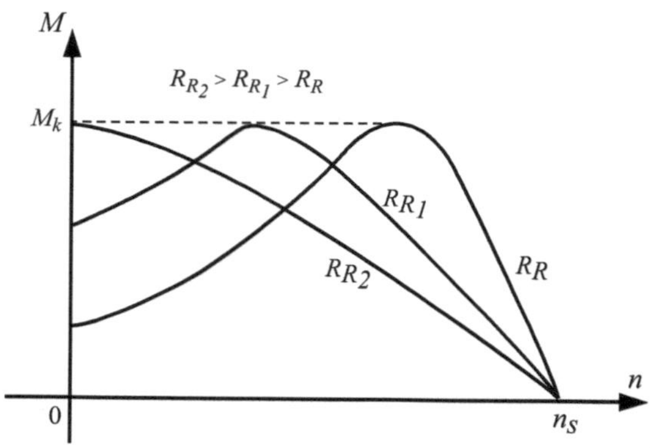

Fig. 16-62: Influence of the rotor resistance on the *M-n* curve of an induction motor

2. Double cage rotor

A disadvantage of the squirrel cage is that the resistance of the cage cannot be adjusted to achieve a higher starting torque for example. A solution was developed by Boucherot in the form of a double cage. In fig. 16-63a the principle is shown of a rotor with two concentric cages.
The outer cage K_1 is composed of bars with a small diameter which are close to the air gap.
The resistance of such a cage is high and the inductance is low (flux passes partly through the air gap). The inner cage K_2 is formed of thicker bars which are almost totally enclosed by the rotor iron. This cage has a low resistance and high inductance.

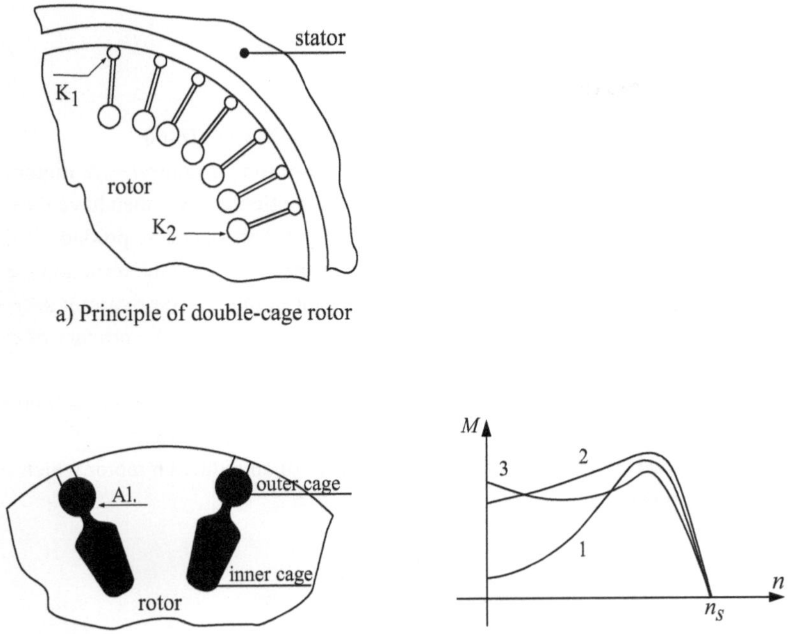

a) Principle of double-cage rotor

b) Modern double cage (cast aluminium)

c) *M-n* curves double-cage motors

Fig. 16-63: Double cage rotor of an asynchronous motor

We now examine the behaviour of a double cage at start up of a motor. At stand still the rotor frequency is large (50 Hz) and the reactance plays a major role, resulting in a small current in the inner cage but a larger current in the outer cage. As the motor increases in speed the rotor frequency decreases to a low value (1 to 4 Hz) and the reactance no longer plays a large role therefore the current will flow in the inner cage with its low resistance. We note that the rotor resistance at start-up is very high (the outer cage) and decreases as the motor accelerates: the inner cage " takes over". The rotor resistance gradually drops and this in a completely electrical manner. The advantages of ease of construction and the reliability of operation of a cage rotor remain valid for the double cage. These motors have an increased starting torque and can start loaded. With different cage shapes and air gap dimensions we are able to produce various *M-n* curves (fig. 16-63c).

Curve 2: the most common type in standard motors. Here the starting torque is between 1 and 1.75 times the nominal torque M_n of the motor.

Curve 1: sometimes chosen for large motors since the lower rotor resistance has a higher efficiency, at the price of lower starting torque.

Curve 3: if a higher start torque is required, then the design of the motor is so that it has curve 3 as its *M-n* curve. This is called the saddle curve.

3.4 Starting current of induction motor

Important issues with starting an induction motor are

1. the starting torque
2. the inrush starting current

The issue of the starting torque has just been addressed in fig. 16-60.

Now we consider the inrush starting current. When starting a squirrel cage motor the power grid voltage is connected to the stator while the rotor is motionless. We then have the equivalent of a transformer with short-circuited secondary (= rotor!). This of course produces a large inrush current. As the motor accelerates the difference in speed between the rotor and the stator fields decreases and the induced emf decreases. As a result of this the rotor current decreases (secondary equivalent of a transformer) and therefore also the stator current (the primary of our equivalent transformer!).

The power grid has to deliver this inrush current and the power utility sets a limit on this current. Normally this inrush current is 3 to 7 times the nominal stator current.

Fig. 16-64 shows the *M-n* curves and the *I-n* curves of an induction motor which is directly connected to the power grid.

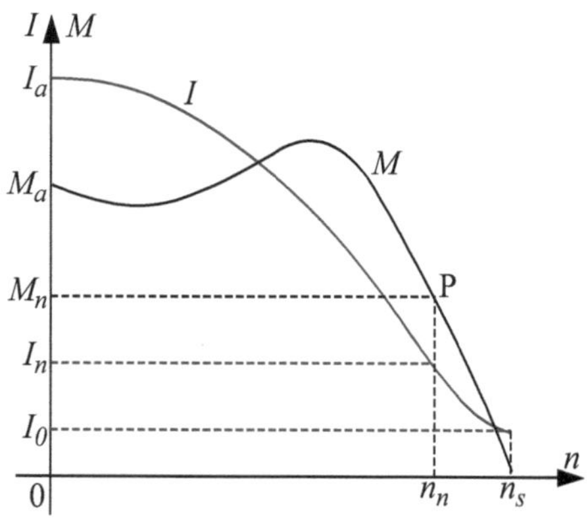

Fig. 16-64: *M-n* and *I-n* curves of an induction motor, directly connected to the power grid

Classical methods to reduce this inrush current include star-delta switching, autotransformer and the soft starter.

1. Star-delta switching

For motors of low power levels (below 10 or 20 kW) star-delta starting can be used whereby the normal configuration of the motor is in delta. With the aid of a contactor the three phases are connected in star when the motor starts. When the motor has achieved nominal speed the connection is changed over to delta. In star the phase voltages are $\sqrt{3}$ lower than normal and the line currents drawn are three times smaller than in the normal delta configuration of the motor. Such a starting method may be used for a motor that starts unloaded (e.g. woodworking machines…).

Fig. 16.65 shows the *M-n* and *I-n* curves for this starting method.

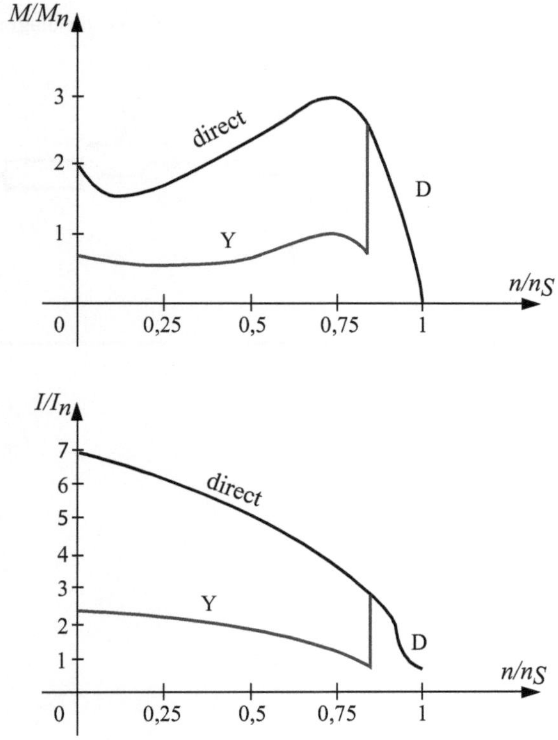

Fig. 16-65: *M-n* curves in the case of star-delta starting

2. Autotransformer

If the power level of the motor is above 10 to 20 kW an autotransformer may be used. The full power grid voltage is applied to the entire winding. Via taps on the autotransformer the voltage to the motor is increased gradually and in steps.

3. Soft starter

With the aid of an electronic device a reduced voltage is applied to the motor at start-up. As the motor speed increases the terminal voltage is raised. For a description and operation of a soft starter you are referred to chapter 20. The price of a soft starter is almost the same as that of a star-delta starter, so that soft starting is taking over from star-delta starting.

3.5 Displacement factor of an induction motor

We draw the T-equivalent of fig. 16-57 in fig. 16-66a. As was the case with the transformer we have included the iron losses ($\vec{I}_V$). In fig. 16-66b we have drawn the vector diagram associated with the configuration of fig.16-66a.

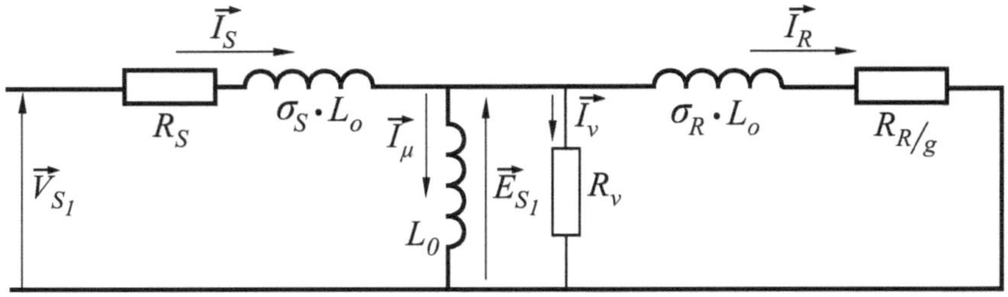

Fig. 16-66a: T-equivalent of an induction motor

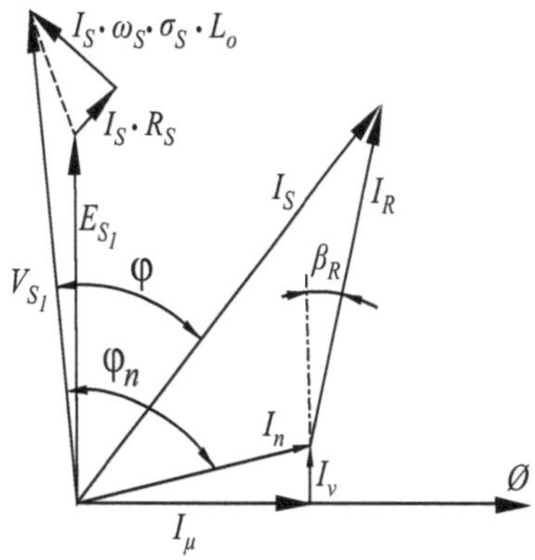

Fig. 16-66b: Vector diagram of an induction motor

For the power drawn from the power grid we can write:

active power: $P = 3 \, . \, V_S \, . \, I_S \, . \, \cos\varphi$ (16-60)

reactive power: $Q = 3 \, . \, V_S \, . \, I_S \, . \, \sin\varphi$ (16-61)

V_S and I_S are phase values of the motor and φ is the displacement angle between V_S and I_S.

At no load $\vec{I}_R = 0$ and $\vec{I}_S = \vec{I}_n$. The displacement factor (cos φ_n) is very small, resulting in the asynchronous motor drawing a large current at no load. This is practically 20 to 30 % of $I_{full\ load}$ (even to 50%!). As the motor load increases then I_R increases (and also I_S) and φ is reduced so that the displacement factor (cos φ) improves for example to cos $\varphi = 0.82$.

Determining the line current of an induction motor per kW of power:

Assume η is the efficiency of the motor and P is the power on the shaft. If the line current and voltage are V and I, then the electrical power drawn from the grid is : $P_{grid} = \sqrt{3} \, . \, V \, . \, I \, . \, \cos\varphi$

and $P = \eta \, . \, P_{grid}$ so that: $I = \dfrac{P}{\eta \, . \, \sqrt{3} \, . \, V \, . \, \cos\varphi}$

With for example $\eta = 87\%$ and $\cos\varphi = 0.82$ we find for a three-phase 1kW motor connected to 3x400V a line current of : $I = \frac{1000}{0.87 \cdot \sqrt{3} \cdot 400 \cdot 0.82} = 2\,\mathrm{A}$.

We find a line current drawn of about 2A/kW when connected to 3x400V. A 4 kW motor at full load draws approximately 8A from a 3x400V power grid. From a 3x230 V power grid the line current is $\sqrt{3}$ greater and is about 3.7A/kW.

3.6 Power diagram

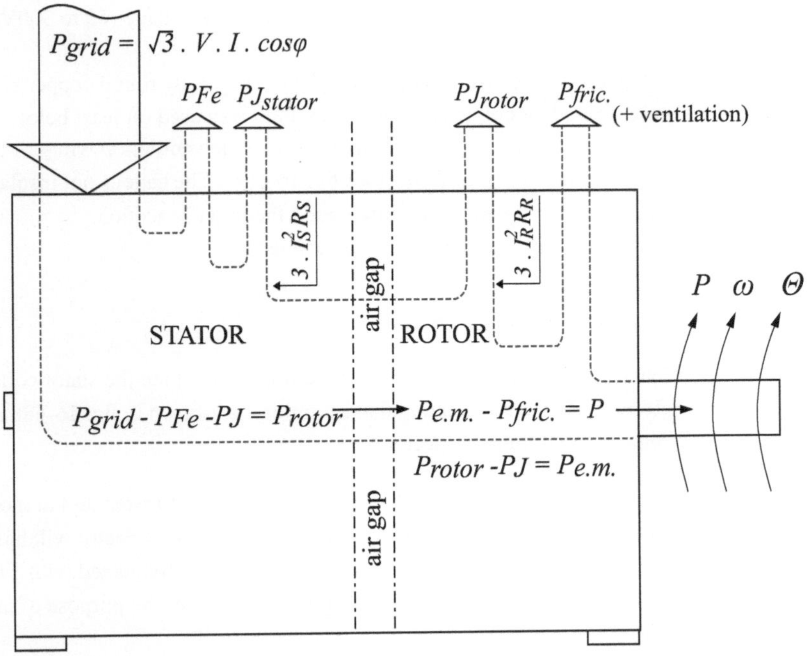

Fig. 16-67: Power of an asynchronous motor

Here we now give an impression of the percentile powers.
The electrical power from the power grid (100%) is reduced by iron losses (2.5%) and the joules losses (4%) in the stator and this difference forms the air gap power (93.5%) that is transmitted to the rotor. The iron losses in the rotor are marginal (low frequency) and are neglected here. If we subtract the joules losses in the rotor (3.5%) then what remains is the electro mechanical power. We also need to subtract the friction and ventilator losses (3%) and what remains over is the shaft power (87%) as stated on the name plate.

3.7 Construction of induction motors

The magnetic circuit of stator and rotor are constructed from thin metal plates. In an induction motor the air-gap needs to be as small as possible. The purpose is to keep the reluctance as low as possible so that the magnetising current I_μ is as small as possible. A small magnetising current has a positive effect on the $\cos\varphi$ and the efficiency.
For extremely small motors the air gap may be a fraction of a millimeter and can be between 1 and 3 mm for large diameters. The rotor needs to be well balanced to prevent every coincidental contact between rotor and stator.
In high voltage motors the stator windings can be made to be directly connected to the H.V. net (e.g. 6 or 12kV) because the electrical windings of stator and rotor are separate. The construction is similar to that of an alternator, especially as far as the precautions for insulation are concerned.
In the case of a wound rotor this is always dimensioned for low voltage (50 to 300V with open rotor windings).
The construction of squirrel cage rotors is much simpler. Originally round copper bars were used to which rings were welded on both ends. Later aluminum was used (at least below 50kW) which was cast in one go complete with short circuiting rings and the ventilation wings at both ends. This procedure allows for a choice of rotor forms (see fig. 16-63b). The cage is not insulated from the metal so that the contact with the metal assures a good thermal connection.
For high efficiency motors one uses a rotor in cast copper.

3.8 Motor terminals

The motor terminals are normally composed of six terminals to which the stator coils are connected as shown in fig. 16-68a. By interconnecting these terminals as shown in fig. 16-68b and 16-68c we obtain star and delta configurations respectively.

In addition to these six terminals there is also an earthing terminal present and in modern motors there are also two terminals for a PTC resistor connection. This PTC resistor will have a very high resistance if the motor temperature becomes too high. The PTC is connected with a relay which can disconnect the main contactor and switch off the motor. This serves the purpose of protecting the motor against overload.

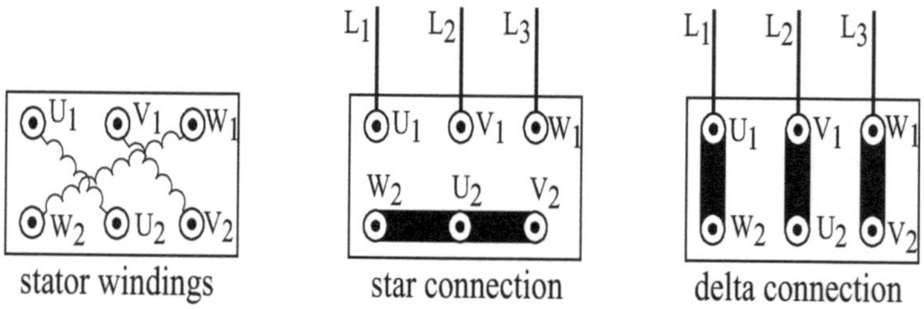

Fig. 16-68: Connection terminals of a three-phase motor

The three-phase stator windings of asynchronous and synchronous motors are identified as U, V, W according to IEC 60034-8. The terminals of these windings are then U_1-U_2, V_1-V_2 and W_1-W_2. The DC excitation winding of the synchronous machine are identified as F_1-F_2.

3.9 Speed and torque control

With the introduction of frequency controllers it became possible to efficiently control the speed and (or) torque of a motor. This will be studied in chapter 20.

Photo Siemens: 1FK7 synchronous servomotor 1LE1 standard induction motor 1PH8 induction servomotor

The **1FK7** redesigned Siemens 1FK7 servo motor for standard motion control applications. Mechanical decoupling of the encoders from the motor shaft increases both the ruggedness and serviceability of the motors, while improved integration of encoder interface electronics makes for an even more compact motor design, thereby simplifying machine installation.

The **ILE1** is a classic induction motor. It is a work horse of industry. Specifications see p. 16.47.

The **IPH8** series by contrast is an induction servomotor intended for converter applications. Supply voltages of 400 to 690 V. Speeds of 400 to 6000 rpm. Protection classes IP23, IP55 and IP65. Powers from 2.8 to 1340kW. Encoders: resolvers, incremental encoder (sin/cos, 1Vpp), EnDat absolute encoder, HTL impulse encoder. Dynamic performance, compact build. Intended for extruders, print machines, main drives of mechanical machines, etc.

The Siemens 1PH8 range of motors for main drives has models equipped with specially energy-efficient Electronically Commutated (EC) separate fans.

3.10 Simple mathematical model of an induction motor

The mathematical model of an AC induction motor is not as simple as that of a DC commutating machine. Especially complicated is the expression for the momentum of torque. An exact mathematical model which is required for the study of speed and torque will be dealt with in chapter 20. In the mean time a number of simplifications can be made:

1. **airgap flux:** the total linked air gap flux Φ can be written with the help of (16-49) as:

$$\Phi = \frac{3}{2} \cdot \Phi_{S1} = \frac{3}{2} \cdot \frac{E_{S1}}{\omega_S} \approx k_1 \cdot \frac{V_s}{f_s} \rightarrow \Phi = k_1 \cdot \frac{V_S}{f_S} \quad (16\text{-}62)$$

2. **pull out torque:** from $M_{max.} = \frac{3 \cdot p \cdot \Phi_{S1}^2}{2 \cdot \sigma_R \cdot L_0}$ follows: $M_{max.} = k_3 \cdot (V_S/f_S)^2$ (16-63)

3. **nominal torque:** we can state that the torque is provided by the operation of the rotor current I_R on the rotating field Φ. The orthogonal component of the rotor current with Φ as shown in fig. 16-66b can be calculated as $I_R . cos\ \beta_R$. The electro mechanical torque is therefore also calculated as: $M_{em} = k \cdot \Phi \cdot I_R . cos\ \beta_R$. At nominal load and neglecting the friction and ventilation losses a simplified formula can be written for the nominal torque:

$$M_n \approx M_{em} = k_4 \cdot \Phi_n \cdot I_n \quad (16\text{-}64)$$

Here in: k_4 = machine constant
I_n = nominal load current of motor
Φ_n = nominal rotating flux

With these formulas a simplified model of a three-phase induction motor can be constructed.

Table

$P = \eta \cdot \sqrt{3} \cdot V \cdot I \cdot \cos\varphi$	$\Phi = k_1 \cdot \frac{V_S}{f_S}$
$n_S = \frac{60 \cdot f_S}{p}$	$M_n = k_4 \cdot \Phi_n \cdot I_n$
$g = \frac{n_S - n}{n_S}$	$M_{po} = M_{max.} = k_3 \cdot (V_S/f_S)^2$
$f_R = g \cdot f_S$	$P = \omega \cdot M_n$
$n = (1 - g) \cdot n_S$	$M_n = \frac{9550 \cdot P}{n}$ (Nm) with P in kW
$M = M_l + M_v$ and $M_v = J_m \cdot \frac{d\omega}{dt}$	

16.4 Synchronous Machines

CONTENTS

4.1 Synchronous three-phase machine

4.1.1 General

The stator winding of a three-phase synchronous machine is comparable to that of a three-phase induction motor. The rotor of a synchronous machine produces a constant magnetic field by means of a DC field winding. In the stator we have a rotating field with synchronous speed $\omega = \frac{\omega_S}{p}$

Here in: ω_S = stator pulsation
p = number of pole pairs in the stator winding

Synchronous machines are mainly used as alternators for generating electrical energy. They are driven by fast rotating steam turbines in electric power stations or by means of hydraulic turbines in water powered generating stations. In the case of steam turbines 2 or 4 pole alternators are used. With (slow rotating) hydraulic turbines a large number of pole pairs (e.g. 40) are required to produce the 50Hz power grid frequency.

Alternators of low power levels are used in cars, airplanes, emergency generators,...

Large power synchronous motors (e.g. up to 12 MW) are also used to improve the power factor of the plant.

Small synchronous motors are often used as servomotors in electrical positioning systems where they are of course controlled by an electronic converter.

4.1.2 Cylindrically wound rotor

Since the stator is identical to that of an asynchronous motor the text of nrs. 3.1.2 to 3.1.4 (sinusoidally distributed winding, stator rotating field, magnetising inductance) are also valid here for the synchronous machine.

The rotor of a synchronous machine is usually built up as shown in fig. 16-69

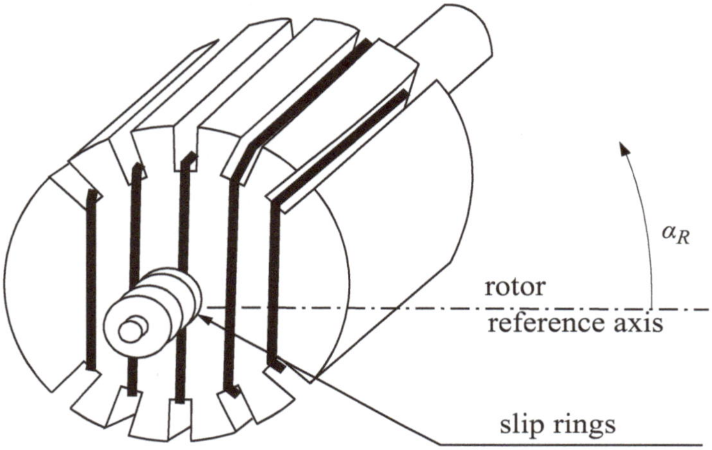

Fig. 16-69: Wound rotor of a synchronous machine

The stator windings of synchronous machines are also sinusoidally distributed.
A DC excitation current flows in the rotor winding.

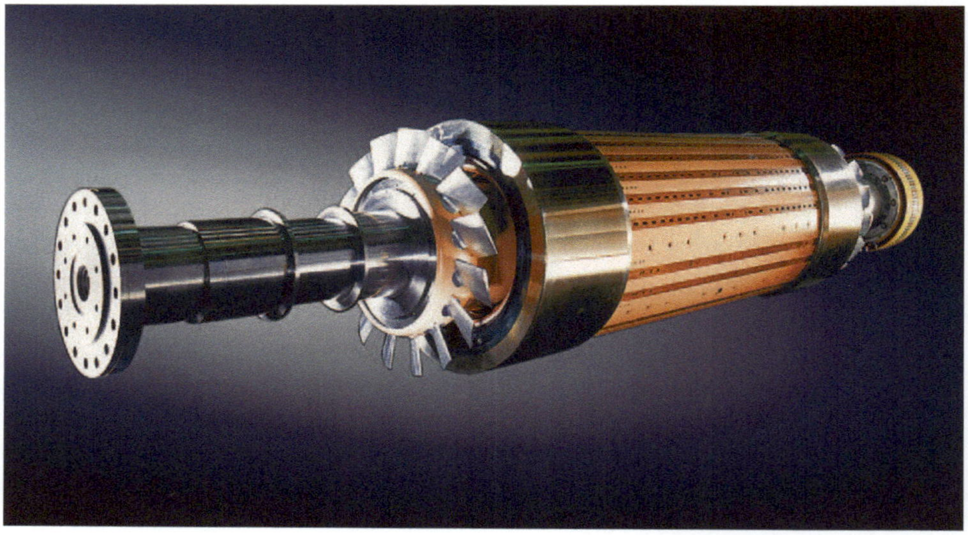

Photo Siemens: Rotor of a 2 pole H-modyn synchron high speed machine. Power:3.3-50 MW (2 to 6 pole).
Standard versions: 6 kV-50 Hz ; 10 kV-50 Hz; 13.2 kV-60 Hz (air to water cooling)

4.1.3 Equivalent diagram of a synchronous machine in nominal service

In the equivalent diagram of an induction motor (fig. 16-57) we have introduced the phasor $\vec{I}_S$ as being responsible for the stator emf. If we adopt the stator winding from the induction motor and add the rotor influence of the synchronous machine, then we arrive at the simple equivalent diagram for a three-phase synchronous machine. Here the (equivalent) rotor current $\vec{I_m}$ is responsible for the rotor-mmf. The total flux in the machine is formed by the stator mmf ($\vec{I_S}$) and the rotor-mmf ($\vec{I_m}/\beta_0$).

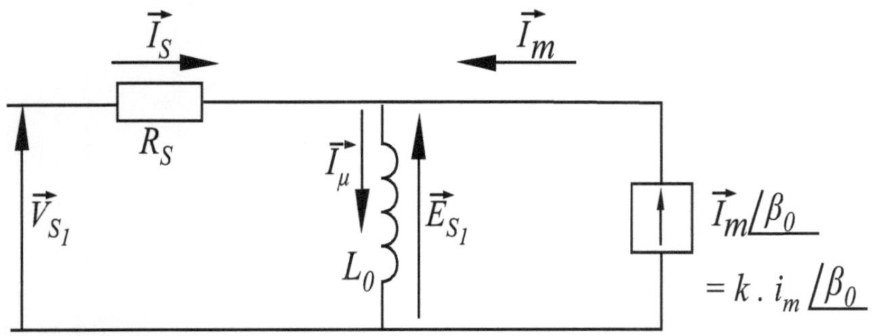

Fig. 16-70: Equivalent diagram of a synchronous machine

Since the rotating flux rotates at the same speed as the rotor there will be no induced emf in the rotor windings. In nominal service the rotor current is calculated with

$$i_m = \frac{V_m}{R_m} \tag{16-65}$$

V_m = DC-supply voltage to the rotor
R_m = resistance of rotor windings
The stator sees a constant current source. A rotor leakage flux is also present which does not link with the stator. This leakage flux is in series with the current source $\vec{I_m}$ and plays no role during nominal service ($\omega = \omega_S$) of the machine.

4.1.4 Leblanc damper windings

These damper windings are composed of a number of copper bars in slots on the pole pieces and parallel with the shaft of the rotor. At both front and back ends these bars are connected with a ring and form a kind of cage similar to a squirrel cage rotor. If the angular velocity of the motor is different from synchronous speed then induced voltages and currents are generated in the cage in a similar fashion to the asynchronous motor. If the rotor speed is lower (higher) than the synchronous speed then the induced currents produce a positive (negative) torque. The cage therefore dampens oscillation of the rotor. This is known as a damper winding (Leblanc "amortisseur").

4.1.5 Equivalent diagram during transient state

The equivalent diagram in fig. 16-70 is a simplified diagram of the motor in nominal service. During a transient state we need to account for the leakage reactance's. For the induction motor we introduce L_{lS} for the stator leakage reactance. For the synchronous machine we also have the excitation winding of the rotor so that we need to take the leakage reactance L_{lm} and the resistance R_m into account.

In addition during the transient state an emf $e = \omega.\Phi_S$ is induced in the magnetizing inductance L_0. We can also replace the current source $k.i_m$ with a voltage source v_m. The damper winding is around the excitation winding and has a linked inductance L_{kS}. The damper winding has a leakage inductance L_D and a resistance R_D. If we add the stator leakage reactance and the damper winding to the diagram of fig. 16-70 then we obtain the equivalent diagram of fig. 16-71.

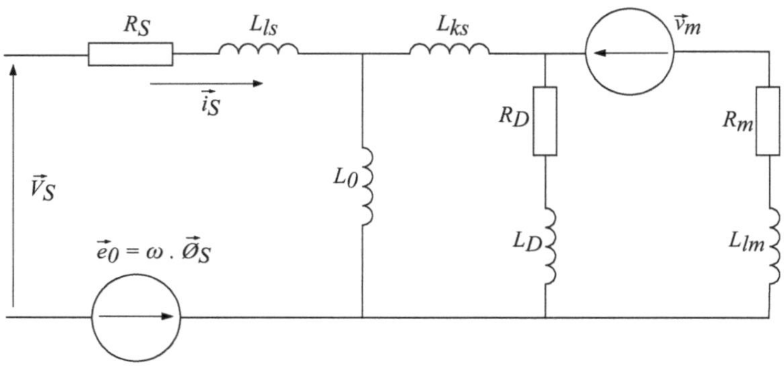

Fig. 16-71: Equivalent diagram of a synchronous machine with stator reference frame

4.1.6 Excitation of a synchronous machine

For large alternators the excitation power is about 1% of nominal power. For a 250 MW alternator that is 2.5MW. It is impossible to transmit this via slip rings. Therefore a second alternator is used for excitation. The field winding of this alternator is wound on the stator of the main alternator and the phase coils are on the rotor of the main alternator. A bridge rectifier mounted on the shaft provides the DC excitation of the main alternator.

4.1.7 Starting a synchronous machine

If the stator coils of a synchronous machine with energized excitation winding are connected to the power grid the machine will not start. As a result of inertia the rotor will not be able to follow the rapidly changing stator field: the rotor will start to vibrate.

To start a synchronous motor there are a number of possibilities:

1. Starting as an induction motor: this is the easiest method. Only applicable for salient pole machines. Via an autotransformer a reduced voltage is applied to the stator winding. The motor starts as an induction motor with no load and about 95% of the synchronous speed is reached. At that instant the excitation current is switched on and regulated to the desired power factor and the stator voltage is increased to the nominal value.

2. Starting with a DC motor: in the case where the synchronous motor includes a DC (excitation-) generator. This dynamo is operated as a motor with a DC supply. The DC machine functions as a start up motor.

3. A small asynchronous motor as start motor: this induction motor needs to have one pole pair less than the synchronous machine in order to reach the synchronous speed of the main machine. Once the synchronous speed is reached the main machine is treated as an alternator and connected in parallel with the power grid.

4.Variable frequency drive: with a frequency drive the synchronous machine can be started from stand still and brought to synchronous speed.

4.2 Electronically controlled synchronous motors

A motor in which the rotor rotates at the same speed as the magnetic field of the stator is called a synchronous machine. In addition to the previously studied AC synchronous motor there are other motor types that comply with this definition. Here they are described in short.

4.2.1 Permanent magnet synchronous machine (PMSM)

Instead of a DC-excited rotor, for small machines permanent magnets are used. These motors are implemented as " stand alone" and mainly as servomotors in an electrical positioning system.

4.2.2 Switched brushless DC-motor

These BLDC motors are also used as servomotors. They will be studied in chapter 22.
BLDC = brushless DC-motor.

4.2.3 Switched reluctance motor (SRM)

See chapter 21

4.2.4 Synchronous reluctance motor (syncRM)

See chapter 21

4.2.5 Stepper motor

See chapter 22

4.2.6 Efficiency

Classification from most to least efficient: PMSM, BLDC, synRM, SRM, ACIM (AC induction motor)

16.5 Small appliance motors

CONTENTS

5.1 Single phase induction motor

5.1.1 Principle

This motor has a cage just like the three-phase induction motor. The stator winding has only one sinusoidally distributed winding. As a result this motor does not start in the same manner as the three-phase induction motor.

In fact two equal but opposite fields are produced in the stator iron. At stand still these fields produce two equal but opposite torques. Indeed, for a sinusoidally distributed stator winding it follows from (16-41): $F_{\mu,ag,\alpha} = \frac{N_{Se} \cdot i_{S1}}{2} \cdot \cos\alpha$

With $i_{S1} = \hat{i}_{\mu} \cdot \cos\omega_S t$ we find: $F_{\mu,ag,\alpha} = \frac{N_{Se}}{2} \cdot \hat{i}_{\mu} \cdot \cos\omega_S t \cdot \cos\alpha$

$$F_{\mu,ag,\alpha} = \frac{N_{Se}}{4} \cdot \hat{i}_{\mu} \cdot \left[\cos(\alpha + \omega_S t) + \cos(\alpha - \omega_S t) \right]$$

We now draw the field strength in the air gap for α = 0 in fig. 16-72. We see a stationary field composed of two opposing rotating stator fields Φ_1 and Φ_2. In fig. 16-73 the *M-n* curves are drawn which both fields should produce. The resulting *M-n* curve shows that the working torque at stand still is zero: the motor will not start by itself. If we assist the motor manually in one of the two directions and exceed the opposing torque M_t, then the motor will start and move to operating point P (fig. 16-73).

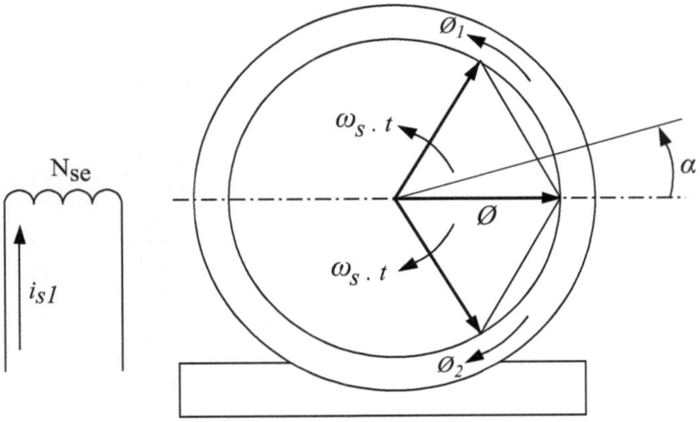

Fig. 16-72: Alternating fields in a single phase induction motor

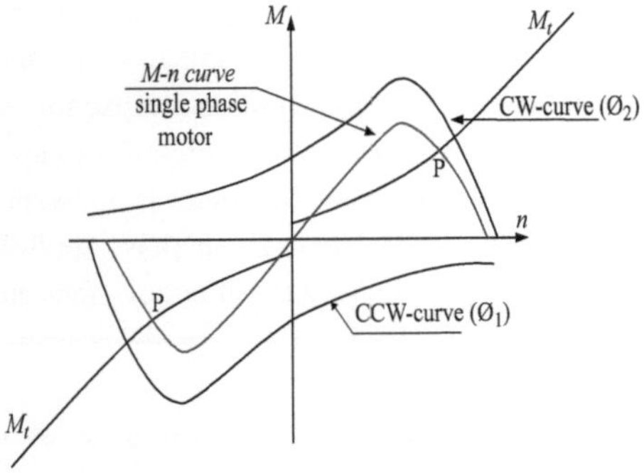

Fig. 16-73: *M-n* curves of a single phase induction motor

5.1.2 Capacitor start motor

In the study of the three-phase asynchronous motor we have seen that a rotating field is created if the three windings which are displaced physically by 120 electrical degrees are fed with three currents which are displaced with respect to each other by a third of a period. The same happens if two windings on the stator are separated by 90° spatially and fed by two currents which are displaced in time by 90°. This system of two single phase voltages displaced from each other by 90° is referred to as a two phase system.

A (two) pole motor with two windings spatially displaced by 90° is called a two phase motor. The operating principle of such a motor can be found in chapter 20 under nr. 7.1.

We can develop the operating principle of a two phase power grid and a two phase motor ourselves by including an auxiliary winding B in the stator of a single phase induction motor (fig. 16-74). This auxiliary winding is displaced spatially by 90° with respect to the main winding A. When the main winding is directly connected to the AC power grid and the auxiliary winding is connected via a capacitor, the current in the auxiliary winding is 90° out of phase with that of the main winding: an artificial two-phase net has been created.

Such a motor starts automatically. In most cases a centrifugal switch disconnects the auxiliary winding and the capacitor once the motor is up to speed so that the motor continues to rotate as an (automatically started) single phase motor.

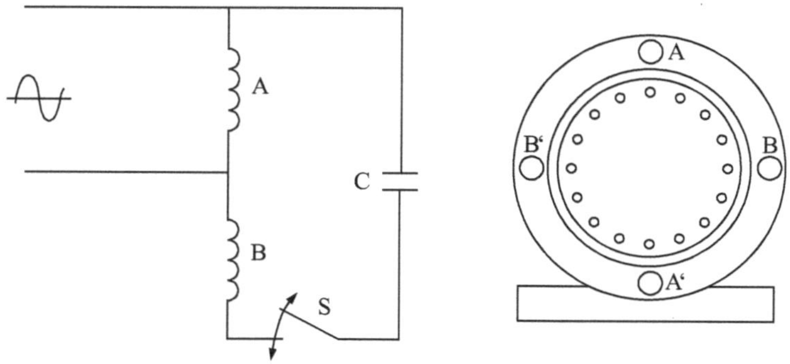

Fig. 16-74: Single phase induction motor with auxiliary winding

The starting torque of such a motor can be 2.8 times the nominal torque. The starting current is relatively large (up to three times the nominal current). A disadvantage of this motor is the relatively high cost price due to the auxiliary winding and additional capacitor. An advantage (in addition to the automatic start) is that the power factor can be high if the capacitor is split into two parts of which only one part is switched out when the motor is up to speed.
This type of motor is only available in the power range up to about 1kW. Applications for the capacitor start motor include household appliances, driving small pumps, fans etc.

Remark

Auxiliary winding: instead of a capacitor a coil can also be switched in series with the auxiliary winding.

5.1.3 Shaded pole motor

Once again a cage rotor is used but now both stator poles are split and on one leg of each pole a ring is placed: see fig. 16-75.

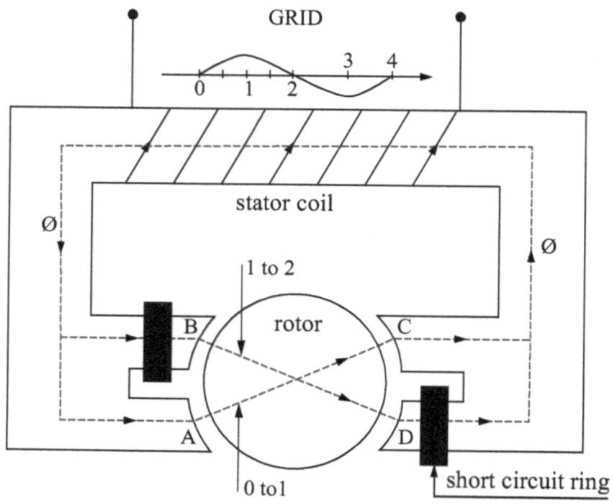

Fig. 16-75: Principle of a shaded pole motor

Since the rings find themselves in a variable magnetic field an induced voltage is produced. According to Lenz's law these rings will produce a magnetic field that oppose the one causing the induced voltage. The causing field was the varying magnetic field in the poles. Assume a main flux Φ as shown in fig. 16-75 and the flux is increasing (supply current between times 0 an 1). The rings produce a flux so that the poles B and D are opposed and the main flux flows almost totally via poles A and C. Between the times 1 and 2 the current reduces and the flux in B and D increases. The north pole is displaced from A to B and the south pole from C to D. After this the supply current reverses and between 2 and 3 it increases with the effect that the flux flows in C and A. The north pole is displaced from D to C and the south pole from B to A. During the last quarter of the cycle the current reduces and poles B and D are strengthened. D is the north pole (coming from C) and B the south pole (coming from A). We can see that there is a rotating field in the stator iron and the cage rotates. These motors are used for small power applications (small ventilators, timer relays,…). This automatic starting asynchronous motor has only one direction of rotation. To reverse the direction the rings would need to be placed on the other pole piece (A and C). The shaded pole motor is the simplest single phase induction motor. It is manufactured up to about 150 W. Power factor and efficiency are poor (often less than 0.4). Low starting torque. The torque on the shaft varies with twice the net frequency centred on an average value.

5.2 Universal motor

5.2.1 Principle

If we connect a (DC-) series motor to an AC supply it should in theory rotate since the excitation and the field currents reverse at the same instant. The torque operates continually in the same direction.

5.2.2 Practical implementation

The configuration of a universal motor is similar to a DC-series motor. Since the stator is now subject to an alternating magnetic field it should be laminated to reduce iron losses (in contrast to the DC-series motor). To limit sparking of the collector special high resistance brushes are used. This motor is called a universal motor because it can operate with AC and DC current.

5.2.3 Properties

The universal motor has practically the same properties as the series motor: large starting torque, speed dependent on the load, runaway under no-load conditions. For the same power level a universal motor is smaller in volume and weight than a cage induction motor. The disadvantage is the maintenance of brushes and collector. For the motor curves see fig. 16-41.

5.2.4 Applications

The universal motor is primarily used in domestic applications up to a power level of 1kW such as coffee grinders, vacuum cleaners, hand drills, angle grinders,... It's use is especially interesting in the speed range where an AC induction motor is not applicable, namely above 3000rpm. If an electric motor with brushes and collector is present in a domestic appliance then it is almost always a universal series motor.

5.2.5 Interference suppression

As a result of the sparking of the brushes high frequency currents may be produced resulting in radio and TV interference. In order to limit this interference as much as possible interference suppression capacitors are used connected in parallel with the brushes and frame.
Fig. 16-76a shows an example. The capacitors may be placed in a single housing as shown in fig. 16-76b. Since the field winding in series with the armature is a coil (ωL) this serves to reduce interference. To reduce interference in every supply conductor the field winding is usually split into two halves (see fig. 16-76a).

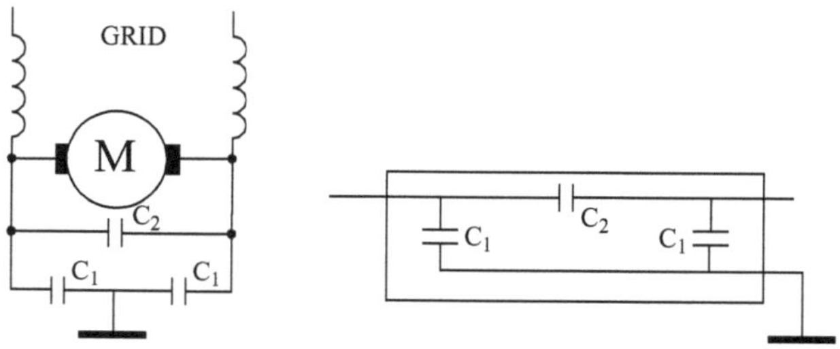

(a) Universal motor with RFI suppression

(b) Suppression capacitors in one case

Fig. 16-76: Universal motor

For electronic control of a universal motor you are referred to chapter 21.

17 Drive Systems

CONTENTS

1. HISTORY

In ancient times the Greeks and Romans used slaves to construct their buildings and works of art. In the best case these slaves had simple mechanical tools at there disposal for moving large loads. Slavery disappeared but slave like labour remained since it took many centuries before more powerful mechanical tools became available. The power became available at the end of the eighteenth century with the first steam engine (1775) of James Watt. This steam machine powered the pumps that were needed to drain the flood water from the deep shafts of the coal mines.
The steam machine was responsible for the birth of the industrial revolution. It was the first time in the history of mankind that people had such large mechanical power available to produce rotating motion at the time and place of their choice!
When electricity became the primary form of energy the electric motor replaced the steam machine. The electric motor experienced little change or development during the first half of the twentieth century up until the second world war. At that time the servo mechanisms were developed. These mechanisms enabled large powers to be controlled together with very accurate positioning.
The technical systems that produce an accurate position and (or) speed of a "load" are referred to as a drive system. Theses drive systems can refer to linear or rotational movement (translation and rotation). By means of the servo mechanisms intelligence was added to the drive system.

Power Electronics: DRIVE TECHNOLOGY and MOTION CONTROL
http://dx.doi.org/10.1016/B978-0-12-814641-5.50002-0,

2. CONTROL THEORY

2.1 Open circuits and closed loops

Consider a power plant where a steam turbine drives an alternator. The steam feed and as a result the speed of the turbine can be controlled by a valve in the supply line.

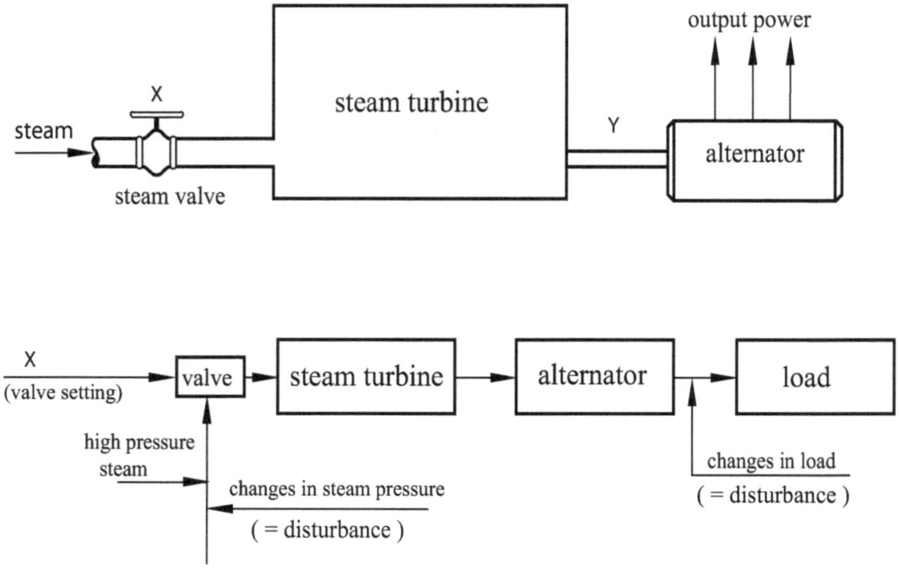

Fig. 17-1: Example of an open loop

If there is no change in the operating conditions of the whole system the position of the valve X could be calibrated as a function of the generator speed. The speed of the turbine is given by Y= k.X, whereby k is a constant that is determined by the operating conditions of the entire system. It is clear that a change in steam pressure or in the load of the alternator will result in a different turbine speed and consequently a different alternator frequency. The system shown in fig. 17-1 is unreliable if we want to maintain a constant frequency of the power grid.
Fig. 17-1 shows a block diagram of an **open loop control system**.
In order to ensure that the speed of the turbine achieves and maintains a predetermined value a configuration as shown in Fig 17-2 is used. The output speed of the turbine is measured using a tachometer. An operator reads the speed compares it with the desired value and subsequently regulates the steam flow with a hand wheel until the output speed is at the desired level. Fig. 17-2 is an example of a **closed loop control system**. The human operator functions not only as an error detector but also as a power amplifier and motor. In a servo system the operator is replaced by a pneumatic, hydraulic or electric system. One theoretical method of operating the steam valve is to use an electric motor as shown in fig. 17-3. Later in this book we will learn about practical solutions.

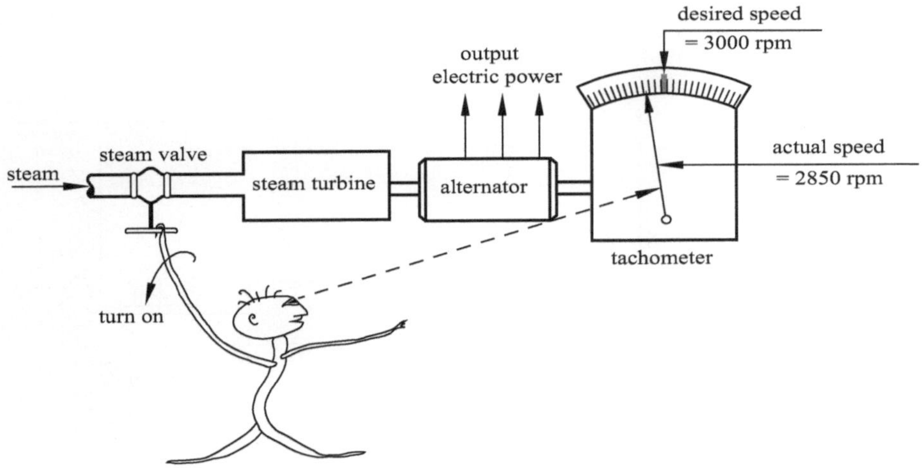

Fig. 17-2: closed loop control (via human operator)

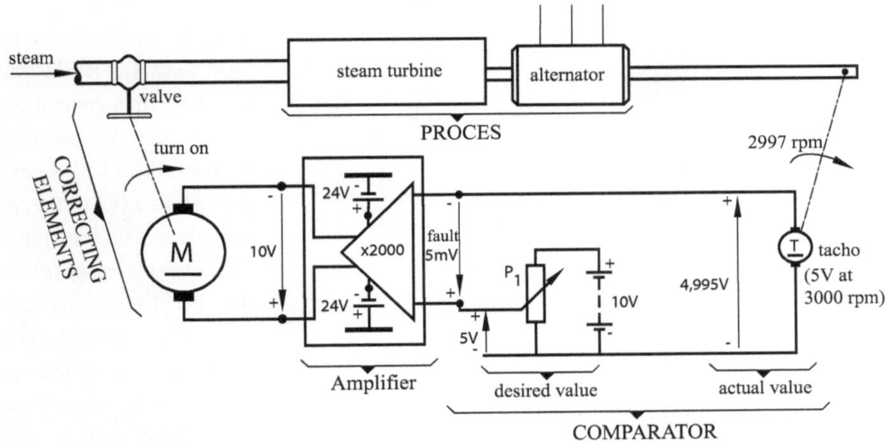

Fig. 17-3: Theoretical speed controller of a steam turbine

The speed is measured via a tacho-generator. Assume that nominal speed of the turbine should be 3000 rpm and the tacho-generator produces 5V at 3000 rpm. The potentiometer P_1 is set to 5V (set point). This set point is then compared with the tacho voltage (measured process value).
If the turbine rotates either too quick or too slow an error voltage results. The polarity of the error voltage is directly related to whether the turbine runs too quick or too slow. This determines the polarity of the armature voltage and thus the direction of rotation of the motor. We need to ensure that when the turbine speed is too high (low) the steam valve is closed (opened).
The block diagram of this closed loop controller is shown in fig. 17-4.

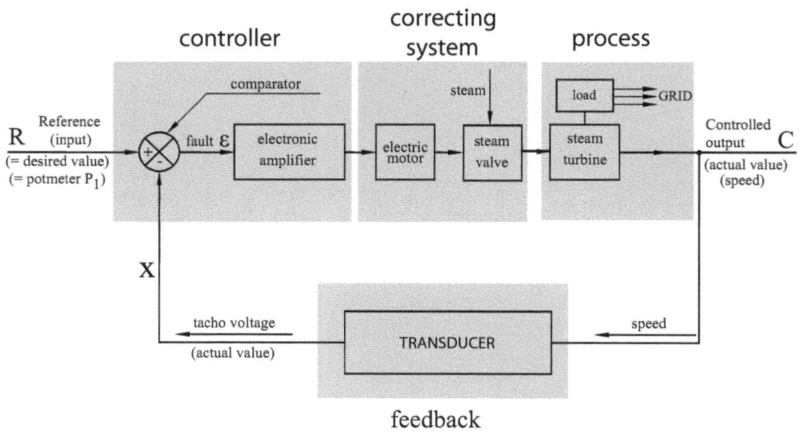

Fig 17-4: Block diagram of configuration of fig. 17-3

2.2 General block diagram of a control loop

In fig.17-4 a block diagram of a closed loop controller is illustrated. Based on this diagram a general block diagram for all closed loop controllers has been derived (fig. 17-5). At every instant the actual value (C) of the output value is compared with the set point value (R). Any eventual error (ε) is corrected. The feed back of the output signal to the comparator at the input is referred to as feedback. In the example considered here the output signal is subtracted from the input signal R, this is referred to as negative feedback.

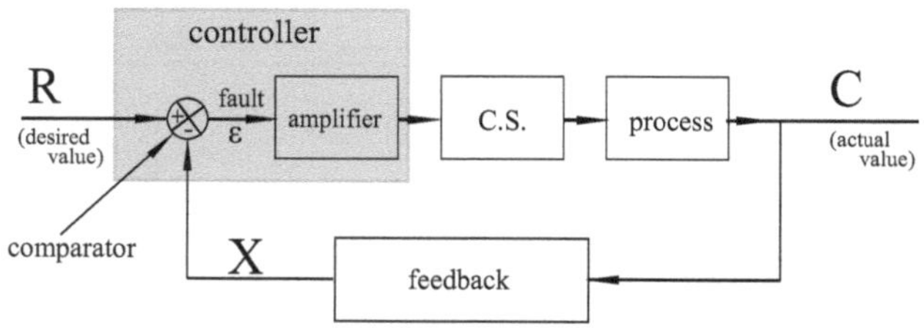

(a) Block diagram feedback system

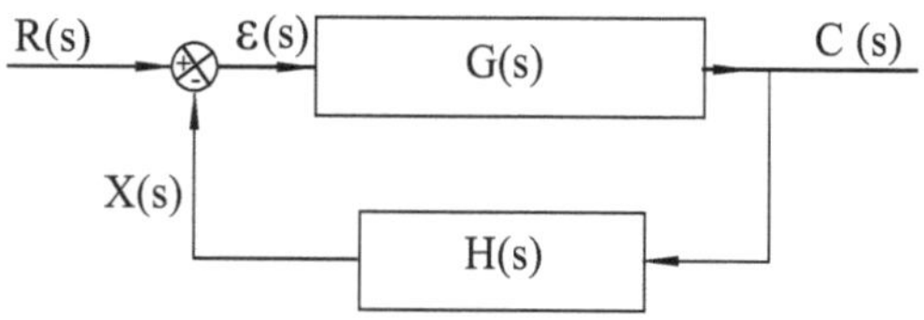

(b) Canonical form feedback system

Fig. 17-5: General block diagrams of a closed loop controller

In fig. 17-5b we can write (with Laplace-notations and with ∓ feedback):

$$C(s) = \varepsilon(s) \,.\, G(s) = G(s) \,.\, [\,R(s) \mp X(s)\,] = G(s) \,.\, [\,R(s) \mp H(s) \,.\, C(s)\,]$$

$$\rightarrow\rightarrow\; C(s) \pm H(s) \,.\, G(s) \,.\, C(s) = G(s) \,.\, R(s)$$

So that:

$$\frac{C(s)}{R(s)} = \frac{G(s)}{1 \pm H(s) \,.\, G(s)} \qquad (17\text{-}1)$$

G(s)	=	*forward transfer function*	*H(s)*	=	*feedback transfer function*
G(s).H(s)	=	*loop gain*	*C(s)/R(s)*	=	*closed loop transfer function*
ε(s)/R(s)	=	*error ratio*			

For a practical application of (17-1): see fig. 19-36 and number 6.1.1 on page 19-35 !

Remark

With positive feedback the comparator produces the sum of the input and output signal. This is undesirable in a control loop. In fig. 17-3 this would mean for example that with a low turbine speed the steam valve would be closed even more. It is clear that this does not result in a well controlled turbine speed.

2.3 History of power electronics

During the second World War a group of engineers and scientists (including the mathematician Norbert Wiener) worked on a project for the US government to develop a fireline control system to combat German aircraft. Firelines are radar controlled cannons. In implementing this project the designers concluded that the heart of the problem lay in comparing what happened and what should have happened. This is the principle of feedback as described above in controlling the speed of a steam turbine. The next step in their research was confirming the similarity between a control process of machines and higher developed organisms. Wiener developed these thoughts in two important books: “Cybernetics” and “The human use of human beings”. Cybernetics was born and in the technical realm the foundation was laid for the theory of servomechanisms. Application of this theory of servomechanisms after the second World War saw the birth of industrial automation. The term cybernetics is derived from an old Greek word for “steersman”.

In solving automation problems the controlling element of the control loop often uses electrical energy. In order to control the electrical energy use was made of electronics. A new technology was born: industrial or power electronics.

From this point in time the three pillars of electrical engineering were **power engineering** for the generation and distribution of electrical energy, **electronic engineering** for processing data and signals and the third was **control or automation** working with closed loop control loops.

An important application of power electronics is the control of motors in electrical drive trains.

2.4 Energy flow in a controlled system

The arrows in the drawing of fig. 17-5 represent the flow of control energy or information and not the supply energy of the system to be controlled (not the steam energy of the turbine example).

In control system block diagrams the energy flow is not included but as an exception this will be included in fig. 17-6 to highlight in which parts of the automated system the drive energy is added.

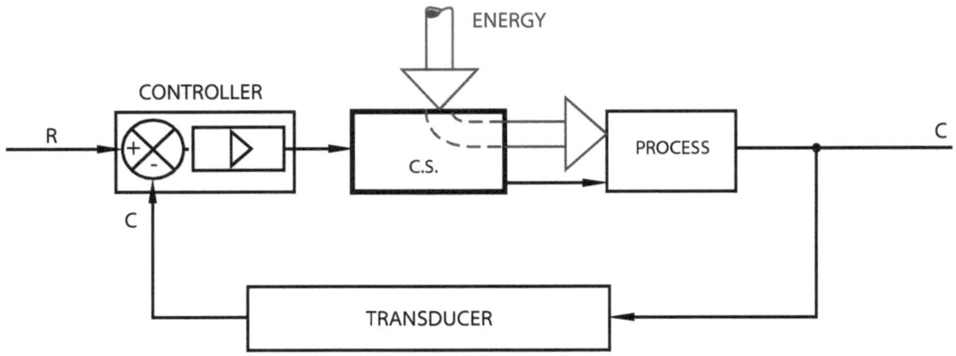

Fig.17-6: Energy flow with automatic control

The energy that flows via the corrective device to the process is not necessarily electrical but can also be hydraulic or pneumatic energy. We then refer to it as a hydraulic or pneumatic drive system respectively rather than an electrical drive system. If the controller and converter are of different physical types to the corrective device then it is referred to as a combined system. In this manner we refer to electro-hydraulic drive systems whereby the control is electrical and the corrective device is hydraulic.

3. TYPES OF DRIVE SYSTEMS

In previous paragraphs we have seen that drives are primarily either hydraulic, pneumatic or electrical. The following table gives an idea of the advantages and disadvantages of the different drive systems.

The advantage of hydraulics is the availability of large force compared to volume and weight of the drive system. Operating canal lock doors is a typical application.

Pneumatics is interesting from the perspective of safety with respect to overload (air is compressible!). The compressibility of air is then a disadvantage when it comes to accurate positioning. Pneumatics also have an advantage in explosive atmospheres. Typical applications are operating doors on trains and department stores, packing machines etc...

Due to the recent applications of micro electronics electrical drives have become more intelligent and this for a very low price. This intelligence makes electrical drives even more attractive. Applications include mechanical machines, pumps, ventilators, bridges, cranes, electrical traction, etc. The intelligent control has as a consequence been responsible for the replacement of hydraulic drives in heavy duty robots with electrical. Arguments for the use of electrical drives include:

- environmentally friendly: electrical energy is “proper”
- energy saving: the energy conversion should have a high efficiency. Of the different drive types electrical drives have the best efficiency.

- optimising the production process: industrial applications require more and more speed, accuracy and reliability. These properties are inherent in electrical systems
- user friendly: thanks to micro electronics drive systems are very user friendly
- easy control and reproduction of speed, torque, acceleration, deceleration …

Table 7-1

SYSTEM PROPERTY	DRIVE SYSTEM		
	Hydraulic	Pneumatic	Electric
• continuous control	• yes	• yes	• yes
• slow linear movement (with large force)	very good	possible	possible
• fast linear movement	possible	very good	good to very good
• rotating movement	easy to realise	easy to realise	very good
• can system operate safely in explosive atmosphere ?	possible	very good	possible
• positional accuracy	good	poor	very good
• cost price energy and installation	high	very high	low
• efficiency	poor	poor	very high
• how is the energy availability ?	very poor (one high pressure supply per installation)	good (central compressor per plant)	very good (electricity is available everywhere)
• energy density	very high	poor	poor
• safety	good	good	good
• intelligent control	moderate	moderate	very good and economical

4. ELECTRONIC DRIVE TECHNOLOGY

4.1 Generalities

In an electric drive system the mechanical energy required to move a load (position and speed) is delivered by an electric motor.

Depending on the type of current we have DC or AC motors. The available energy source is an AC or DC supply.

In order to control the energy flux from electrical power supply to motor with high efficiency, electrical energy converters are used. We distinguish four types of electrical energy converters:

- **Controlled rectifier:** controls the energy flow between AC power grid and DC consumer
- **AC controller:** controls the energy flow between AC power grid and AC consumer
- **Chopper:** allows the energy to be regulated between a DC power source and a DC consumer
- **Inverter:** able to regulate the energy between a DC source and an AC consumer.

Every one of these converters together with an electric motor will play the role of corrective device in the configuration shown in fig. 17-6.

Remarks

1. In the control and regulating of electric motors we distinguish between:
   - speed control (drive system)
   - position control (motion control)
2. According to power level we distinguish between:
   - small power motors: up to 20 kW
   - medium power motors: 20 to 100 kW
   - large power motors: above 100 kW

Motors between 10 and 500 W are extremely low power and motors above 1 MW are extremely large.

3. The following factors influence motor choice:
   1. total cost price of motor + control electronics
   2. maintenance and energy costs
   3. dynamic behaviour.

4.2 Choice of drive train

In the choice of drive train the most important properties of the driven system are:

1. *M-n* curves of the mechanical system
2. Required power on the shaft of the motor
3. Required control properties such as dynamic behaviour, control range, speed.

In addition the properties of the available energy source play a role in the chosen solution.

4.3 *M-n* curves

In drive technology a distinction is made with respect to the torque and speed. One classification can look like this:

1. Torque independent of speed:
   - applications with constant power (winding and unwinding of material, drills,...)
   - applications with constant torque (conveyer belts, lifting equipment with constant load,...)
   - torque varies quadratically with speed (fans, centrifugal pumps,...)
   - torque is proportional to speed (take off roller,...)

2. torque is a function of time: (rolling mill, CNC machines,…); CNC = computer numerical control
3. torque dependent on the angular position (metal cutters, presses,…)

Remarks

1. It is clear that the motor torque at every instant needs to be at least as large as the counter torque of the driven load. In addition some applications require a large starting torque to get moving. Examples of this are plunger pumps, mixers with material that has hardened, equipment whereby at low temperature the lubricant has a higher resistance than during normal service, etc…
   In the choice of drive train account needs to be taken of the required (large) starting torque.
2. We classify the following practical groups of load torque (M_t)
   1. $M_t = K$ = constant (rolling mills, cranes, conveyer belts, mixers, escalators, mills,…)
   2. $M_t = K.\omega$ (paper machines, screw jacks, …)
   3. $M_t = K.\omega^2$ (centrifugal pumps and ventilators)
   4. $M_t = \frac{K}{\omega}$ (wrapping machines, sanders)

   In the groups 1, 2 and 3 an electric motor with constant flux will be used. For group 4, often in part of the operating region, the motor is controlled via flux control.

3. $P = \omega \,.\, M_t$ provides the respective types of load as power:
   1. M_t = constant = K → $P = K \,.\, \omega$
   2. $M_t = K \,.\, \omega$ → $P = K \,.\, \omega^2$
   3. $M_t = K \,.\, \omega^2$ → $P = K \,.\, \omega^3$
   4. $M_t = \frac{K}{\omega}$ → P = constant (K)

4.4 Four quadrant service

If the motor speed needs to be controlled in both directions and in addition electrical braking is possible we refer to four quadrant service. Examples are:

- lifting equipment (lifting and lowering with a crane, draw-bridges, elevators,…)
- loads with a large moment of inertia where rundown without braking would take too much time (centrifuges,…)
- rolling mills (also have a large moment of inertia): in addition it can occur that the material to be rolled will have to be rolled several times backwards and forwards resulting in fast braking being required
- metal working machines where time loss (production!) will not allow to switch to slower speed.

4.5 Types of electric motor used with electronic drives

Until 1975 DC motors were used almost exclusively for speed control. Due to the development of the microcontroller, digital signal processor (DSP) and the IGBT, speed control of asynchronous and synchronous motors is now possible for a large range of applications. These days an AC drive is the normal solution. Since the existing DC drives will not be put onto the scrapheap any time soon it is advisable to study speed control of both DC and AC motors. In addition the knowledge of DC drives is useful in understanding vector control of AC motors.

The term DC or AC refers to the type of voltage to which the motor may be directly connected. In addition to the large power synchronous motor (MW!) many small synchronous motors are used. Specifically this includes stepper motors, switched reluctance motors, the brushless DC motor and the AC servo motor, which are discussed in chapters 21 and 22. These motors operate from a DC supply with the inclusion of an electronic converter. We do not call them DC motors since they cannot operate directly from a DC supply. It is more accurate to categorize them with synchronous motors. Fig. 17-7 shows the motors used for speed and position control

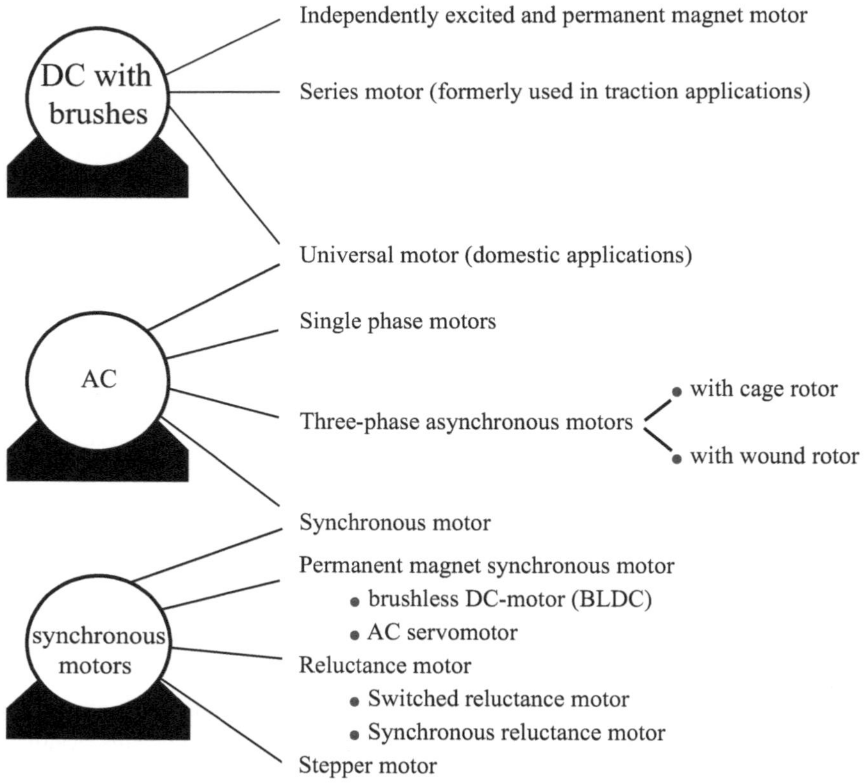

Fig. 17-7: Types of electric motors used in electronic drive systems

Table 17-2 provides an overview of the most commonly used motors and their advantages, disadvantages and typical applications.

Table 17-2

MOTOR TYPE	Advantages	Disadvantages	Typical applications
DC-MOTOR WITH BRUSHES (independently excited type and permanent magnet motor)	- can rotate extremely slowly - small torque ripple - over loadable - simple, inexpensive controllers	- speed limited by counter emf - small power density - brushes and collector (maintenance) - poor thermal performance (losses central in motor!)	- extreme heavy duty compressors and fans - automotive (seats, windows, mirrors,...)
UNIVERSAL MOTOR	- high start torque - small in comparision with single phase motor	- brushes and collector (maintenance)	- household applications (vacuum cleaner, drill,...)
THREE-PHASE ASYNCHRONOUS MOTOR (SQUIRREL CAGE)	- simple construction - good efficiency at full load - quite, smooth operation, low torque ripple - easy to use above "basic" speed	- poor efficiency with small load - poor thermal overload behavior (not suitable for servo applications) - average power density - needs vector control to be competitive	- pumps and ventilators - compressors - general industrial speed control - cheap low power industrial and domestic applications (fridges, air conditioners, washing machines, dish washers, machine tools)
BRUSHLESS DC MOTOR (BLDC)	- controllable across a large speed range - fast acceleration and deceleration - high efficiency at all speeds (if loaded) - best overload properties - quiet, smooth, extremely low torque ripple - large mechanical power density - long lasting reliability	- requires more complex control compared to a classic DC machine - requires vector control to be competitive with a DC machine - poor efficiency at low load - limited choice of application above basic speed - not suitable for AC power, requires a converter	- audio equipment for Cd's (CD-ROM player ,bar-code reader, ticket printer, etc...) - robotics - cooling - electric cars - active auto suspension - electrical servo steering - in standalone industrial drives
SWITCHED RELUCTANCE MOTOR (SRM)	- cheap motor - operates well in rough situations - high starting torque - long lasting reliability - extremely flexible - extremely high speeds	- high torque ripple at low speed - noise caused by torque ripple - the drive design has to be adjusted	- automotive - fans - domestic applications.. - cheap brush-less drive applications with large speed range - hard disk drive
SYNCHRONOUS RELUCTANCE MOTOR (syncRM)	- high efficiency (even at partial load) - efficiency better than IE4 level (IM) - short run-up time (low inertia rotor)	- requires frequency converter (FOC: with special algorithms) - low power factor	- fans and pumps - conveyers - cranes - compressors
STEPPER MOTOR	- quick angular position - accurate operation in open loop - large speed range - reliable operation - no maintenance, long lifespan	- the complete stepper motor with control card is not cheap	- robotics - coordinate tables - recorders and plotters - printers, type writers, cash registers, disc drives

4.6 Comparison table of frequency controlled and DC drive trains

Table 17-3

Property	Frequency control		DC controller
	Scalar	Closed loop vector control	
MOTOR			
. dimensions	. small		. normal
. price	. cheap		. expensive
. protection level	. Ex-protection level		. low
. maintenance	. minimal		. collector and brushes
CONVERTER			
. dimensions	. large		. small
. price	. high		. low
. "electronics"	. complicated		. extremely simple
DRIVE			
. torque control	. no	. extremely good	. extremely good
. torque at stand still	. no	. yes, limited by $I_{inv.}$	. yes, limited by brushes and commutator
. dynamic behaviour		. reasonably good	. extremely good
. constant shaft torque		. extremely good	. extremely good
. "fan" characteristic	. extremely good		. not good
. control of speed:			
. constant torque	. 1-100%		. 0-100%
. constant power	. up to 200%		. up-to 500%
. accuracy	. 2-3%	. 0.5...0.01%	. 0.5...0.01%
maximum speed (50 Hz net)			
. small power	. 6000 rpm		. 5000 rpm
. large power	. 3000 rpm		. 1000 rpm
. efficiency	. high		. extremely high
. power factor	. extremely high		. low to high
. noise	. high		. low
. EMC	. many preventative measures required		. no problems

5. USEFUL MECHANICAL FORMULAS

5.1 Basic equations

5.1.1 Force of inertia

The first basic mechanical formula is : $\Sigma \vec{F} = \int_0^m \vec{a} \, . \, dm = m \, . \, \vec{a}$ = force of inertia.

If a force f_a is applied to a mass *m,* and the opposing force is f_t as shown in fig. 17-8 , then the relationship between force and acceleration *a* is given by:

$$\Sigma \vec{F} = f_a - f_b = m \, . \, a = m. \frac{dv}{dt} = m \, . \frac{d^2x}{dt^2}$$

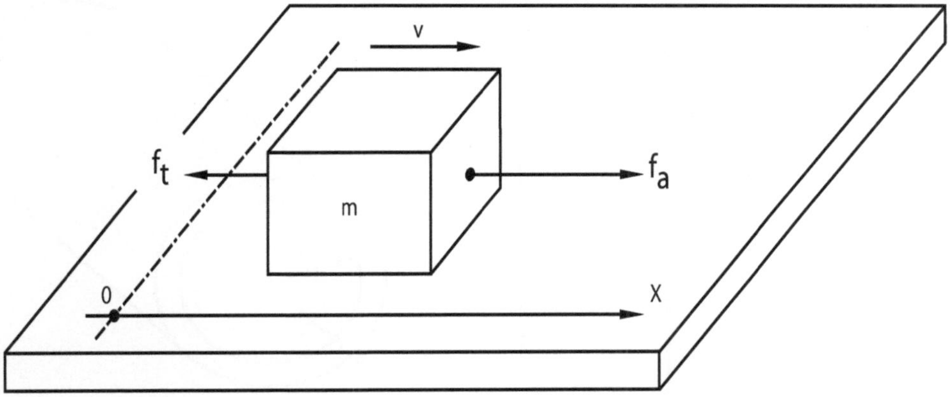

Fig. 17-8: Acceleration by translation

5.1.2 Moment of inertia

A second basic mechanical equation is that of the moment of inertia: $\vec{M} \, \Sigma F = \vec{M} \int_0^m a \, . \, dm$

In the case of rotation around the axis of rotation and with M = driving moment and M_t = opposing moment (= load moment) these two basic formulae become:

$$M - M_t = \int a \, . \, dm \, . \, r = \int dv/dt \, . \, r \, . \, dm = \frac{d\omega}{dt} \int r^2 \, . \, dm$$

The term $\int r^2 \, . \, dm$ is in similar fashion to the force of inertia of a translation called the moment of inertia, so that: $J = \int r^2 . \, dm$

5.1.3 Moment of inertia of cylinders

1. Full cylinder

When the mass is distributed in a homogeneous cylinder (fig. 17-9) with radius *R*, height *h* and specific density ρ, then we can determine the moment of inertia *J* of this cylinder as follows. We call *dm* the mass of an infinitely thin tube with thickness *dR* and height *h* and we find:
$dm = \rho \,.\, 2 \,.\, \pi \,.\, R \,.\, h \,.\, dR$.
The moment of inertia of the entire cylinder is given by:

$$J = \int_0^R R^2 \,.\, dm = \int_0^R R^2 \,.\, \rho \,.\, 2 \,.\, \pi \,.\, h \,.\, R \,.\, dR = \rho \,.\, 2 \,.\, \pi \,.\, h \,.\, \int_0^R R^3 \,.\, dR$$

$$= \rho \,.\, 2 \,.\, \pi \,.\, h \,.\, \frac{R^4}{4} = \rho \,.\, \pi \,.\, R^2 \,.\, h \,.\, \frac{R^2}{2} = m \cdot \frac{R^2}{2} \quad \rightarrow\rightarrow \quad J = m \cdot \frac{R^2}{2} \qquad (17\text{-}2)$$

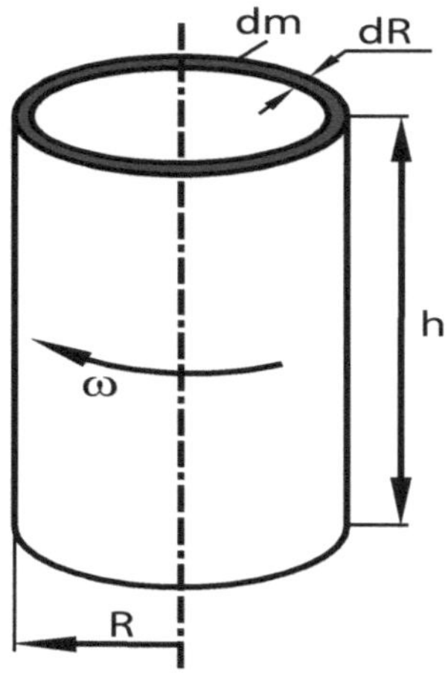

Fig. 17-9: Mass distributed across the homogenous cylinder

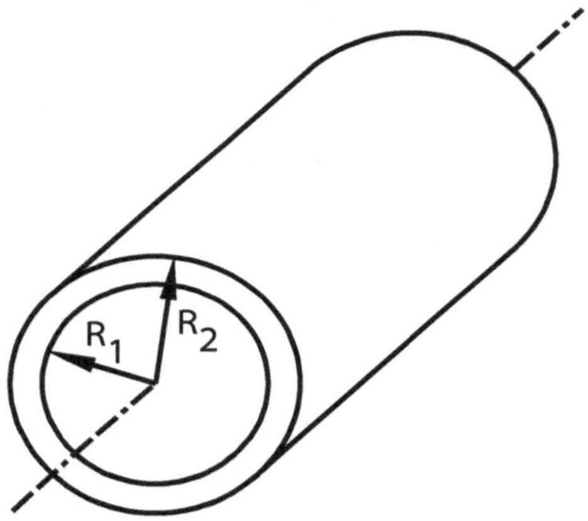

Fig. 17-10: Hollow cylinder

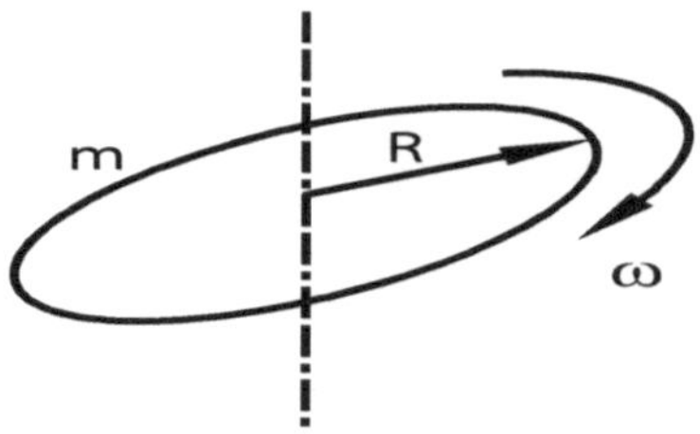

Fig. 17-11: Mass concentrated in thin ring

2. Hollow cylinder

For a hollow cylinder (fig.17-10) we find : $J = \int_{R_1}^{R_2} \rho \,.\, 2 \,.\, \pi \,.\, h \,.\, R^3 \,.\, dR = \frac{\rho \,.\, \pi \,.\, h}{2} \,.\, (R_2^4 - R_1^4)$

with the mass $m = \rho \,.\, \pi \,.\, h \,.\, (R_2^2 - R_1^2)$, gives:

$$J = \frac{m}{2} \cdot (R_2^2 + R_1^2) \qquad (17\text{-}3)$$

If the mass is concentrated in a thin ring ($R_1 = R_2 = R$) as in fig. 17-11, then (17-3) becomes: $J = m \,.\, R^2$.

5.2 Moment of inertia of several mechanisms

5.2.1 Equivalent moment of inertia

Assume that the driving force f_a is produced by an electric motor with a belt as shown in fig 17-12.

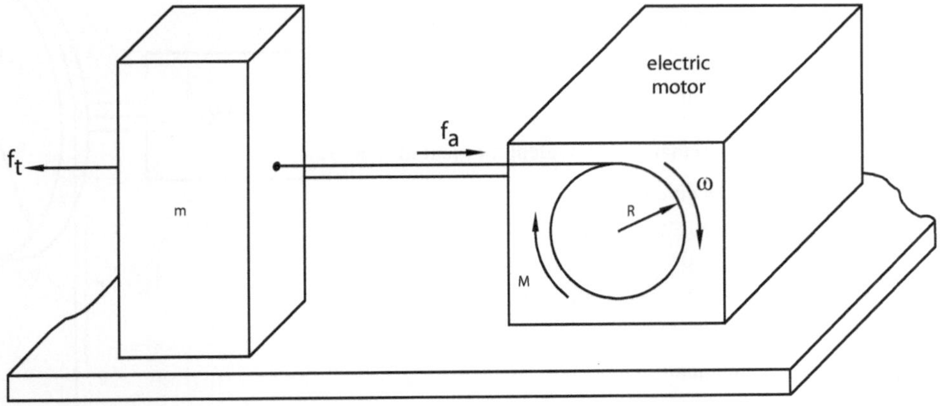

Fig. 17-12: Drive with electric motor

For fig. 17-12 we can write :

$$f_a - f_t = m \,.\, \frac{dv}{dt}\,;\; (f_a - f_t) \,.\, R = m \,.\, R \,.\, \frac{dv}{dt} = m \,.\, R \,.\, \frac{d\omega R}{dt} = m \,.\, R^2 \,.\, \frac{d\omega}{dt}$$

$$\overrightarrow{M}\, \Sigma F = M - M_t = m \,.\, R^2 \,.\, \frac{d\omega}{dt}$$

Here in: M = driving torque (Nm)

M_t = opposing torque (Nm)

From the equation it can be seen that the term $m \cdot R^2$ is present even though the mass m is not rotating. Therefore: $m \,.\, R^2 = J_{eq.}$ = the "equivalent" moment of inertia of the linearly moving mass, reflected to the shaft of the motor (kgm^2).

We find therefore:

$$M - M_t = J_{eq.} \cdot \frac{d\omega}{dt} \qquad (17\text{-}4)$$

Remarks

1. $J_{eq.} = m \, . \, R^2 = m \frac{D^2}{4}$ .Sometimes the inertia is quoted as $m \, . \, D^2$, in which case it needs to be divided by 4 to find $J_{eq.}$
2. The kinetic energy $W = \frac{1}{2} \, . \, m \, . \, v^2$ may also be written as:
   $W = \frac{1}{2} \, . \, m \, . \, v^2 = \frac{1}{2} \, . \, m \, . \, R^2 \, . \, \omega^2 = \frac{1}{2} \cdot J_{eq.} \cdot \omega^2$
   This expression is similar to (electrical) energy:
   $W = \frac{1}{2} \cdot L \cdot i^2 \;\; and \;\; W = \frac{1}{2} \cdot C \cdot V^2$

5.2.2 Pulley

A load with mass m (kg) operates on the circumference of a wheel R, so that $J_{1eq.} = m_1 \, . \, R^2$ (kgm²⁾ .

The wheel R with mass m_2 (kg) also needs to be accelerated. We introduce a moment of inertia $J_2 = \frac{m_2 \, . \, R^2}{2}$ so that the equivalent moment of inertia of the pulley is given by

$$J_{eq.} = m_1 \cdot R^2 + \frac{m_2 \cdot R^2}{2} \quad (17\text{-}5)$$

The moment equation is: $M - M_t = J_{eq.} \cdot \frac{d\omega}{dt}$.

M is the motor torque and M_t is the load torque:

$$M_t = F \cdot R \quad (17\text{-}6)$$

Here by: $F = m_1 \, . \, g$.

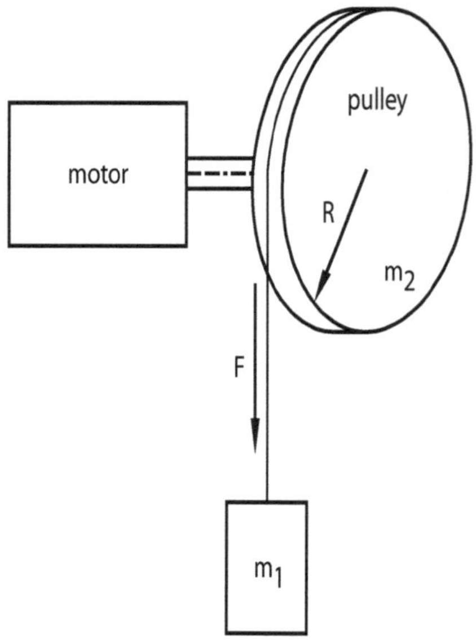

Fig. 17-13: Pulley

The equivalent moment of inertia calculated on the shaft of the machine is also valid for the setup shown in fig. 17-14 and may be expressed by expression (17-5).

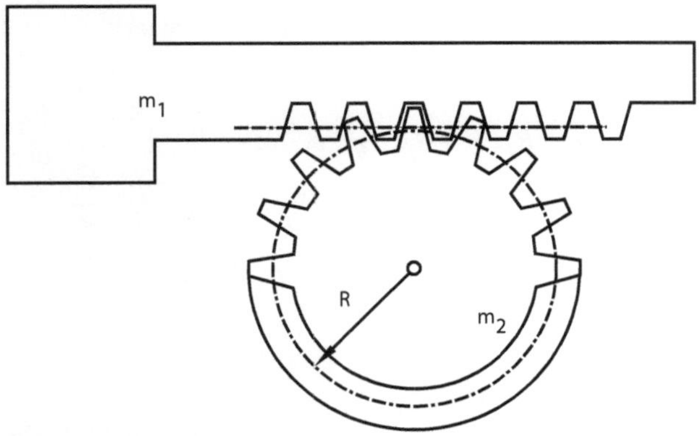

Fig. 17-14: Rack and pinion

5.2.3 Gear wheel transmission

Assume a reduction whereby gearwheel R_2 is driven by gearwheel R_1 with a torque M_a.

The driving power of gearwheel R_1 is F_1. The opposing force $F_2 = F_1 = F$.

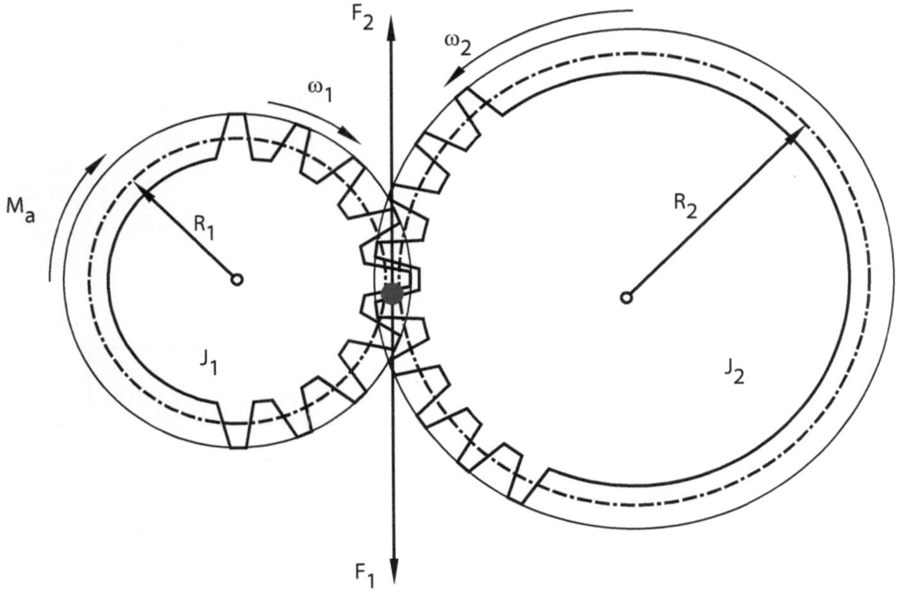

Fig. 17-15: Gearwheel reduction

Problem: Which moment is needed to bring the unloaded gearwheels up to speed? To solve this problem we separate each gearwheel and determine the rotation balance.
By neglecting every form of friction we can write:

- for the left gearwheel: $M_a - R_1 \,.\, F = J_1 \,.\, \frac{d\omega_1}{dt}$ or: $M_a = R_1 \,.\, F + J_1 \,.\, \frac{d\omega_1}{dt}$
- for the right gearwheel : $R_2 \,.\, F = J_2 \,.\, \frac{d\omega_2}{dt}$ then: $R_1 \,.\, F = \frac{R_1}{R_2} \,.\, J_2 \,.\, \frac{d\omega_2}{dt}$

With $r = \frac{R_1}{R_2}$ is $R_1 \,.\, F = r \,.\, J_2 \,.\, \frac{d\omega_2}{dt}$ and: $M_a = r \,.\, J_2 \,.\, \frac{d\omega_2}{dt} + J_1 \,.\, \frac{d\omega_1}{dt}$

From $\omega_2 = \omega_1 \,.\, \frac{R_1}{R_2} = \omega_1 \,.\, r$

it follows: $M_a = J_1 \,.\, \frac{d\omega_1}{dt} + r \,.\, J_2 \,.\, \frac{d(\omega_1 \,.\, r)}{dt} = (J_1 + r^2 \,.\, J_2) \,.\, \frac{d\omega_1}{dt}$

The equivalent moment of inertia reflected to the shaft of the gearwheel R_1 is:

$J_{eq.} = J_1 + r^2 \,.\, J_2$

Calculating with revolutions and a transmission ratio N:

$N = \frac{\text{input shaft speed}}{\text{output shaft speed}} = \frac{1}{r}$, then :

$$J_{eq.} = J_1 + \frac{J_2}{N^2} \qquad (17\text{-}7)$$

Similarly valid for a reduction (fig. 17-16):

$$J_{eq.} = J_r + \frac{J_b}{N^2} \qquad (17\text{-}8)$$

J_r = inertia of the reductor (kgm^2)

J_b = inertia of the load connected to the output shaft

N = transmission ratio $= \frac{\omega_1}{\omega_2}$

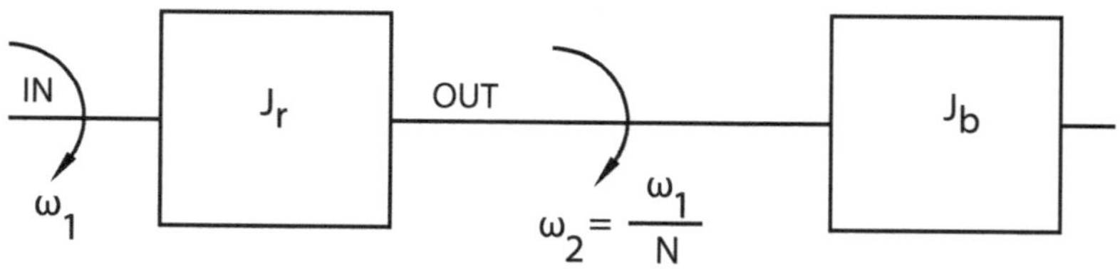

Fig. 17-16: Reduction (N) loaded with moment of inertia

5.2.4 Belt or chain transmission

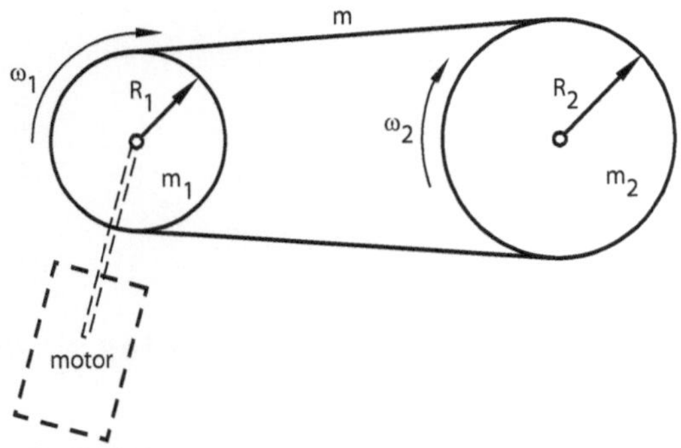

$$J_{eq.} = \frac{(m_1 + m_2)}{2} \cdot R_1^2 + m \cdot R_1^2 \qquad (17\text{-}9)$$

Fig. 7-17: Belt or chain transmission

$J_{eq.}$ = total equivalent inertia on motor shaft (kgm^2) . If the pulley wheel R_2 is also loaded with inertia J_b then:

$$J_{eq.} = \frac{(m_1 + m_2)}{2} \cdot R_1^2 + m \cdot R_1^2 + \frac{J_b}{N^2} \qquad (17\text{-}10)$$

Here in N = transmission ratio

5.2.5 Worm wheel transmission

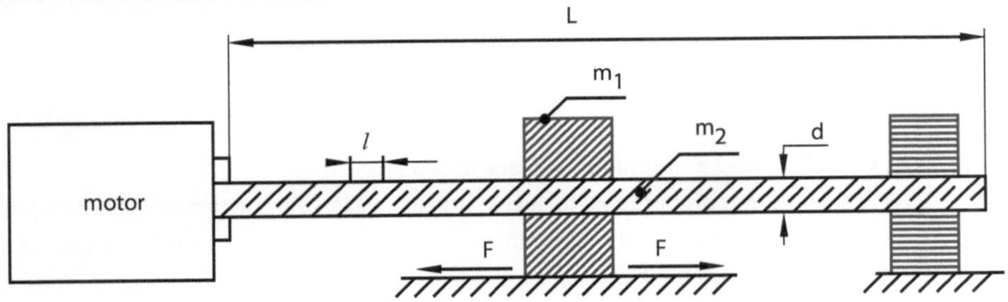

Fig. 17-18: Motor with worm wheel transmission

A rotation of one revolution ($2\ \pi\ r$) of the motor shaft corresponds to a displacement l of the load with mass m_1, so that $J_{1eq.} = m_1 \cdot r^2 = m_1 \cdot \left[\frac{l}{2 . \pi}\right]^2$. The rotation of the worm wheel with mass m_2 requires a moment of inertia $\frac{m_2 \cdot r^2}{2} = \frac{m_2 \cdot d^2}{8}$. If we express the pitch of the screw and the average diameter of the worm in mm, then the total equivalent inertia moment is:

$$J_{eq.} = \frac{m_1 \cdot l^2}{4 \cdot \pi^2} \cdot 10^{-6} + \frac{m_2 \cdot d^2}{8} \cdot 10^{-6} \quad (\text{kgm}^2) \qquad (17\text{-}11)$$

Remark

The worm and gear produce a friction torque, quantified by:

$$M_f = \frac{F \cdot l \cdot 10^{-3}}{2 \cdot \pi \cdot \eta} \quad \text{(Nm)} \qquad (17\text{-}12)$$

F = force of friction (N)
l = screw pitch (mm)
η = transmission efficiency to the load. This efficiency is between 0.3 and 0.9. A ball screw spindle can achieve an efficiency of 0.9.

5.3 Counter torque on the motor shaft

There are four groups of torques to be overcome:

a. Inertia J_m
Only has affect during a change of motor speed. The total moment of inertia consists of the sum of the moments of inertia of the rotor with the equivalent inertia of the complete load reflected to the shaft of the motor.

b. Counter torque produced by forces
In the example of the pulley (fig. 17-13) the counter torque is $M_t = F\,x\,R$.
If there is a reduction (N) between motor shaft and counter torque then the equivalent torque is :

$$M_{eq.} = \frac{M_{load}}{N} \qquad (17\text{-}13)$$

c. Counter torque as a result of viscous friction
This torque is proportional to speed. It is the result of the action of a gas or liquid on a solid object that moves in this environment. Internally in the motor we find a viscous friction between the rotor and air.

d. Counter torque as a result of dry friction
This torque is the result of movement of one object against another fixed object.
In addition to external friction the internal friction in the motor also need to be included (friction of the motor shaft in its bearings!).

Remark

As already mentioned on p. 17.9 we categorize different load torques (M_t):

1. M_t = constant (rolling mills, cranes, conveyor belts, mixers, escalators, mills,...)
2. $M_t = K.\omega$ (paper machines, screw jacks, ...)
3. $M_t = K.\omega^2$ (centrifugal pumps and fans)
4. $M_t = \frac{K}{\omega}$ (wrapping machines, sanders)

6. MOMENT AND POWER OF A MOTOR

The required torque is indicated by:

$$M = M_t + J_m \cdot \frac{d\omega}{dt} \qquad (17\text{-}14)$$

M = required motor torque (Nm)

M_t = sum of all friction torques and other force torques. For example in a pulley transmission M_t = total torque produced by friction + $F \times R$ (pulley)

J_m = sum of all the moments of inertia reflected to the shaft of the motor. In J_m the moment of inertia of the rotor is included. (kgm^2)

$\frac{d\omega}{dt}$ = angular acceleration motor (rad/s^2)

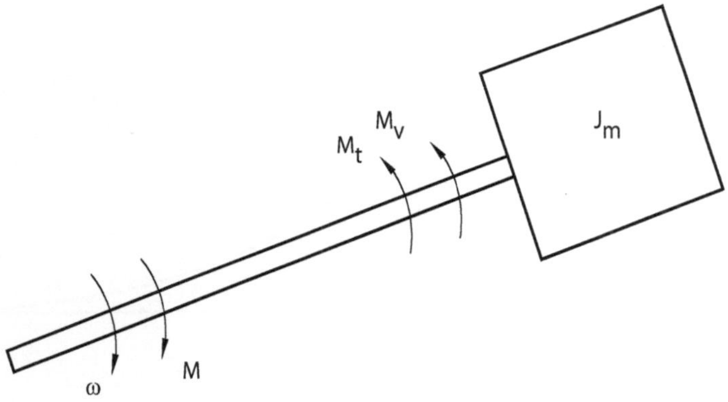

Fig. 17-19: Drive system

For fig. 17-19 we can write: $M = M_t + M_v$ with : $M_v = J_m \cdot \frac{d\omega}{dt}$

Here M_v is the acceleration or deceleration torque, depending on the sign of $\frac{d\omega}{dt}$

Multiplying (17-14) with ω produces: $\omega \cdot M = \omega \cdot M_t + \omega \cdot J_m \cdot \frac{d\omega}{dt}$

$$P_{sh.} = P_t + \omega \cdot J_m \cdot \frac{d\omega}{dt} \qquad (17\text{-}15)$$

$P_{sh.} = \omega \cdot M$ = driving power (on the shaft)

$P_t = \omega \cdot M_t$ = load power

$\omega \cdot J_m \cdot \frac{d\omega}{dt}$ = change of kinetic energy, stored in the rotating masses

7 RUN DOWN TEST TO DETERMINE THE INERTIA OF A DRIVE SYSTEM

From (17-15) follows a method to determine the inertia of a complete drive.
The input power P_i of the motor is measured at different steady state speeds ω. Since these measurements were in steady state (ω = constant) $\omega \,.\, J_m \cdot \frac{d\omega}{dt} = 0$ and $P_{shaft} = P_t$.

If the copper losses P_{Cu} in the motor are subtracted from P_i, then $P_t = P_i - P_{Cu}$ remains and this for different values of ω. In this way the effective load torque $M_t = \frac{P_t}{\omega}$ for different motor speeds may be determined. Fig. 17-20 shows an example.

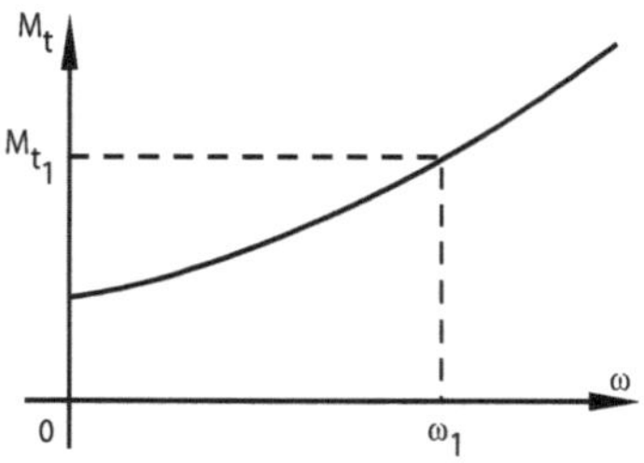

Fig. 17-20: Motor counter torque

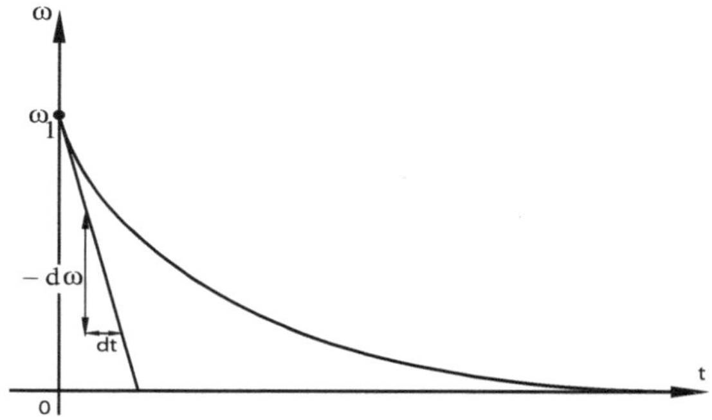

Fig. 17-21: Run down speed of motor

The motor is now accelerated to a value ω_1 with a load torque M_{t1} (fig. 17-20). Then the motor is disconnected from the electrical supply. It will run down to stand still, we note the speed as function of time. (fig. 17-21)

During run down: $0 = M_{t1} + J_m \cdot \frac{d\omega}{dt}$ so that:

$$J_m = -\frac{M_{t1}}{[d\omega/dt\,]_{\omega_1}} \tag{17-16}$$

If we can determine the steepness $\frac{d\omega}{dt}$ of the rundown curve, then from (17-16) the inertia of the drive can be calculated. If the ω-t curve in fig. 17-21 is a straight line, then J_m is easily determined. In the case of a non linear curve we can calculate for tangents at different points in order to determine the average value of J_m.

8. NUMERIC EXAMPLES

1. We call P_{sh} and M_{sh} the nominal values of power and moment of torque on the shaft of the motor.
   Determine the start time of a drive with the following details:
   $P_{sh} = 4$ kW ; $n = 3000$ rpm ; $M_{start} = 1.5 \cdot M_{sh}$; $M_t = 0.75 \cdot M_{sh}$; $J_{rotor} = 0.06$ kgm² ; $J_{eq.load} = 0.25$ kgm² .

   SOLUTION:

   Total inertia: : $J = 0.25 + 0.06 = 0.31$ kgm²

   $$\omega = \frac{2 \cdot \pi \cdot n}{60} = 314 \text{ rad/s}$$

   $$P_{sh} = M_{sh} \cdot \omega \rightarrow M_{sh} = \frac{4500}{314} = 14.32 \text{ Nm}$$

   $M_{sh} = M_t + M_v$ produces at start: $M_{start} = M_t + M_v$, so that the acceleration torque is:
   $M_v = 1.5 \cdot M_{sh} - 0.75 \cdot M_{sh} = 10.74$ Nm

   From $M_v = J \cdot \frac{d\omega}{dt}$ it follows that the start time:

   $$dt = t \rightarrow t = \frac{J \cdot d\omega}{M_v} = \frac{0.31 \cdot 314}{10.74} = 9 \text{ seconds}$$

2. Determine the nominal power of a motor, if the drive has the following properties:
   $m \,.\, D^2 = 10$ kgm² ; $\Delta n = 0$ to 1500 rpm in 4 seconds ; $M_{start} = 2 \,.\, M_{sh}$;
   $M_t = 0.6 \text{ x } M_{sh}$

   SOLUTION:
   $M_{start} = M_t + M_v = 2 \cdot M_{sh} = 0.6 \cdot M_{sh} + M_v$, so that: $M_v = 1.4 \cdot M_{sh}$

   $$M_v = J \cdot \frac{d\omega}{dt} = \frac{m \cdot D^2}{4} \cdot \frac{d\omega}{dt} = \frac{10}{4} \cdot \frac{2 \cdot \pi \cdot 1500}{60 \cdot 4} = 98.17 \text{ Nm}$$

   $$M_{sh} = \frac{M_v}{1.4} = 70.12 \text{ Nm}$$

   $$P_{sh} = M_{sh} \cdot \omega = 70.12 \cdot \frac{2 \cdot \pi \cdot 1500}{60} = 11 \text{ kW}$$

Photo Maxon Motor Benelux:

ESCON 36/2 DC: first product in the new servo controller range by maxon motor. Compact, powerful 4-quadrant PWM servo controller offers efficient control of brushed permanent-magnet DC motors up to 72 W.

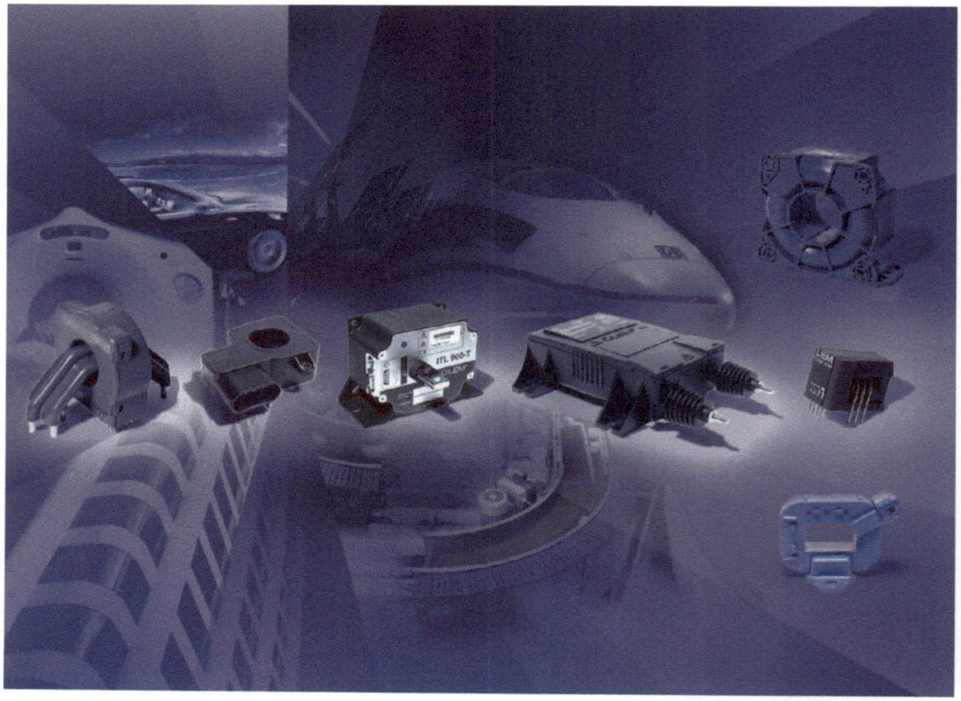

Photo LEM: Isolated current and voltage measurement in the industry (this is the cover photo from LEM)

18 CURRENT - , ANGULAR POSITION - , SPEED TRANSDUCERS

CONTENTS

For speed and position control we need as a minimum to measure the speed and position. In addition the motor current needs to be known for amongst other things the protection of the power bridge and for the torque controller. In this chapter we study the operating principles and properties of the standard current, angular position and speed sensors.

Power Electronics: DRIVE TECHNOLOGY and MOTION CONTROL
http://dx.doi.org/10.1016/B978-0-12-814641-5.50003-2,

1.CURRENT SENSORS

1.1 Instrument current transformer

Via current transformers (fig. 18-1) and a three-phase diode bridge the AC currents in the lines (L_1, L_2, L_3) are converted to a DC voltage. With a suitable choice of *R* we can convert 0-500A input into an output voltage of 0-10V.

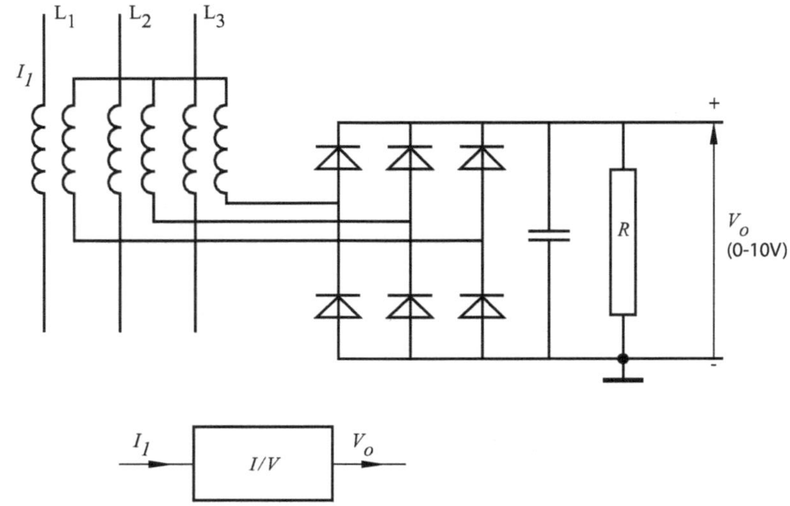

Fig. 18-1: Instrument current transformer

In an AC motor drive using this technique we therefore have a means of judging the motor current. In a DC drive this circuit is also applicable since the AC supply (L_1, L_2, L_3) to the controlled rectifier is also a measure of the motor current on the output of the converter.
The current transformers also implement galvanic separation between power and control circuit.

1.2 Shunt converter

The DC voltage drop across a shunt-resistor is chopped, amplified and modulated. Between amplifier and demodulator, galvanic separation is implemented (transformer or opto-coupler).
The advantage of the measurement converter in fig. 18-2 with respect to the current transformer is that the circuit (with the exception of the shunt) is dimensioned independent of the current.
The polarity of the output voltage depends on the direction of the measured motor current. A shunt converter is therefore suitable for four quadrant drive trains.

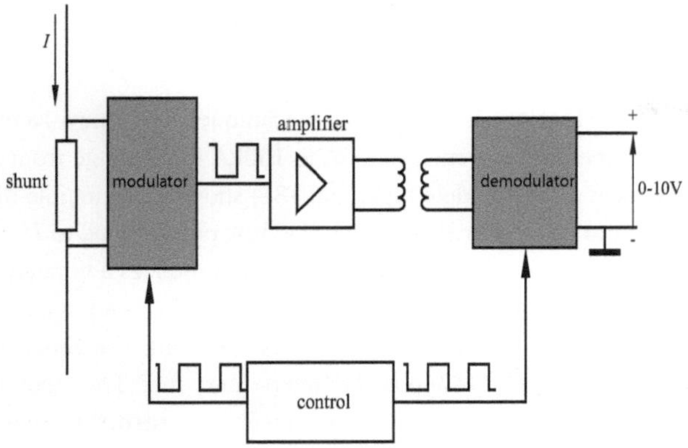

Fig. 18-2: Shunt converter

1.3 Hall sensor open circuit

The current to be measured flows through an inductor (fig. 18-3). In the air gap of the coil a Hall generator is located which is supplied with a constant current I_C . The Hall voltage V_H is proportional to the flux (Φ) and therefore to the measured current I . Especially large currents (e.g. 30 to 1000 A) can be measured using this method. From the graphic of fig. 18-3 it is clear that this principle may also be used in four quadrant drive trains. For the operation of a Hall sensor you are referred to chapter 5.

The Hall voltage can, after amplification via an opamp be transported to the required location.

Advantages of open loop sensor (OL = open loop):

Simple electronics; good price performance ratio; low power consumption (opamp); small dimensions in relation to the large currents it measures.

Disadvantages:

Small bandwidth (up to 50 kHz) as a result of the quality of the magnetic circuit and (to a lesser degree) the electronics; high offset voltage and drift of the amplification factor.

Fig. 18-3: Hall measurement sensor

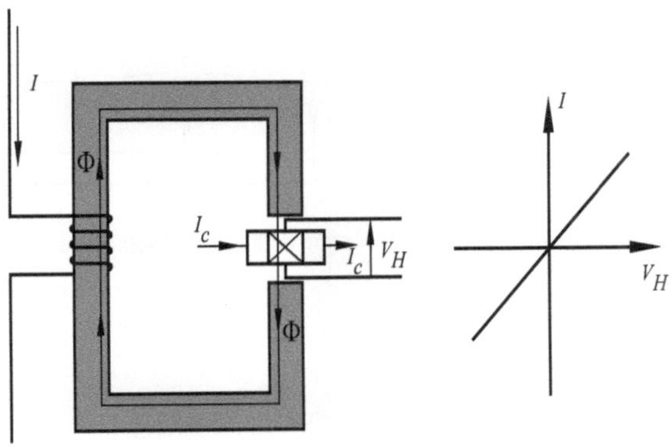

1.4 Closed loop sensor with Hall transducer

1.4.1 The zero flux current transformer

The Swiss company LEM (Liaisons Electroniques Mécaniques SA Genève) supplies the market with modules which can measure currents from 5A to 100kA in the range from DC to 500kHz. The maxima are dependent on the module type. Fig. 18-4 shows the principle of operation. Feedback is used. The current to be measured I_1 produces a flux, proportional to $N_1 \cdot I_1$ ampere windings. The Hall generator supplied with a constant current I_C produces a voltage V_H which after amplification (A) in an opamp and via booster transistor T_1 (or T_2) sends a current I_2 through a coil with N_2 windings. A flux, proportional with $N_2 \cdot I_2$ is the result. The connections are made in such a way that this (secondary) flux is opposed to the primary flux. The secondary flux ($N_2 \cdot I_2$) attempts to equal the primary flux ($N_1 \cdot I_1$) so that the result is: a **zero flux transformer**.
With a large amplification A of the opamp it is almost so that

$N_2 \cdot I_2 = N_1 \cdot I_1$, and: $I_2 \cdot R = \frac{N_1}{N_2} \cdot R \cdot I_1$.

The voltage across the resistor R is a measure of the current I_1 .
Due to the nature of the configuration there is galvanic separation between input and output.
The insulation voltage of the LEM-module varies between 2.5 and 12kV. LEM modules have an output current of 25 to 400mA and an accuracy of 0.2 to 1% of the nominal current. All these figures depend of course on the chosen module. Modules are available to measure currents from a few amps to tens of kA!.

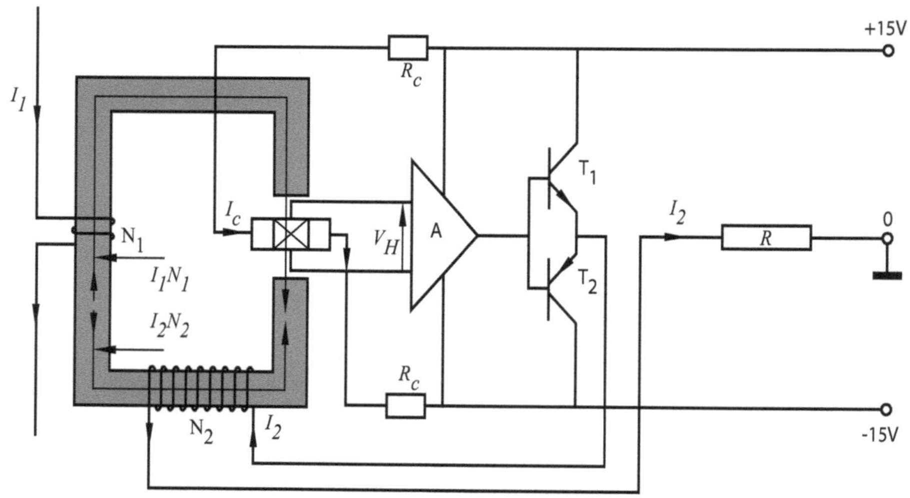

Fig. 18-4: Principle of a current sensor with closed loop: the zero flux transformer from LEM

Advantages of closed loop sensor (CL = closed loop):

- large bandwidth (up to 200 kHz)
- high accuracy
- low drift.

Disadvantages:

- high current usage (mainly due to the compensation of the secondary winding)
- expensive current amplification stage

1.5 ETA-technology (LEM)

With this technology LEM combines the principle of an open loop sensor with those of a CL-sensor. This is only possible through the use of an ASIC (Application Specific Integrated circuit) that improves the accuracy of the open loop. An ASIC is located in the air gap with a temperature compensated Hall element. With DC and low frequency an open loop is used. In the case of AC the secondary winding provides the output voltage. Both signals (OL and CL) are electronically added and form a common output voltage.
ETA stands for the Greek letter η (= efficiency!).

Photo LEM: LA306-S: Closed loop "high performance" current sensor for 400A

1.6 Closed loop sensor with flux-gate technology

1.6.1 Flux gate sensor

The flux gate principle is familiar from the flux-gate magnetometers. These were developed during the second world war and used by low flying aircraft to detect submarines. These days flux-gate sensors are used in gyro compasses and in lab equipment to measure remanent magnetism for example. The flux-gate magnetometer can measure the individual components of the earth's magnetic field from 0.5 to 1 nT accuracy. The flux gate sensor is comprised of a core of minimal cross sectional area and made from material with high permeability and steep and small magnetic curves. Around this plate a coil is wound. The small cross sectional area is saturated by a small current in the coil. Fig. 18-5 shows the hysteresis curve of the flux-gate.

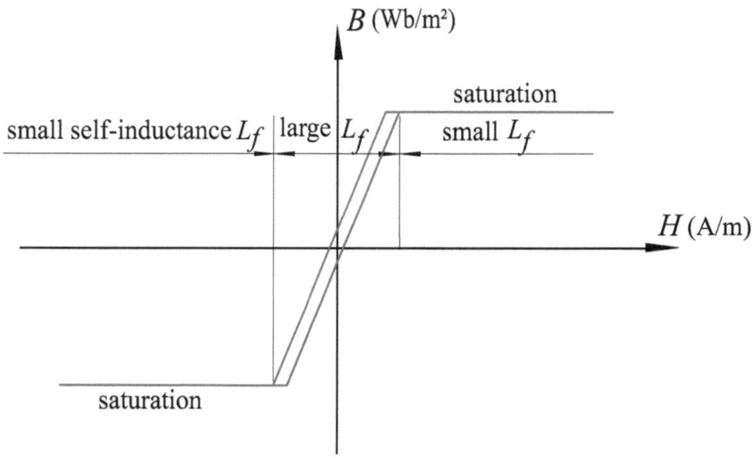

Fig.18-5: Flux gate hysteresis loop

In fig 18-6 we see the location of the flux gate in the magnetic circuit of our future current sensor. Assume I_1 is zero and therefore there is no flux in the core of fig. 18-6. We now apply a block wave to the coil of the flux-gate so that it becomes saturated. The current is as shown in fig. 18-7. The current will exponentially rise, but in a short time the core is saturated and the self induction L_f can be neglected so that only the resistance R_f of the coil limits the current. The average value of the current in N_f is zero.

If there is a current I_1 present and therefore a flux in the core of the current sensor, then the flux-gate in one direction will be more quickly saturated resulting in the current in the coil looking like that shown in fig. 18-8. The average value of the current will be positive or negative corresponding to the direction of the current I_1 .

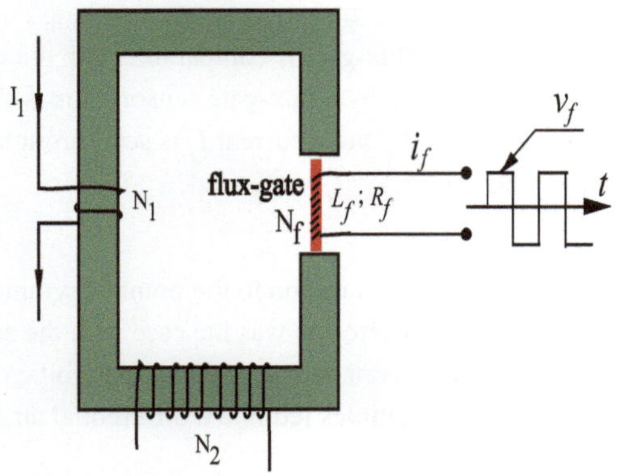

Fig. 18-6: Magnetic circuit current sensor

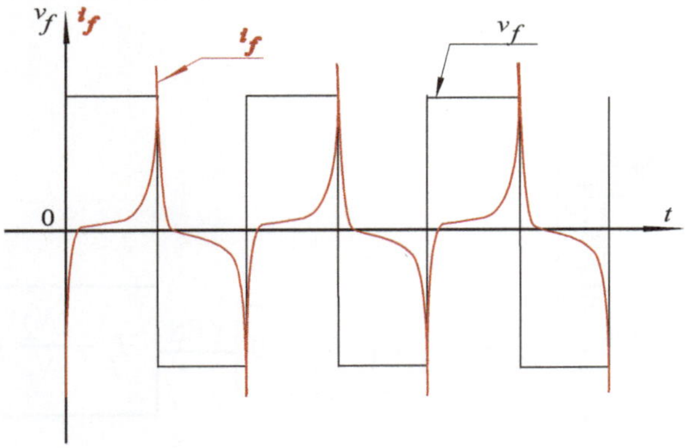

Fig 18-7: Current and voltage in flux-gate coil when $I_1 = 0$

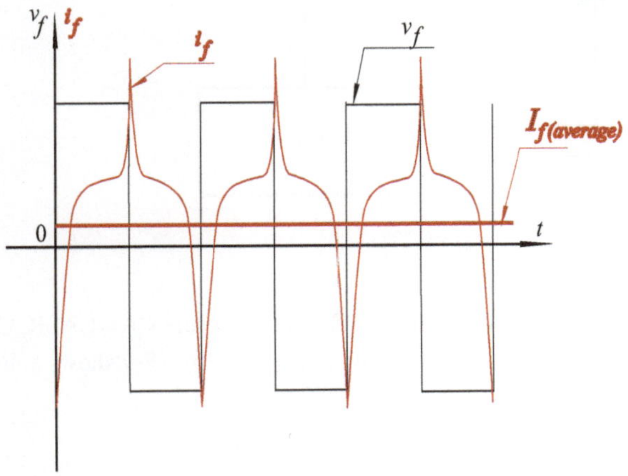

Fig. 18-8: Current and voltage in flux-gate coil with current I_1

1.6.2 Closed loop sensor

In fact the construction of a closed loop sensor with flux-gate is comparable with the magnetic circuit in fig. 18-4 in which a Hall sensor is now replaced by a flux-gate sensor. This is shown in fig. 18-9. The current to be measured I_1 produces a flux Φ_1 and a current I_2 is sent through the secondary coil (flux Φ_2) in such a way that $\Phi_2 = -\Phi_1$. Here $N_1 \cdot I_1 = N_2 \cdot I_2$.

It follows that: $I_1 = \frac{N_2}{N_1} \cdot I_2$.

We are continually trying to bring the flux-gate out of saturation to the point of symmetry of the hysteresis loop so that the resulting flux ($\Phi_1 + \Phi_2$) is zero. As was the case with the zero flux transducer (fig.18-4), the voltage drop of I_2 over a resistor produces the output voltage of the current sensor. Since this is a floating output it is sometimes fed into a differential amplifier with opamps.

The signal generator which supplies the winding N_f is comprised of a comparator circuit with hysteresis (Schmitt-trigger). The current change in I_f produces noise in the primary of the current sensor as a result of the transformer operation. A filter is required to remove the noise.

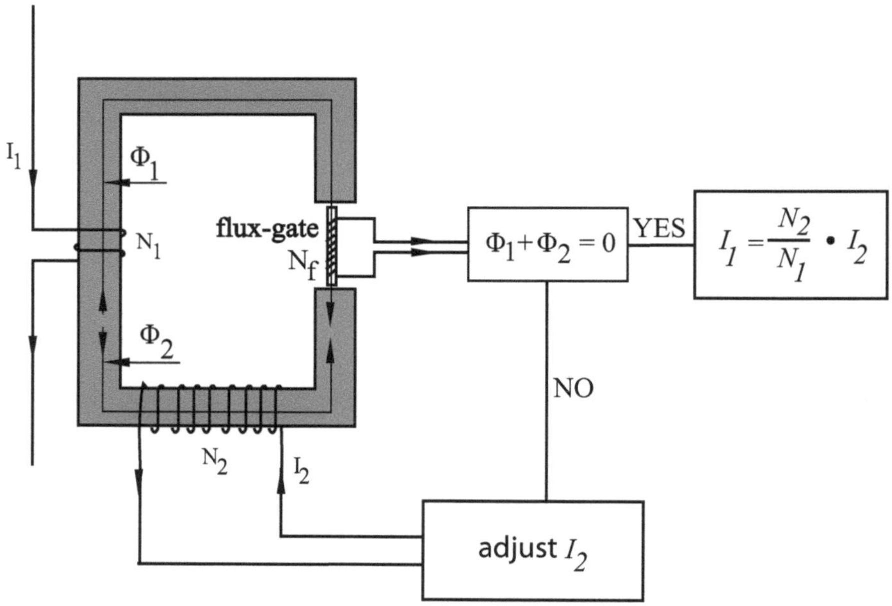

Fig. 18-9: Adjusting the closed loop

The firm LEM supplies the market with the following flux-gate sensors CAS/CASR, CFSR and CTSR (see photo's on p. 20.13 and p.20.82).The photo at the top of p. 18-9 shows a didactic model of a flux-gate current sensor.

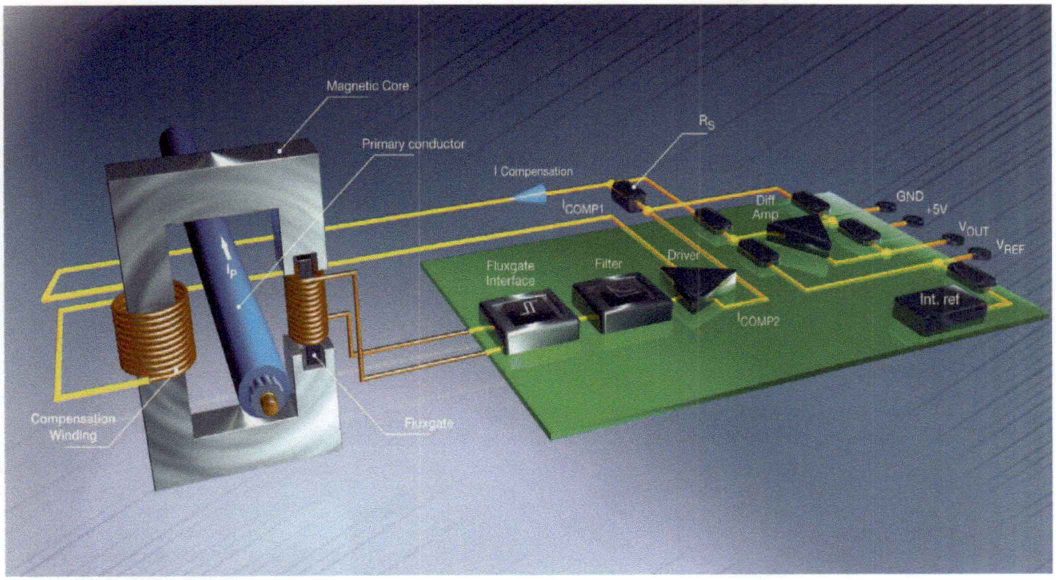

Photo LEM: Didactic example of a flux-gate current sensor

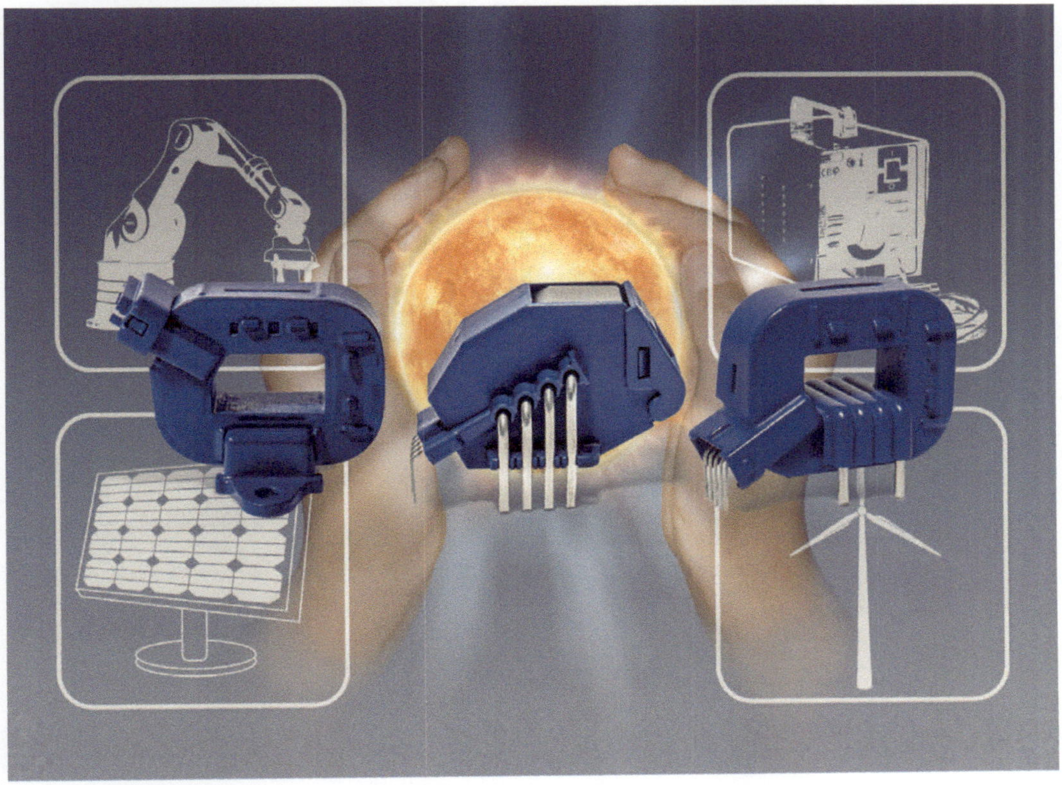

Photo LEM: HO series: High performance Open Loop ASIC based current transducers up to 250A

LEM

Current Transducer LTS 25-NP

I_{PN} = 25 At

For the electronic measurement of currents: DC, AC, pulsed, mixed with galvanic isolation between the primary circuit (high power) and the secondary circuit (electronic circuit).

5 Year Warranty · CE · RoHS COMPLIANT 2002/95/EC 16054 · c UL us

Electrical data

I_{PN}	Primary nominal current rms		25	At
I_{PM}	Primary current, measuring range		0 .. ± 80	At
V_{OUT}	Output voltage (Analog) @ I_P		$2.5 \pm (0.625 \cdot I_P/I_{PN})$	V
	$I_P = 0$		2.5 [1]	V
G	Sensitivity		25	mV/A
N_S	Number of secondary turns (± 0.1 %)		2000	
R_L	Load resistance		≥ 2	kΩ
R_{IM}	Internal measuring resistance (± 0.5 %)		50	Ω
TCR_{IM}	Temperature coefficient of R_{IM}		< 50	ppm/K
V_C	Supply voltage (± 5 %)		5	V
I_C	Current consumption @ V_C = 5 V	Typ	$28 + I_S$ [2] $+ (V_{OUT}/R_L)$	mA

Accuracy - Dynamic performance data

X	Accuracy @ I_{PN}, T_A = 25°C			± 0.2	%
	Accuracy with R_{IM} @ I_{PN}, T_A = 25°C			± 0.7	%
ε_L	Linearity error			< 0.1	%
			Typ	Max	
TCV_{OUT}	Temperature coefficient of V_{OUT} @ I_P = 0	- 10°C .. + 85°C	50	100	ppm/K
		- 40°C .. - 10°C		150	ppm/K
TCG	Temperature coefficient of G	- 40°C .. + 85°C		50 [3]	ppm/K
V_{OM}	Magnetic offset voltage @ I_P = 0, after an overload of 3 x I_{PN}			± 0.5	mV
	5 x I_{PN}			± 2.0	mV
	10 x I_{PN}			± 2.0	mV
t_{ra}	Reaction time @ 10 % of I_{PN}			< 100	ns
t_r	Response time to 90 % of I_{PN} step			< 400	ns
di/dt	di/dt accurately followed			> 60	A/µs
BW	Frequency bandwidth (0 .. - 0.5 dB)			DC .. 100	kHz
	(- 0.5.. 1 dB)			DC .. 200	kHz

General data

T_A	Ambient operating temperature	- 40 .. + 85	°C
T_S	Ambient storage temperature	- 40.. + 100	°C
m	Mass	10	g
	Standards	EN 50178: 1997 IEC 60950-1: 2001	

Notes: [1] Absolute value @ T_A = 25°C, $2.475 < V_{OUT} < 2.525$
[2] $I_S = I_P/N_S$
[3] Only due to TCR_{IM}.

Features

- Closed loop (compensated) multi-range current transducer using the Hall effect
- Unipolar voltage supply
- Isolated plastic case recognized according to UL 94-V0
- Compact design for PCB mounting
- Incorporated measuring resistance
- Extended measuring range.

Advantages

- Excellent accuracy
- Very good linearity
- Very low temperature drift
- Optimized response time
- Wide frequency bandwidth
- No insertion losses
- High immunity to external interference
- Current overload capability.

Applications

- AC variable speed drives and servo motor drives
- Static converters for DC motor drives
- Battery supplied applications
- Uninterruptible Power Supplies (UPS)
- Switched Mode Power Supplies (SMPS)
- Power supplies for welding applications.

Application domain

- Industrial.

1.6.3 Current measurement in a photovoltaic installation

As discussed in chapters 5 and 15 we want a PV-installation to operate at the MPP (maximum power point). This means for every PV topology we need to measure the DC-output (current and voltage) of the solar panel. In addition a current measurement in the input of the control loop is necessary for protection against short circuit and over current. In installations without transformer the maximum DC-current that the PV installation can send into the power grid is a maximum of 10mA to 1A.

The value depends on the standard used in the different countries. (IEC 61727,...VDE 0126-1).
From all these requirements current sensors used need to have:
an accuracy better than 1%, low offset, low drift in amplification, operate well with DC and low frequencies.

A sensor with closed loop flux-gate technology meets the specified requirements. The firm LEM has developed their CTSR-series for this type of application.

In addition to the mentioned requirements (accuracy, low drift,...) the CTSR-series has a number of additional functions such as: reference pin (for self testing), demagnetizing function (also via the reference pin). The CTSR can be used for single and three-phase voltages. The CTSR also has a version with four individual primary conductors. Three conductors are used for three-phase systems and the fourth conductor is used for testing the operation or as neutral conductor of the three-phase net.

In addition the magnetic core of the CTSR has two magnetic screens to protect the flux-gate from external magnetic fields. The reference pin provides access to the 2.5 V reference voltage. This $V_{ref.}$ can be used as a reference for an AD-converter.

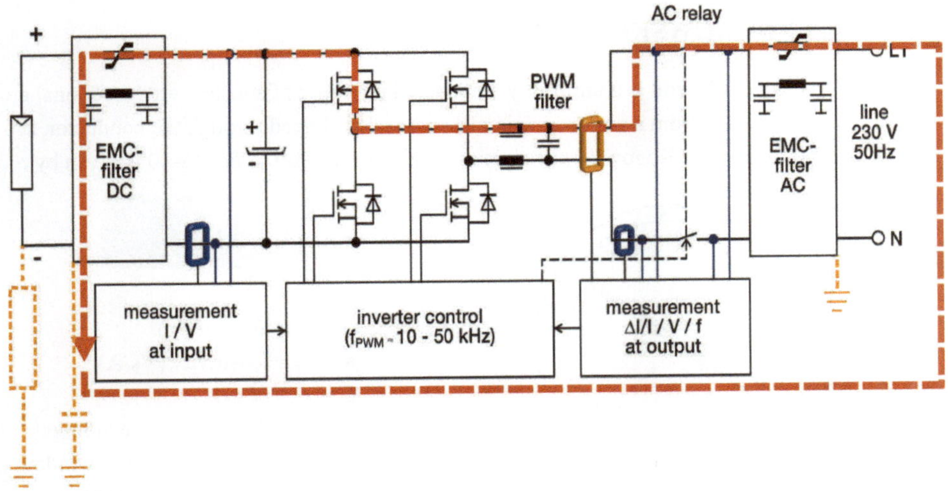

Fig. 18-10 (photo LEM): Example of a leakage current

Fig. 18-10 shows an example of how a leakage current can flow in a PV-installation without transformer. The capacitance between solar panel and the roof can be the cause of the leakage current which can result in the solar panel rising to the net potential. It is therefore necessary that any leakage currents are measured contact free and in a galvanically isolated manner.

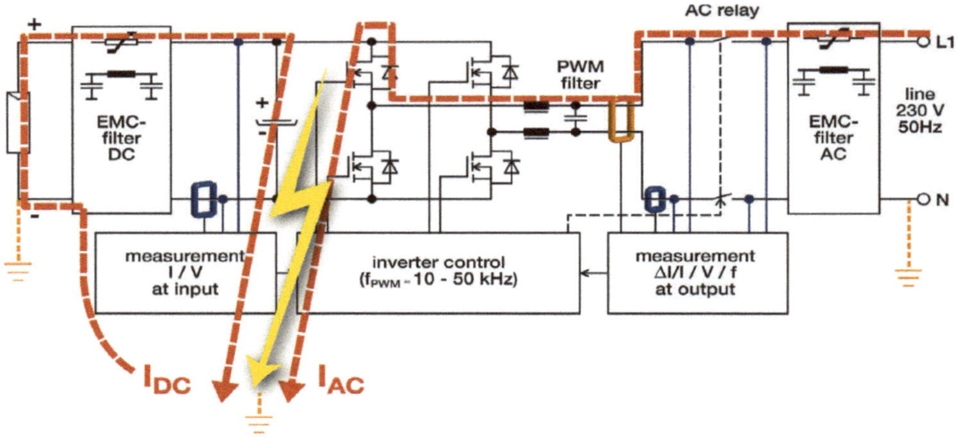

Fig. 18-11 (Photo LEM): Example of a fault current caused by an earth fault

In the case of an insulation fault an earth current can flow. It is also necessary to control (monitor) this. The sensors used for this should be able to measure AC and DC currents since the fault current can be either AC or DC, depending on the fault location. Fig. 18-11 shows an example of how to measure (Δi) in order to detect the fault current. It is clear that the current sensors required in a transformer free PV-installation do need to meet the specifications already mentioned (low offset, accuracy, DC-signals,…). The current measurement on the output measures the current difference as the result of a leakage current.

1.7 ROGOWSKI-measurement converter

A Rogowski-coil is a uniformly wound coil on a ring of constant cross sectional area. This ring is constructed from a non magnetic material and is slipped through the conductor, see fig. 18-12. According to the Rogowski principle the output of the coil in fig. 18-10 is given by :

$$e = K \cdot \frac{dI}{dt} \qquad (18\text{-}1)$$

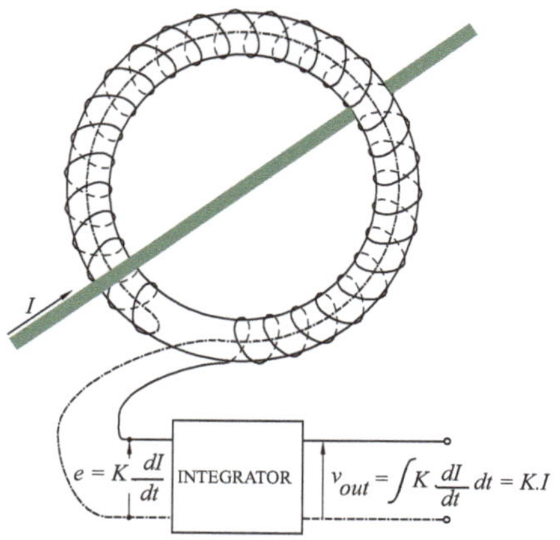

e = induced emf (V)
K = coil sensitivity (Vs/A)
$K = \mu_0 \cdot N \cdot A$
N = winding density = windings/meter
A = cross sectional area of winding (m²).
Integrating the voltage e produces a voltage v_{out} which is a measure of the current strength I.

Fig. 18-12: Measurement configuration using a Rogowski-coil

Remarks

1. The Rogowski-transducer is (typically) suited to high current levels (hundreds to thousands of amps) but is also used to measure tens of amps
2. The Rogowski-transducer has no DC component in its output and from this viewpoint is comparable to an AC coupled meter
3. This measurement converter has galvanic separation between the power circuit (I) and the output measurement signal (v_{out})
4. There is no saturation with large currents
5. The transducer does not load the power circuit
6. A bandwidth of 50Hz to 1MHz is possible with a Rogowski coil
7. Typical Rogowski coils have between 20 and 50 windings per cm. The ring has for example a cross sectional area of 10 to 15 mm^2 with a "coil length" of 20 to 30cm

1.8 Comparative table

The following table provides an overview of standard current sensors.
CL = closed loop.

Property	Current transformer	Shunt	CL with Hall sensor	CL with flux gate sensor	Rogowski sensor
AC? DC?	AC	AC/DC	AC/DC	AC/DC	AC
large current	average	poor	average	extremely good	good
temperature dependent	low	average	large	low	extremely low
linearity	large	large	average	extremely large	large
bandwidth	low	low	average	large	average
galvanic separation	yes	none	yes	yes	yes
power consumption	low	high	low	low	extremely low
dimensions	small	extremely small	small	average	average

2. ANGULAR POSITION SENSORS (SHAFT ANGLE TRANSDUCER) - (2.7) SPEED SENSORS

Almost every machine or industrial process contains one or more rotating shafts. It is therefore important in most instrumentation systems or control systems to be able to measure the exact shaft angle of a mechanical shaft. This angular data can be used to control position, speed or acceleration of a mechanism.

Standard shaft angle transducers include:

- potmeters
- encoders (optical or magnetic)
- resolver

Since control systems operate digitally it is necessary to have the data in digital form.
Potentiometers need to be followed by an A/D converter while resolvers require an RDC (resolver to digital converter).
The sensors discussed here may be implemented with a linear or angular displacement.
Linear displacements are usually obtained via a motor and worm wheel (fig. 18-13) so that a shaft angle transducer which is coupled to the motor can give an indication of linear displacement.

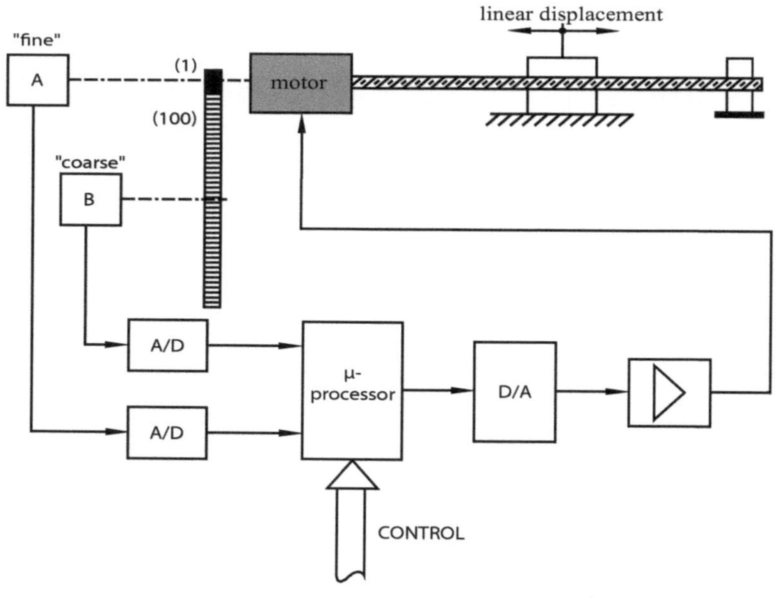

Fig. 18-13: Use of shaft angle transducers to control linear movement

Per revolution of B, sensor A produces 100 revolutions. Sensor B is the coarse and sensor A the fine shaft angle transducer.
Sensor A could be a resolver and sensor B a low resolution sensor such as an absolute encoder or potentiometer. In the case of a revolver the associated A/D converter is an RDC.
In certain applications in order to accurately measure a linear displacement it may be safer to monitor the linear displacement rather than the angular position of a motor. We return to this point under nr. 2.5.

2.1 Potentiometer

The wiper of the potmeter is connected with the mechanical shaft of which we want to know the position. A stabilised voltage is connected across the potmeter. The voltage between wiper and one end of the potmeter is an electrical indication of the angular position. To be useful in a digital control system the output voltage has to be digitised. Due to the presence of the connection terminals the potmeter has only a useful rotation angle of between 300 and 340°. Typical resistor values lie between 5 and 50 kΩ. By turning the wiper the output impedance is changed (of our signal source) between zero and maximum. This can affect the accuracy of the conversion.
The most accurate potmeters are wire wound and the maximum resolution is normally 12 bit. By moving the wiper over the wire wound potmeter body noise is produced which cannot be completely filtered out. Wear and tear reduces the total number of movements of the wiper.
The most important advantages of potmeter sensors are:

- suitable for use with high temperatures (up-to several hundred degrees Celsius)
- can handle vibrations and shocks

2.2 Optical encoders

We distinguish between:

- incremental optical encoder
- absolute optical encoder

The incremental encoder gives a specific number of impulses per revolution (e.g. 500 impulses) while the absolute encoder only gives a unique code which corresponds to a specific angular position. Fig. 18-14 shows the principle configuration of an optical encoder.

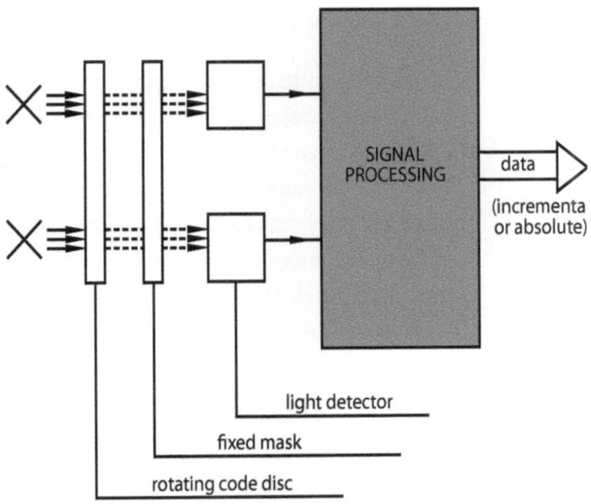

Fig. 18-14: Principal configuration of an optical encoder

The encoder is comprised of:

- a light source (e.g. LED)
- code disc, with fields that are darkened and allow light through.
- light detector, mostly comprised of photo-transistors.

2.2.1 Incremental encoder

Incremental encoders have three output signals as standard:

- a signal A consisting of n pulses per revolution (this signal can be a block or sine-wave)
- a signal B, identical to A but 90° displaced
- a signal Z (= zero marker output)

From the combination A,B and Z we can:

1) determine the shaft position. For this the Z pulse, the initialisation pulse, is used and there after the pulses A or B are counted.
2) determine the direction of rotation (comparing A with B!)

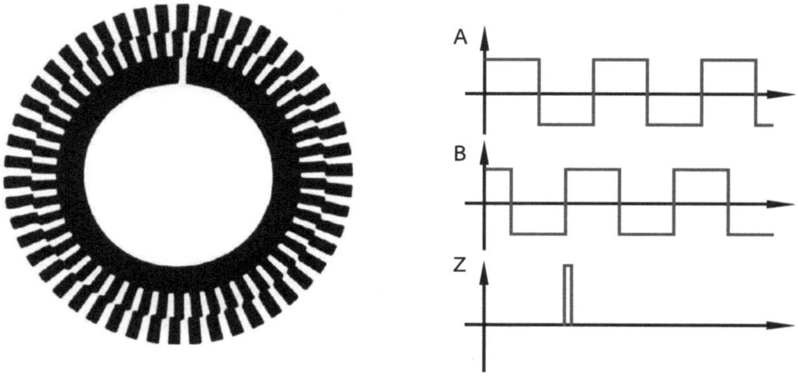

Fig. 18-15: Code disc of incremental encoder together with the resulting waveforms

The rotating code disc of an incremental encoder consists of a disc that allows light through and which has a number of darkened strips to prevent light shining through. Fig. 18-15 shows what this involves. Notice also how A and B channels and the Z pulse are formed with the transparent parts. The code disc can be made of glass, metal or plastic.

If the resolution has to be increased (more pulses per revolution) then the diameter of the disc (and thus the encoder) must be larger by necessity.

A standard "trick" to increase the number of pulses per revolution involves differentiating the block wave. Take an encoder with 500 pulses per rev. Differentiation of the positive flank of the block wave in fig. 18-15 results in 1000 pulses per rev, and differentiation of the positive and negative flanks results in 2000 pulses per rev.

The encoder is usually specified as : n pulses per rev with the options x2 and x4.

The photo below from the firm Heidenhain shows once again the transparent principle of an optical encoder.

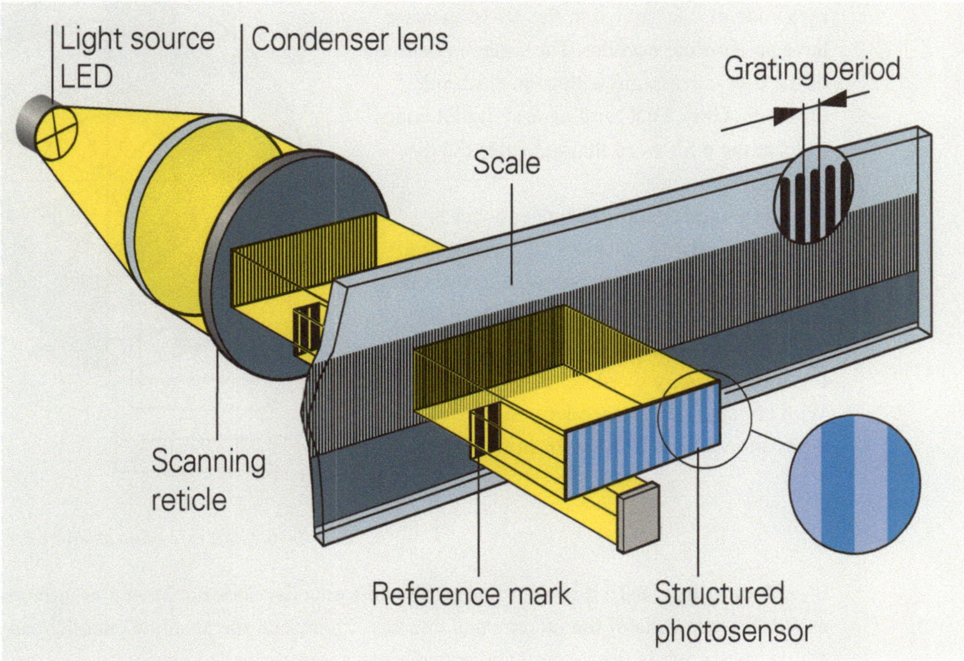

Photo HEIDENHAIN: Principle configuration optical encoder

The light beam emitted by the LED is concentrated by the condenser lens and thereafter one large area is taken out by the detection plate.This detection plate has a grating of its own which differs slightly from the grating used on the glass ruler. The glass scale moving with respect to the detection plate causes a type of filtering that creates homogenous signals of a shape very close to a sine wave. A special structured photo sensor is used to generate four 90° electrically phase-shifted scanning signals. Since the four sinusoids are not symmetrical with respect to the zero line two resulting symmetrical sinusoids S1 and S2 are produced via a push pull circuit which are 90° out of phase with each other. This is the SIN/COS output. A typical amplitude is 1 V peak to peak. The sinusoids can also be converted internally to block waves (as in fig. 18-15) which then appear on the output at TTL or HTL level. The second field serves as a reference R (one per revolution of the shaft).

Encoders may be single turn and multi turn types. Single turn systems provide the actual position within one revolution. Multi-turn pulse sensors operate with multiple revolutions.

2.2.2 Absolute encoder

If the stripe code disc of fig. 18-15 is replaced by a code disc as shown in fig. 18-16 then we have an absolute encoder. The sixteen positions of the disc correspond with a specific code. Usual are Gray, BDC and natural digital codes to form the 4 bit word that indicates the disc position.

The advantage of the absolute encoder is amongst others its "memory". In this way a short or long duration power outage will not interfere with the output signal. Indeed for every position of the encoder disc there is only one unique binary word.

With an incremental encoder you have to wait every time for the Z pulse for initialisation.

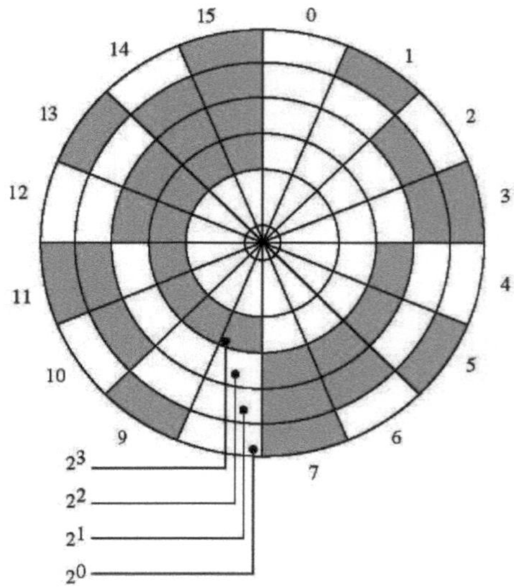

Fig 18-16: Code disc of an absolute encoder

If we look at fig. 18-16 it is clear that an absolute encoder does not have 500 different "positions" as did the strip code of the incremental encoder. Therefore, the absolute encoder can serve as a course sensor while the incremental encoder can serve as the fine sensor.

The picture on p.18.20 shows code discs for absolute encoders.

2.3 Magnetic encoder

We discuss the case of an incremental magnetic ERM encoder from HEIDENHAIN. It consists of a scanning disk with alternating positive and negative local magnetic fields along the circumference of the disk and a scanning reticle with magnetoresistive sensors.

2.3.1 Measuring principle

Magnetic encoders use a graduation carrier of magnetizable steel alloy. A write head applies strong local magnetic fields in different directions to form a graduation consisting of north poles and south poles (MAGNODUR process). The following grating periods are possible on the circumference:

- approx. 200 μm for ERM 2200
- approx. 400 μm for ERM 200, ERM 2400, ERM 2410
- approx. 1000 μm for ERM 2900

Due to the short distance of effect of electromagnetic interaction and the very narrow scanning gaps required, finer magnetic graduations have significantly tighter mounting tolerances.

The permanently magnetic MAGNODUR graduation is scanned by magnetoresistive sensors. They consist of resistive tracks whose resistance changes in response to a magnetic field. When a voltage is applied to the sensor and the scale drum moves relative to the scanning head, the flowing current is modulated according to the magnetic field.

The special geometric arrangement of the resistive sensors and the manufacture of the sensors on glass substrates ensure a high signal quality. In addition, the large scanning surface allows the signals to be filtered for harmonic waves. These are prerequisites for minimizing position errors within one signal period.
A magnetic structure on a separate track produces a reference mark signal. This makes it possible to assign this absolute position value to exactly one measuring step.

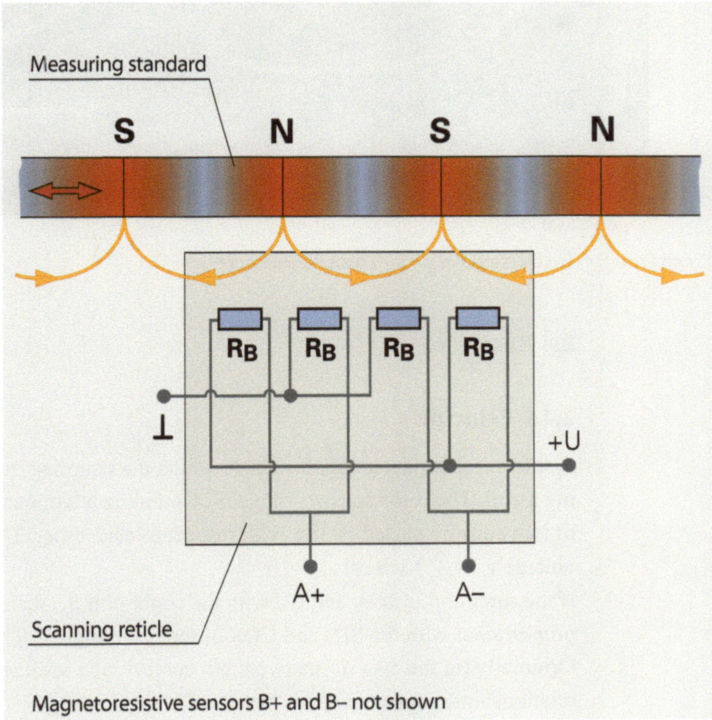

Fig. 18-17: Magnetoresistive principle

2.3.2 Remarks

Magnetoresistive scanning is typically used for medium-accuracy applications, or for where the diameter of the machined part is relatively small compared to the scale drum. It can be exposed to heavy loads of cooling lubricants and operate under high humidity, heavy dust loads and in oily atmospheres.

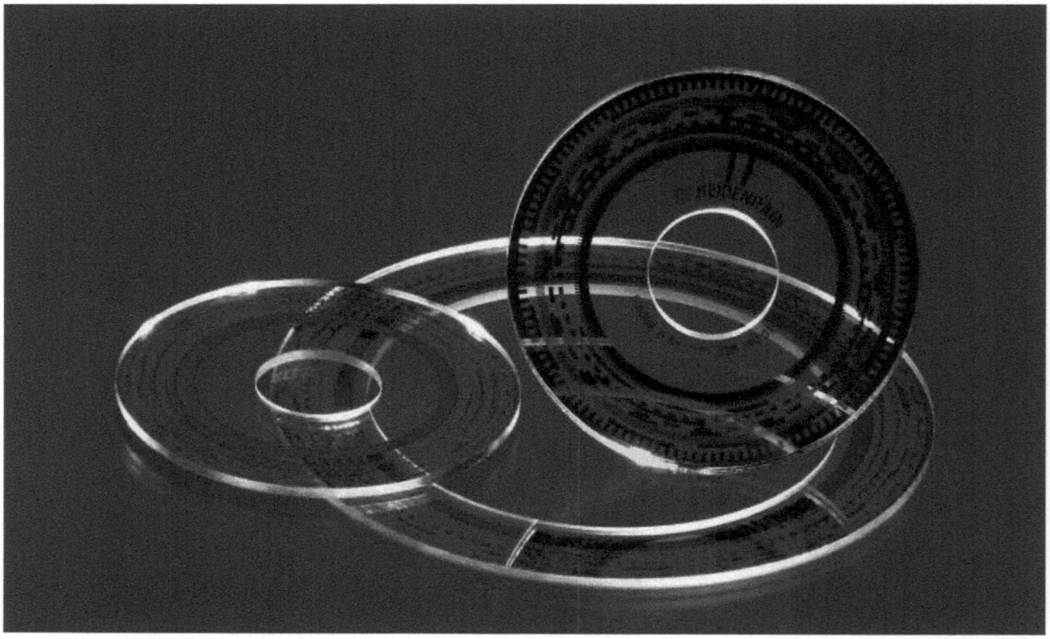

Photo HEIDENHAIN: Code discs for absolute encoders

2.4 Resolver

2.4.1 Principle

A resolver is in fact a transformer in which the coupling between the coils can be changed by turning a coil. The rotor consists of a coil, wound on a laminated iron core, The stator winding consists of two coils displaced by 90° with respect to each other. The rotor is supplied with a sinusoidal voltage $v_i = \hat{v}_i \cdot \sin \omega t$.
If the rotor coil is at an angle θ with the stator coil S_1, then the stator voltages are respectively proportional with the SIN and COS of the rotor angle θ. This is indicated in fig. 18-18.
Originally (in the 50's of the twentieth century) the resolver was used to solve trigonometric relationships, hence the name "resolver".

Applications of resolvers

A modern application is for example a robot for which the angular positions (of the various degrees of freedom) are determined with resolvers. Other applications: the angular position of a radar antenna or remotely controlling the course of a ship (gyro-compass), CNC machines, etc.

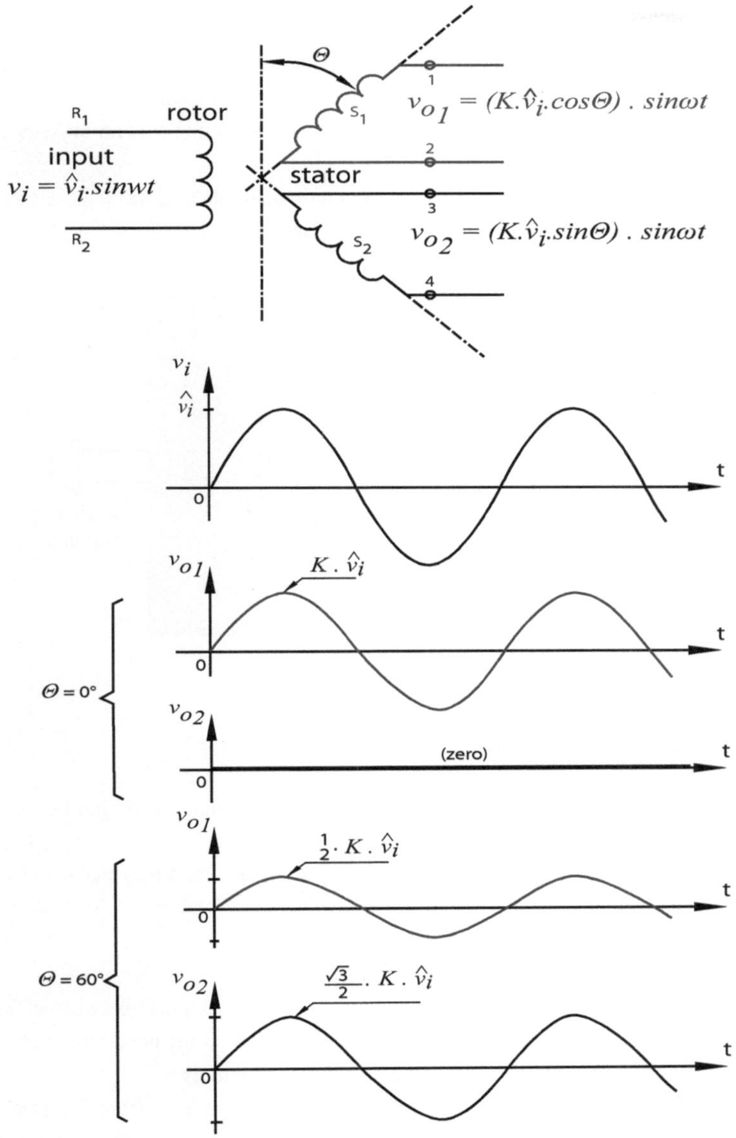

Fig. 18-18: Resolver

2.4.2 Synchro converters

Synchros are small machines which were developed during the second world war. The resolver is a derivative type which is practically as old as the synchro. These devices were used in analogue servo systems. Due to their excellent properties resolvers continue to be widely used.
Control systems have become completely digital so that a converter is required between resolver and micro-controller. These converters are available hybrid micro-electronic modules. It is referred to as an RDC (resolver-to-digital converter).

2.4.3 Brushless resolvers

In most cases "brushless" resolvers are used. At one end of the rotor the secondary of a transformer is wound which delivers the reference voltage for the rotor.
The primary of this supply transformer is wound on the stator. In this way no brushes are required to supply the rotor.
The output coils of the resolver are on the stator and with an angle α of the rotor we obtain the voltages $v_i .sin\alpha$ and $v_i .cos\alpha$ across the stator coils S_1 and S_2 .

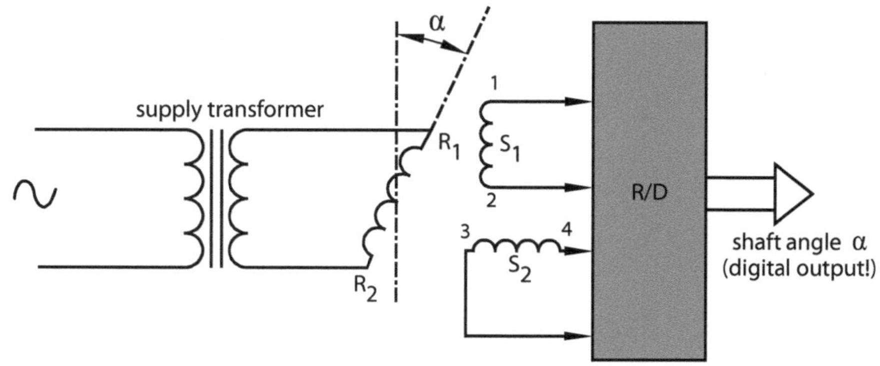

Fig. 18-19: Resolver configuration followed by a converter

Observing the ratio $\frac{v_i \,.\, sin\ \alpha}{v_i \,.\, cos\ \alpha} = tan\ \alpha$, then we notice supply voltage variation has no influence on the measured angular position α.
The distance between resolver and converter (RDC- or RD-converter) may be tens of meters.

2.5 Linear measuring systems

In addition to rotating encoders linear encoders also exist. We distinguish between closed and open linear measuring systems. The closed systems are resistant to the penetration of dust, moisture and shavings and are predominately used for machine tools.
There are also closed linear encoders which can measure linear distance from 5 cm to for example 30m. Accuracy classes of for example ± 3μm/m to ± 5μm/m are possible. This means that within 1 m of the ruler the error is maximum 3 to 5μm. Typical measurement increments are 0.1μm / 0.5μm / 1μm / 10μm.
Both optical and magnetic measurement systems are in operation. Also here we see the distinction between absolute and incremental encoders.
The top photo on p. 18-23 shows an optical linear encoder from the firm Heidenhain. The ruler has a Diadur-scale (see p. 18.23). Measurement spans may be from 140 to 4240 mm. Accuracy ± 3μm to ± 5μm. Sinusoidal signal 1V p-p. Frequency limit 150 kHz.
On p 18.24 a magnetic measurement system from MTS is shown.

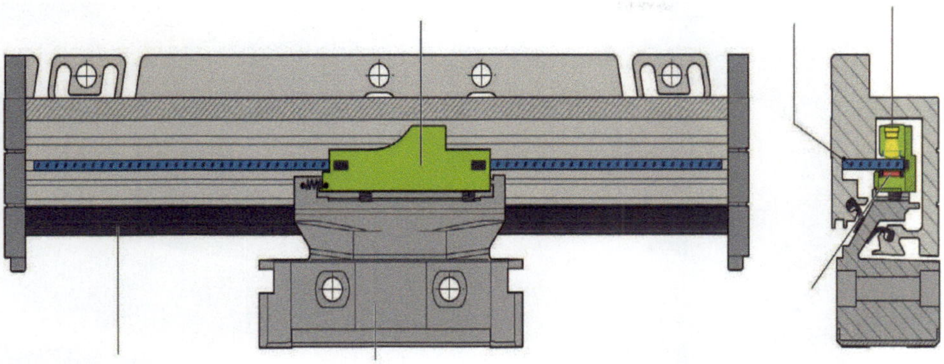

Photo HEIDENHAIN: Linear encoder LC115

The most important parts of the measuring system from HEIDENHAIN are the scale carriers, mostly in the form of a graduated scale. HEIDENHAIN manufactures the precision graduations in specially developed, photolithographic processes:

- AURODUR: Highly reflective gold lines and matte etched gaps. AURODUR graduations are usually on steel carriers and have a typical grating period of 40µm
- METALLUR: With its special optical composition of reflective gold layers, METALLUR graduations show a virtually planar structure. They are therefore particularly tolerant to con tamination and have a typical graduation period of 40µm.
- DIADUR: Precision graduation composed of an extremely thin layer of chromium on a sub strate usually of glass or glass ceramic. The accuracy of the graduation structure lies within the micron and submicron range, the graduation period is typically 20µm.
- SUPRADUR: The graduations act optically like three-dimensional phase gratings, but they have a planar structure and are therefore particularly insensitive to contamination. The graduation period is 8µm or finer.
- OPTODUR: an optically three dimensional, planar structure with particularly high reflectance and a typical graduation period of 2 µm and finer.
- MAGNODUR: Thin magnetically active layers in the micron range are structured for very fine, magnetized graduations.

Along with these very fine grating periods, these processes permit a high definition and homogeneity of the line edges. Together with the photoelectric scanning method, this high edge definition is a precondition for the high quality of the output signals. The master graduations are manufactured by HEIDENHAIN on custom-built high-precision dividing engines.
The photo on p.18.24 shows carriers of both linear and rotating encoders.

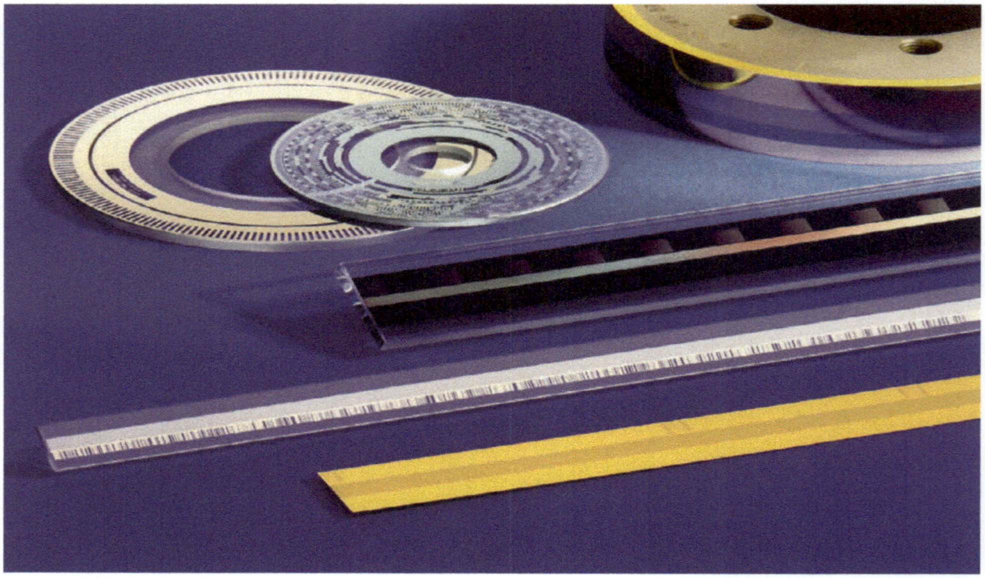

Photo HEIDENHAIN: DIADUR and AURODUR scales on different carriers

An example of a magnetic system from MTS is shown in the photo below.

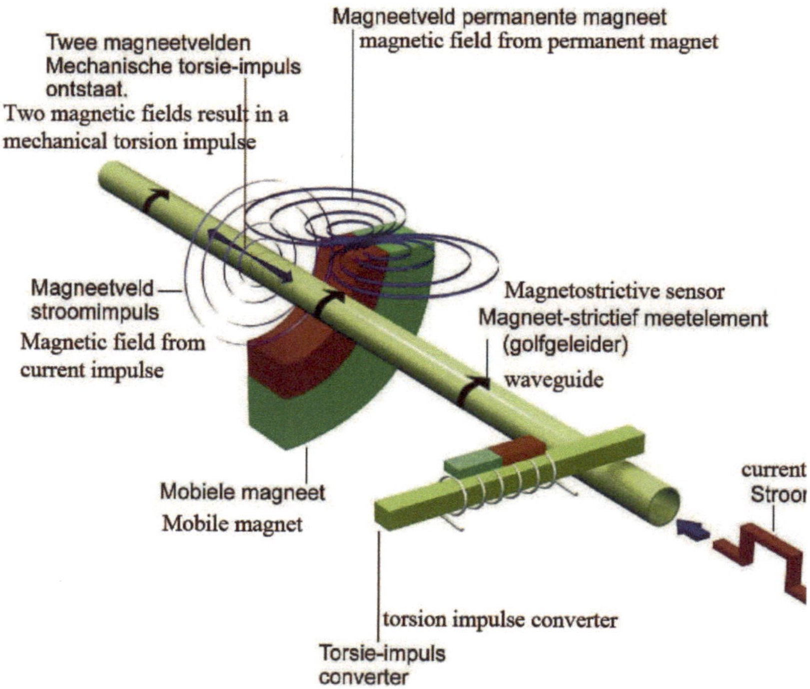

Photo MTS (Multiprox): MTS-sensor for linear positioning

Some ferromagnetic materials change their length under the influence of a magnetic field and conversely there magnetic state can change under the influence of mechanical force.
This is called the magnetostrictive property of the material. Central in the MTS-sensor is such a magnetostrictive waveguide.
A movable position magnet which is connected to the position to be determined produces a magnetic field in the longitudinal direction of the waveguide. A current impulse through the waveguide creates a radial magnetic field around the waveguide. At the instant that both magnetic fields coincide, a torsion impulse is produced in the waveguide. This impulse flows as a sound wave with the constant speed of the sound from the measurement point to the end of the waveguide and is converted in the sensor head to a distance proportional output signal.

2.6 Comparison of the various angular position sensors

2.6.1 Operating environment

Dirt, oil, salt atmosphere, high temperature, shocks and vibrations have only a small effect on resolvers and magnetic encoders. Optical encoders are sensitive to vibration and shocks. Potmeters can display “bounce “ as the wiper vibrates as a result of shocks or movement. Resolvers can operate in ambient temperatures up to 200°C and higher. The weak point of an optical encoder can be that a part of the accompanying electronics is placed close to the sensor so that temperature influence can be important. By contrast a resolver can have its electronics tens of meters away. High temperatures will also result in oxidation and wear of moving parts of a potmeter.

2.6.2 Accuracy and resolution

The **resolution** of an angular position sensor is the weight of the smallest movement in the digital scale or expressed another way, the least significant bit (LSB): $LSB = \frac{\text{full angular range}}{2^N}$.
Resolvers have an infinite resolution and it is mainly the converter (RDC) that determines the resolution of the measurement. An RDC typically delivers 10, 12, 14 or 16 bit output depending on the type of converter. A 14-bit RDC has a positional resolution of $\frac{360°}{2^{14}} = 1.32$ arc minutes.
The resolution of an optical encoder is determined by the amount of transparent windows.
The measurements of optical encoders are smaller when a laser light source is used.

Accuracy is the maximum error of the digital code with respect to the original value.
Optical encoders have the best accuracy. Resolvers have an accuracy of about 10 arc minutes.
Here of course we need to include the accuracy of the RDC in the calculation with its accuracy of a few arc minutes.

2.6.3 Maximum speed

In the case of a resolver we usually obtain one output period per revolution resulting in an absolute position measurement. With an encoder, the maximum speed is mainly limited by the frequency response of the encoder input of the positioning system.

Example:

An encoder has 2000 pulses/rev and the maximum encoder input of the drive train is 150 kHz. The motor can then have a maximum speed of 4500 rev/min in order not to miss a pulse.

Indeed: $\frac{2000 \times 4500}{60} = 150$ kHz.

2.7 Speed sensors

2.7.1 DC tachogenerator

A small DC generator with permanent magnets has an emf: $E = k.n.\Phi = k_1 \, . \, n$.

The polarity of E is determined by the direction of rotation so that this type of tachogenerator is suitable for four quadrant drives.

The constant of such a tachogenerator is typically 60V/1000 rpm or 20V/1000 rpm.

2.7.2 Angular encoder

With the present technology tachogenerators have been totally replaced by angular encoders. Differentiation of the angular position Θ (of the motor shaft) gives us $d\Theta/dt = \omega$.

This provides the option to also measure speed with an encoder.

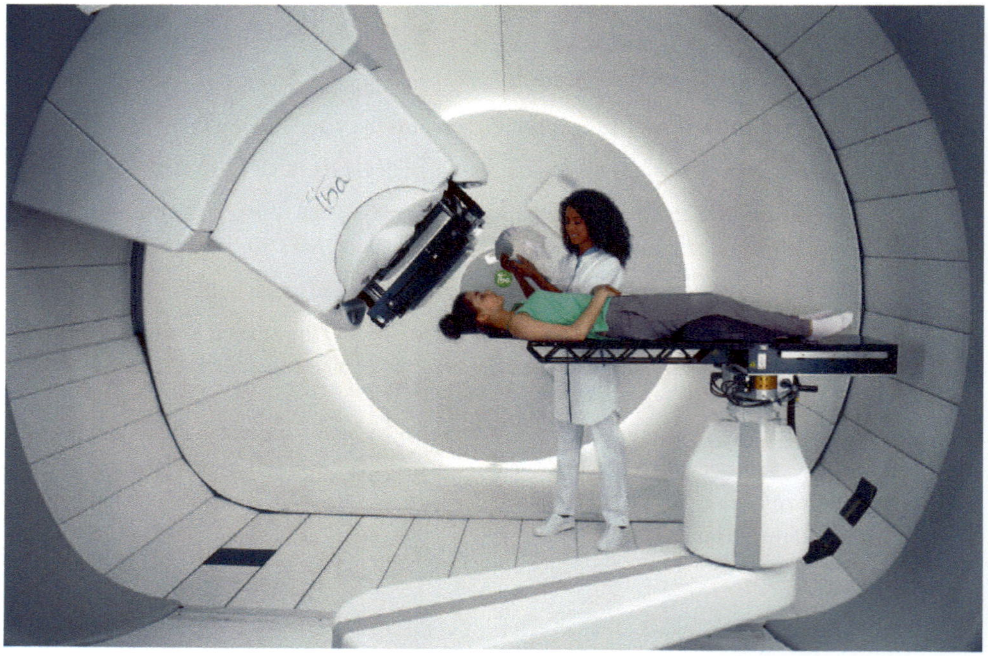

Courtesy of IBA: Picture of the Proteus ® PLUS. This scanner requires highly accurate measurement of angles which is achieved by angle encoders of HEIDENHAIN

2.8 Angular measurement systems

2.8.1 Measuring principles

The term angle encoder is typically used to describe encoders that have an accuracy of better than ± 5" and a line count above 10,000. These angle encoders are found in applications that require the highly accurate measurement of angles in the range of a few angular seconds, e.g. in rotary tables and swivel heads on machine tools, C-xes on lathes, but also in measuring equipment and telescopes. Other applications, such as scanners, positioning systems, printing units or beam deflection systems, require high repeatability and/or a high angular resolution. Encoders for such applications are likewise referred to as angle encoders. In contrast, rotary encoders are used in applications where accuracy requirements are less stringent, e.g. in automation, electrical drives, and many other applications. Angle encoders can have one of the following mechanical designs:

A. ***Angle encoders with integral bearing, hollow shaft and stator coupling***
Because of the design and mounting of the stator coupling, it must absorb only that torque caused by friction in the bearing during angular acceleration of the shaft. These angle encoders therefore provide excellent dynamic performance. With a stator coupling, the stated system accuracy also includes deviations from the shaft coupling. The RCN, RON and RPN angle encoders have an integrated stator coupling, whereas the ECN has a stator coupling mounted on the outside.
Other advantages:

- compact size for limited installation space
- hollow shaft diameters up to 100 mm to provide space for power lines, etc.
- simple installation

B. ***Angle encoders with integral bearing, for separate shaft coupling***
ROD angle encoders with solid shaft are particularly suited to applications where higher shaft speeds and/or larger mounting tolerances are required. The shaft couplings allow axial tolerances of ± 1 mm.

Photo HEIDENHAIN: ROD 880 incremental angle encoder with K16 flat coupling

Photo HEIDENHAIN: RCN8580 absolute angle encoder

C. *Angle encoders without integral bearing*

The ERP, ERO and ERA angle encoders without integral bearing (modular angle encoders) are intended for integration in machine elements or apparatuses. They are designed to meet the following requirements:

Large hollow shaft diameters:

- (up to 10 m with a scale tape)
- high shaft speeds up to 20 000 min–1
- no additional starting torque from shaft seals
- segment versions

Photo HEIDENHAIN: ERA4000 incremental angle encoder

D. *Modular magnetic encoders*

The robust ERM modular magnetic encoders are especially suited for use in production machines. The large inside diameters available, their small dimensions and the compact design of the scanning head predestine them for the C axis of lathes, simple rotary and tilting axes (e.g. for speed measurement on direct drives or for integration in gear stages), and spindle orientation on milling machines or auxiliary axes.

2.8.2 Accuracy

The accuracy of angular measurement is mainly determined by:

- the quality of the graduation,
- the stability of the graduation carrier,
- the quality of the scanning process,
- the quality of the signal processing electronics,
- the eccentricity of the graduation to the bearing,
- the error of the bearing,
- the coupling to the measured shaft.

These factors of influence are comprised of encoder-specific error and application dependent issues. All individual factors of influence must be considered in order to assess the attainable total accuracy.

A. *Encoder-specific error*

The encoder-specific error is given in the accuracy of graduation and the position error within one signal period.

1. Accuracy of graduation

The accuracy of the graduation ± a results from its quality, this includes:

- the homogeneity and period definition of the graduation,
- the alignment of the graduation on its carrier,
- for encoders with massive graduation carriers: the stability of the graduation carrier, in order to also ensure accuracy in the mounted condition,
- for encoders with steel scale tape: the error due to irregular scale-tape expansion during mounting, as well as the error at the scale-tape butt joints of full-circle applications.

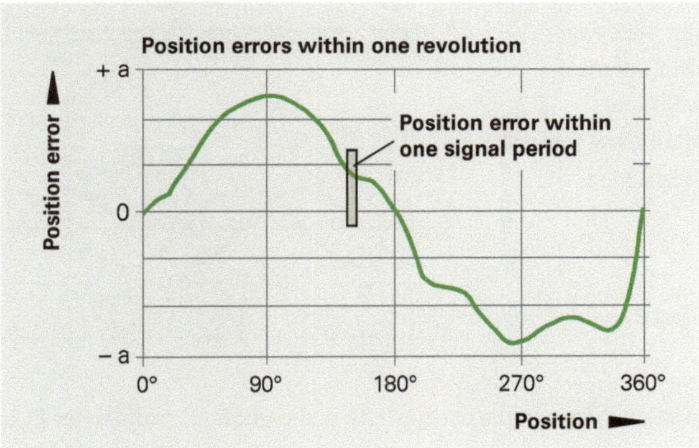

The accuracy of the graduation ± a is ascertained under ideal conditions by using a series-produced scanning head to measure position error at positions that are integral multiples of the signal period.

Fig. 18-20: Accuracy of graduation

2 Position error within one signal period

The position error within one signal period ± u results from the quality of the scanning and, for encoders with integrated pulse shaping or counter electronics, the quality of the signal-processing electronics. For encoders with sinusoidal output signals, however, the errors are determined by the signal processing electronics of the subsequent electronics.
The following individual factors influence the result:

- the length of the signal period
- the homogeneity and period definition of the graduation
- the quality of scanning filter structures
- the characteristics of the detectors
- the stability and dynamics of further processing of the analog signals

These factors of influence are to be considered when specifying position error within one signal period. Position error within one signal period ± u is specified in percent of the signal period. For modular angle encoders without integral bearing the value is typically better than ± 1 % of the signal period (ERP 880: ± 1.5 %). Position errors within one signal period already become apparent in very small angular motions and in repeated measurements. They especially lead to speed ripples in the speed control loop.

B. ***Application-dependent error***

The mounting and adjustment of the scanning head, in addition to the given encoder-specific error, normally have a significant effect on the accuracy that can be achieved by encoders without integral bearings. Of particular importance are the mounting eccentricity of the graduation and the radial runout of the measured shaft. The application-dependent error values must be measured and calculated individually in order to evaluate the total accuracy.

1. Errors due to eccentricity of the graduation to the bearing

Under normal circumstances, the graduation will have a certain eccentricity relative to the bearing once the disk/hub assembly, scale drum or steel scale tape is mounted. In addition, dimensional and form deviations of the customer's shaft can result in added eccentricity.

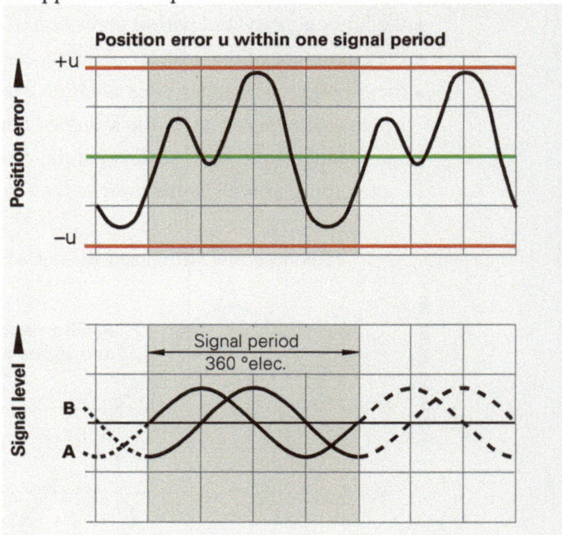

Fig. 18-21: Position error for encoder with sinusoidal output

The following relationship exists between the eccentricity e, the graduation diameter D and the measuring error Δφ (see fig. 18-22):

$\Delta\varphi = \pm 412 \cdot \frac{e}{D}$

- Δφ = measurement error in angular seconds (")
- e = eccentricity of the scale drum to the bearing in μm (1/2 the radial deviation)
- D = mean graduation diameter in mm

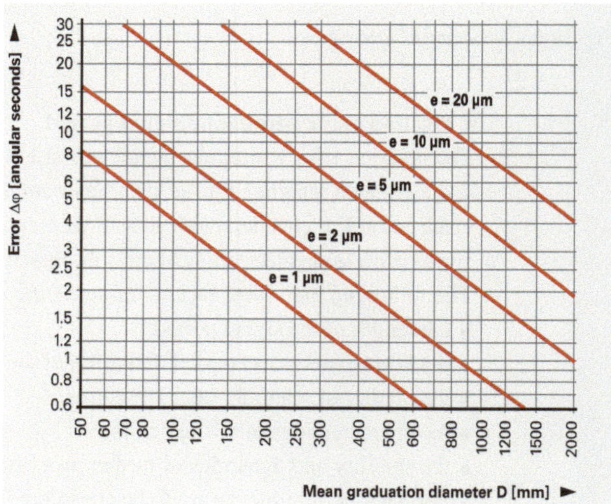

Fig. 18-22: Errors due to eccentricity

2. Error due to radial runout of the bearing

The equation for the measuring error Δφ is also valid for radial error of the bearing if the value e is replaced with the eccentricity value, i.e. half of the radial error (half of the displayed value). Bearing compliance to radial shaft loading causes similar errors.

3. Compensation possibilities

The mounting eccentricity of the graduation and the radial runout of the measured shaft cause a large share of the application dependent errors. A common and effective method of eliminating these errors is to mount two or even more scanning heads at equal distances around the graduation carrier. The subsequent electronics mathematically combine the individual position values. The EIB 1500 from HEIDENHAIN is an electronics unit suitable for mathematically combining the position values from two scanning heads in real time, without impairing the control loop.

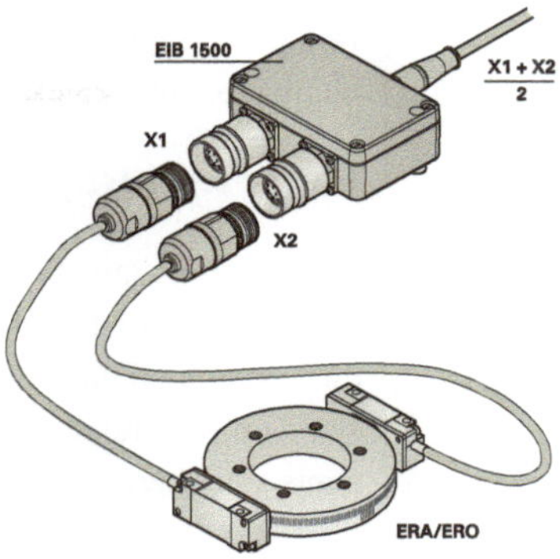

Fig. 18-23: Position calculation of two scanning heads

The accuracy improvement actually attained by this in practice strongly depends on the installation situation and the application. In principle, all eccentricity errors (reproducible errors due to mounting errors, non-reproducible errors due to radial eccentricity of the bearing) as well as all uneven harmonics of the graduation error are eliminated.

2.8.3 Calibration chart

The calibration chart documents the graduation accuracy including the graduation carrier. It is ascertained through a large number of measuring points during one revolution. All measured values lie within the graduation accuracy listed in the specifications. The deviations are ascertained at constant temperatures (22 °C) during the final inspection and are indicated on the calibration chart. The calibration standard indicated in the manufacturer's inspection certificate documents and guarantees traceability to recognized national and international standards. The accuracy data of the calibration chart do not include the position error within one signal period and any error resulting from mounting.

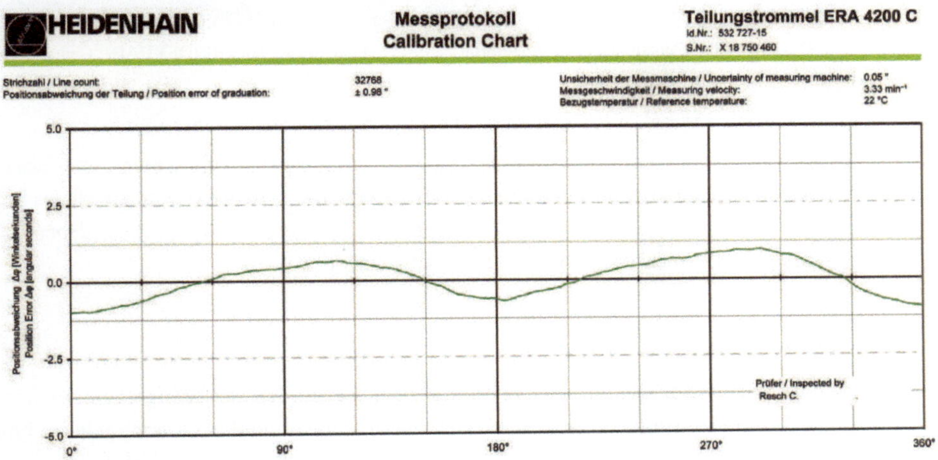

Fig. 18-24: Calibration chart example ERA 4200C scale drum

2.8.4 Reliability

Exposed angle encoders without integral bearing from HEIDENHAIN are optimized for use on fast, precise machines. In spite of the exposed mechanical design, they are highly tolerant to contamination, ensure high long-term stability, and are quickly and easily mounted.

Lower sensitivity to contamination

Both the high quality of the grating and the scanning method are responsible for the accuracy and reliability of the encoders. Encoders from HEIDENHAIN operate with single-field scanning. Only one scanning field is used to generate the scanning signals. Local contamination on the measuring standard (e.g. fingerprints or oil accumulation) influences the light intensity of the signal components, and therefore of the scanning signals, in equal measure. The output signals do change in their amplitude, but not in their offset and phase position. They remain highly interpolable, and the position error within one signal period remains small.
The large scanning field additionally reduces sensitivity to contamination. In many cases this can prevent encoder failure. Even if the contamination from printer's ink, PCB dust, water or oil is up to 3 mm in diameter, the encoders continue to provide high-quality signals. The position errors within one revolution remain far below the specified accuracy.
Despite significant contamination, the specified value of ± 1 % of maximum position errors within one signal period is exceeded only slightly.

Fig. 18-25: Photo HEIDENHAIN: Contamination by fingerprint

Fig. 18-26: Photo HEIDENHAIN: Contamination by toner dust

2.9 Interfaces

As well-defined transitions between encoders and subsequent electronics, interfaces ensure the reliable exchange of information. HEIDENHAIN offers encoders with interfaces for many common subsequent electronics. The interface possible in each respective case depends, among other things, on the measuring method used by the encoder.

2.9.1 (1Vpp)

Encoders with 1 Vpp interface provide voltage signals that can be highly interpolated. The sinusoidal incremental signals A and B are phase-shifted by 90° and have amplitudes of typically 1 VPP. The illustrated sequence of output signals, with B lagging A, applies for the direction of motion shown in the dimension drawing (fig. 18-27).

11μAmp is an older version of this 1Vpp interface, it uses similar signals with lower amplitudes. These are more susceptible to interference and less suitable for longer cable lengths (e.g. 150m for 1Vpp and 30m for 11μAmp)

2.9.2 TTL

HEIDENHAIN encoders with TTL interface incorporate electronics that digitize sinusoidal scanning signals with or without interpolation. The incremental signals are transmitted as the square-wave pulse trains Ua1 and Ua2, phase-shifted by 90° elec. (fig.18 -28).

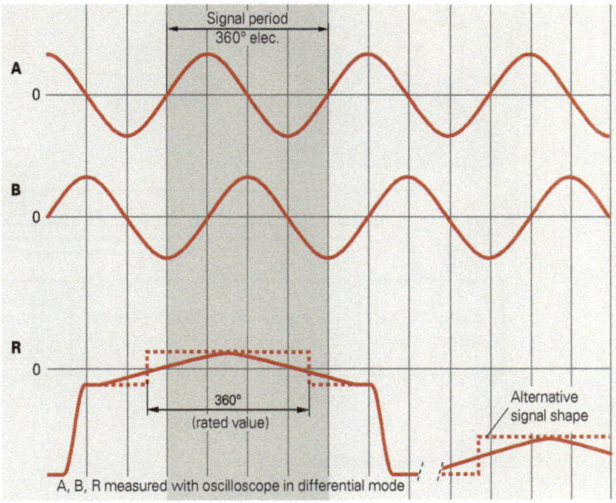

Fig. 18-27: Signals from 1Vpp interface

The reference mark signal consists of one or more reference pulses Ua0, which are gated with the incremental signals. In addition, the integrated electronics produce their inverted signals $\overline{U_{a1}}$, $\overline{U_{a2}}$ and $\overline{U_{a0}}$ for noise-proof transmission. The illustrated sequence of output signals, with Ua2 lagging Ua1, applies to the direction of motion shown in the dimension drawing. The fault-detection signal $\overline{U_{as}}$ indicates fault conditions such as breakage of the power line or failure of the light source. It can be used for such purposes as machine shut-off during automated production.

HTL is a variant of the TTL interface where the signal amplitude depends on the voltage supply. HTL is more suited for long cable lengths or high noise conditions.

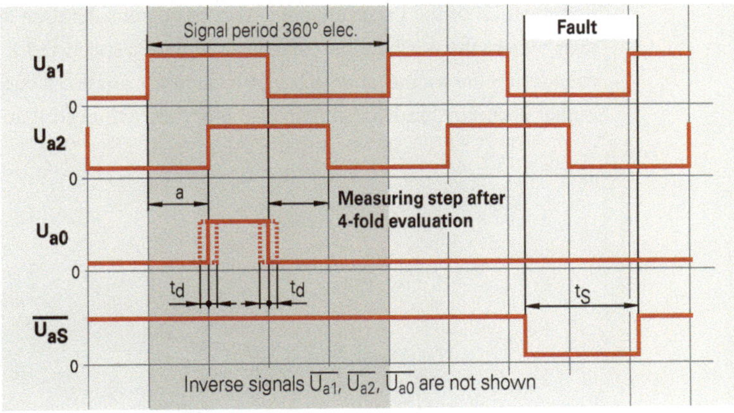

Fig. 18-28: TTL-interface signals

2.9.3 ENDAT

The EnDat interface is a digital, bidirectional interface for encoders. It is capable both of transmitting position values as well as transmitting or updating information stored in the encoder, or of saving new information. Thanks to the serial transmission method, only four signal lines are required. The data is transmitted in synchronism with the clock signal from the subsequent electronics. The type of transmission (position values, parameters, diagnostics, etc.) is selected through mode commands that the subsequent electronics send to the encoder. More information on this protocol can be found on www.endat.de.

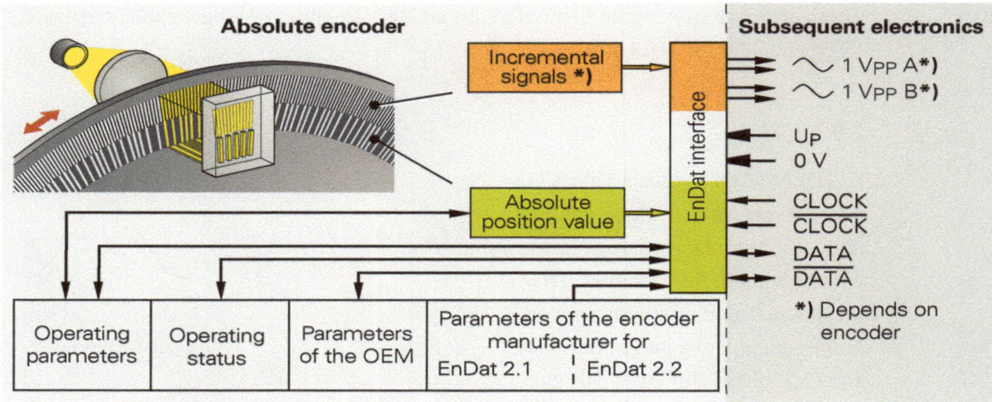

Fig. 18-29: ENDAT serial interface

The ENDAT serial interface (fig. 18-29) is developed by HEIDENHAIN. Several other companies use their own serial interface, for instance Drive-CliQ from Siemens, FSIα and FSIαi from Fanuc, Mitsu from Mitsubishi, YASK from Yaskawa and Profibus which is a noproprietary open fieldbus.

2.10 Closed loop position measurement on feed drives

If a linear encoder is used for measurement of the slide position, the position control loop includes the complete feed mechanics. This is therefore referred to as a closed-loop operation. Play and inaccuracies in the transfer elements of the machine have no influence on the accuracy of the position measurement. Measurement accuracy depends almost solely on the precision and installation location of the linear encoder. This basic consideration applies both for linear axes and rotary axes, where the position can be measured with a speed-reduction mechanism connected to a rotary encoder on the motor, or with a highly accurate angle encoder on the machine axis. Significantly higher accuracy grades and reproducibility are achieved if angle encoders are used.

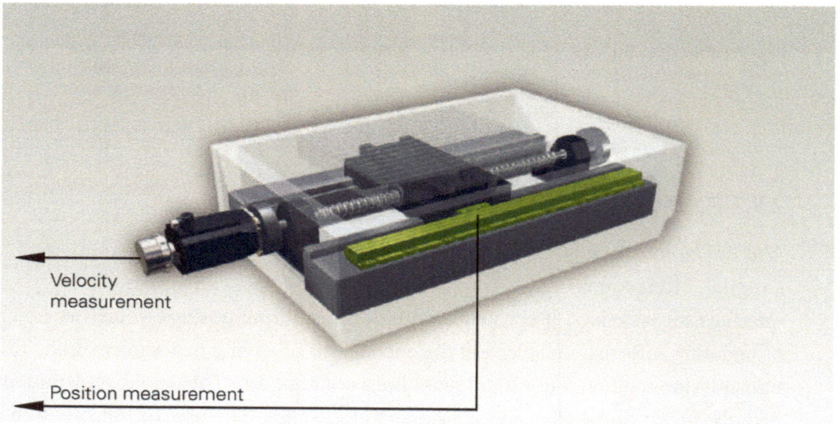

Photo HEIDENHAIN: Position and velocity measurement on feed drives

19 SPEED- and (or) TORQUE CONTROL of a DC-MOTOR

CONTENTS

From an industrial viewpoint AC power grids dominate so that regulation or control of a DC motor for industrial applications necessarily requires a thyristor bridge. In rolling mills, wrapping machines, lifting machines, etc...when a DC machine is used it is almost invariably an independently excited motor. It is the regulation of this motor, supplied from an AC power grid which is studied in part A of this chapter. Note that as far as the armature control is concerned, the theory discussed here is applicable to the permanent magnet motor.

DC drives are very dynamic.

A number of performance indicators of a DC drive are:

- speed accuracy better than 0.5%
- torque response about 10ms
- speed response : 28 to 40 ms.

If the supply is DC then we place a chopper between source and motor.

DC supplies appear as batteries or as overhead wire + rails in electric traction.

Batteries are important if portable apparatus are involved. In these applications mostly a PM-motor with chopper is used.

Power Electronics: DRIVE TECHNOLOGY and MOTION CONTROL
http://dx.doi.org/10.1016/B978-0-12-814641-5.50004-4,

A. DC-MOTOR supplied from an AC POWER GRID

1. CONTROL OF AN INDEPENDENTLY EXCITED MOTOR

1.1 Generalities

A classic DC motor is constructed from a stator with field winding F_1F_2 and a rotor with commutator, brushes and an armature winding A_1A_2: see fig. 19-1. The armature winding sits in the groves of the laminated rotor. The field winding F_1F_2 may also be replaced by permanent magnets. To counteract the armature action machines have auxiliary poles n_h-z_h (for power levels from 1kW) and a compensation winding C_1C_2 (above 100 kW). This compensation winding sits in the teeth of the main poles and is connected in series with the armature winding. The flux Φ in the machine is proportional with the excitation current I_m and depends upon the magnetic induction and the cross sectional area of the iron.

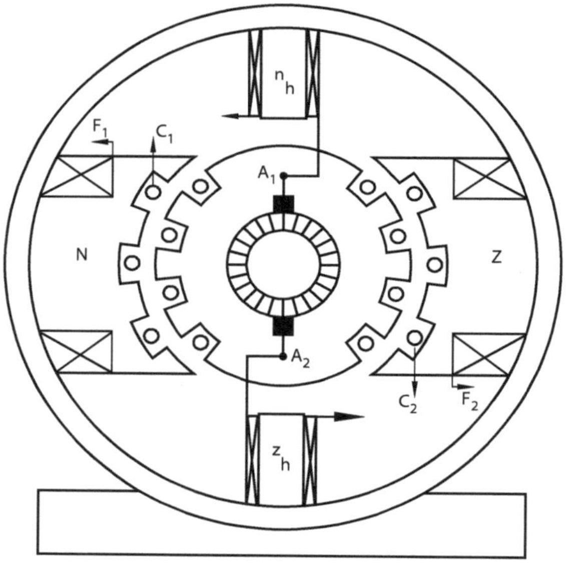

Fig. 19-1: Motor cross section

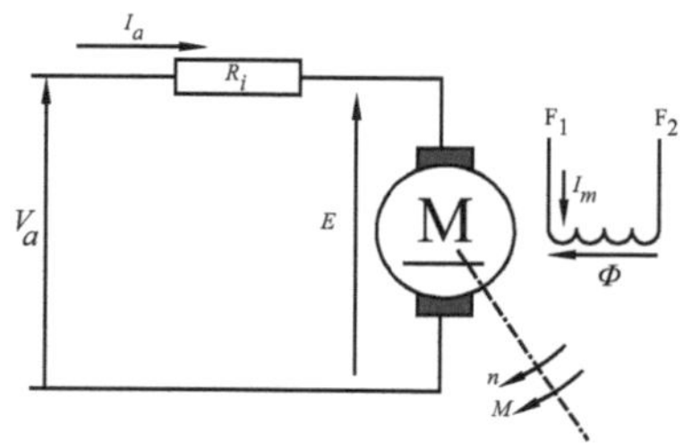

Fig. 19-2: Independently excited motor

If an excitation motor current I_a is applied to the armature then as a result of the Lorentz force the armature will rotate and torque is developed with moment $M_{em} = k_2 \cdot I_a \cdot \Phi$.
In the armature an emf E is produced, indicated by: $E = k_1 \cdot n \cdot \Phi$. The difference between applied armature voltage V_a and counter emf E in the stationary situation is given by $I_a \cdot R_i$. Here R_i is the total resistance of the armature circuit.
Fig. 19-2 shows the principle configuration of an independently excited motor.

In the formulas for E and M_{em} the machine constants k_1 and k_2 are given by:

$$k_1 = \frac{p}{a} \cdot \frac{N}{60} \quad \text{and} \quad k_2 = \frac{p}{a} \cdot \frac{N}{2.\pi}$$

with:
- p = number of pole pairs
- 2·a = number of parallel armature branches
- N = total number of wires on the armature circumference

Sometimes E and M_{em} are written as:

$$E = k_1 \cdot n \cdot \Phi = K_G \cdot \omega \qquad (19\text{-}1)$$

$$M_{em} = k_2 \cdot I_a \cdot \Phi = K_M \cdot I_a \qquad (19\text{-}2)$$

with:
- $K_G = k_1 \cdot \Phi \cdot \frac{30}{\pi} = \frac{p}{a} \cdot \frac{N}{2.\pi} \cdot \Phi$ = generator constant
- $K_M = k_2 \cdot \Phi = \frac{p}{a} \cdot \frac{N}{2 \cdot \pi} \cdot \Phi$ = motor constant
- $K_M = K_G$ as apparent above. Both constants have a different dimension (see chapter 16).

If the current I_a is not constant, for example because of acceleration of the motor, then we also need to include the self induction of the armature circuit L_a into account. This is shown in fig. 19-3. Then we can write:

$$v_a = i_a \cdot R_i + L_a \cdot \frac{di_a}{dt} + e \qquad (19\text{-}3)$$

with I_a constant $\frac{di_a}{dt} = 0$ and therefore $V_a = I_a \cdot R_i + E$ (nominal service) (19-4)

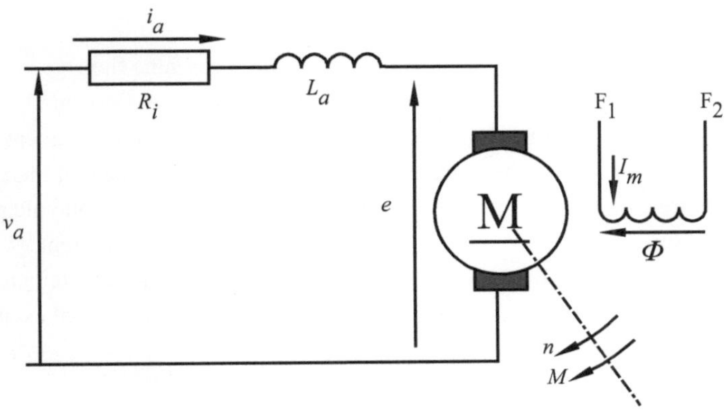

Fig. 19-3: Independently excited motor in transient state

1-2 Equivalent circuit with independently excited motor

The ideal DC-motor is described by two equations:

$$E = k_1 \cdot n \cdot \Phi = K_G \cdot \omega \ \text{(V)} \tag{19-1}$$

$$M \approx M_{em} = k_2 \cdot I_a \cdot \Phi = K_M \cdot I_a \ \text{(Nm)} \tag{19-2}$$

In fig. 19-4 we have just made the field supply and the armature supply variable by implementing controlled rectifiers. For simplicity these converters are represented as one SCR. In reality the motor will be fed from a B_6-bridge while for the field circuit normally a single phase thyristor bridge is sufficient. The motor drives a mechanical load so that the properties of this load (moment of inertia J_m, counter torque M_t) also need to be taken into account. Fig. 19-4 shows the equivalent circuit of a loaded DC-commutator motor.

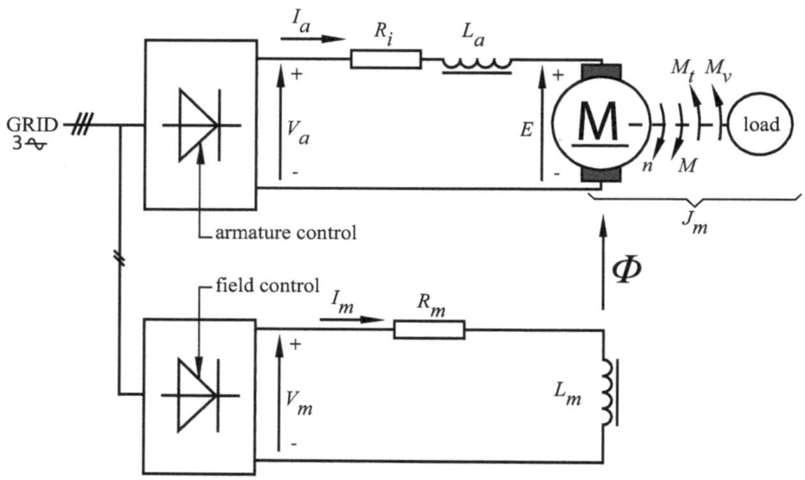

Fig. 19-4: Electronic control of an independently excited motor

In fig. 19-4 we write:

V_a	=	terminal voltage	Φ	=	machine flux
I_a	=	armature current	M_t	=	counter moment
R_i	=	total resistance of armature circuit	M_v	=	acceleration moment
L_a	=	total self induction of the armature circuit	J_m	=	total moment of inertia
E	=	armature emf	V_m	=	field excitation voltage
n	=	motor speed	I_m	=	excitation current
ω	=	angular velocity of shaft	R_m	=	resistance of excitation circuit
M	=	moment of torque on motor shaft	L_m	=	self induction of excitation circuit
M_{em}	=	electromagnetic motor torque			

1.3 Mathematical model

The torque M on the shaft of the motor serves to eliminate the counter torque M_t. Any difference $M - M_t = M_v$ will accelerate or decelerate the moment of inertia (rotor + load), depending on the polarity of M_v :

$$M_v = J_m \cdot \frac{d\omega}{dt} \tag{19-6}$$

As a result of friction and iron losses we have $M < M_{em} = k_2 \cdot I_a \cdot \Phi$, but at low speeds the difference is minimal. If the friction and iron losses are shifted to M_t then written exactly we have:

$$M = k_2 \cdot I_a \cdot \Phi = M_t + M_v \tag{19-7}$$

In fig. 19-4 we also see that:

$$\Phi = k_3 \cdot I_m \tag{19-8}$$

We reconsider here table 16-1 (mathematical model) as table 19-1.

Table 19-1

$v_a = E + i_a \cdot R_i + L_a \cdot \frac{di_a}{dt}$	(19-3)	$M_t = J_m \cdot \frac{d\omega}{dt}$ (Nm)	(19-6)
nominal service: $V_a = I_a \cdot R_i + E$	(19-4)	$M \approx M_{em} = k_2 \cdot I_a \,.\, \Phi = M_t + M_v$	(19-7)
$\Phi = k_3 \cdot I_m$ (Wb)	(19-8)	$P = \omega \cdot M = (V_a \cdot I_a - I_a^2 \cdot R_i) - P_{Fe} - P_{fric.}$	(19-9)
$E = k_1 \cdot n \cdot \Phi$ (V)	(19-1)	$M = \frac{9550 \cdot P}{n}$ (Nm), with P in kW !	(19-5)

Remark

In (19-9) and (19-5) P and M are the available power and torque on the shaft of the motor. The expression $M_{em} = k_2 \cdot I_a \cdot \Phi$ (19-2) describes the internally produced electromechanical torque of the motor. This is obviously larger than the torque on the shaft of the machine since in (19-2) no iron losses and friction losses have been accounted for. For simplicity for the following study we assume (with a small error until otherwise informed) that the available moment on the shaft of the machine is indicated by $M = k_2 \cdot I_a \cdot \Phi$.

2. *M-n* CURVES

2.1 Constant work torque

If we maintain a constant flux Φ and vary the armature voltage V_a we can derive from (19-1) and (19-4): $n = \frac{E}{k_1 \cdot \Phi} = \frac{V_a - I_a \cdot R_i}{k_1 \cdot \Phi} \approx k \cdot V_a$.

By neglecting the $I_a \cdot R_i$ voltage drop the speed n is directly proportional to the armature voltage V_a if Φ is maintained constant. When $\Phi = \Phi_{nom.}$ and $V_a = V_{a\,nom.}$ then $n = n_{nominal}$.
Consider a motor with nominal values:
$V_a = 440$ V; $I_a = 13$A; $P = 4.7$kW; $n = 1540$ rpm; $R_i = 0.68\Omega$; $L_a = 30$mH; $J_{rotor} = 0.02\text{kgm}^2$.
Such a motor requires approximately 10V (transitional resistance brushes, $I_a \cdot R_i$ - losses) to start rotating. Assume that I_a remains constant, then we find:
$M \approx M_{em} = k_2 \cdot I_a \cdot \Phi$ = constant and $P \approx k_4 \cdot V_a \cdot I_a \approx k_5 \cdot V_a$.
Both expressions determine the curves for armature control in fig. 19-5.
From (19-5): $M = 29.14$ Nm.

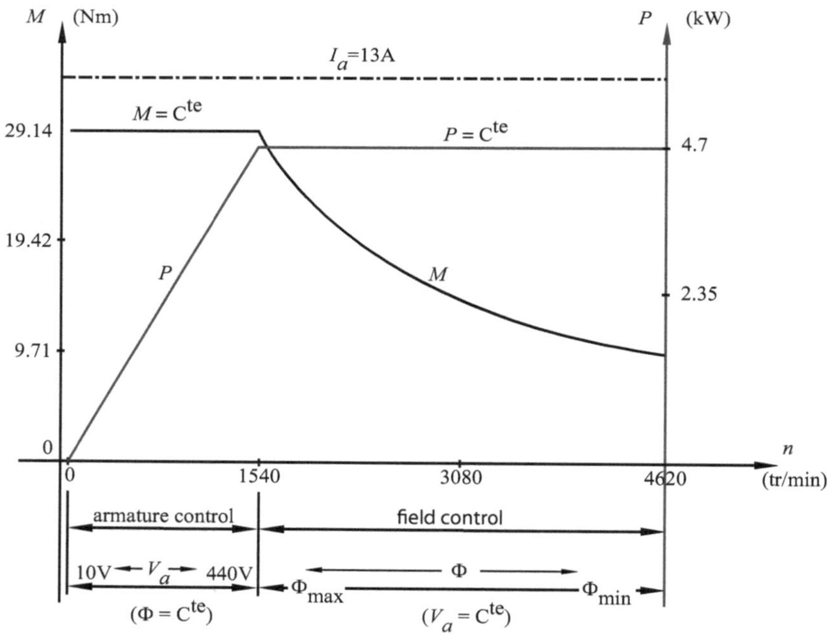

Fig. 19-5: Curves of independently excited motor

Remark
At stand still and with full armature current the collector lamella will be damaged. That is the reason the manufacturer specifies a minimum speed of for example 20 rpm.

2.2 Constant Power

A common task for a motor is winding or unwinding wire or cable, rolling paper, textile, etc. In fig. 19-6 we see such a machine (wrapping machine). Evident conditions are a constant supply of material (v) and a constant pull (F).

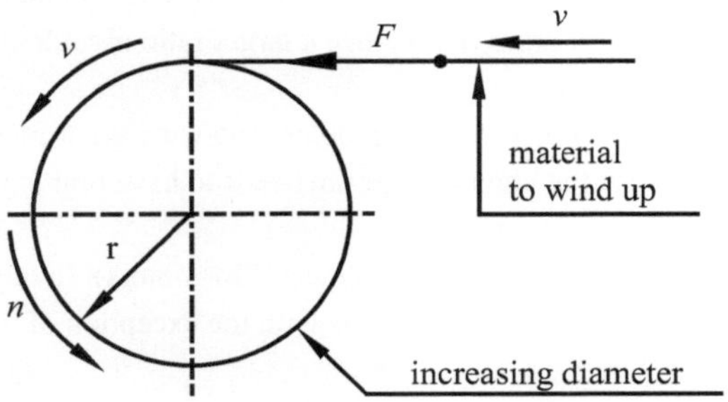

From $M = F \cdot r$

and $v = \frac{2 \cdot \pi \cdot r \cdot n}{60}$

it follows: $M \cdot n = F \cdot r \cdot \frac{60 \cdot v}{2 \cdot \pi \cdot r} = \text{constant}$

Fig. 19-6: Winding-drum

The product $M.n$ is constant and the M-n curve is hyperbolic.

In addition: $P = \omega \cdot M = M \cdot \frac{2 \cdot \pi \cdot n}{60} = \frac{2 \cdot \pi}{60} \cdot M \cdot n = \text{constant}$

To wind and unwind correctly, speed control with ***constant power*** is required.

Electronic solution:

We maintain V_a and I_a constant so that $P = V_a \cdot I_a = \text{constant}$.

V_a and Φ are set for the average winding speed: $n = \frac{V_a}{k_1 \cdot \Phi}$.

If the diameter of the winding drum increases then M has to increase proportionally.
Since $M \approx M_{em} = k_2 \cdot I_a \cdot \Phi$, this can be achieved by increasing the flux (therefore via I_m) proportionally via a controlled rectifier.
It boils down to maintaining V_a and I_a constant and varying I_m proportionally with the diameter of the winding drum in order to achieve speed control with constant power.

Remark

The right hand side of the graphic in fig. 19-5 (field control) is also clear now. Since V_a and I_a are constant, the power is constant and with reducing flux the speed increases and the torque decreases.

Remarks

1. The mechanical construction of a motor allows a maximum speed of three times $n_{nom.}$ (sometimes exceptionally to $5 \cdot n_{nom.}$).
   The speed variance via field control is therefore maximum 1:3 (up to 1:5). This limitation is also valid for the ratio between empty and full drum (in practice a larger ratio is seldom required!).
2. In practice wind up and off drums can each be driven by one motor. The wind up drum can be controlled as described and at the same time the unwinding drum needs to have braking to maintain the correct tension on the material being wound.
3. Smaller motors (up to tens of watts) often have permanent magnets (PM-motors). In this case we are limited to armature control. Everything discussed here is with the exception of field control applicable to PM-motors.

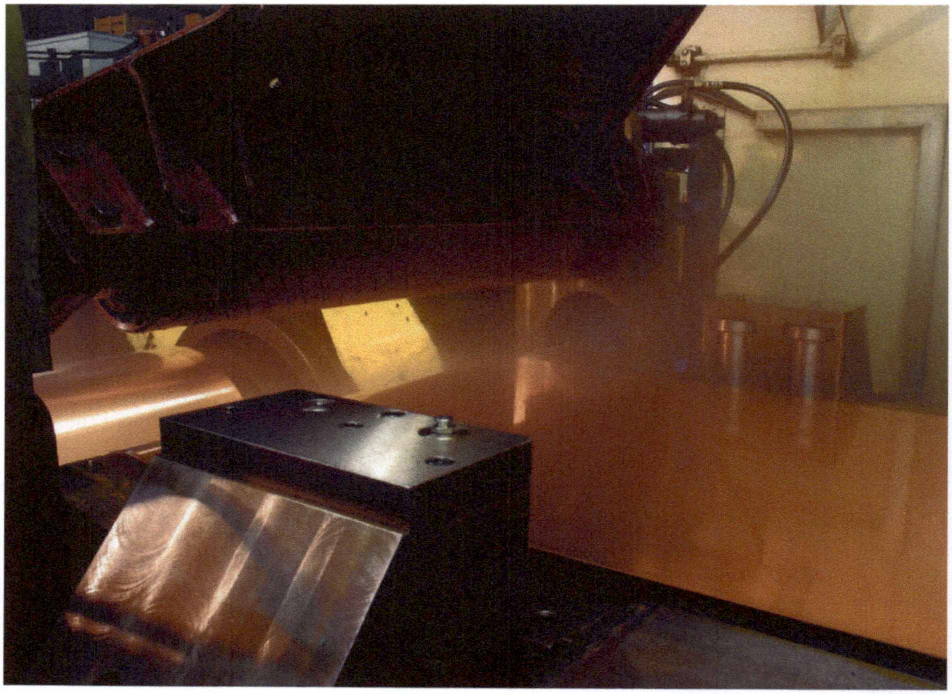

Photo Siemens: Cold rolling reversing stand in the Voehringen factory. Siemens modernized the Voehringen plant for cold rolling of non ferrous metals in the Wielandwerke AG (Ulm Germany). This involved a world leader in the production of half finished and special products made from copper and copper alloys. The project included the basic automation, process automation and drive systems, including the commissioning

Numeric example 19-1:

Given:

The motor with its specifications from numeric example 16-7.

Problem:

Determine the speed of the motor in question if the armature voltage is respectively 110V and 440 V and the motor in both cases is 40% and 100% loaded.

Solution:

1. From numeric example 16-7 we know that:
   - generator constant: $K_G = 2.6738\ \mathrm{V/_{rad/s}}$
   - normalised emf: $e_N = 0.28$ V/rpm
   - nominal momentum motor torque: $M = 29.14$ Nm

2. $V_a = 110\text{V}$
   a) 40% motor load: $M_1 = 0.4 \cdot 29.14 = 11.656$ Nm
   with $M_{em} \approx M$ is: $I_a = 0.4 \cdot 13 = 5.2\text{A}$
   $V_a = 110\text{V}$ results in an emf: $E = V_a - I_a \cdot R_i = 110 - 5.2 \cdot 0.68 = 106.464\text{V}$
   so that: $n = \dfrac{E}{e_N} = \dfrac{106.464}{0.28} = 380.22$ rpm

   b) 100% motor load: $M_2 = 29.14\text{Nm}$ and $I_a = 13\text{A}$
   $V_a = 110\text{V}$ results in an emf: $E = 110 - 13 \cdot 0.68 = 101.16\text{V}$
   so that: $n = \dfrac{E}{e_N} = \dfrac{101.16}{0.28} = 361.28$ rpm

3. $V_a = 440\text{V}$
   a) 40% motor load: $I_a = 5.2\text{A}$
   $E = V_a - I_a \cdot R_i = 440 - 5.2 \cdot 0.68 = 436.464\text{V}$
   $n = \dfrac{E}{e_N} = \dfrac{436.464}{0.28} = 1558.8$ rpm

   b) 100% motor load: $I_a = 13\text{A}$
   $E = V_a - I_a \cdot R_i = 440 - 13 \cdot 0.68 = 431.16\text{V}$
   $n = \dfrac{E}{e_N} = \dfrac{431.16}{0.28} = 1540$ rpm

This is the name plate data which was the starting point to determine M, K_G, e_N.

2.3 *M-n* curves with armature control

We go deeper with the most common situation, namely "armature control". As was apparent from numeric example 19-1 the voltage drop $I_a \cdot R_i$ has an effect on the speed with changing motor load.

Indeed from $E = k_1 \cdot n \cdot \Phi = V_a - I_a \cdot R_i$ it follows that the speed is: $n = \dfrac{V_a - I_a \cdot R_i}{k_1 \cdot \Phi}$

With constant flux it is then: $$n = k_4 \cdot [\, V_a - I_a \cdot R_i] = k_4 \cdot E \qquad (19\text{-}10)$$

In addition (19-2) leads to: $$M = K_M \cdot I_a \qquad (19\text{-}11)$$

Expression (19-11) allows us to draw fig. 19-7. The torque is independent of the speed, and depends only on the actual motor current I_a. We have taken the numeric data of the motor of fig. 19-5, which corresponds with the motor in numeric example 19-1.
With the aid of expression (19-10) we can now draw figure 19-8 for the same motor. The figures 19-7 and 19-8 are known as the speed control characteristics.

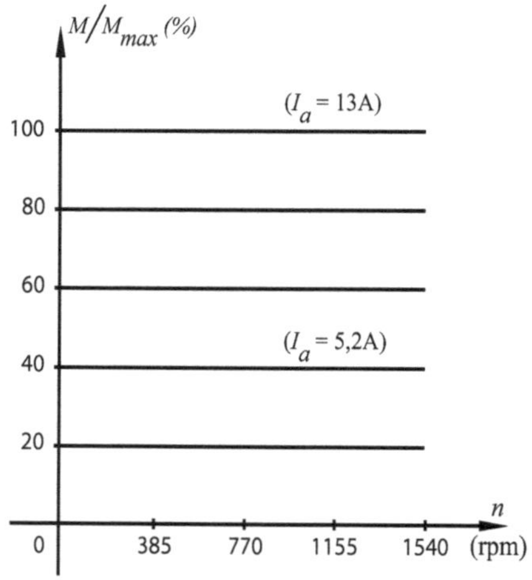

Fig. 19-7: Speed control characteristic (M = *constant*)

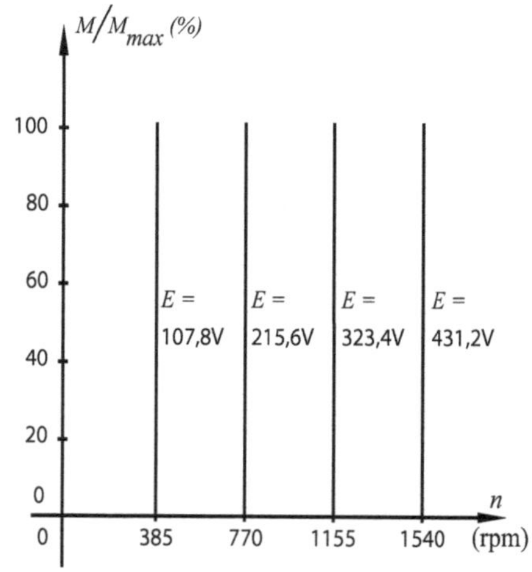

Fig. 19-8: Speed control characteristic (n = *constant*)

In fig. 19-7 the torque remains practically constant as the speed reduces from nominal speed to zero. Keeping the torque constant is important for the drive train when multiple motors are involved (rolling mills, wire pulling machines, textile industry,…).
Normally the main motor determines the speed at which the material is moved and the other motors have to adjust to this speed to prevent unnecessarily stressing the material. The auxiliary motors are set for a constant torque that corresponds to the acceptable elongation of the material.

Remark
The line M/M_{max} = 100% in fig.19-7 is the line at M = 29.14 Nm = constant in fig. 19-5!

2.4 Single quadrant operation

Consider the previous motor as drawing 5.2A and having a counter emf of 215.6V. In fig. 19-7 we see that the operating point lies on the horizontal at 40% . M_{max} . In fig. 19-8 the operating point lies somewhere on the vertical at E = 215.6V. The operating point P is the intersection of both lines. This is shown in fig. 19-9.

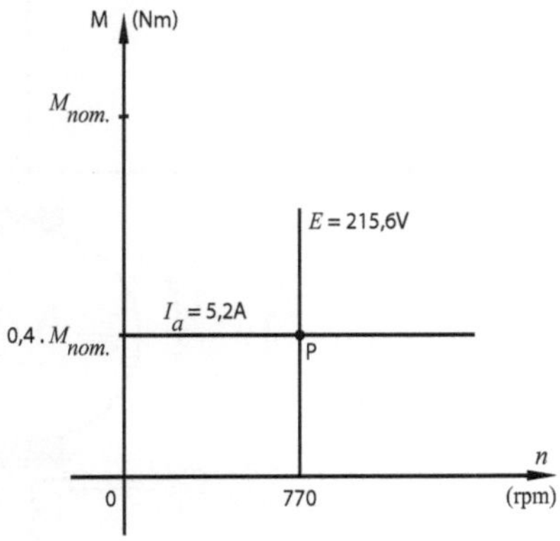

Fig. 19-9: Operating point of DC-motor

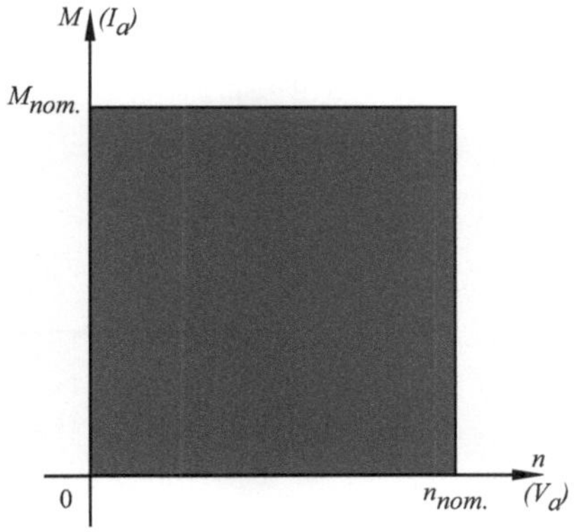

Fig. 19-10: Single quadrant service

With armature control all possible operating points of the motor (in one direction of rotation) are somewhere in the rectangle enclosed by the intersecting M-n lines of M_{nom} and n_{nom} .
Here we see that M_{nom} and n_{nom} are the nominal values as found on the motor data plate.
The value of M_{nom} is first calculated with (19-5): $M_{nom} = \frac{9550 \cdot P}{n}$.
The collection of all operating points lie in one quadrant: we speak of single quadrant operation. This is drawn in fig. 19-10.
Since $M = K_M \cdot I_a$ and $n = k_4 \cdot E \approx k_4 \cdot V_a$ we refer to the graphic as an I_a-V_a coordinate system.

3. CONTROLLED SINGLE QUADRANT DRIVE

3.1 Operation

A good speed controller must meet two requirements:

1. n should be independent of the load
2. I_a should be limited to an adjustable maximum.

The block diagram of the control circuit for the speed control with armature control is shown in fig. 19-11.

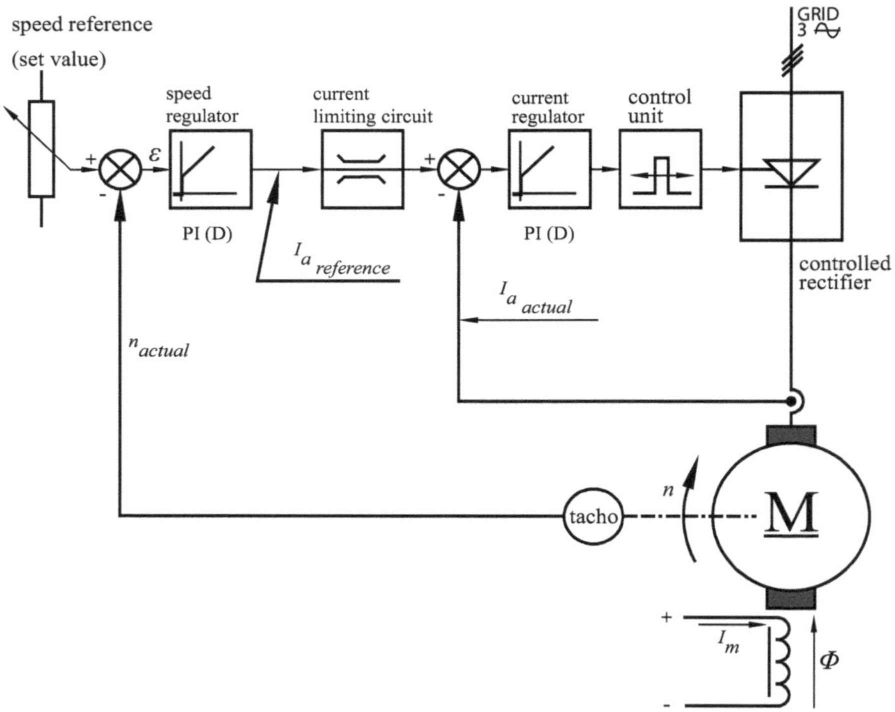

Fig. 19-11: Block diagram of a controlled single quadrant drive

With a potmeter the desired speed is set. A tachometer produces a voltage proportional with the actual speed of the motor. Set-point and process value in the form of voltages are compared with each other in a comparator. At the output of this comparator the difference ε appears.
If $\varepsilon \neq 0$ the speed of the motor is not equal to the set-point, in other words there is an error.
The fault ε is referred to as the error signal. The error ε is transferred to the first controller. If this controller is a proportional amplifier, then the output signal will be proportional to the difference between set-point and the actual speed of the motor. In other words the output of the speed controller indicates if the motor should accelerate or decelerate. For a positive ε the motor should accelerate and for a negative ε the motor should decelerate. An acceleration or deceleration corresponds with an increased or decreased motor torque. This is achieved by an increased or decreased current in the motor.

The output of the speed regulator is therefore a measure of the desired armature current of the motor. We compare this desired armature current with the actual armature current in a current comparator.
The output of this comparator is now the input for a second controller to achieve an as exact as possible current control. The output of this second controller sends a pulse (control signal), which for its part controls the thyristor bridge. The output voltage of the bridge is controlled as a function of the input voltage of the pulse generator.
As we shall see later the controllers are not just proportional controllers but normally PI-controllers. In exceptional cases PID-controllers may be encountered.

3.2 Advantages of internal current control

The ***current control circuit*** is especially active with ***changes*** in the ***motor load torque***. The short control time (7 or 10 ms) in the current circuit do not allow the load variations to affect the speed control since the armature current is quickly adjusted to the required value by the new load torque.
With this internal current control the maximum motor current is easily controlled. If we limit the input current controller to for example 120% of the nominal value of the armature current, then thyristor bridge and motor are protected against overload and against a high start current surge of the motor.

Current limitation is especially important when ***quick speed changes*** are ***required*** from the machine. If the motor load has a large moment of inertia then large (required) speed changes cannot be followed. If the speed set-point is suddenly raised, then the controller output will want a much higher armature current than the nominal armature current. The current limitation sees to it that the nominal value is not exceeded. With constant flux and nominal armature current during acceleration the motor will maintain an almost constant torque $M \approx M_{em} = k_2 \cdot I_a \cdot \Phi$.
If the counter torque of the mechanical load is M_t, then the difference $M_v = M - M_t$ will cause the machine to accelerate according to $M_v = J_m \cdot (d\omega/dt)$.
Here $d\omega/dt$ = acceleration; J_m = moment of inertia. If M_v is constant then the machine will accelerate linearly to the desired speed.
If the desired speed is less than the actual speed then with single quadrant operation the motor will run till the desired speed is reached and this with zero armature current.

Remark

Why is the armature current limit set to 120% of the nominal armature current?
Very simply since a fully loaded motor ($I_a = I_{a\ nominal}$) could no longer accelerate if we should limit the current to 100%.

Numeric example 19-2:

Given:

- DC-motor: 220V / 10A / 1.76kW / 1000 rpm / $R_i = 1\Omega$
- tacho generator: constant is 10V / 1000 rpm
- B_6 - controlled rectifier with impulse module: $V_{control} = 0$ to 10V gives $V_{di\alpha} = V_a = 0$ to 220V
- control characteristic: fig. 19-12

  To enable maximum use of this numeric example a number of values are indicated in this characteristic.

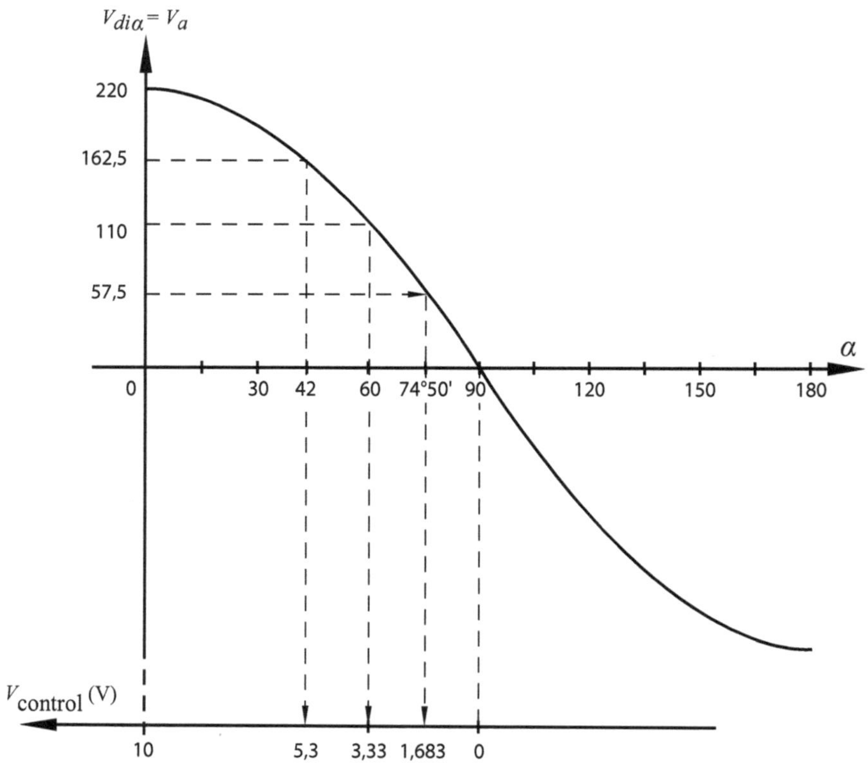

Fig. 19-12: Control characteristic of a B_6 - Bridge

Required:

1. With $n = 250$ rpm and with 50% motor load determine the input and output values of the different "blocks "in fig. 19-11. The internal current control and current limitation are inactive.

2. The same question but now with $n = 500$ rpm, 50% motor load and active current control as in fig. 19-11.

Solution:

To make the solution as short as possible we will ignore the transient behaviour, and only concentrate on the final steady state values. The transient behaviour of a single quadrant drive is studied under optimising the controllers (p.19.34).

From the name plate of the motor we obtain the following:

at full load, $V_a = 220\text{V}$ and $n = 1000$ rpm:

$$I_a = 10 = \frac{V_a - E}{R_i} = \frac{220 - E}{1} \quad \text{or: } E = 210\text{V at 1000 rpm}$$

The normalised emf e_N is: $e_N = 0.21\text{V/rpm}$

The nominal torque is: $M = \frac{9550 \cdot P}{n} = \frac{9550 \cdot 1.76}{1000} = 16.8\text{ Nm}$

1. $n = 250$ rpm - 50% motor load

- $M_t = 50\% \cdot 16.8 = 8.4\text{Nm}$
- $I_a = 50\% \cdot 10 = 5\text{A}$
- $n = 250$ rpm $\rightarrow\rightarrow$ $E = 250 \cdot e_N = 250 \cdot 0.21 = 52.5\text{V}$
- $V_a = E + I_a . R_i = 52.5 + 5 \cdot 1 = 57.5\text{V}$
- Fig. 19-12: $V_{d\alpha} = 57.5 = 220 \cdot \cos\alpha \rightarrow\rightarrow \alpha = 74°50' \rightarrow\rightarrow V_{control} = 1.683\text{V}$
- The single quadrant controller without internal current loop is drawn in fig. 19-13.

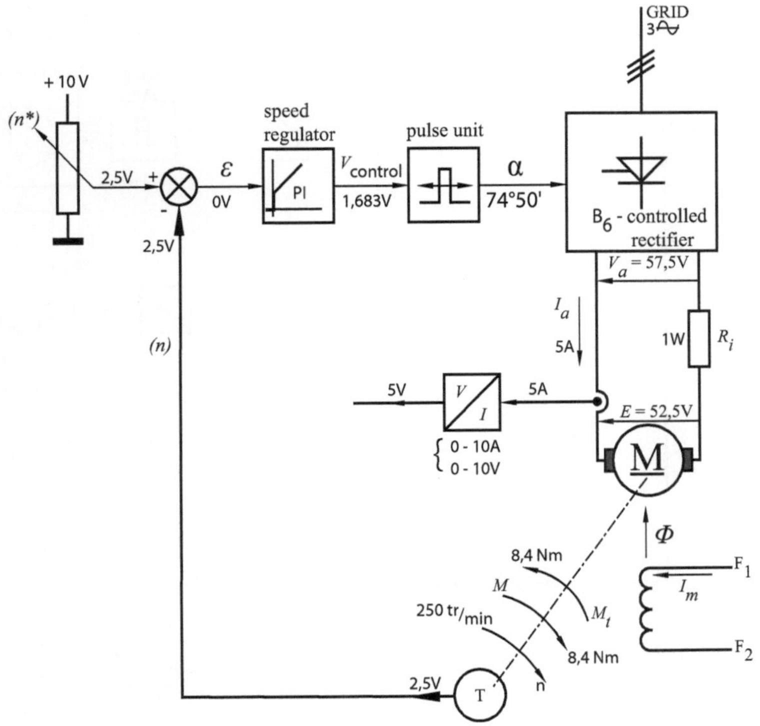

Fig. 19-13: Speed controlled DC-motor at 250 rpm

2. $n = 500$ rpm - 50% motor load

To set the desired speed $n^* = 500$ rpm we adjust the wiper of the potmeter from 2.5 V to 5V. In the first instant the motor remains rotating at a speed of 250 rpm and the tachometer produces 2.5 V. The error voltage ε becomes $5 - 2.5 = 2.5$ V. The n-controller immediately makes a jump at its output (proportional part of the PI controller) and at the same time the controller begins to integrate (I - part) so that the output voltage of the n-controller reaches 10V in a number of ms. With $V_{control}$ =10 V then $\alpha = 0°$ and $V_{di\alpha} = V_a = 220$ V. Due to the inertia of the motor with its drive in the first instant E remains practically unchanged at $E = 52.5$V so that:

$$I_a = \frac{V_a - E}{R_i} = \frac{220 - 52.5}{1} = 167.5\text{A !}$$

Too much for our 10 A motor and also for the associated SCR bridge . We are compelled to limit the armature current and maintain it within safe operating values. The only way to do this is to measure the current I_a with a current sensor and to control this current to below an adjustable (acceptable) maximum. For this reason the internal current loop of fig. 19-11 is now active. This is shown in fig. 19-14 with the current limited to 9 A, and speed set-point n^*= 500 rpm with the motor 50 % loaded. The current sensor is followed by an I/V converter that for example converts 0-10 A to 0-10 V.

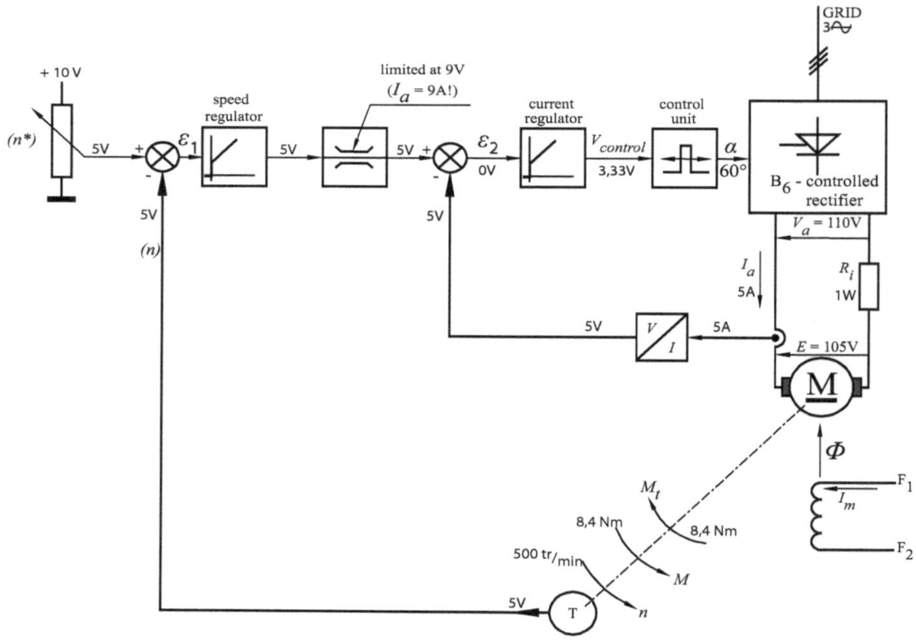

Fig. 19-14: Single quadrant DC-motor at $n = 500$ rpm, $M_t = 0.5\, x\, M = 8.4$ Nm and $I_{a\,max} = 9$A

$$E = 500 \cdot e_N = 500 \cdot 0.21 = 105\text{V};\; V_a = E + I_a \cdot R_i = 105 + 5 \cdot 1 = 110\text{V}$$
$$V_{di\alpha} = 110 = 220 \cdot \cos\alpha \longrightarrow\!\!\longrightarrow \alpha = 60° \longrightarrow\!\!\longrightarrow V_{control} = 3.33\text{V}$$

Remark

If the desired speed is 750 rpm then the potmeter wiper is set to7.5V. In the first instant the motor remains at 500 rpm and the error voltage ε is 2.5V. The output of the n-controller becomes quickly 10V as a result of integration. The "current limiting" block is set to $I_{a\,max}$ = 9A, that means the output can (with 1V/A) reach a maximum of 9V.

The comparator of the current controller receives 9 V (setpoint) and 5V (for the actual current). The error ε_2 = (9V − 5V) = 4V causes $V_{control}$ to rise so that α falls and $V_{d\alpha} = V_a$ rises.

The current $I_a = \frac{V_a - E}{R_i}$ increases, but is limited to 9A by the controller.

Indeed , if I_a > 9A then the error ε_2 would be negative, the PI controller integrates down, $V_{control}$ decreases, V_a and I_a decrease.

With I_a = 9A the driving torque of the motor is: $M = 90\% \cdot 16.8 = 15.12$ Nm.

$M = M_t + M_v$ becomes: $15.12 = 8.4 + M_v$. There is an acceleration torque $M_v = 6.72 = J_m \cdot \frac{d\omega}{dt}$.

The acceleration $d\omega/dt$ depends upon the value of J_m.

If the motor exceeds 750 rpm then ε_1 is slightly negative and the output of the n-controller drops and after a few short fluctuations it sets the output to 5V and the armature current I_a is then 5A (which corresponds to 50% load) and the values of fig. 19-14 are then:

- n = 750 rpm ; I_a = 5A
- $E = 750 \cdot e_N = 750 \cdot 0.21 = 157.5$V
- $V_{d\alpha} = V_a = E + I_a \cdot R_i = 162.5$V
- $\alpha = 42°$
- $V_{control}$ = 5.3V
- $\varepsilon_2 = 0$
- output armature current converter = 5V
- output current limiter = output n-controller = 5V
- output tacho = 7.5V.

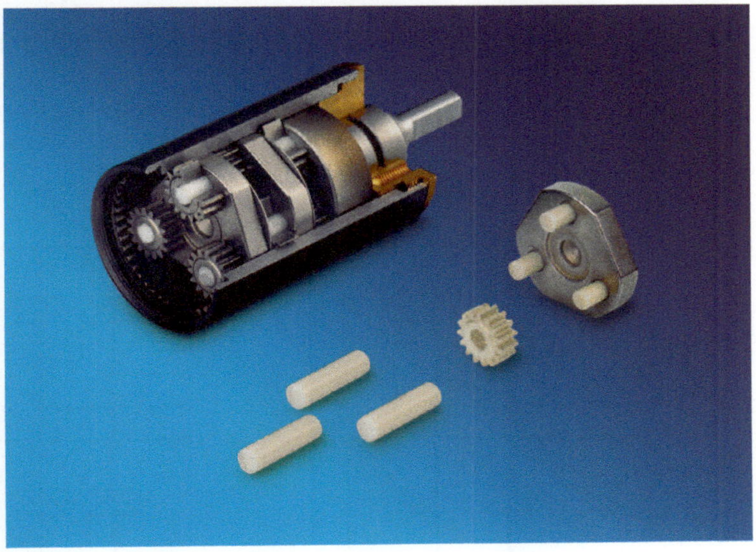

Photo Maxon Motor Benelux: Ceramic is better than steel. Possible applications are where steel reaches the limit of its properties. Properties such as resistance to abrasion and good conduction prove that the lifespan are dramatically extended. Thanks to these properties a ceramic shaft of zirconiumdioxide can be used in planetary gears up to 10,000 rpm

4. TWO QUADRANT AND FOUR QUADRANT OPERATION

4.1 Reversing direction of rotation and braking of independently excited motor

Fig. 19-15 shows a machine operating as generator or as motor. In the case of the generator the current leaves the terminals via the + terminal and power is generated: the machine is operating as a ***transmitter***. In the case of the motor the current leaves via the – terminal, power is received: the machine operates as a ***receiver***.

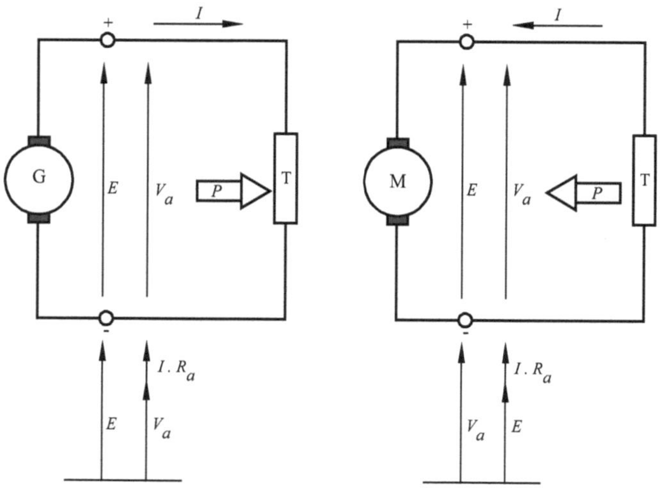

Fig. 19-15: DC-machine operating as generator or as motor

If we look at a fully controlled three-phase rectifier (fig. 19-16/17), we see that the current I can only flow in one direction through the SCR's.

If current leaves via the + terminal, then power is transferred from the power grid to the load: the bridge operates as ***rectifier***. If I leaves via the – terminal then power is transferred from the load to the power grid: the bridge operates as an ***inverter***.

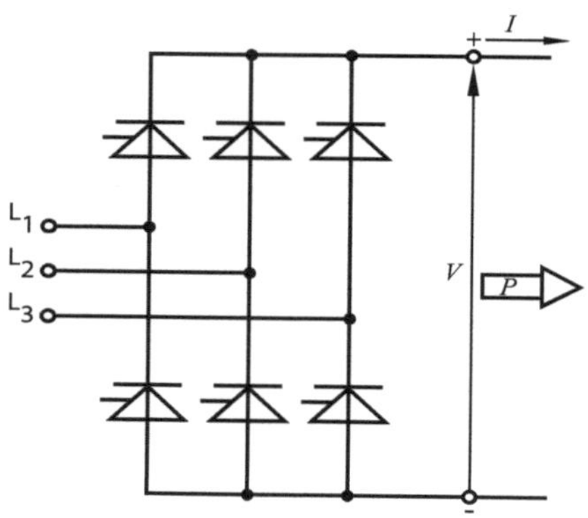

Fig. 19-16: B_6 bridge in rectifier mode

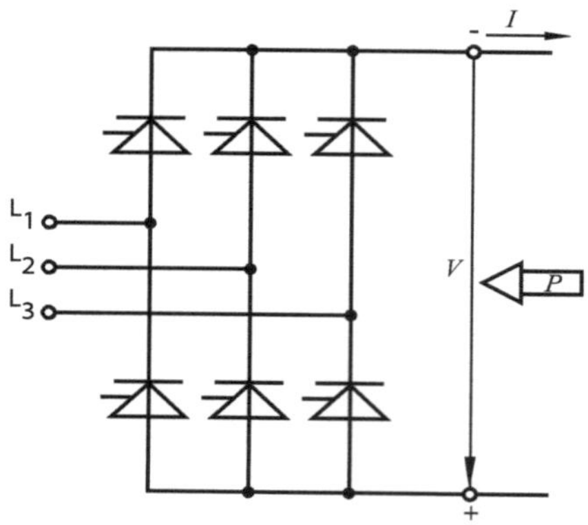

Fig. 19-17: B_6 bridge in inverter mode

From fig. 19-15 we can derive two practical rules:

1. In a power supply the current leaves via the plus terminal and in a consumer the current leaves via the minus terminal.
2. With a DC-machine the polarities of V_a and E are the same, in other words the plus terminal of V_a is also the plus terminal of E.

These two practical rules will be useful in the study of two quadrant and four quadrant operation of a DC-machine.

Photo Siemens: Typical roughing stand with edger. Siemens VAI Metals Technologies has received an order from Thyssen Krupp Steel Europe AG to completely modernize its hot strip rolling mill no. 1 in Duisburg-Bruckhausen (Germany). The objectives of the project are to further improve product quality and broaden the range of products. The Bruckhausen location of Thyssen Krupp Steel Europe produces more than three million tons of cast steel slabs per annum. The plant can produce not only high quality steels but also high-silicon electri cal strip grades. The plant has a compact hot strip mill and a wide hot strip mill. The finishing mill has seven, four-high rolling stands. All the stands in the finishing mill will be equipped with Smart Crown technology to improve the profile and flatness control. The profile and flatness control will be matched to the associated new actuators, and optimized. The same will apply to the controllers of the new hydraulic actuators in the roll screw, loopers and side guides

4.2 Motor braking - Two quadrant operation in the same direction

In this two quadrant service the machine operates as motor in one direction of rotation and brakes as a generator in the same direction of rotation.
We consider for example a rolling mill connected to two anti-parallel bridges (fig. 19-18).
Note the indicated polarity of V_1 and V_2 corresponds with the rectifier operation of the respective bridges.

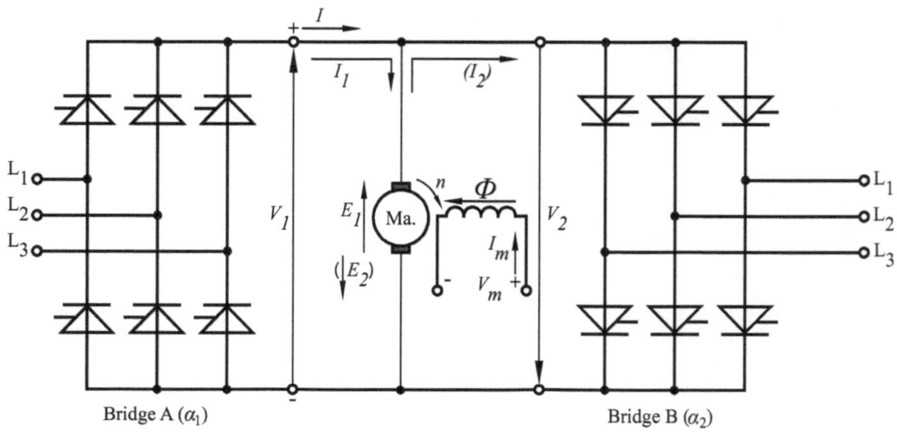

Fig. 19-18: DC-motor controlled using two thyristor bridges

Assume bridge A is operating as a rectifier ($\alpha_1 < 90°$) so that with V_1 and I_1 the machine rotates clockwise as a motor. This is called quadrant 1.
We disconnect bridge A and control bridge B as an inverter ($\alpha_2 > 90°$ with V_2 negative). Due to the mechanical inertia the machine can now operate as a generator. With an unchanged direction of rotation (n) and flux (Φ) the polarity of $E_1 = k_1 \cdot n \cdot \Phi$ do not change.
With $|-V_2| < E_1$ the current I_2 flows through bridge and generator. The machine generates power and brakes until it reaches stand still in quadrant 4. This two quadrant mode is shown in fig. 19-19.
It is clear that $M (= k_2 \cdot I_a \cdot \Phi)$ has the same polarity as I_a since the direction of the flux is unchanged.
Note that the polarity of the counter emf corresponds with that of the armature voltage V_1.

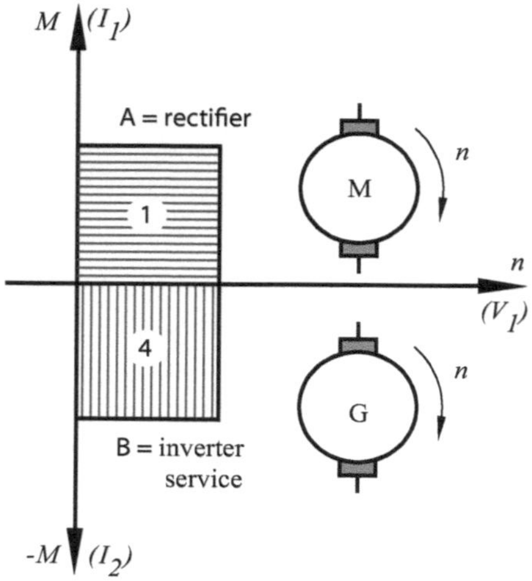

Fig. 19-19: Two quadrant-“rolling mill” service of a DC-machine

4.3 Motor braking -Two quadrant operation with the same torque direction

A lifting motor can also operate in two quadrant mode but then in quadrant 1 and 2 (or in quadrant 3 and 4!).

Consider a lifting motor that lifts a load with a positive torque and rotates clockwise (fig. 19-21). This is motor mode in quadrant 1 (fig. 19-20) together with the rectifier operation of bridge A in fig. 19-18 (applied voltage V_1 ; armature current I_1 ; counter emf E_1) .

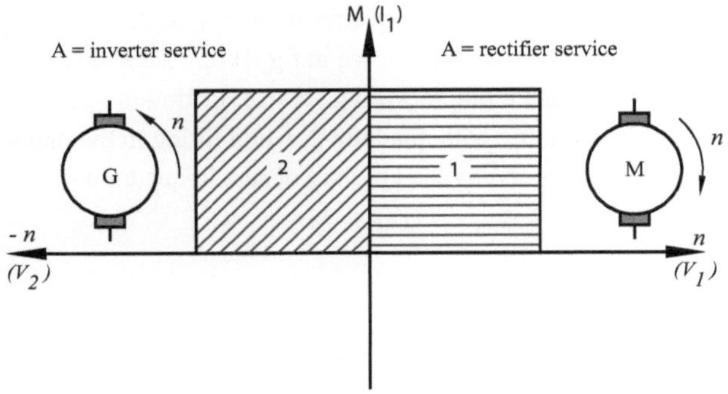

Fig. 19-20: Two quadrant -"lifting" operation of a DC-machine

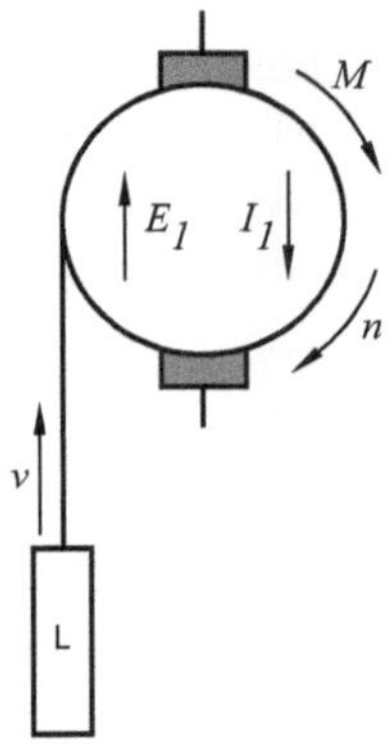

Fig. 19-21

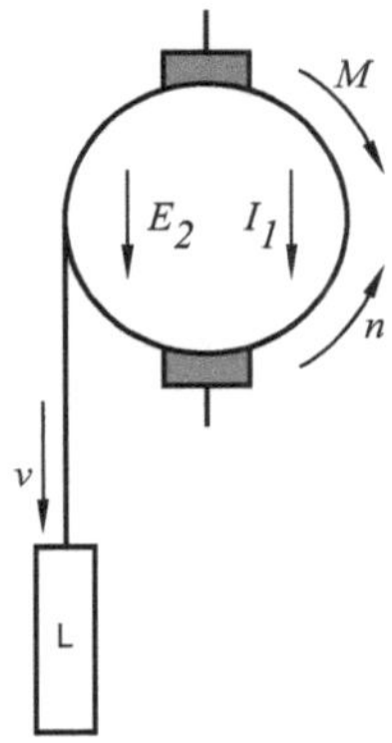

Fig. 19-22

If we reduce the firing of rectifier A so that V_1 decreases, then at a certain instant the motor torque will be insufficient to raise the load and the load will begin to drop. (fig. 19-22). This is known as load pulling. The direction of rotation of the motor reverses and since the flux has not changed, the polarity of $E = k_1 \cdot n \cdot \Phi$ reverses from what it previously was. The emf is now E_2 (fig. 19-18) and if bridge A is controlled as an inverter the polarity of V_1 can reverse.

With $|V_1| < E_2$ the machine will deliver energy to the power grid. This is operation in the second quadrant as represented in fig. 19-20. The load L has to be sufficiently large for load pulling to occur otherwise operation in quadrant 2 is not possible. Note that one bridge (A or B) is sufficient for this type of two quadrant operation.

4.4 Four quadrant mode

We reconsider the rolling mill as presented in fig. 19-19 and to the right of fig. 19.23. Controlling bridge A as rectifier allows the motor to rotate clockwise. Bridge B as inverter allows the machine to brake (still in a clockwise direction!) until it is at stand still.
If we now control bridge B in fig. 19-18 as a rectifier ($\alpha_2 < 90°$) then the motor will rotate counter clockwise (with V_2 and I_2). The flux Φ is indeed unchanged, but the direction of I_2 is opposite to I_1 from quadrant 1 so that the motor turns counter clockwise. This is quadrant 3 in fig. 19-23. Disconnecting bridge B and controlling bridge A as inverter produces quadrant 2 whereby the machine runs to stand still as a generator. This is shown in fig. 19-23. This figure is equally the graphical representation of four quadrant mode for lifting service. Lifting in one direction corresponds with quadrant 1 and load pulling with quadrant 2, while lifting in the opposite direction corresponds to quadrant 3 and corresponding load pulling occurs in quadrant 4.

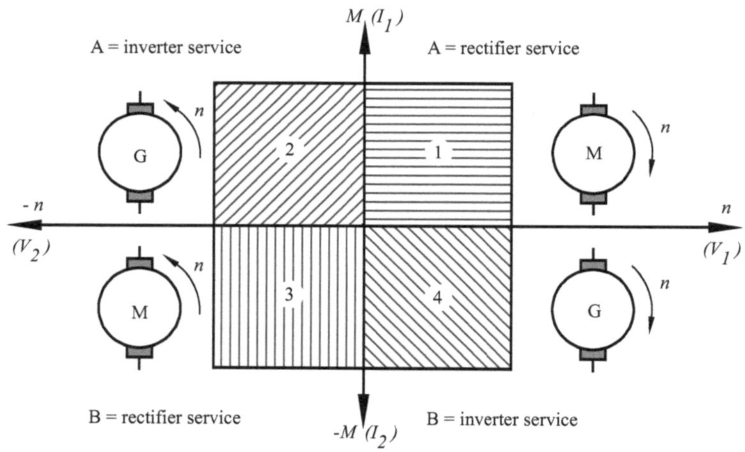

Fig. 19-23: Four quadrant operation of the configuration from fig. 19-18

Example

Consider a DC-machine operating an elevator. The counter weight is equal to the weight of the lifting cage increased by half the maximum load. The motor load is in any case a maximum of the half load of the lift. We assume:

a) that the fully loaded motor raises the load by rotating clockwise and we call this "quadrant 1"
Determine in which quadrant the lift is operating when:

b) the lift is fully loaded when it is lowered
c) the lift is empty when lowered
d) the lift is empty when raised.

Solution:

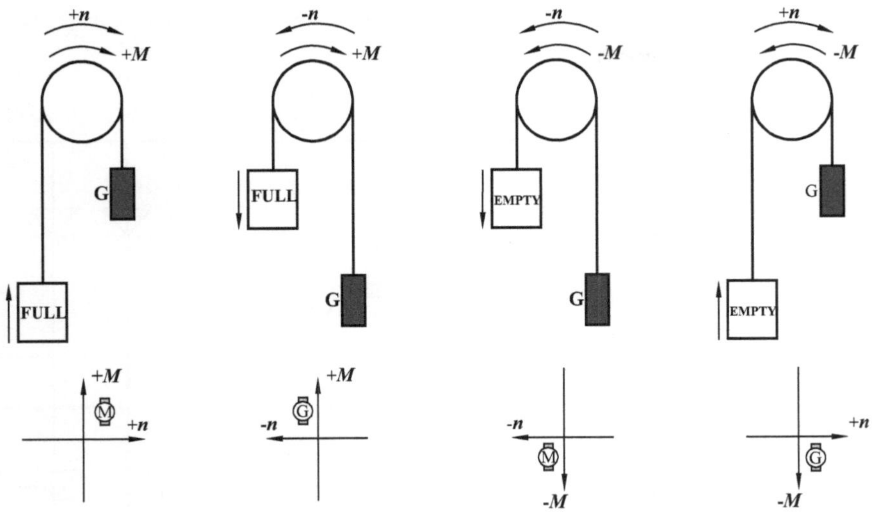

Fig. 19-24

Remark

Fig. 19-25 shows the control circuit of a four quadrant configuration. We refer to this circuit as circulating current free as opposed to the configuration in fig. 19-26 and 19-27.

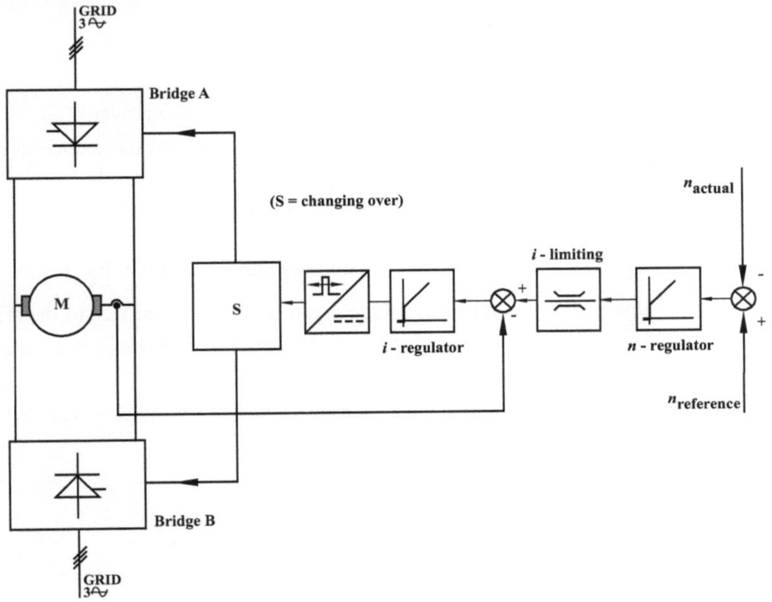

Fig. 19-25: Four quadrant drive with anti-parallel bridges circulating current free

4.5 Four quadrant mode with circulating current

To ensure a speedy transition between the different quadrants both bridges are operated simultaneously. This is studied in fig. 19-26. At every instant we ensure that $\alpha_1 + \alpha_2 = 180°$.

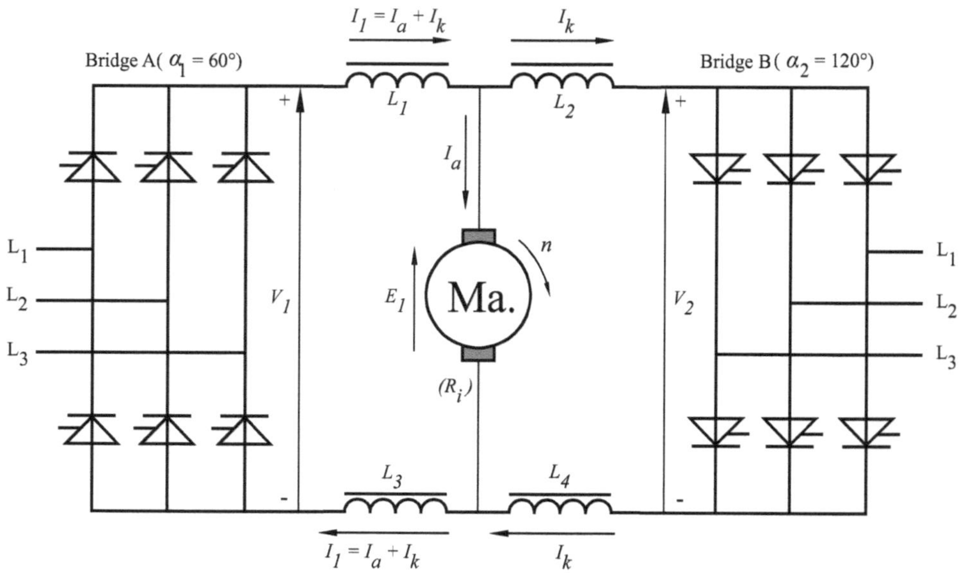

Fig. 19-26: Anti-parallel circuit with circulating current

In quadrant 1 bridge A is operated as rectifier and delivers current I_1 while bridge B is operated as inverter with no current. If bridge A has for example $\alpha_1 = 60°$, then the output voltage is:

$$V_1 = \frac{3 \cdot \hat{v}}{\pi} \cdot \cos \alpha_1 = \frac{3 \cdot \hat{v}}{2 \cdot \pi}.$$

The amplitude $\hat{v}$ here is that of the line voltage of the three-phase power grid.
The motor is supplied with voltage V_1, draws a current $I_1 = I_a$ and has a counter emf E_1 with the same polarity as V_1 and in addition $E_1 = V_1 - I_a \cdot R_i$.
If bridge B is operated simultaneously as inverter with $\alpha_2 = 120°$ ($= 180° - \alpha_1$) then the average voltage is $V_2 = \frac{3 \cdot \hat{v}}{\pi} \cdot \cos 120° = -\frac{3 \cdot \hat{v}}{2 \cdot \pi}$ with the polarity shown in fig. 19-26.
Since V_2 is as large and opposed to V_1, bridge A can only deliver current to the armature of the motor. Bridge B cannot supply current to the motor since the thyristors are polarized in the wrong direction.
During braking with the same direction of rotation $V_1 < E_1$ and bridge A (rectifier) is without current. With $E_1 > V_2$ bridge B will carry current.

Example
With $\alpha_1 = 45°$ (bridge A = rectifier) then $\alpha_2 = 135°$ (B = inverter).
Bridge A supplies the machine as a motor.
In a 3x400 V net then $\hat{v} = \sqrt{2} \cdot 400 = 565$V.
Assume that $R_i = 1\Omega$. With $\alpha_1 = 45°$ then $V_1 = 382$V and if for example $I_1 = 50$A, then $E_1 = 332$V.

Since $\alpha_2 = 135°$ then $V_2 = 382V$.
Bridge B cannot supply the motor since the thyristors are blocking.
A quick change of α_1 to 60° ($\alpha_2 = 120°$) gives: $V_1 = 269.8V = V_2$. Since the machine does not slow down instantly, E_1 does not change much and as long as $E_1 > V_2$ bridge B will operate as an inverter and the machine will brake and operate as a generator. E_1 cannot send current through bridge A since the thyristors are blocking. At the instant that $E_1 < V_1$ bridge A will operate as rectifier and the machine will operate as a motor.

Numeric example 19-3:

Given:

The configuration in fig. 19-26 whereby the firing angle α_1 of bridge A is 60° and α_2 of bridge B is 120°. The supply voltage is 3x230 V - 50 Hz.

Required:

Determine the instantaneous output voltages v_1 and v_2 of both bridges at the times corresponding with $\omega t = 135°$ and $\omega t = 165°$.

Solution:

The reference voltage is always v_{13}, it goes through zero at $t = 0$.

1. $\omega.t = 135°$
   Fig. 8-21: $v_1 = v_{13} = 230 \cdot \sqrt{2} \cdot \sin 135° = 230V$
   Fig. 8-25: $v_2 = v_{12} = 230 \cdot \sqrt{2} \cdot \sin (135° + 60°) = -84.18V$

2. $\omega.t = 165°$
   Fig. 8-21: $v_1 = v_{13} = 230 \cdot \sqrt{2} \cdot \sin 165° = 84{,}18V$
   Fig. 8-25: $v_2 = v_{12} = 230 \cdot \sqrt{2} \cdot \sin (165° + 60°) = -230V$

Circulating current

In reality the average output currents of both bridges can be equally large but the instantaneous values of the output voltages are different. This is clearly seen in numeric example 19-3.
The result of this is that equalizing or circulating current I_k will flow. To limit these circulating current, chokes are used. In fig. 19-26 the current $I_a + I_k$ will flow through L_1 and L_3. To avoid saturation large and expensive coils would need to be used. Small coils are used however which saturate for $I_a + I_K$ but not for I_k. The unsaturated coils L_2 and L_4 limit I_k. In the third quadrant L_1 and L_3 limit the circulating current while the other coils are saturated. The advantage of speed control with circulating current is that the current in the armature can change direction quickly. The advantage of this is good dynamic control behaviour.
If the bridge is only fired when the current almost goes to zero the circulating current will be limited: this is called a limited circulating current system. In this way the chokes can be even much smaller. Fig. 19-27 shows a control block diagram of a four quadrant drive with circulating current.

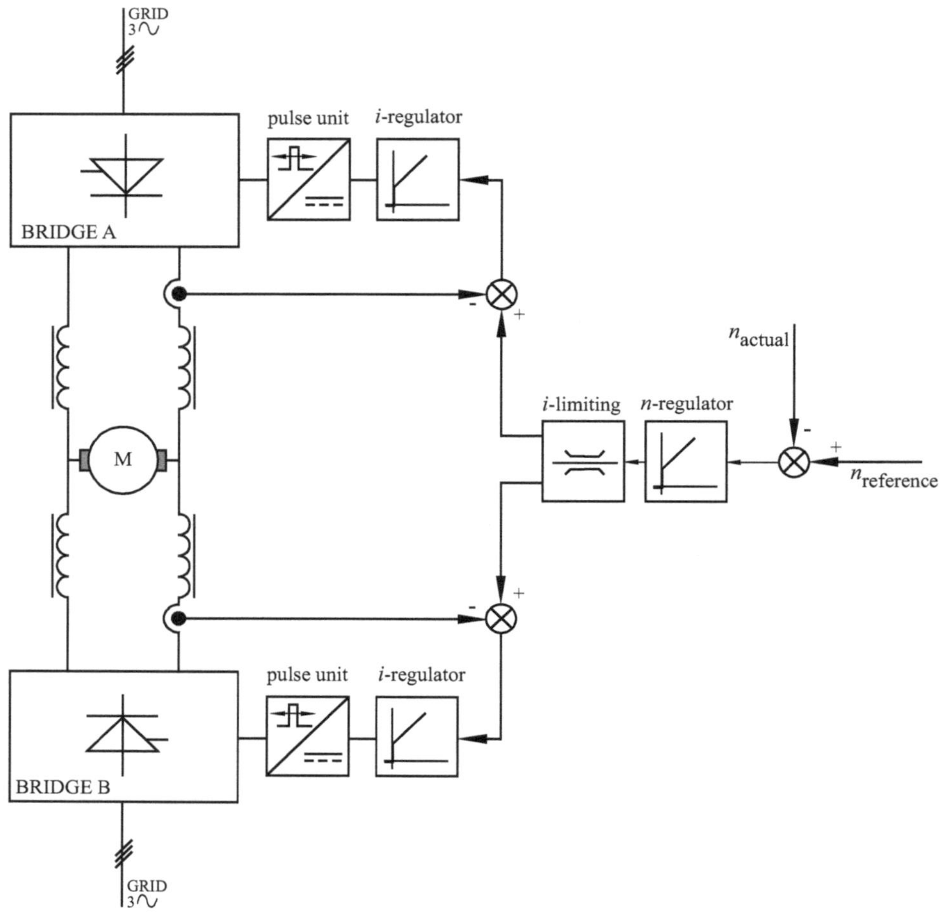

Fig. 19-27: Four quadrant drive with anti-parallel bridges and circulating current

Remark

Advantages of the non circulating current configuration:

- no expensive chokes required
- one pulse generator, one current controller and one speed controller are sufficient. In a circuit with circulating current two current controllers and two pulse generators are required.

Disadvantages of non circulating current:

- an additional circuit is required for the pulse generator to fire the combined bridge from the unique trigger circuit.
- there is a dead time with reversal compared to the circulating current circuit.

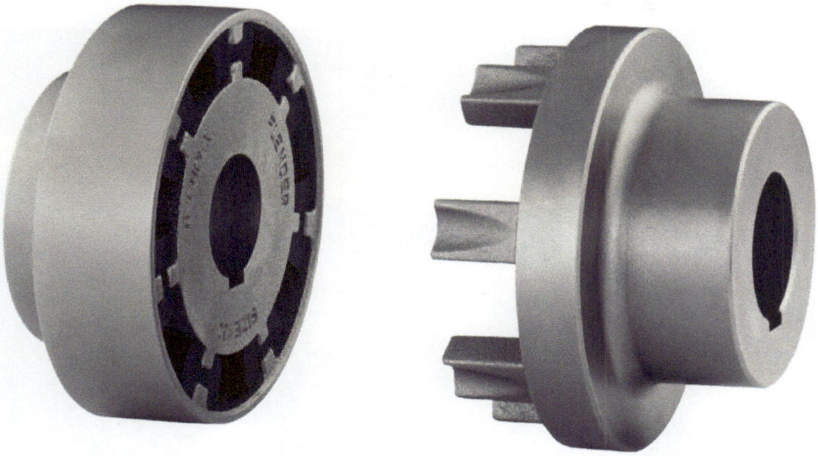

Photo Siemens: The Siemens N-EUPEX coupling is a universal coupling made of high-quality cast iron GG-25. The flexible elements are resistant to many media. The metal pins and the flexible elements are designed so that only minor wear occurs at permissible misalignment.
A distinction is made between the overload-keeping fail-safe series (N-EUPEX) and the overload disconnecting series without fail-safe device (N-EUPEX DS)

Photo Siemens: Typical application of a DC drive train is a hot rolling mill in a steel factory.
For further explanation about the DCM-series you are referred to p.19.28

Photo Siemens: Siemens modernised the paper machine at SCA packaging Containerboard GmbH in Aschaffenburg (North Bavaria) with a multi-motor drive system. The project goal was to increase production

4.6 Block diagram of digital DC drive train

The SINAMICS DCM-series from Siemens is a microprocessor controlled speed controller for DC motors from 6 kW to 25 MW (in parallel even up to 30 MW). With the Sinamics DC Master (DCM) the SINAMICS functions and tools such as SIZER and STARTER can also be used for DC applications. The SINAMICS DC master integrates the control (open loop and closed loop) with a supply unit in one device and is therefore very compact and space saving.
Applications of the DCM can be found in rolling mills, wire pulling machines, cable cars and lifts.
Speed response 40 ms. Torque response 6 to 9 ms. Protection category IP00 to IP20.
Energy recovery possible with certain configurations.
Communication RS232, RS485, PROFIBUS DP, PROFINET.
Fig. 19-28 shows a block diagram. A full controlled B_6 - bridge is responsible for the armature supply while a single phase SCR-bridge supplies the field control. With the aid of current transformers the armature current is measured, this is on the AC side of the converter. Transducers make digital signals of the armature voltage, AC supply voltage and field current.
The tachogenerator voltage is connected to the terminals XT1 (terminal 103 and 104).
The AC voltage connected to XP1 is rectified and serves as a supply for the internal electronics of the DCM. In emergency's the emergency stop will switch off the contacter.

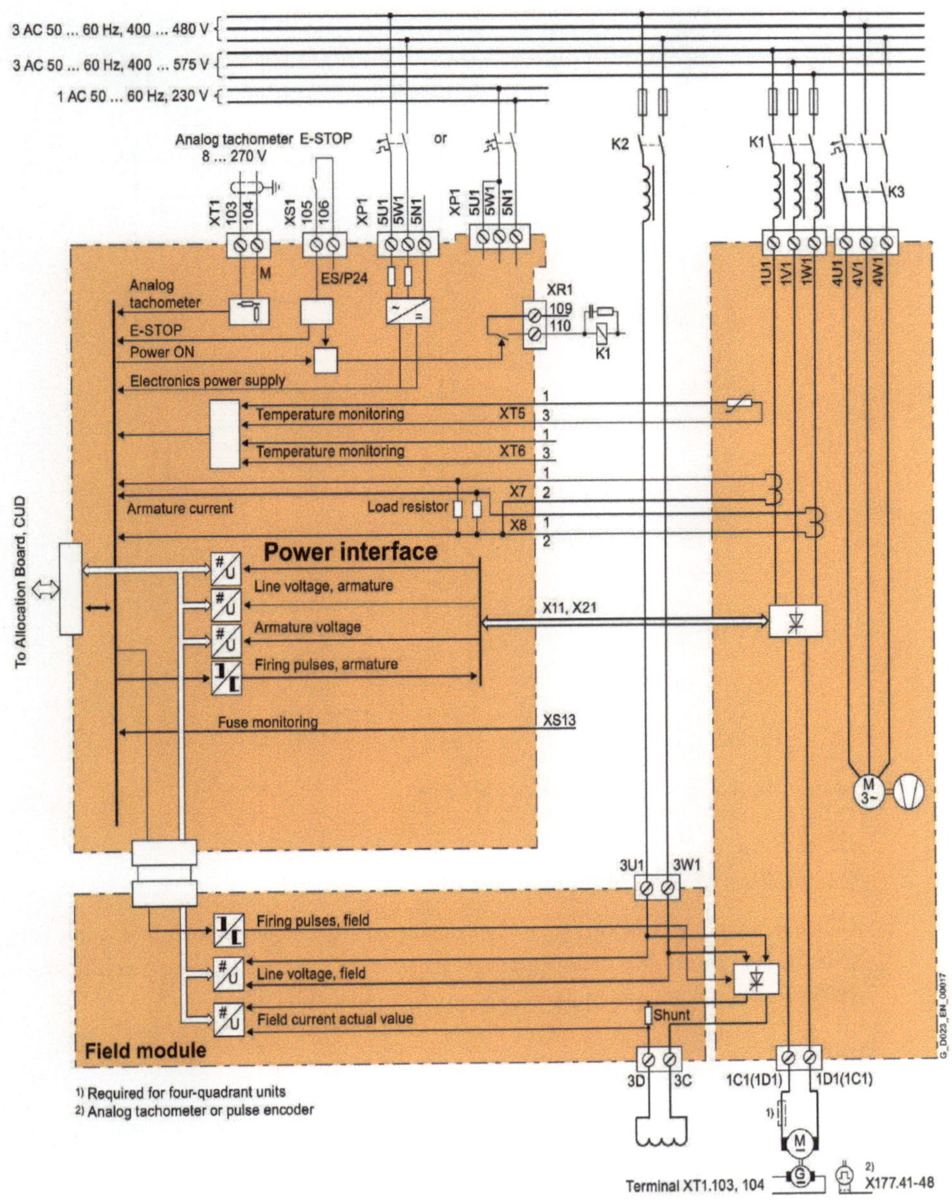

Fig. 19-28: Block diagram of a DCM-drive train from Siemens

5. FUNCTIONAL CONTROL DIAGRAM OF A SINGLE QUADRANT DRIVE

5.1 Control diagram

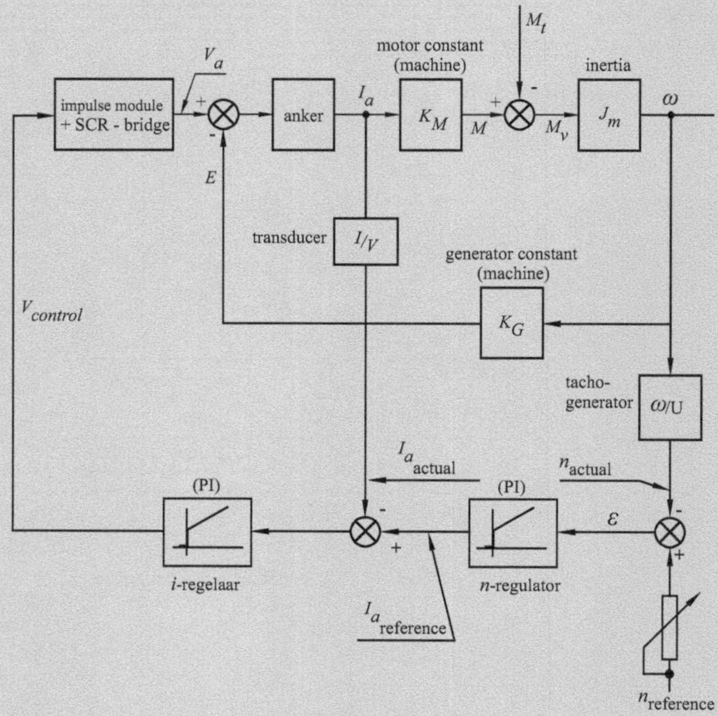

Fig. 19-29: Functional control diagram of an independently excited motor

5.2 Analysis of control block diagram

We examine the "blocks" in fig. 19-29 from the control perspective in other words in dynamic state. At the same time the measurements are provided from an existing speed controller of a 220V- 3.3kW – 3000 rpm motor.

5.2.1 Armature circuit

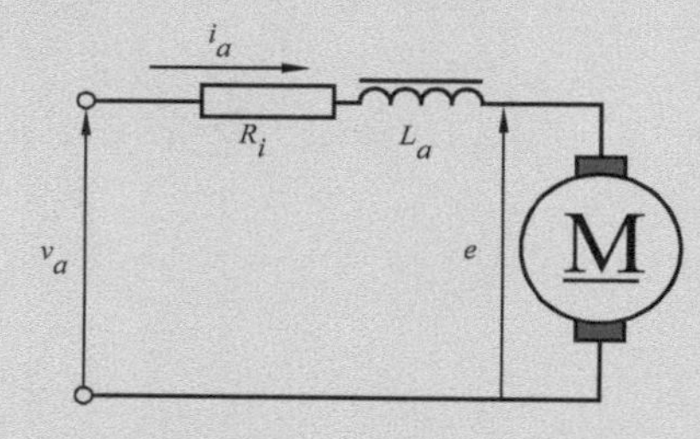

Fig. 19-30: Armature circuit

$$v_a = e + i_a \cdot R_i + L_a \cdot \frac{di_a}{dt}$$

$$V_a = E + I_a \cdot R_i + s \cdot L_a \cdot I_a$$

$$\frac{I_a}{V_a - E} = \frac{1}{R_i + s \cdot L_a} = \frac{1}{R_i} \cdot \frac{1}{1 + \frac{L_a}{R_i} \cdot s}$$

$$\frac{I_a}{V_a - E} = k_1 \cdot \frac{1}{1 + s \cdot \tau_a}$$

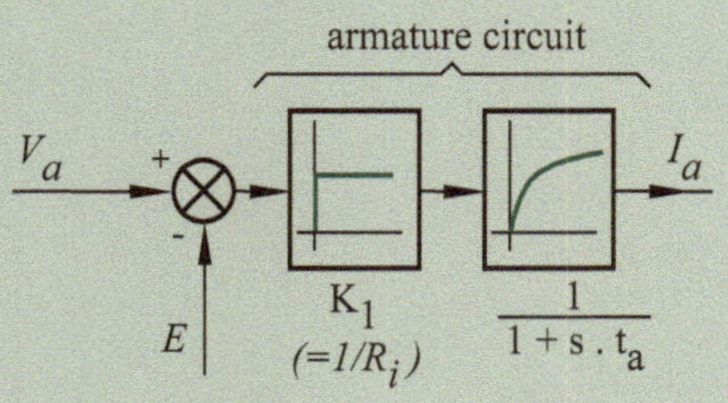

Measures results: $R_i = 1.12\Omega$

$L_a = 30\text{mH}$

From this: $K_1 = \frac{1}{R_i} = 0.893$

$\tau_a = \frac{L_a}{R_i} = 26.8\text{ms}$

Fig. 19-31: Armature circuit

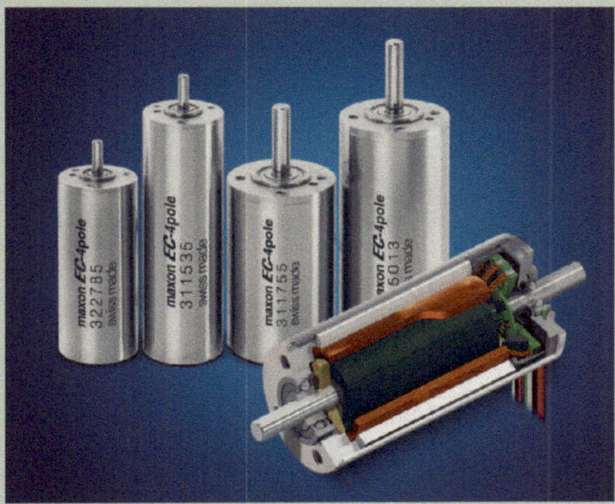

Photo Maxon Motor Benelux: EC-4-pole 30 Ø30mm: innovative, maximum performance from minimum dimensions.
EC (electronically commutated) motor from 4.5 to 48 V.
Type EC-powermax 30: 24 V / 100 W / stand still torque: 1.240 mNm / to 17.800 rpm.
Type EC-powermax 30: 48 V / 200 W / stand still torque: 3.430 mNm / to 16.500 rpm.

5.2.2 Machine constants

1. Motor constant K_M

$M \approx M_{em} = k_2 \cdot I_a \cdot \Phi$. With $K_M = k_2 \cdot \Phi$ is approximately:

$$M = K_M \cdot I_a \qquad (19\text{-}11)$$

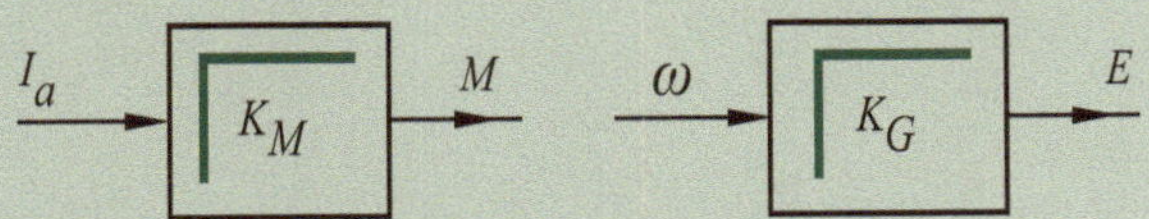

Fig. 19-32: Machine constants K_M and K_G

2. Generator constant K_G

$$E = k_1 \cdot n \cdot \Phi . \frac{\pi}{30} \cdot \frac{30}{\pi} = k_1 \cdot \Phi \cdot \omega \cdot \frac{30}{\pi}$$

With $K_G = k_1 \cdot \Phi \cdot \frac{30}{\pi} \left(= \frac{30}{\pi} \cdot \frac{E}{n} \right)$ becomes: $E = K_G . \omega$ (19-1)

Motor constant = generator constant: $K_M = K_G$

5.2.3 Inertia

- available torque on the shaft: M
- counter torque (mechanical load, friction in bearings, resistive torque of fan): M_t
- acceleration torque: $M_v = M - M_t$

If M_v is constant then the motor will accelerate linearly according to the following expression:

$$M_v = J_m \cdot \frac{d\omega}{dt}$$

$M - M_t = s \cdot \omega \cdot J_m$ so that: $\frac{\omega}{M - M_t} = \frac{1}{s \cdot J_m}$

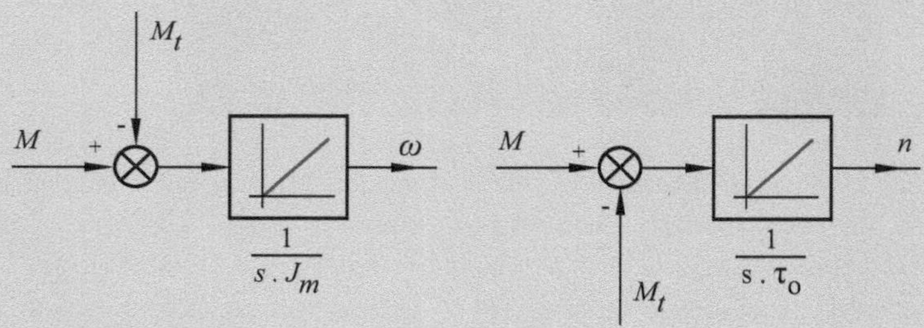

Fig. 19-33: Integrating action (mechanical inertia)

Remark

We can also consider the speed as output:

$$\frac{\omega}{M - M_t} = \frac{1}{s \cdot J_m} \longrightarrow \frac{\pi \cdot n}{30 \cdot (M - M_t)} = \frac{1}{s \cdot J_m}$$

$$\frac{n}{M - M_t} = \frac{1}{s \cdot \frac{J_m \cdot \pi}{30}} = \frac{1}{s \cdot \tau_0} \quad \text{with} \quad \tau_0 = \frac{J_m \cdot \pi}{30}$$

5.2.4 Control module + SCR bridge

We know from chapter 8 that the output voltage of a three- phase converter obeys a cosine law. The module that generates the pulse train (variable in time) for the gates of the thyristors is controlled by a standard voltage (0 to 10V). This 0 to 10 V is the output voltage from the current controller. At the output of the thyristor bridge a variable voltage V_a for the motor is produced. The voltage is variable between 0 and 220 V. If $V_{control}$ is the output voltage of the current controller, then for a certain configuration $K_2 = \left(\frac{\Delta V_a}{\Delta V_{control}}\right)_{max}$ can be determined.

We can calculate that $K_2 \approx 0.19\, x. V_{line}$. With a 3x230 V supply for the SCR bridge $K_2 \approx 43$ and with 3x400V $K_2 \approx 76$. We also know that the (static) average dead time of the three-phase SCR bridge ≈ 1.7ms (= t_d), so that we can draw fig. 19-34.

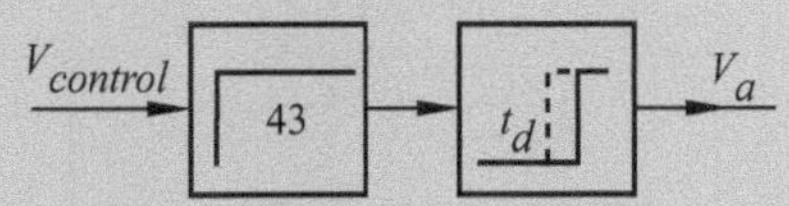

Fig. 19-34: Equivalent diagram of an SCR bridge and control module

5.2.5 Armature current-converter

The armature current is converted (galvanic isolation via opto-coupler) to a standard signal of 0 to 10 V. Assuming a maximum armature current of 25A then the constant of the transducer is : K_3 = 10V/25A = 0.4V/A.
Due to the non ideal rectification of the thyristor bridge the armature current has a ripple. (fig. 8-21). Since the current controller would not interpret this ripple as an error but as a load fluctuation, a filter with time constant t_{vi} (= 1 to1.2 ms) needs to be used, see fig. 19-39. This filter eliminates the 300 Hz current ripple of the SCR bridge.

5.2.6 Tachogenerator

The speed range is for example 0 to 3000 rpm and we want a standard signal of 0 -10 V.
The transfer constant of the *n/V* converter is then: $K_4 = \frac{10V}{3000\ rpm}$ = 0.0033 V/rpm = 0.2 V/rev/s.
If the tachogenerator produces for example 60 V/1000 rpm then a voltage divider is required between tachogenerator and controller (10 V/180V =1/18). The voltage divider has been included in the transfer constant K_4 .
As a result of the non ideal behaviour of the tachogenerator the output voltage contains a ripple. This can be produced by the collector lamella of the DC tacho. The speed controller can interpret this ripple as speed changes. To eliminate this error a filter with time constant t_{vn} (e.g. 0.5 ms) is included in the circuit, see fig. 19-46.

5.2.7 Controllers

Current controllers and speed controllers are generally of the PI-type.
For speed control via armature control of a DC-motor the following rules are often used to optimise the controller:

1. Amount optimum: used for the current controller
2. Symmetrical optimum: used for the speed controller.

6. OPTIMISING THE CONTROLLERS

6.1 Optimum amount

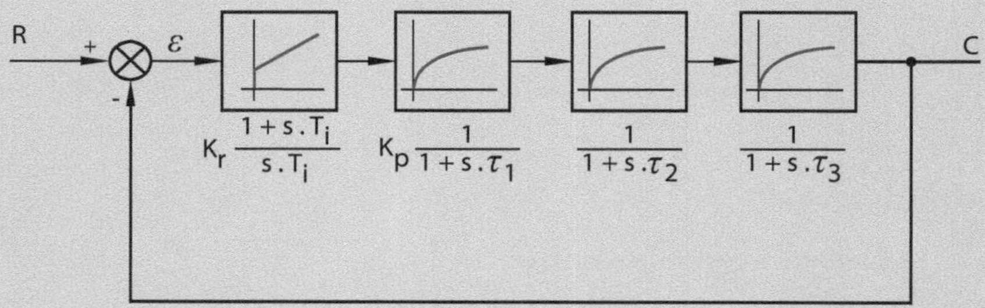

Fig. 19-35: PI-control of a cascade of first order systems

Consider a control circuit with a PI-controller and a process that is comprised of multiple first order systems. The gain (K_p) of the process can be concentrated in one of the first order systems. If the time constant of one of the combined systems τ_1 is much larger than the remaining time constants τ_2, τ_3, ... then without much error these time constants can be replaced by a first order system $\sigma = \tau_2 + \tau_3$ (fig. 19-36).

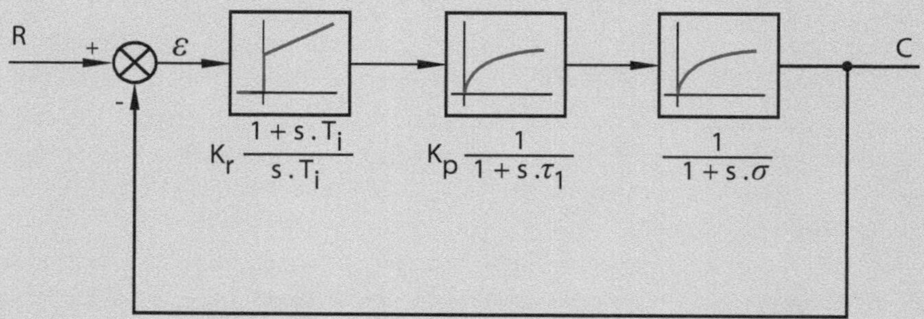

Fig. 19-36: Equivalent diagram of fig. 19-35

6.1.1 Determining the controller parameters

$$\frac{C(s)}{R(s)} = F(s) = \frac{G(s)}{1 + G(s)\,.\,H(s)} = \frac{K_r \,.\, \dfrac{1 + s\,.\,T_i}{s\,.\,T_i} \,.\, K_p \,.\, \dfrac{1}{1 + s.\,\tau_i} \,.\, \dfrac{1}{1 + s\,.\,\sigma}}{1 + K_r \,.\, \dfrac{1 + s\,.\,T_i}{s\,.\,T_i} \,.\, K_p \,.\, \dfrac{1}{1 + s\,.\,\tau_i} \,.\, \dfrac{1}{1 + s\,.\,\sigma}}$$

Setting the integration time (T_i) of the controller equal to the large time constant τ_I we find:

$$F(s) = \frac{K_r \,.\, K_p \,.\, \dfrac{1}{s\,.\,T_i\,.\,(1 + s\,.\,\sigma)}}{1 + K_r \,.\, K_p \,.\, \dfrac{1}{s\,.\,T_i\,.\,(1 + s\,.\,\sigma)}} = \frac{K_r \,.\, K_p}{K_r \,.\, K_p + s\,.\,T_i + s^2\,.\,T_i\,.\,\sigma}$$

so that: $F(s) = \dfrac{C(s)}{R(s)} = \dfrac{K_r \,.\, K_p}{s^2\,.\,T_i\,.\,\sigma + s\,.\,T_i + K_r\,.\,K_p}$

In the frequency domain this becomes:

$$F(j\omega) = \frac{K_r \,.\, K_p}{-\omega^2\,.\,T_i\,.\,\sigma + j\omega\,.\,T_i + K_r\,.\,K_p} \quad ; \quad |F(j\omega)| = \frac{K_r \cdot K_p}{\sqrt{\left[K_r \cdot K_p - \omega^2 \cdot T_i \cdot \sigma\right]^2 + \omega^2 \cdot T_i^{\,2}}}$$

Ideally $|F(j\omega)| = 1$. The **value** $|F(j\omega)| = 1$ is called in German "Der **Betrag**" which means "the amount". This explains the term amount optimum (as translation for the German "**Betragsoptimum**"). We find the amount optimum for:

$$(K_r\,.\,K_p)^2 = (K_r\,.\,K_p)^2 + \omega^4\,.\,T_i^2\,.\,\sigma^2 - 2\,.\,\omega^2\,.\,T_i\,.\,\sigma\,.\,K_r\,.\,K_p + \omega^2\,.\,T_i^2$$

or: $\omega^2\,.\,(\omega^2\,.\,T_i^2\,.\,\sigma^2 + T_i^2 - 2\,.\,T_i\,.\,\sigma\,.\,K_r\,.\,K_p) = 0$

This is the case when $\omega = 0$. If we want that $|F(j\omega)|$ remains close to unity for as many frequencies as possible we make the term

$T_i^2 - 2 \cdot T_i \cdot \sigma \cdot K_r \cdot K_p$ equal to zero $\longrightarrow$ $K_r = \dfrac{T_i}{2 \cdot K_p \cdot \sigma}$

For the settings of the PI-controller of fig. 19-35 we then use:

$$\boxed{T_i = \tau_i} \qquad (19\text{-}12)$$

$$\boxed{K_r = \frac{T_i}{2 \cdot K_p \cdot \sigma}} \qquad (19\text{-}13)$$

K_r	=	proportional gain of the controller (= ***r***egulator)
T_i	=	integration time constant
τ_i	=	large time constant of process
σ	=	sum of the smaller process time constants
K_p	=	gain factor of the process to be controlled

6.1.2 Transfer function optimised control loop

$K_r = \dfrac{T_i}{2 . K_p . \sigma}$ gives:

$$F(s) = \frac{K_r . K_p}{s^2 . T_i . \sigma + s . T_i + K_r . K_p} = \frac{\frac{T_i}{2 . \sigma}}{s^2 . T_i . \sigma + s. T_i + \frac{T_i}{2 . \sigma}} = \frac{1}{2 . \sigma^2 . s^2 + 2 . \sigma . s + 1}$$

In the frequency domain: $F(j\omega) = \dfrac{1}{-2 . \sigma^2 . \omega^2 + 2 . \sigma j\omega + 1}$

$$| F(j\omega) | = \frac{1}{\sqrt{(1 - 2 . \sigma^2 . \omega^2)^2 + 4 . \sigma^2 . \omega^2}}$$. When $\omega . \sigma << 1 \rightarrow\rightarrow | F(j\omega) | \approx 1$

With $\omega . \sigma << 1 \rightarrow\rightarrow F(j\omega) = \dfrac{1}{2 \cdot \sigma \cdot j \cdot \omega + 1}$; $F(s) = \dfrac{1}{1 + 2 \cdot \sigma \cdot s}$

The optimised control circuit behaves as a first order system with equivalent time constant:

$$\tau_{eq.} = 2 \cdot \sigma \qquad (19\text{-}14)$$

6.1.3 Step response of optimized control circuit

$$F(s) = \frac{C(s)}{R(s)} = \frac{1}{2 . \sigma^2 . s^2 + 2 . \sigma . s + 1} = \frac{\frac{1}{2 . \sigma^2}}{s^2 + \frac{1}{\sigma} .s + \frac{1}{2 . \sigma^2}} \left[= \frac{\omega_0^2}{s^2 + 2\, \beta\, \omega_0\, s + \omega_0^2} \right]$$

$\omega_0 = \dfrac{1}{\sigma \sqrt{2}}$; $2 . \beta . \omega_0 = \dfrac{1}{\sigma} \rightarrow\rightarrow \beta = \dfrac{1}{\sqrt{2}}$

The step response of a second order system is given by:

$$y(t) = R . \left[1 - \frac{e^{-\beta \omega_0 t}}{\sqrt{1 - \beta^2}} \cdot \cos\left(\sqrt{1 - \beta^2} \cdot \omega_0 t - \Phi \right) \right]$$

With: $\tan \Phi = \dfrac{\beta}{\sqrt{1 - \beta^2}} = 1$ and overshoot : $D = R \cdot e^{-\frac{\pi \cdot \beta}{\sqrt{1 - \beta^2}}} \cdot 100\ \%$

The step response is drawn in fig. 19-37. We can now determine t_1/σ , t_2/σ and D . Each $y(t) = R$ we find an intersection of the step response and the input step $r(t)$.

t_1/σ ? $\rightarrow\rightarrow \dfrac{e^{-\beta \omega_0 t}}{\sqrt{1 - \beta^2}} \cdot \cos\left(\sqrt{1 - \beta^2} \cdot \omega_0 t - \Phi \right) = 0$

$\sqrt{2} \cdot e^{-t_1/2\sigma} \cdot \cos\left(\dfrac{t_1}{2\sigma} - \dfrac{\pi}{4} \right) = 0$; $\dfrac{t_1}{2\sigma} - \dfrac{\pi}{4} = \dfrac{\pi}{2} \rightarrow\rightarrow \dfrac{t_1}{\sigma} = 4.71$

t_2/σ ? $\rightarrow\rightarrow \dfrac{t_2}{2\,\sigma} - \dfrac{\pi}{4} = \dfrac{3 \cdot \pi}{2}$; $\rightarrow\rightarrow \dfrac{t_2}{\sigma} = 10.99$

D ? $\rightarrow\rightarrow D = R \cdot e^{-\pi \frac{\beta}{\sqrt{1 - \beta^2}}} \cdot 100\ (\%)$; with $\beta = \dfrac{1}{\sqrt{2}} \rightarrow\rightarrow D = 4.33 \cdot \% \cdot R$

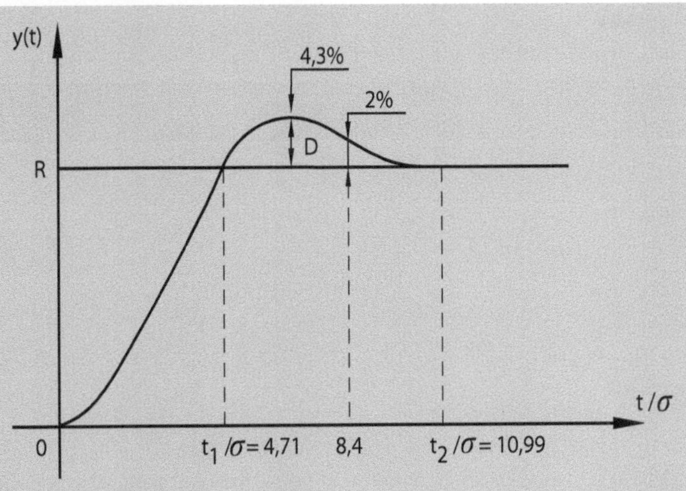

Fig. 19-37: Step response second order system

6.1.4 Amount optimum applied to armature control of DC-motor

In fig. 19-38 we redraw the internal current control loop of fig. 19-11.

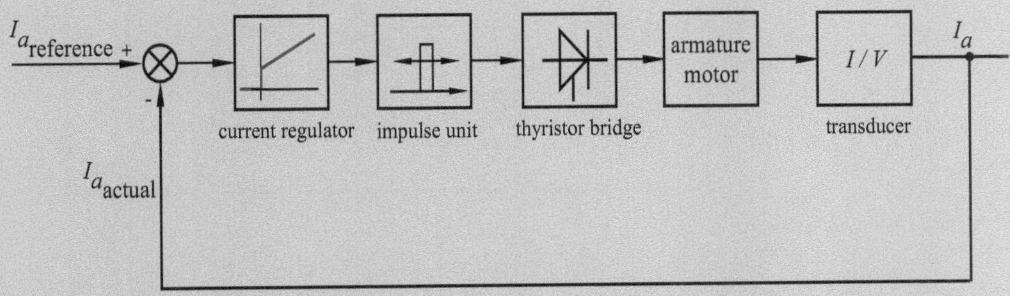

Fig. 19-38: Internal current controller for an independently excited motor

For every block we can now draw the equivalent diagram in fig. 19-39.

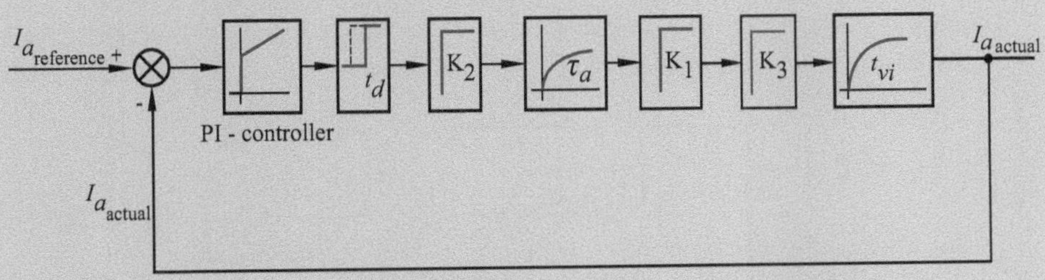

Fig. 19-39: Equivalent diagram of fig. 19-38

Table 19-1

part of process	time constant	gain
controlled rectifier	t_d = dead time	K_2
armature	τ_a	K_1
transducer (current)	t_{vi} = delay due to current smoothing filter (e.g. 1.1ms)	K_3

In practice:
- τ_a = largest time constant
- $\sigma = t_d + t_{vi}$ = sum of the smaller time constants
- For applications with an amount optimum the large time constant should be $> 4.\sigma$ (guideline: $\tau_a \geq 20 . \sigma$)

If we assume: $K_1 \cdot K_2 \cdot K_3 = K_p$

Fig. 19-39 can be approximated in fig. 19-40, from which we recognise the configuration of fig. 19-36.

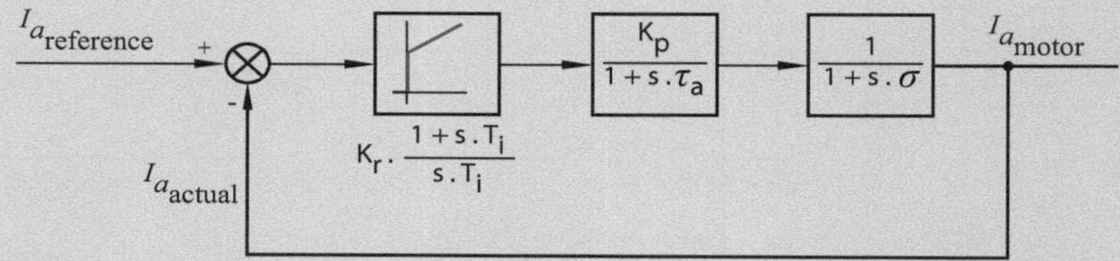

Fig. 19-40: Equivalent diagram for fig. 19-39

6.2 Symmetrical optimum

6.2.1 Integrator in the process

If in addition to a first order process an integrator is also present in the process then a controller setting using the amount optimum will not suffice.

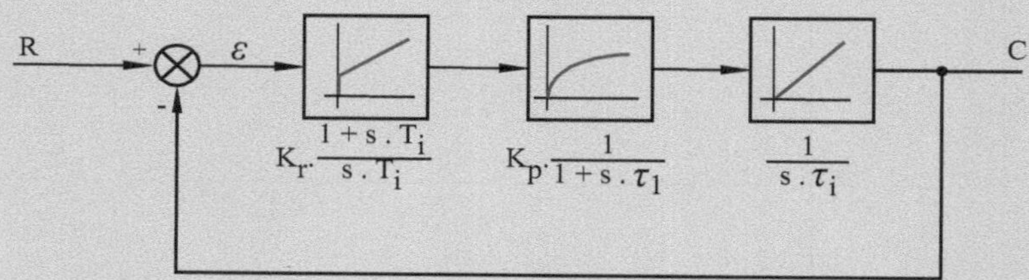

Fig. 19-41: Process with integrator

$$F(s) = \frac{C(s)}{R(s)} = \frac{G(s)}{1 + G(s) \cdot H(s)} = \frac{\frac{K_r \cdot K_p (1 + s \cdot T_i)}{s \cdot T_i} \cdot \frac{1}{(1 + s \cdot \tau_1)} \cdot \frac{1}{s \cdot \tau_i}}{1 + \frac{K_r K_p}{s \cdot T_i} \cdot \frac{1 + s \cdot T_i}{1 + s \cdot \tau_1} \cdot \frac{1}{s \cdot \tau_i}}$$

If we have to optimise this to the amount then we set $T_i = \tau_1$, so that:

$$F(s) = \frac{C(s)}{R(s)} = \frac{\frac{K_r \cdot K_p}{s^2 \cdot T_i \cdot \tau_i}}{1 + \frac{K_r \cdot K_p}{s^2 \cdot T_i \cdot \tau_i}} = \frac{1}{1 + \frac{T_i \cdot \tau_i}{K_r \cdot K_p} \cdot s^2}$$

This corresponds to integrating control of an integrating process. We know that this is unstable. Indeed in the equation of F(s) we see the expression for a second order system with:

- natural resonance $\omega_0 = \sqrt{\frac{K_r \cdot K_p}{T_i \cdot \tau_i}}$
- damping: $\beta = 0$

With a step $R(s) = \frac{R}{s}$ there is a corresponding response : $C(s) = \frac{R}{s} \cdot \frac{1}{1 + \frac{s^2}{\omega_0^2}} = \frac{R}{s} \cdot \frac{\omega_0^2}{s^2 + \omega_0^2}$

Splitting into partial fractions gives: $C(s) = \frac{R}{s} - \frac{R \cdot s}{s^2 + \omega_0^2}$

Inverse Laplace transformation produces for the time function:

$c(t) = R - R \cdot \cos \omega_0 \cdot t = R \cdot (1 - \cos \omega_0 \cdot t)$. This is drawn in fig. 19-42.

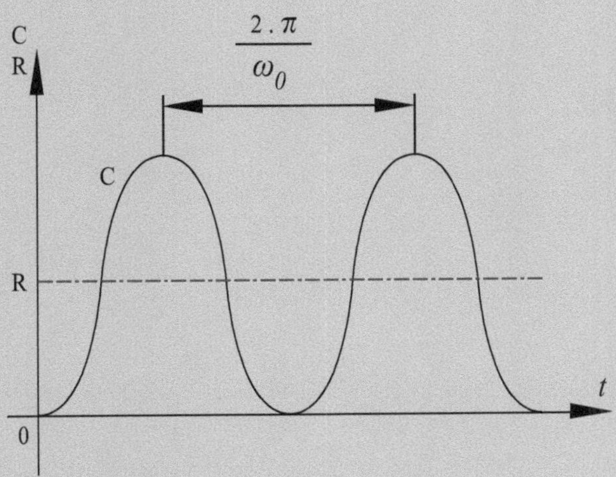

Fig. 19-42: Response of an integrating system

6.2.2 Symmetrical optimum

We assume a control loop comprised of an integrator and multiple first order systems with small time constants ($\tau_1 + \tau_2 + ... = \varepsilon$).

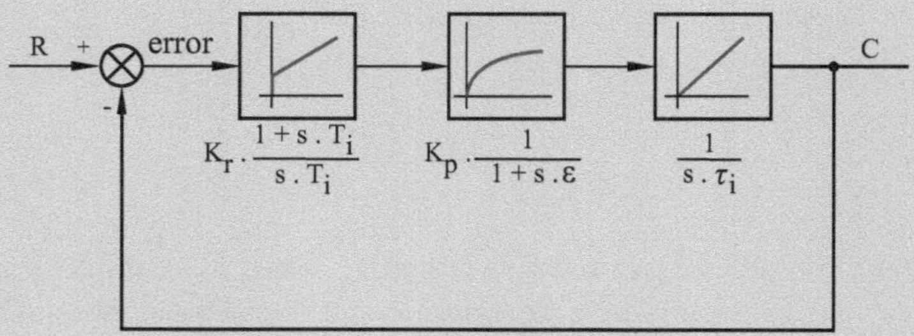

Fig. 19-43: Control loop with PI-controller and a process with an integrator

$$F(s)=\frac{C(s)}{R(s)}=\frac{K_r K_p \cdot \frac{1+s\,T_i}{s\,T_i}\cdot\frac{1}{1+s.\varepsilon}\cdot\frac{1}{s.\tau_i}}{1+K_r K_p \cdot \frac{1+s.T_i}{s.T_i}\cdot\frac{1}{1+s.\varepsilon}\cdot\frac{1}{s.\tau_i}}=\frac{K_r K_p.(1+s.T_i)}{s^3\,T_i\tau_i\varepsilon+s^2\,T_i\tau_i+K_rK_p\,T_i.\,s+K_rK_p}$$

If:

$$\begin{aligned} T_i\tau_i\varepsilon &= T_1\\ T_i\tau_i &= T_2\\ K_rK_p &= K_1\\ K_rK_pT_i &= K_2\end{aligned}$$

then: $$F(s)=\frac{K_1+s.K_2}{s^3\,T_1+s^2\,T_2+s\,K_2+K_1}$$

In the frequency domain:

$$F(j\omega)=\frac{K_1+j\omega K_2}{-j\omega^3\,T_1-\omega^2\,T_2+j\omega K_2+K_1}=\frac{K_1+j\omega K_2}{j\omega\,(K_2-\omega^2\,T_1)+(K_1-\omega^2\,T_2)}$$

Amplitude: $$|F(j\omega)|=\sqrt{\frac{K_1^2+\omega^2\,K_2^2}{K_1^2-\omega^2\,(2.K_1.T_2-K_2^2)-\omega^4\,(2.T_1.K_2-T_2^2)+\omega^6.\,T_1^2}}$$

$|F(j\omega)|=1$ for $\omega=0$

If we want $|F(j\omega)|$ to remain close to unity for as many frequencies as possible, then in the denominator we can:

- $2.K_1\cdot T_2-K_2^2=0 \quad \rightarrow\rightarrow \quad 2\cdot K_rK_p\cdot T_i\cdot\tau_i-(K_rK_p\cdot T_i)^2=0$ (1)
- $2.T_1\cdot K_2-T_2^2=0 \quad \rightarrow\rightarrow \quad 2\cdot T_i\cdot\tau_i\cdot\varepsilon\cdot K_rK_p\cdot T_i-(T_i\cdot\tau_i)^2=0$ (2)

From (1): $T_i=\dfrac{2\cdot\tau_i}{K_rK_p}$ (3)

From (2): $K_r=\dfrac{\tau_i}{2\cdot K_p\cdot\varepsilon}$ so that: $T_i=\dfrac{2\tau_i}{K_rK_p}=\dfrac{2\cdot\tau_i\cdot 2\cdot K_p\cdot\varepsilon}{\tau_i\cdot K_p}=4\,\varepsilon$

Setting of the PI-controller: $$T_i = 4 \cdot \varepsilon \qquad (19\text{-}15)$$

$$K_r = \frac{\tau_i}{2 K_p \varepsilon} \qquad (19\text{-}16)$$

With this optimization it becomes:

$$F(s) = \frac{K_r K_p (1 + s \cdot T_i)}{s^3 T_i \tau_i \varepsilon + s^2 T_i \tau_i + K_r K_p T_i \cdot s + K_r \cdot K_p} = \frac{1 + 4 \cdot s \cdot \varepsilon}{s^3 8 \varepsilon^3 + s^2 8 \varepsilon^2 + 4 s \varepsilon + 1}$$

In the frequency domain we have:

$$F(j\omega) = \frac{1 + j4\varepsilon\omega}{-j\omega^3 8\varepsilon^3 - \omega^2 8 \varepsilon^2 + j\omega 4 \varepsilon + 1}$$

Calculation for the amplitude leads to: $|F(j\omega)| = \sqrt{\dfrac{1 + 16(\varepsilon\omega)^2}{1 + 64(\varepsilon\omega)^6}}$

6.2.3 Step response

$$F(s) = \frac{C(s)}{R(s)} = \frac{1 + 4 s \varepsilon}{s^3 8 \varepsilon^3 + s^2 8 \varepsilon^2 + s 4 \varepsilon + 1}$$

With $R(s) = \frac{R}{s}$ we find: $C(s) = \frac{R}{s} \cdot \dfrac{1 + 4 s \varepsilon}{s^3 8 \varepsilon^3 + s^2 8 \varepsilon^2 + s 4 \varepsilon + 1}$

If $2 \varepsilon s = x \quad \rightarrow\rightarrow \quad C(s) = \dfrac{R \cdot 2 \cdot \varepsilon}{x} \cdot \dfrac{1 + 2x}{x^3 + 2x^2 + 2x + 1}$;

$$x^3 + 2x^2 + 2x + 1 = (x + 1)(x^2 + x + 1)$$

$$\frac{1}{x} \frac{1 + 2x}{x^3 + 2x^2 + 2x + 1} = \frac{1 + 2x}{x \cdot (x + 1)(x^2 + x + 1)} = \frac{a}{x} + \frac{b}{x + 1} + \frac{c x + d}{x^2 + x + 1}$$

Solution: $a = 1$; $b = 1$; $c = -2$; $d = -1 \rightarrow\rightarrow \frac{1}{x} \cdot \dfrac{1 + 2x}{x^2 + x + 1} = \frac{1}{x} + \frac{1}{x+1} - \dfrac{2x + 1}{x^2 + x + 1}$

So that: $C(s) = 2 \varepsilon R \left[\dfrac{1}{2 \varepsilon s} + \dfrac{1}{1 + 2 \varepsilon s} - \dfrac{4 \varepsilon s + 1}{4 \varepsilon^2 s^2 + 2 \varepsilon s + 1} \right]$

$$C(s) = 2 \varepsilon R \left[\frac{1}{2 \varepsilon s} + \frac{1}{1 + 2 \varepsilon s} - \frac{2(2 \varepsilon s + \frac{1}{2})}{(2 \varepsilon s + \frac{1}{2})^2 + \frac{3}{4}} \right]$$

$$= R \cdot \left[\frac{1}{s} + \frac{1}{s + \frac{1}{2\varepsilon}} - 2 \frac{s + \frac{1}{4\varepsilon}}{(s + \frac{1}{4\varepsilon})^2 + \frac{3}{16 \cdot \varepsilon^2}} \right]$$

$$\mathscr{L}^{-1} \rightarrow\rightarrow c(t) = R \cdot \left[1 + e^{\frac{-t}{2\varepsilon}} - 2 \cdot e^{\frac{-t}{4\varepsilon}} \cdot \cos \frac{\sqrt{3}\, t}{4 \varepsilon} \right]$$

The step response is drawn in fig. 19-44.

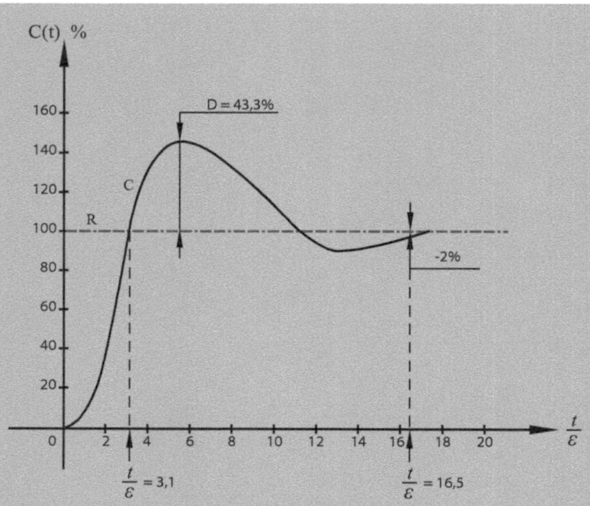

Fig. 19-44: Step response in the case of symmetrical optimum

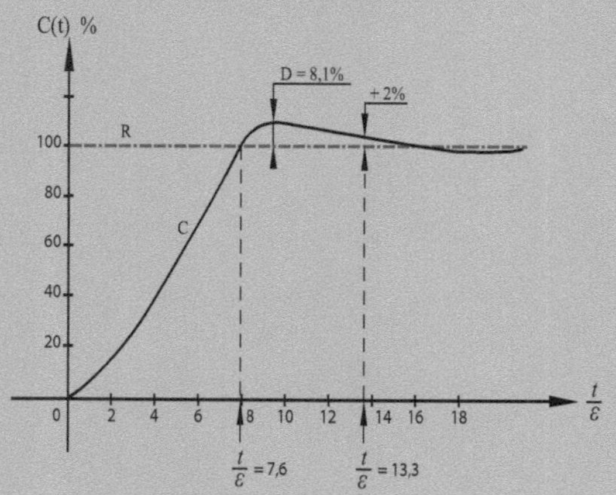

Fig. 19-45: Step response with improved symmetrical optimum

The overshoot of 43% is simply too high.
To improve this a delay network can be included with the transfer function: $\frac{1}{4\varepsilon s+1}$ so that:

$$C(s) = \frac{R}{s} \cdot \frac{1}{s^3 8\varepsilon^3 + s^2 8\varepsilon^2 + s 4\varepsilon + 1}.$$

The step response is now drawn in fig. 19-45.

Remarks

1. If we draw the Bode diagram of the open control loop in which the controller is set-up according to the rules for the symmetrical optimum we see that the amplitude response is symmetrical with respect to the 0 dB axis, which explains the name symmetrical optimum
2. $\boxed{\varepsilon = \tau_{eq} + t_{vn}} = 2 \cdot \sigma + t_{vn}$ with $t_{vn} \approx 0.5\text{ms}$ (19-17)

6.2.4 Symmetrical optimum applied to speed control of a DC-motor

In fig. 19-46 the speed control circuit is drawn. The internal current controller is now replaced by an equivalent time constant. This internal current controller is set-up according to the amount optimum.

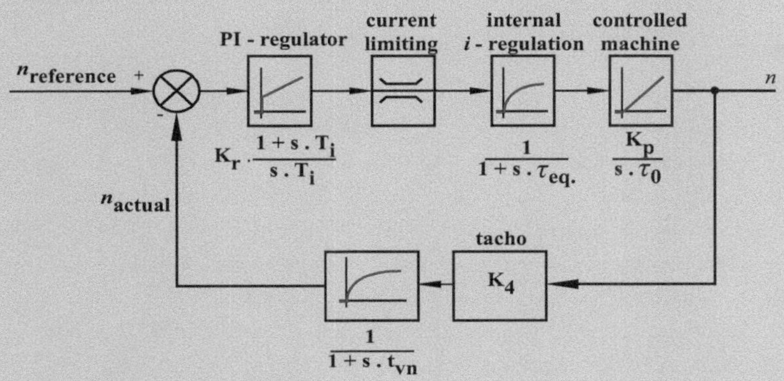

Fig. 19-46: Equivalent circuit speed control

6.3 Commissioning rules

Table 19-2:

controller	set up method	K_r	T_i	$t_{100\%}$	t_{out}	D %	τ_{eq}
armature control	A.O.	$\frac{\tau_a}{2 \cdot K_p \cdot \sigma}$	τ_a	4.7 σ	8.4 σ	4.3	2 σ
speed control	S.O.	$\frac{\tau_0}{2 \cdot \varepsilon \cdot K_p}$	4 ε	3.1 ε 7.6 ε	16.5 ε 13.3 ε	43.3 8.1	4 ε —

Remark

Digital controller: These days the controllers are digital as a result of which a number of small time delays are included in the controller. We find additional delays as a result of:

- sample time (sample period) of the speed and current measurements
- calculation of the desired armature current in the speed controller
- delay in the current controller.

Due to these additional delays the optimum settings of a digital controller will only differ slightly from what we have concluded in the previous pages.

7. NUMERIC EXAMPLE 19-5

Given:

DC-motor 220 V; 3.3kW; 3000 rpm; $R_i = 1.12\Omega$; $L_a = 30\text{mH}$.
$J_m = 0.07\ \text{kgm}^2$ for the entire drive (run down test).
Tachogenerator: 60V/1000 rpm. Armature current converter: 0.4V/A

Question:

1. In practice how can you determine the time constant of the armature?
2. For this motor we measure at $n = 1440$ rpm that $V_a = 134\text{V}$ and $I_a = 13.3\text{A}$. Determine the motor and generator constants. What is the moment of the work torque?
3. Determine the parameters (K_r and T_i) of the PI-controller in fig. 19-40, taking account of the measurements mentioned above and including the data of the SCR control module (fig. 19-34). Time constant of the smoothing filter: $t_{vi} = 1.1\text{ms}$
4. Use the result up to now to determine the parameters of the speed controller in fig. 19-46. Time constant of the smoothing filter of the n-controller: $t_{vn} = 0.36\text{ms}$.

Solution:

1. Set a step voltage on the armature circuit of the unexcited motor and save the exponential current response with a scope. We can determine τ by drawing the tangent to this exponential. For a permanent magnet motor the rotor should be locked during this measurement....

2. $E = 134 - 1.12 \cdot 13.3 = 119.1\text{V}$; $K_G = \frac{E}{\omega} = \frac{119.1}{\frac{\pi \cdot 1440}{30}} = 0.79\ \text{Vs/rad}$

   $K_M = (K_G) = 0.79\ \text{Nm/A}$; $M = K_M \cdot I_a = 0.79 \cdot 13.3 = 10.5\ \text{Nm}$

3. $K_1 = \frac{1}{R_i} = 0.89\ \text{A/V}$; $K_2 = 43$ (fig. 19-34!) ; $K_3 = 0.4\ \text{V/A}$; $K_p = K_1 \cdot K_2 \cdot K_3 = 15.3$

   $\tau_A = \frac{L_a}{R_i} = \frac{30 \cdot 10^{-3}}{1.12} = 26.8\ \text{ms}$; $\sigma = t_d + t_v = 1.7 + 1.1 = 2.8\ \text{ms}$ (<<< 26.8 ms)

   amount optimum controller: $T_i = \tau_a = 26.8\text{ms}$; $K_r = \frac{T_i}{2 \cdot K_p \cdot \sigma} = 0.31$

4.
   - Optimized current controller: $\tau_{eq} = 2 \cdot \sigma = 5.6\ \text{ms}$
   - Sum of smaller time constants speed controller: : $\varepsilon = \tau_{eq} + t_{vn} = 5.96\ \text{ms}$
   - Equivalent integrator time constant: $\tau_0 = \frac{J_m \cdot \pi}{30} = \frac{0.07 \cdot \pi}{30} = 7.33\ \text{ms}$
   - Gain factors:
     - motor: $K_M = 0.79\ \text{Nm/A}$
     - tachogenerator: $K_4 = 0.2\ \text{V/rpm}$
     - process: $K_p = 0.79 \cdot 0.2 = 0.158$
   - Symmetrical optimum speed controller:
     - $T_i = 4 \cdot \varepsilon = 23.84\ \text{ms}$
     - $K_r = \frac{\tau_0}{2 \cdot K_p \cdot \varepsilon} = \frac{7.33 \cdot 10^{-3}}{2 \cdot 0.158 \cdot 5.96 \cdot 10^{-3}} = 3.89$

Photo Siemens: The photo shows a paper rolling machine. Such installations often operate with linear speeds of between 30 and 300 rpm, with paper drums of 2m and bobbins that often weigh 1000 kg or more
This picture serves also as cover photo from Siemens

B. DC-MOTOR supplied from a DC SOURCE

8. CHOPPER CONTROLLED DRIVE

In the past traction applications used ***open loop*** chopper ***control*** of DC series motors. In new traction projects only the asynchronous motor is used.
With smaller power levels and especially servo applications ***closed loop*** chopper ***control*** can be encountered. The typical motors are then permanent magnet motors (PM-motor).
Also independently excited machines may be used. This type of machine has just been dealt with so that we limit our study here to the PM-motor.

8.1 Permanent magnet motor

With small DC-motors the magnetic field is usually produced by ceramic permanent magnets.
It is clear that the characteristics of such motors are very similar to DC-motors with field windings.
A PM-motor does have better properties:

1. practically linear *M-n* curves (fig. 19-47). The armature reaction has less influence on the flux than in the case of wound poles since:
   a) the permeability of ceramic material is extremely low (almost the same as air)
   b) the coercitive force of PM-material opposes change as a result of field and armature action.

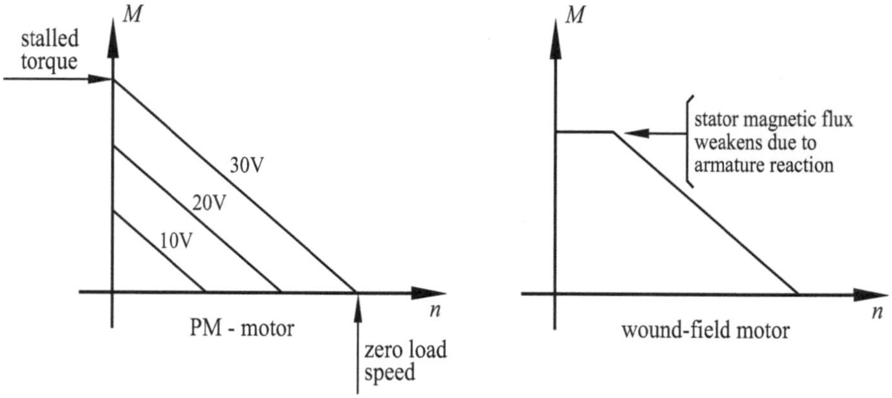

Fig. 19-47: *M-n* curves PM-motor

2. small dimensions of the PM-motor.
   As a result of the high coercitive force of permanent magnets the radial dimensions of the poles are much smaller than for wound pole pieces which results in a PM-motor being smaller and lighter than a classic DC-motor of equivalent power
3. no power loss in the field winding
4. high starting torque.

Remark: *M-n* curves (fig. 19-47)

For a DC-motor we can write: $E = k_1 \cdot n \cdot \Phi$ and $M = k_2 \cdot I_a \cdot \Phi$;

$$I_a = \frac{V_a - E}{R_i} = \frac{V_a - k_1 \cdot n \cdot \Phi}{R_i} \text{ ; so that: } M = k_2 \cdot \Phi \left[\frac{V_a - k_1 \cdot n \cdot \Phi}{R_i} \right]$$

For a PM-motor Φ is constant and therefore: $M = k_3 \cdot V_a - k_4 \cdot n$

1. Torque in blocked state:
   With $n = 0$ is $M = k_3 \cdot V_a$ = torque in blocked state (stalled torque)
2. No load speed: at $M = 0$ is $n = \frac{k_3}{k_4} \cdot V_a$ = no load speed
3. The *M-n* curves:
   With a constant supply voltage V_a and increasing speed *n*, torque *M* decreases linearly.

8.2 Single quadrant drive

The thyristor bridge and the pulse generator of fig. 19-11 are now replaced by a chopper and its controller. The chopper is PW-modulated and since the current is controlled this is called a CR-PWM chopper. PW = pulse width; CR-PWM = current regulated pulse width modulation.
For the optimum configuration of the controllers see the study under number 6.

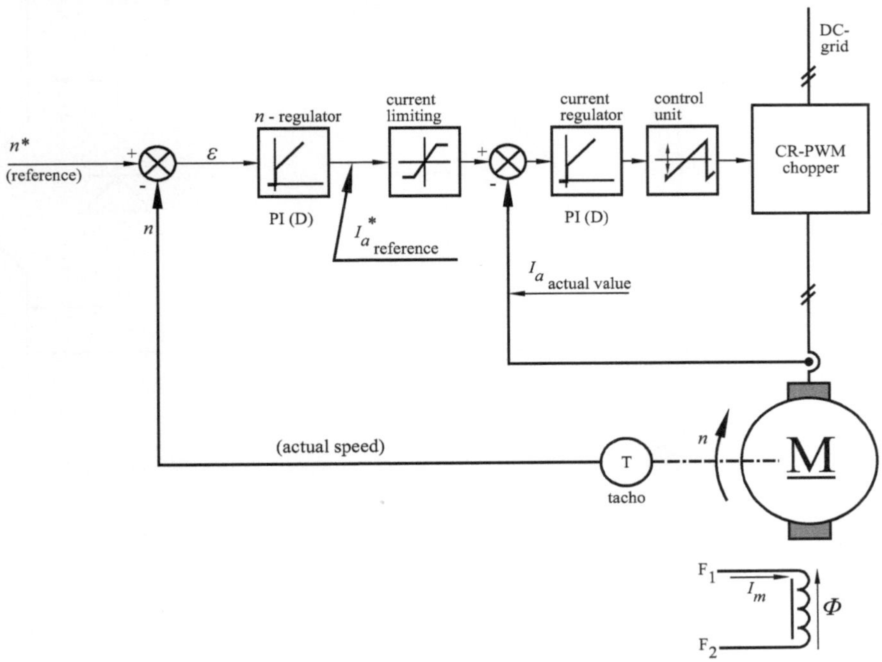

Fig. 19-48: Controlled single quadrant drive

8.3 Two quadrant operation

Fig. 19-49 shows a H-bridge. This is often used in a power stage for the control of servo-motors and stepper motors. The H-bridge can also be used as a linear servo amplifier but for reasons of efficiency this will only be implemented for very small motor powers.

Mostly the transistors are operated as switches and by controlling the ratio a PWM servo amplifier results. The (transistor) switches work in pairs: T_1T_4 and T_2T_3.
If T_1T_4 are closed and T_2T_3 are open, then the armature current flows to the right. The motor turns clockwise for example. With T_2T_3 closed and T_1T_4 open the motor will turn counter clockwise. The bridge can operate in ***drive mode*** in two directions.

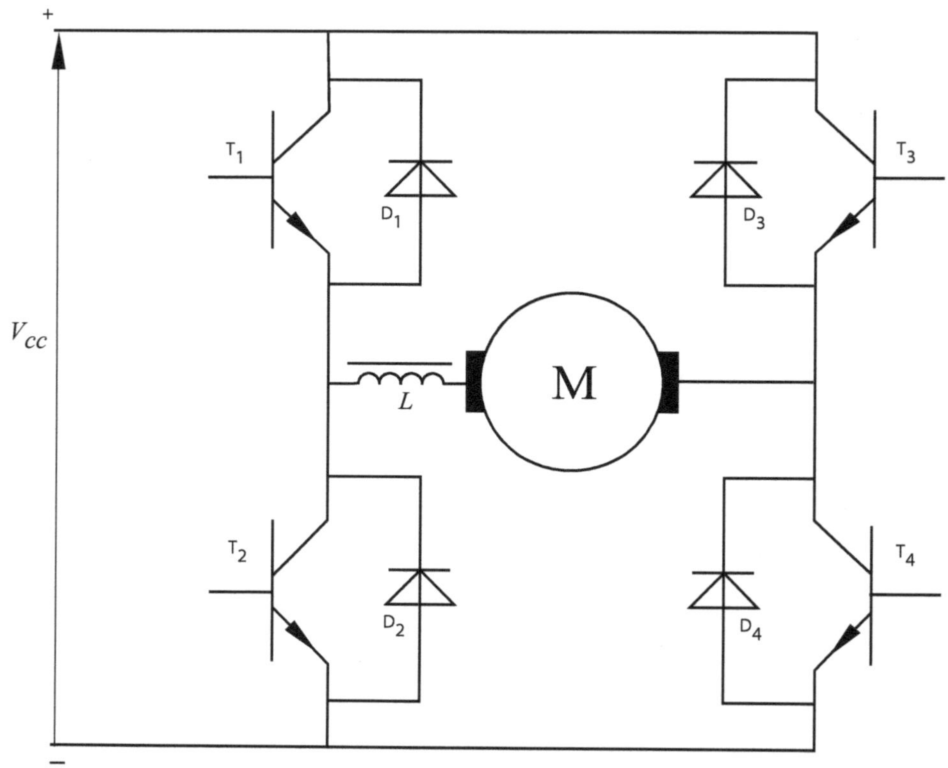

Fig. 19-49: H-bridge control

In principle the H-bridge has two operating modes which could be called the unipolar and bipolar PWM. Fig. 19-50 shows a possible waveform for unipolar PWM.
The motor terminal voltage varies during one working cycle between 0 and V (0 to $+V$ or 0 to $-V$).
Use is made of two switches: T_1T_4 **OR** T_2T_3.
In the bipolar PWM in fig. 19-51 four switches are used for one direction of rotation of the motor. The motor voltage changes from $+V$ to $-V$ and it is the average voltage which determines the direction of rotation of the motor

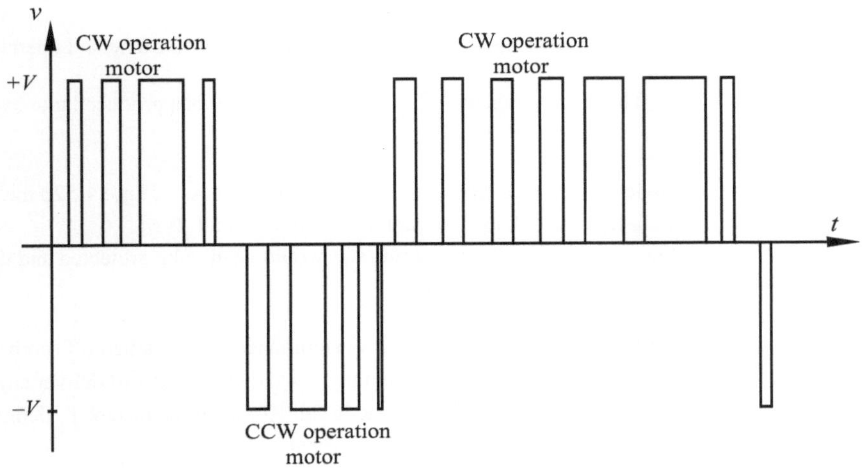

Fig. 19-50: Unipolar PWM

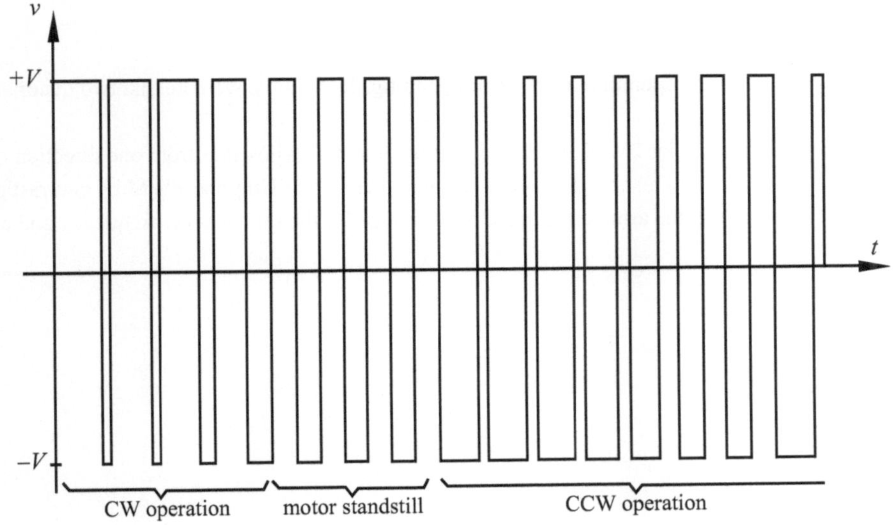

Fig. 19-51: Bipolar PWM

As an example we consider the operation of a H-bridge in the commonly used unipolar PWM. Fig. 19-52a shows T_1T_4 closed with motor rotating clockwise. There are now two possibilities for transistor control: either one switch stays closed (e.g. T_1) and the other PW-controlled (here T_4), or both switches (T_1 and T_4) are PW controlled as in fig. 19-52c.
We first consider the method whereby T_1 remains closed and only T_4 is PW controlled.
When T_4 opens in fig. 15-52b we have:

- induction voltage $e_l = L \cdot \frac{di_a}{dt}$ caused by the inductance of the armature circuit!
- emf $E = k \cdot n \cdot \Phi$: due to the inertia the machine continues to rotate and operate as generator!
- $e = e_l - E$. This e (in series with V_{CC} since T_1 is closed) can produce large overvoltage across T_4!

A free wheel diode is needed to protect the transistor. In the case of fig. 19-52b the emf e will deliver current via D_3 and T_1. Diode D_3 protects the transistor T_4!
In other switching situations a different transistor would need to be protected and therefore all four have free wheel diodes $D_1/D_2/D_3/D_4$.

Another possibility is that T_1 and T_4 are always simultaneously switched off (both PW controlled!). During blocking of the transistors in fig. 19-52c the emf e will deliver energy via D_2 and D_3 to the V_{CC}-source. This is also the case in fig. 19-52b at the instant T_1 definitely opens (together with T_4!). Obviously D_2 needs to be present.
The operation of the motor in the other direction is similar. Of course then we are working with transistors T_2T_3 instead of T_1T_4.

Remarks

1. From the operation of the H-bridge in fig. 19-52 a/b/c we note that two quadrant operation is possible.
2. With bipolar PWM a fast and supple transition is possible from one direction of rotation to the other. This results in good dynamic behaviour. Unipolar PWM by contrast produces less ripple in the armature current of the motor for the same carrier frequency and average current.

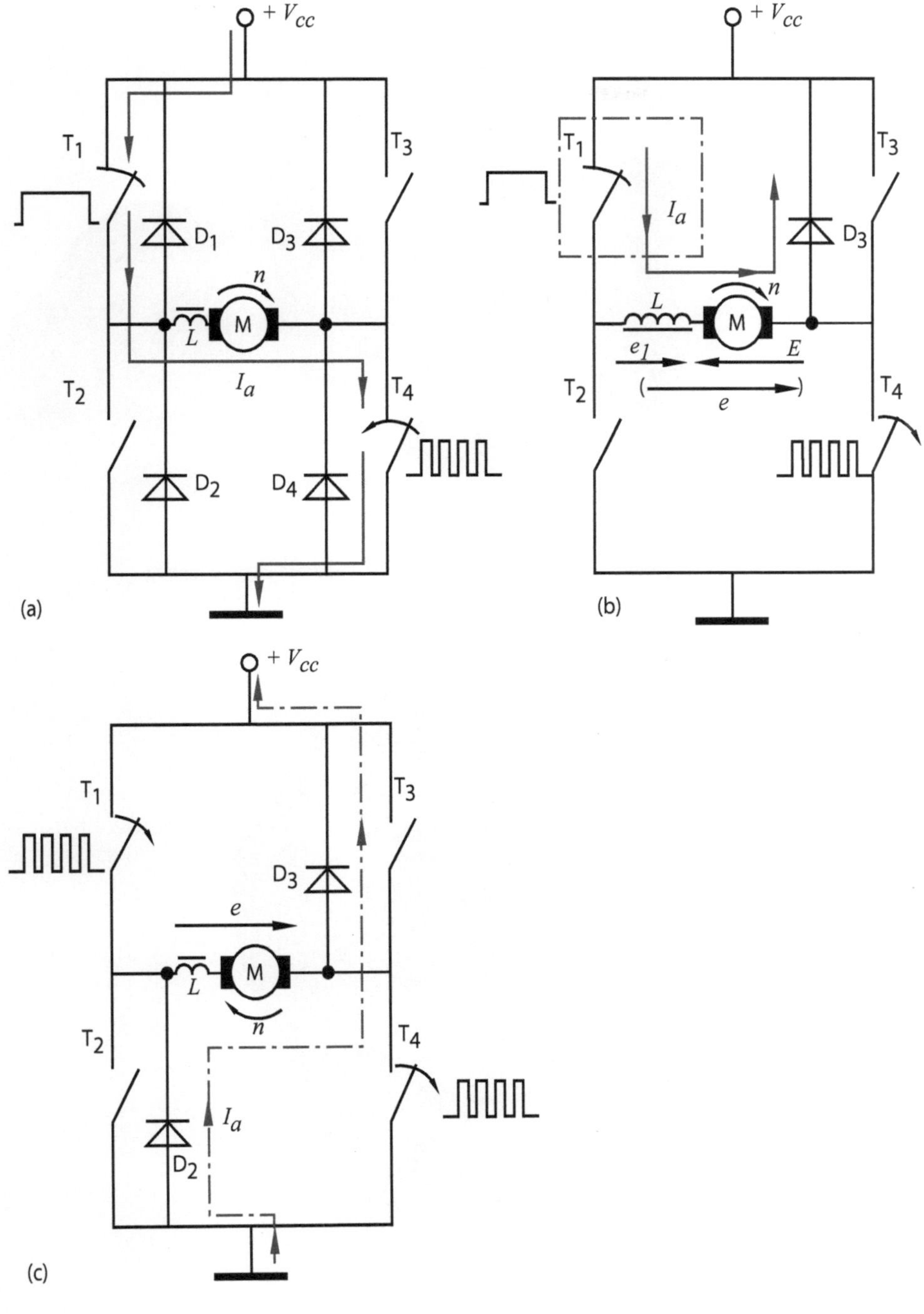

Fig. 19-52: H-bridge operation with unipolar PWM

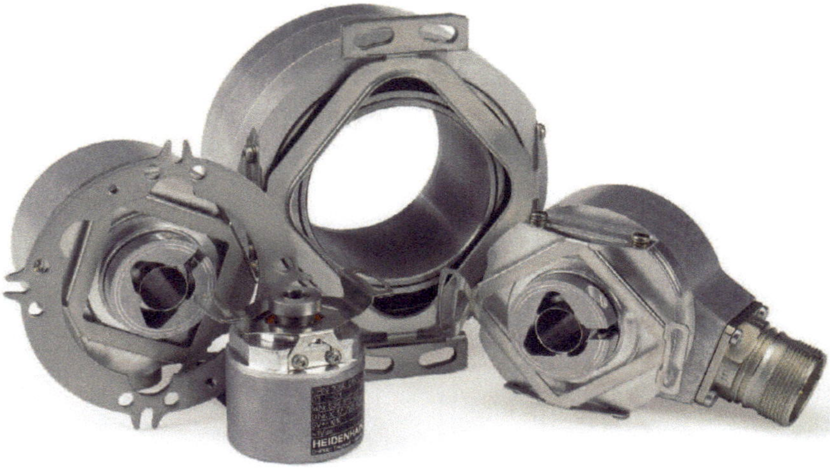

Photo Heidenhain: Impulse transmitters with their own bearings and built-on stator coupling

Photo Rotero: Exploded view of reducer for high accuracy positioning systems

9 CHOPPER CONTROL OF A SERIES MOTOR

Up to about 1990 the DC series motor was used in many countries for traction (train, tram, metro). With DC supplies a chopper was used and with an AC supply a controlled rectifier was used.
In addition to controlling the main (traction) motors, choppers were also used for auxiliary equipment such as the fans for the traction motors.
The power level of the choppers for the traction motors varied from hundreds of kW to many MW. Recent developments in traction sees the use of IGBT's as a power switch. The control is implemented using micro-controllers and DSP's and predominantly three-phase asynchronous motors are used (see chapter 23).

9.1 Traction service

In fig. 19-53 we represent the chopper as a mechanical switch Ch. The duty cycle δ of the chopper will determine the average value of $v_2 \rightarrow\rightarrow V_{2av.} = \delta \cdot V$.
The value of $V_{2\,av.}$ determines the speed of the motor.

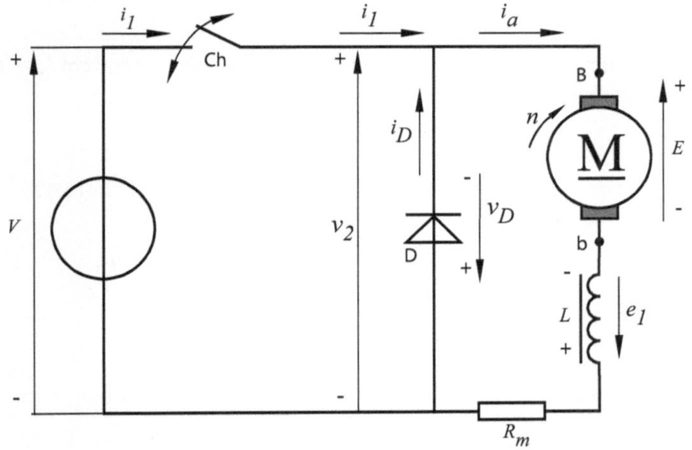

Fig. 19-53: Principle configuration of a chopper controlled DC-motor

The change in current Δi_a is given by expression (12-6) on p. 12.5:

$$\Delta i_a = \frac{\delta\,.\,(1-\delta)\,.\,T\,.\,V}{L}$$

. It is clear that $\Delta i_a = 0$ for $\delta = 0$ or for $\delta = 1$.

The maximum of Δi_a, with δ as independent variable is found with:

$$\frac{d(\Delta i_a)}{d(\delta)} = \frac{T\,.\,V}{L} - \frac{2\,.\,\delta\,.\,T\,.\,V}{L} = \frac{(1-2\,.\,\delta)\,.\,T\,.\,V}{L} = 0 \rightarrow\rightarrow \delta = 0.5$$

With $\delta = 0.5$ and $\frac{1}{T} = f_c =$ chopper frequency then becomes: $(\Delta i_a)_{max.} = \frac{V}{4 \cdot L \cdot f_c}$ (19-18)

From (19-18) it follows that the peak to peak motor ripple current $(\Delta i_a)_{max.}$ is smaller as:

1. the chopper frequency becomes larger
2. the self induction is larger.

If a too low chopper frequency is used then a large and expensive choke coil needs to included.
A high chopper frequency will increase the losses in:

- the semiconductors from which the chopper is built
- protection circuits for these semiconductors
- motor (losses due to the AC component of the current)

As a thyristor chopper is normally used the OFF-time should be at least five times the dead time of the thyristor.

By making the chopper frequency too large the maximum of δ is limited. In this case a large part of the power of the supply could not be transferred to the motor.

Practical chopper frequencies for traction motors were 100 Hz for SCR's and 200 to 600 Hz for GTO's.

Numeric example 19-5:

Given:

In the configuration of fig. 19-53 is $V = 600\text{V}$, full load armature current $I_a = 100\text{A}$, chopper frequency = 100 Hz and motor speed = 1000 rpm.

Question:

Which value of self induction is required to limit the current ripple to 15A?

Solution:

We know that $(\Delta i_a)_{max.}$ occurs at $\delta = 0.5$ and then $(\Delta i_a)_{max.} = \frac{V}{4 \cdot L \cdot f_C}$.

From this: $L_{min.} = \frac{V}{4 \cdot f_C \cdot (\Delta i_a)_{max.}} = \frac{600}{4 \cdot 100 \cdot 15} = 100\text{mH}$.

This might seem small, but note that this coil may not saturate before $I_a + \frac{(\Delta i_a)_{max.}}{2} = 107.5$ A.

The core will be quite large. Together with the winding wire for this large current then it is clear that chopper commutation coils for traction applications can be substantial.

Remark

Usually when braking a series motor it is operated as an independently excited dynamo!

9.2 Line filter

In the case where the supply is a battery (internal impedance ≈ 0) the chopper can be supplied without problem.
If the supply is via a catenary wire then the self induction L_R of this wire will:
1) severely limit the rise time of the current at switch on of the chopper
2) produce high induction voltages at chopper switch off.

To counteract these negative effects at least one L_1 C_1-filter is included (fig. 19-54).

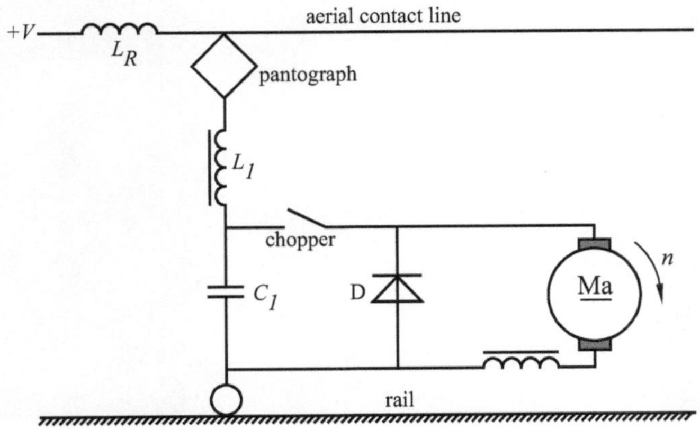

Fig. 19-54: Self induction of catenary wire and input filter of a traction engine

Capacitor C_1 : allows the chopper to absorb current pulses without the self induction of the catenary limiting the current rate of rise. The capacitor operates as an energy reservoir.
In addition the capacitor C dampens the overvoltage on the input of the chopper.
This overvoltage can be produced because of :
1) overvoltage present on the catenary
2) overvoltage as a result of the current interruption of the chopper.

Coil L_1 : will limit the ripple in the catenary wire so that other consumers on this wire experience no problems as a result of the intermittent operating current. These intermittent currents in the catenary wire and rail tracks could cause disturbance in telecom control circuits.
The capacitor C_1 together with $L_R + L_1$ forms a series resonator circuit with resonant frequency

$$f_1 = \frac{1}{2 \cdot \pi \cdot \sqrt{(L_R + L_1) \cdot C_1}} \qquad (19\text{-}19)$$

With a chopper frequency f_C which is lower or equal to the resonant frequency f_1 could produce (high) voltage oscillations. In practice then $f_C > 2 \cdot f_1$ and even $f_C > 3 \cdot f_1$.
Account also needs to be taken of the fact that L_R is variable depending on the distance between substation and carriage.

Photo Siemens: Siemens modernised the main drives of the hot rolling mills of Voestalpine Stahl GmbH (Linz-Austria). This modernisation will be complete in 2013. Voestalpine Steel hot rolling mill will produce 4.8 million tons of steel band between 1.5 and 20 mm thick and 70 to 175 cm wide.
The drive is comprised of cycloconverters, controlled by Simodyn D controllers

Photo Siemens: Finishing of the hot rolling production line at Corus Ijmuiden (The Netherlands)

20 SPEED- and (or) TORQUE-CONTROL of THREE-PHASE ASYNCHRONOUS MOTOR

CONTENTS

Power Electronics: DRIVE TECHNOLOGY and MOTION CONTROL
http://dx.doi.org/10.1016/B978-0-12-814641-5.50005-6,

1. THREE-PHASE ASYNCHRONOUS MOTOR

1.1 Generalities

The asynchronous motor is without doubt the most popular motor in industry.
The stator of this motor is composed of Si-steel plates of between 0.3 and 0.5 mm thick. These plates have slots in which the three-phase windings lie. Every winding has a sinusoidal flux distribution around the stator circumference. For simplicity only one central conductor is shown as in fig. 20-1. If the three-phase winding is connected to the power grid, then a rotating magnetic

field with speed n_S is produced: $$n_S = \frac{60 \cdot f_S}{p} \qquad (20\text{-}1)$$

f_S is the supply frequency of the stator coils and p is the number of pole pairs. n_S is the synchronous speed. The cylinder form of the rotor is also made from laminated Si-steel plates with slots in which the rotor windings are placed. Mostly the rotor winding is formed by bars which are connected at the ends, and this is referred to as a squirrel cage rotor. Sometimes the rotor may be wound. This windings are then connected via slip rings to the end of which resistors are connected. This is known as a slip ring motor.
The rotating field induces voltages in the rotor conductors and with a closed rotor circuit a rotor current flows. The Lorentz force on the rotor conductors causes the rotor to rotate with the rotating field. The speed n of the rotor is less than the stator rotating field, otherwise there would be no induced voltage and current in the rotor and hence no torque. The reduced speed of the rotor with respect to the stator may be expressed as the slip g:

$$g = \frac{n_S - n}{n_S} \qquad (20\text{-}2)$$ or $g = \frac{n_S - n}{n_S} \cdot 100\ (\%)$. From which: $$n = n_S \cdot (1 - g) \qquad (20\text{-}3)$$

Consider a two pole machine. In fig. 20-1 we have drawn the rotating stator field at a certain instant. For simplicity the direction of the induced voltage has only been determined in two windings. If the rotor forms a closed electric circuit, then fig. 20-1 shows the direction of the rotor current in two of the rotor conductors. Note that because of the self induction L_R of the rotor phase, this current heavily lags the induced voltage. The time constant $\tau_R = \frac{L_R}{R_R}$ of a standard induction motor is usually larger than a hundred milliseconds. Here R_R is the resistance of one rotor phase.
In fig. 20-2a we have redrawn a portion of fig. 20-1 in order to determine the force F on the rotor. We see that the rotor is pulled along with the rotating stator field Φ.
If we have to externally drive the rotor with an over synchronous speed ($n > n_S$) then we see in fig. 20-2b that the force F on the rotor acts as a brake. The rotor current reverses polarity and with an unchanged three-phase voltage, energy is delivered to the net. The machine operates as an asynchronous generator.
The frequency f_R of the induced voltage in the rotor is determined by the difference in speed of the rotating stator field (n_S) and the rotor (n) .

$$f_R = \frac{p \cdot (n_S - n)}{60} = \frac{p \cdot g \cdot n_S}{60} = g \cdot \frac{p \cdot n_S}{60} = g \cdot f_S \longrightarrow \quad f_R = g \cdot f_S \qquad (20\text{-}4)$$

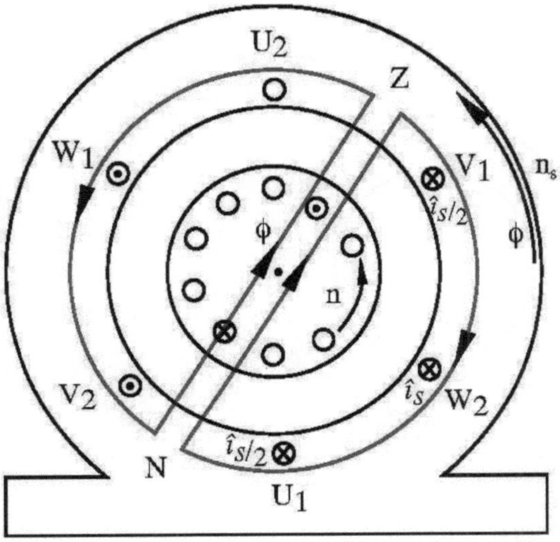

Fig. 20-1: Operation of an asynchronous motor

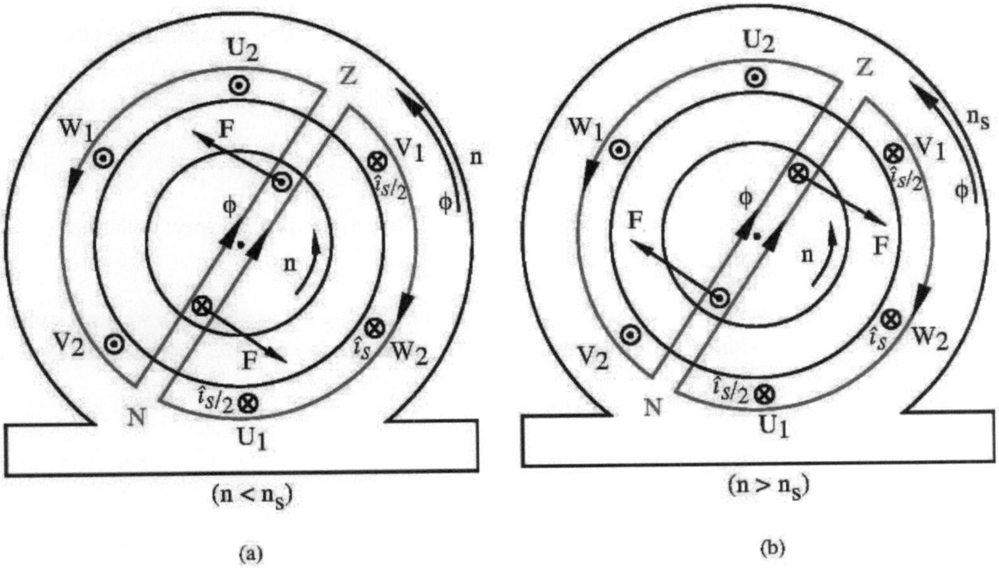

Fig. 20-2: Motor and generator service of an induction machine

1.2 Rotating stator field – rotating rotor field

The stator and rotor are drawn in fig. 20-3, shown spatially and isolated from each other.

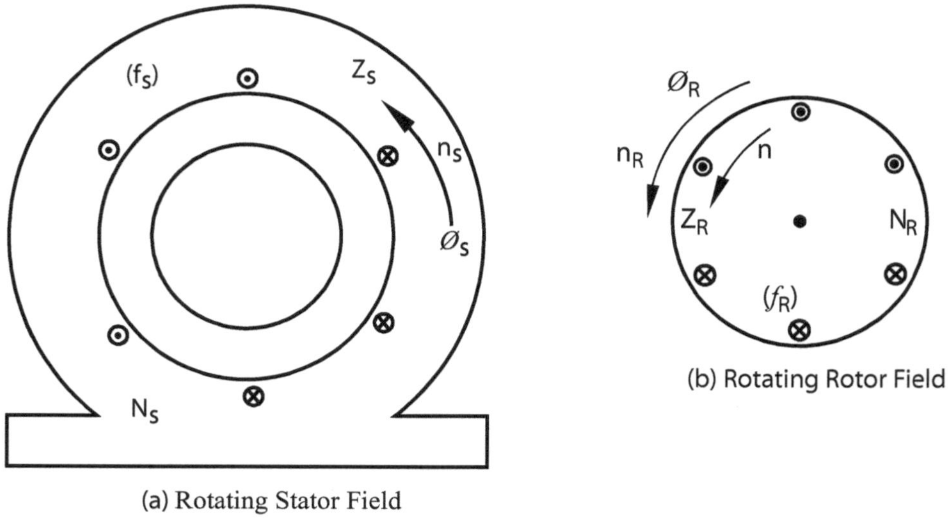

Fig. 20-3: Stator and rotor with respective rotating fields

A three-phase current with frequency f_S in the stator winding creates a rotating magnetic field Φ_S. The speed of this field is $n_S = \frac{60 \cdot f_S}{p}$ with respect to the stator.
The rotor winding behaves as a three-phase winding and similar to the operation in the stator the rotor current produces a rotating magnetic field Φ_R with respect to the rotor. The speed of this rotor field with respect to the rotor is determined by the frequency f_R of the rotor current so that:

$$n_R = \frac{60 \cdot f_R}{p} = \frac{60 \cdot g \cdot f_S}{p} = g \cdot n_S .$$

Since the rotor rotates at n rpm the rotor field will have a speed with respect to the stator of:
$n + n_R = n_S \cdot (1 - g) + g \cdot n_S = n_S$.

Decision: if we take the stator as reference, then in the three-phase machine we have a rotating stator field Φ_S with synchronous speed n_S AND a rotating rotor field Φ_R with the same synchronous speed n_S. Both fields together produce THE rotating stator field Φ !

1.3 Simple equivalent diagram of an induction motor

Fig. 20-4 shows a T-equivalent diagram for one phase of a three-phase motor.

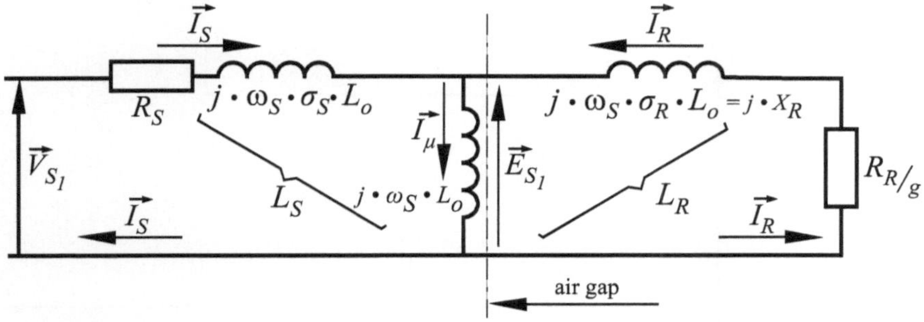

Fig. 20-4: T-equivalent diagram of an induction motor

The symbols in fig. 20-4 have the following meaning:

V_{S1} : voltage on a stator coil
I_S : phase current
L_0 : magnetising inductance, this is the common (coupled) self induction between stator and rotor, and responsible for the flux Φ_{S1} in one phase. As we know the rotating flux is $\Phi = \frac{3}{2}\,\Phi_{S1}$. In addition: $L_0 = \frac{3 \cdot \pi}{8} \cdot N_{Se}^2 \cdot \frac{\mu_0 \cdot l \cdot r}{l_{ag}}$
N_{Se} : equivalent sinusoidal winding. In practice N_{Se} = 97 to 100% of N_S
N_S : number of windings in one stator coil
l : useful axial length of stator
μ_0 : $4 \cdot \pi \cdot 10^{-7}$ H/m
r : internal radius of stator
l_{ag} : length of the air-gap between rotor and stator
L_S : stator self-induction = $L_0 \cdot (1 + \sigma_S)$
L_R : rotor self-induction = $L_0 \cdot (1 + \sigma_R)$

In fig 20-5 the vector diagram associated with fig. 20-4 is shown.

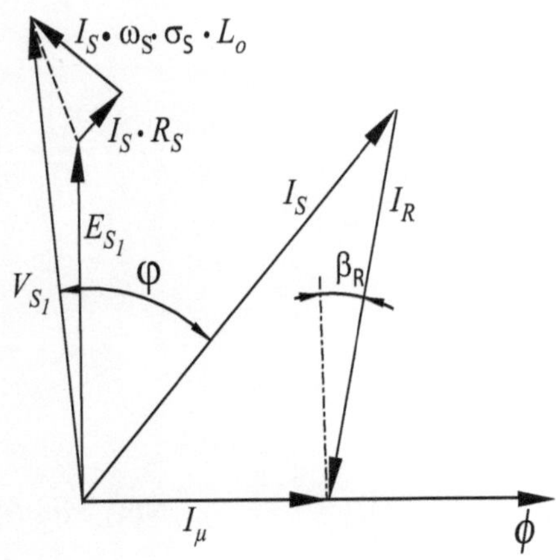

Fig 20-5: Vector diagram for the circuit in fig. 20-4

1.4 M-n curve of an induction motor

The curve which relates torque to speed is important in the study of speed control of motors. This is the so called *M-n* curve, which is shown in fig. 20-6.

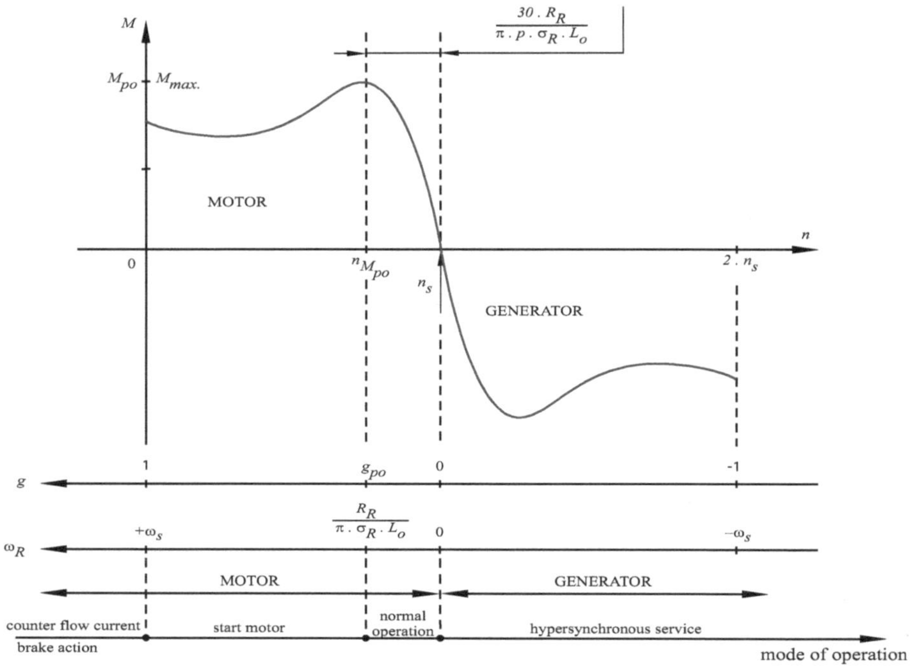

Fig. 20-6: *M-n* curve of an asynchronous motor

1.5 Simple mathematical model of an induction motor

In chapter 16 with a few simplified formulas we constructed a simple mathematical model of a three-phase induction motor. We rewrite this formulas in table 20-1.

Table 20-1

$P = \eta . \sqrt{3} \cdot V \cdot I \cdot \cos\varphi$		$\Phi = k_1 \cdot \dfrac{V_S}{f_S}$	(20-5)
$n_S = \dfrac{60 \cdot f_S}{p}$	(20-1)	$M_{em} = k_4 \cdot \Phi_n \cdot I_n$	(20-6)
$g = \dfrac{n_S - n}{n_S}$	(20-2)	$M_{po} = M_{max} = k_3 \cdot (V_S/f_S)^2$	(20-8)
$f_R = g \cdot f_S$	(20-4)	$P = \omega \cdot M_n$	(20-7)
$n = (1 - g) \cdot n_S$	(20-3)	$M_n = \dfrac{9550 \cdot P}{n}$ (Nm) with P in kW	(20-9)
$M = M_t + M_v$ and $M_v = J_m \cdot \dfrac{d\omega}{dt}$	(20-10)		

Photo Siemens: S110: Line: 1ph 200…240V AC / 3ph 380-480V AC.

The Sinamics S110 is a single shaft drive in the power range of 0.12 to 90kW.

The AC-servo-converter is intended for simple positioning applications with synchronous and asynchronous motors. It has integrated safety functions. The Sinamics S110 can be used for simple pick&place or delivery tasks, for robotics and handling operations, for supply systems in printing and paper machines, in positioning systems in wind and solar applications, etc

Photo Siemens: S120: 0.25 to 4500kW Supply 1ph 230V AC / 3ph 380 – 480V AC/ 3ph 660-690V AC.

The Sinamics S120 drive includes additional safety features with the new version V2.5. They support the simple and norm conform realisation of innovative safety concepts. The safety functions Safe Stop 2, Safe operating Stop, Safe brake ramp, Safely-Limited speed and Safe Speed Monitor complement the existing basic safety functions

2. ELECTRONIC CONTROL OF AN INDUCTION MOTOR

Two important properties on the shaft of a motor are the speed and torque. Application of electronic circuits allow us to alter these values. The simplest circuit will only alter the motor frequency or slip, resulting in a speed change. With more sophisticated electronics we can change and control both speed and torque of the motor. From the control perspective we differentiate between two possibilities:

1. **Open loop control** of the motor speed
2. **Closed loop control** of speed and (or) torque of the induction motor

Open loop control is also known as scalar control. An electronic converter is placed between power grid and motor, in which only the amplitude of flux and frequency are controlled.
If in addition to the amplitude also the phase of the flux is controlled then it is referred to as flux-vector control. An example of this is the field orientated control of an induction motor to produce good dynamic behaviour. A vector controller is (dynamically) better than a scalar controller.
A vector controller with encoder or tacho has a control range of 1:100, a speed accuracy of better than 0.5%, a torque response of 15ms and a speed response of at least 60ms. These statistics almost equal those of a DC-drive: speed accuracy better than 0.5%, torque response 10 ms, speed response 28 to 40 ms.
The characteristics of a high dynamic system include torque control, good control options at very low speeds, four quadrant service, short response times.
A flux-vector controller also includes a torque controller. As a result a simple drive with an asynchronous motor can be very dynamic even at stand still. This is a necessary property for the control of lifting equipment (cranes, elevators).
Dr. F. Blaschke of Siemens-Erlangen formulated in the beginning of the seventies a general theory for the control of AC-motors. Since with this type of control the rotor flux vector serves as a reference, Dr. Blaschke gave it the name “field orientation”.
This type of control requires a lot of fast calculations, which is only economical with the aid of micro-controllers and digital signal processors (DSP). This is the reason that field orientated control was not technologically possible in the seventies.

In summary:

1. **Scalar Regulation**
   - Advantages:
     - cheap converter
     - no feedback required
   - Disadvantages:
     - unknown operating point of motor
     - torque is not controlled.

2. **(Flux) Vector control**
   - Advantages:
     - accurate speed control
     - good torque control
     - full torque at stand still
     - almost equals the dynamic properties of a DC-drive.
   - Disadvantages:
     - feedback required
     - expensive converter.

Fig. 20-7 shows an overview of the types of electronic converters used to regulate and control an induction motor.

As shown under the following nr. 3 we make a distinction between frequency control and slip control. Frequency controllers can be in the form of scalar regulation or vector control.

Slip control is a scalar regulation only.

Modern controls mostly use PWM-converters with DC-voltage link.

To a lesser degree controlling of the stator voltage is also encountered. In a limited number of applications a direct converter using a cyclo-converter (MW-power range) is used or converters with a DC current link (for motors from 2 MW).

Controlling the rotor power is no longer considered for motors below 1 MW.

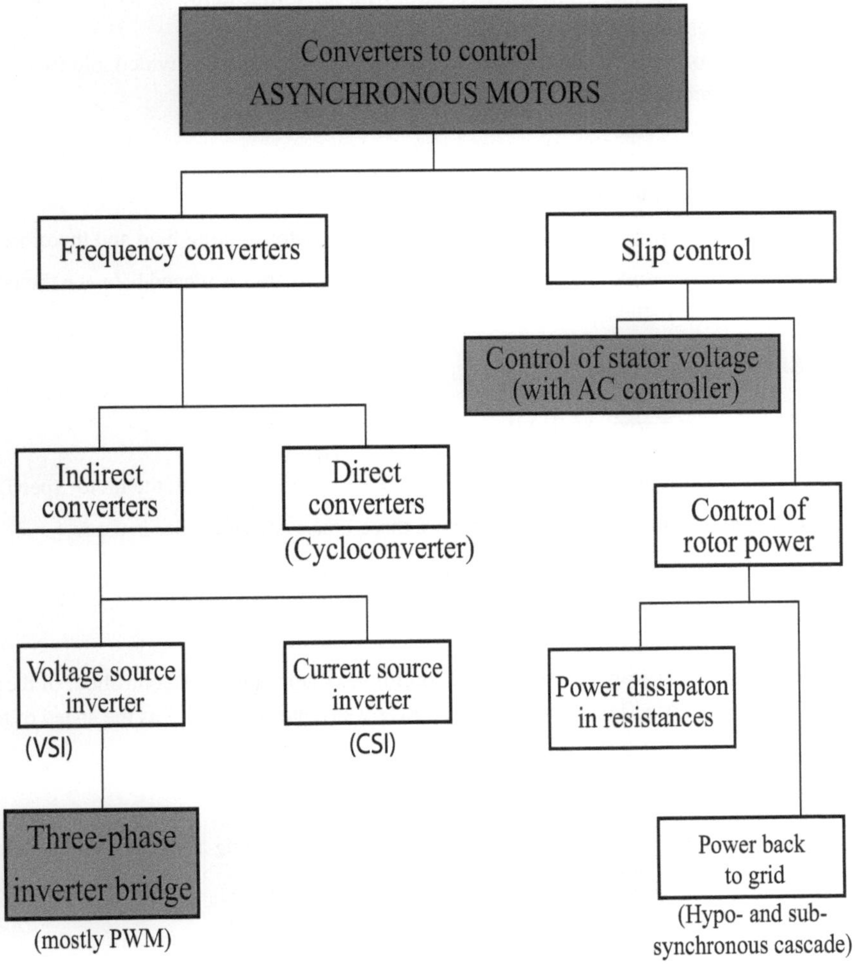

Fig. 20-7: Converters used for regulating and controlling induction motors

3. SCALAR CONTROL OF INDUCTION MOTOR SPEED

The speed of an asynchronous motor is given by $n = n_S(1-g)$ or $n = \frac{60 \cdot f_S}{p}(1-g)$.
The speed is therefore variable by altering f_S, g or p. Altering p will not be considered since switching the number of poles (Dahlandermotor) does not provide continuous speed regulation.
The remaining solution is to control f_S (frequency control) or control g (slip control).
Systems whereby only the amplitude of a variable (flux, slip) are controlled is referred to as a scalar control system. The most common is maintaining a constant voltage-frequency ratio which corresponds to maintaining a constant amplitude of the rotating flux. It is also referred to as a constant flux or *V/f* control. This is especially interesting for the control of pumps and fans to save on energy costs.
Such a constant flux is difficult to control with full torque at low speeds or at stand still and therefore not suitable for servo-drives.
The electronic speed control of an asynchronous motor can be divided into two groups:

- frequency control
- slip control

Frequency control

From $n_S = \frac{60 \cdot f_s}{p}$ we note that the speed of the stator rotating field and therefore the speed of the motor can be varied by varying f_S. A frequency converter is placed between the power grid with it's constant frequency and the motor.

Slip control

The slip $g = \frac{n_S - n}{n_S}$, so that $n = n_S \cdot (1-g)$.
With a fixed supply frequency f_S and thus $n = \frac{60 \cdot f_S}{p}$ = constant, the motor speed n can be changed by adjusting the slip. For this a three-phase AC-controller is used.

4. SLIP CONTROL

In the only real practical case of slip control the stator voltage is controlled. In the past slip ring motors were used for which the rotor power was controlled and thus the speed of the motor.

4.1 Standard motor

In fig. 20-8 we see the principle configuration for controlling a standard motor. Here the stator voltage is controlled with a three-phase AC-controller.

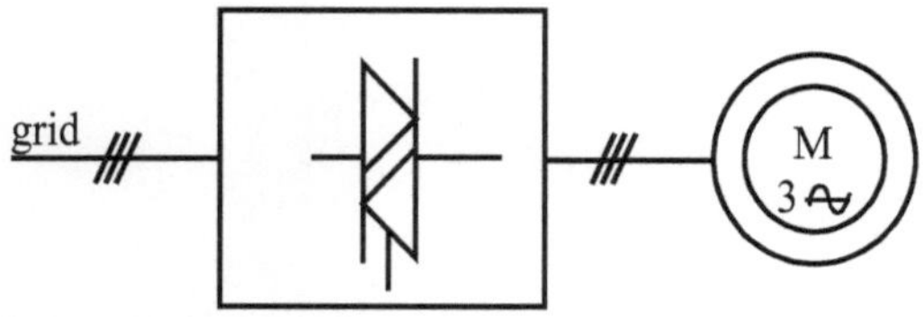

Fig. 20-8: Speed control via a three-phase AC-controller

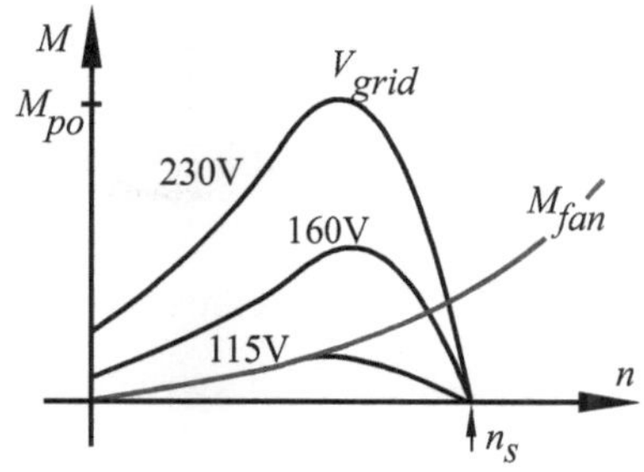

Fig. 20-9: *M-n* curves of motor in fig. 20-8

Since M_{po} is proportional to $[V_S/f_S]^2$ if the frequency f_S is constant then the pull-out torque will approximately decrease proportionally with V_S^2. This is shown in fig. 20-9.
It is clear that this speed regulation is only useful with small asynchronous motors with a so called ventilator load.

Remark: softstarter
To limit the inrush current of an asynchronous motor a soft starter is used. During starting of the motor the stator voltage is controlled with an AC-controller. This is therefore also slip control but it's purpose is not to control speed. The softstarter is discussed under nr. 9 in this chapter.

4.2 Motor with large resistance

In the case of specially constructed motors with large rotor resistance the *M-n* curves are shown in fig. 20-10. The counter torque M_t of the load is also drawn. The advantage is that the motor speed can be controlled easily with the aid of cheap electronic circuits (e.g. AC-controller based on triacs). The disadvantages are the specially constructed motor with large dimensions and the poor efficiency and power factor at low load.

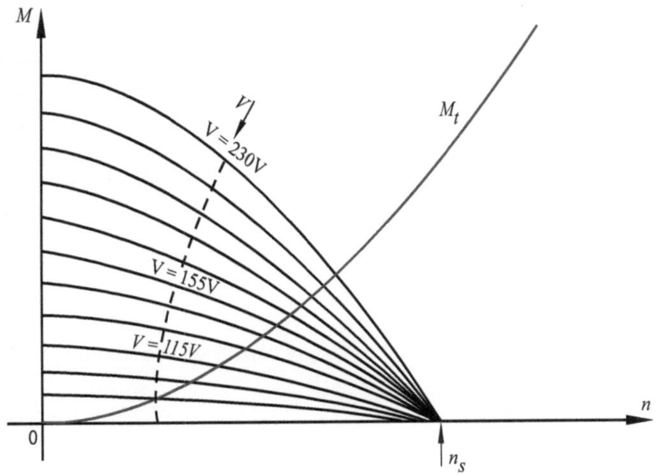

Fig. 20-10: *M-n* curves of induction motor with large rotor resistance

5. SCALAR FREQUENCY CONVERTERS

We make a distinction between the direct and indirect converter.

5.1 Direct converter

This is a cycloconverter. Seldom used. For the study of the cycloconverter see chapter 10.

5.2 Indirect frequency controller

Since the supply frequency of 50 Hz is converted in two steps to a variable frequency we speak of two stage conversion or indirect frequency conversion. In the first step the single or three-phase voltage is converted to a DC-link. The converter produces a DC-voltage or a DC-current.
We make a further distinction between converters with a DC-voltage link and a DC- current link.
In drive jargon we talk about VSI (voltage source inverter) or CSI (current source inverter).
For CSI see nr. 10 of this chapter.
The inverter converts this DC-voltage or current into a three-phase supply which drives the motor (fig. 20-11 or 20-12). The frequency of these three-phase supply is variable.

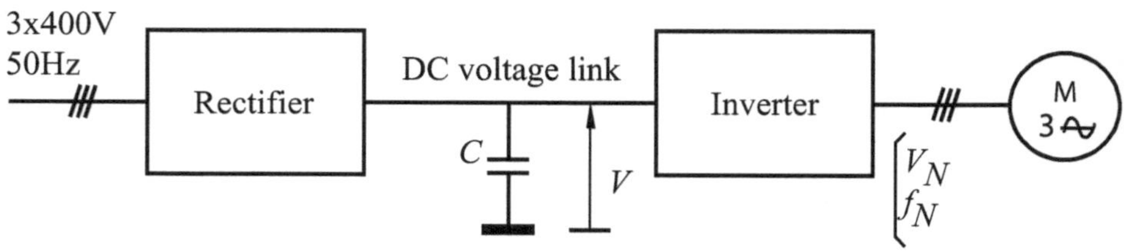

Fig. 20-11: Frequency-converter with DC-voltage link: VSI

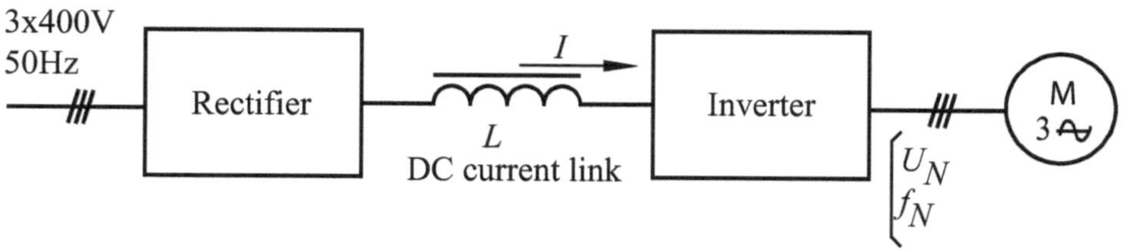

Fig. 20-12: Frequency converter with DC-current link: CSI

Single phase supplies are used to supply converters (for motors) with limited power, up to 2 kW.

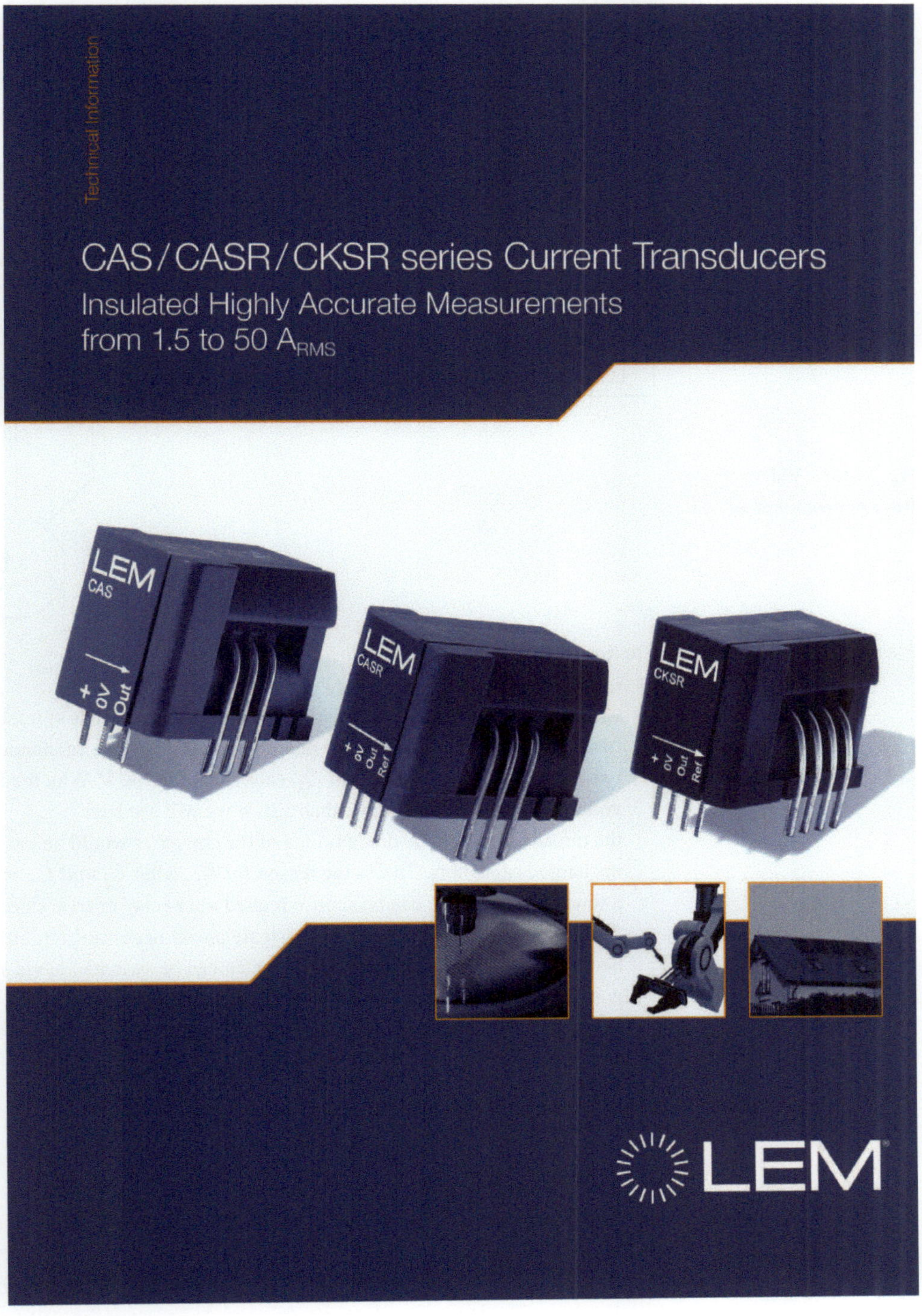

Photo LEM: CAS/ CASR/ CKSR-series of flux-gate current sensors

6. INDIRECT FREQUENCY CONVERTER OF THE VSI-TYPE

A more elaborate diagram than that of fig. 20.11 is shown here in fig. 20-13. Here we extend the configuration of fig. 14-8 with a few elements (*R*, R_1, R_2, T_1). The converter is now a PWM-type.

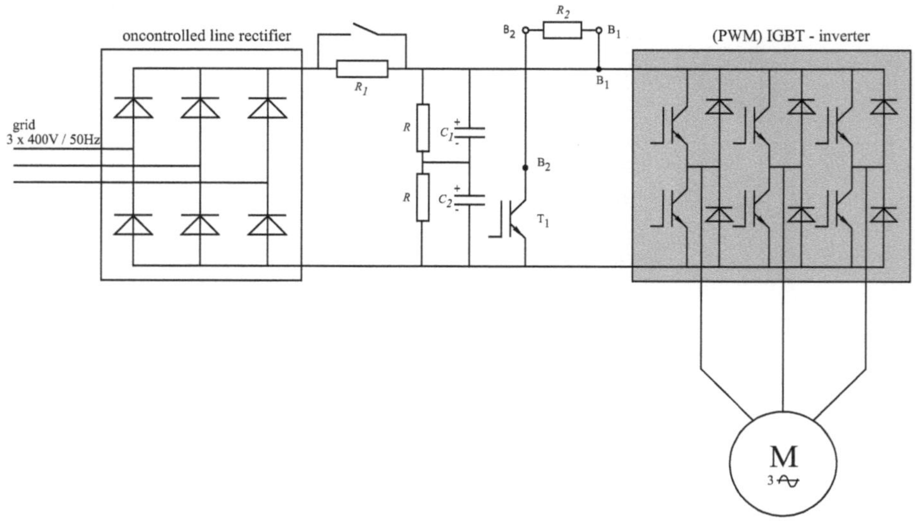

Fig. 20-13: A VSI frequency converter

New compared to fig. 14-8 are:

1) Inverter: is a PWM-type.
2) C_1 *and* C_2: with a power grid voltage of 3x400 V the DC-link voltage is at least 530V. Instead of a smoothing capacitor rated for 900V it is more economical to use two capacitors in series (C_1 and C_2) each rated for 450 V. Why the DC-link voltage can be so much higher than 530 V we will see later.
3) R_1 : the inrush current in the diode bridge of the converter would be too high with uncharged capacitors. That is the reason for R_1 . After C_1 and C_2 are charged to a few tens of volts R_1 can be short-circuited via a relay or triac since R_1 plays no further role but rather would then dissipate power necessarily if left in circuit.
4) R_2 and T_1 : here R_2 is the so called brake resistor which can be switched in via T_1 ? The purpose of R_2 will be explained under nr. 6.5.5 and "braking" of the motor.
5) *R* and *R* : cause a good voltage division between C_1 and C_2 .

6.1 Relationship between torque and supply frequency

6.1.1 Nominal flux

With a fixed number of poles of a machine the speed may be controlled by varying f_S :

From $M_{po} = k_3 \cdot \left[\frac{V_S}{f_S}\right]^2$ we see that to maintain a constant maximum torque the ratio $\frac{V_S}{f_S}$ should be constant. Fig. 20-14 shows what this involves.

The system in which $\frac{V_S}{f_S}$ is kept constant is known as constant flux control. This is a good control system for an asynchronous motor. The system is valid for three-phase squirrel cage motors and for three-phase synchronous motors. This is also called *V/f* control .

We let $\frac{V_S}{f_S} = k =$ V/Hz value of the motor under study.

This k-factor determines the motor flux $\Phi_n = \left[k_1 \cdot \frac{V_S}{f_S}\right] = k_1 \cdot k$

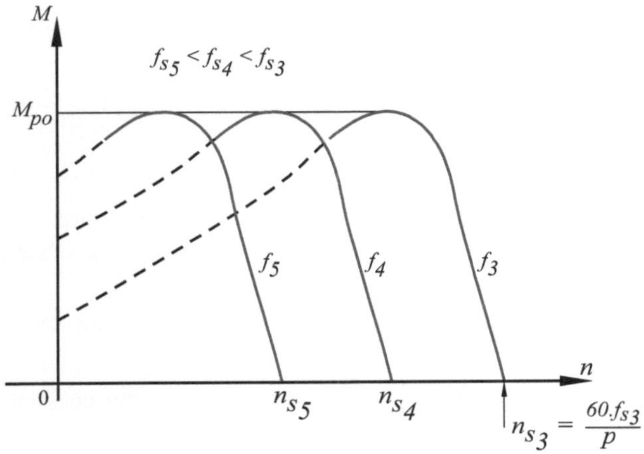

Fig. 20-14: *M-n* curves with constant flux

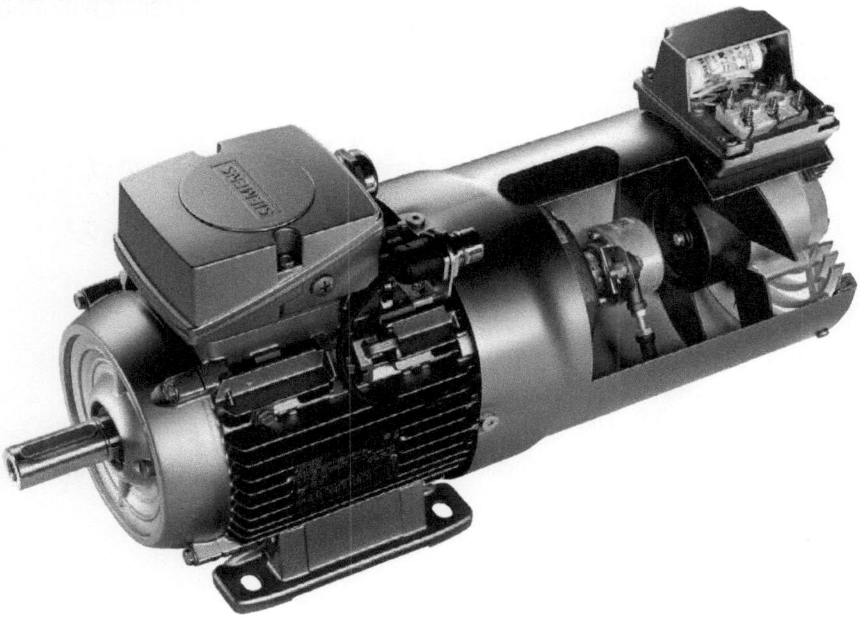

Photo Siemens: Cross section of a 1LE1 asynchronous motor. Encoders, brakes and external fans can easely be added

6.1.2 Field weakening

Assume that we drive a motor from a converter with variable voltage and frequency.

We call: V_N = nominal output voltage of the converter (= maximum output voltage!)
f_N = nominal frequency of the converter (= maximum output frequency!)

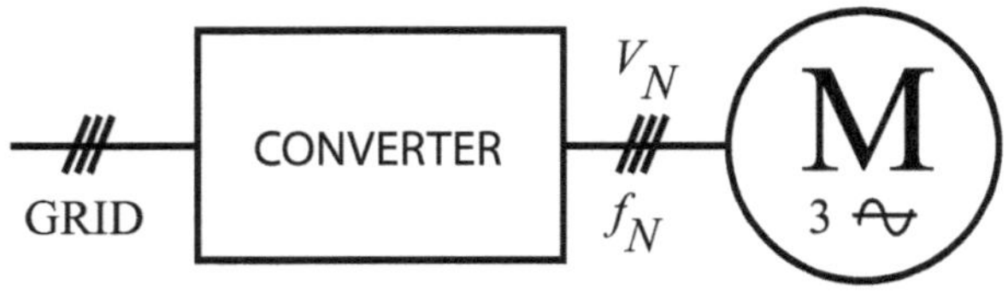

Fig. 20-15: Frequency controller

From a mechanical point of view the maximum frequency that may be applied to a 50Hz motor from a converter is 400Hz. With PWM-converters the lower frequency limit of the converter is at 0.1 Hz.
Assume a motor of 3x230V(Δ) - 50 Hz is connected to the output of a frequency converter where for example V_N = 3x230V and f_N = 100Hz.
The nominal flux of the motor is found by: $\frac{V}{f} = \frac{230}{50}$ = 4.6 V/Hz = k = flux constant.
With a converter frequency of 25 Hz an output voltage of 115V would be sufficient.
Let the frequency and voltage increase, then from 50 Hz and 230V (= $V_N = V_{max.}$) only the frequency of the converter can change up to f_N = 100 Hz.
Between the lowest converter frequency (e.g. 0.1 Hz) and 50 Hz (fig. 20-16a) the motor can operate with its nominal flux $\Phi_n = k_1 \cdot V_S/f_S$ since we can vary both V_S and f_S.
From 50 Hz the maximum output voltage of the converter under discussion is reached and the nominal flux will decrease with increasing frequency f_S.

Indeed, from $V_S = V_N$ is $\Phi = k_1 \cdot \frac{V_S}{f_S} = k_1 \cdot \frac{V_N}{f_S} = \frac{k_5}{f_S}$, so that $M_{em} = k_4 \cdot \Phi \cdot I_n = k_1 \cdot \frac{k_5}{f_S} \cdot I_n$.

If we assume that the load current is nominal I_n = *constant,* then $M_{em} = k_6/f_S$.
This is shown in fig. 20-16a.
The maximum torque is given on p. 20.6 by (20-8) : $M_{max.} = k_3 \cdot (V_S/f_S)^2$.

As long as $V_S < V_N$ is $M_{max.} = M_{po}$ because the ratio V_S/f_S is maintained.

In the region of field weakening ($V_S = V_N$) this becomes $M_{max.} = k_7 \cdot \frac{1}{f_S{}^2}$.
The maximum torque decreases quadratically with increasing frequency (f_S) of the motor.
This is indicated in fig. 20-16b.

6.2 Relationship between power and supply frequency

The dissipated power of a motor is given by: $P = \omega \cdot M = \frac{2 \cdot \pi \cdot n}{60} \cdot M$

6.2.1 Nominal flux

$\Phi_n \rightarrow P = \frac{2 \cdot \pi \cdot n}{60} \cdot M_n = k_8 \cdot n$.

Since (with the exception of slip) speed is proportional to f_S, we find: $P = k_8 \cdot n = k_9 \cdot f_S$
The power increases linearily with f_S, until $V_S = V_N$.

6.2.2 Field weakening

Here is $M_{em} = \frac{k_6}{f_S}$ so that $P = \omega \cdot M = \frac{2 \cdot \pi \cdot n}{60} \cdot \frac{k_6}{f_S} = \frac{2 \cdot \pi}{60} \cdot k_6 \cdot \frac{n}{f_S}$

Since n is proportional to f_S, P will be constant. This is shown in fig. 20-16c.

(a)

(b)

(c)

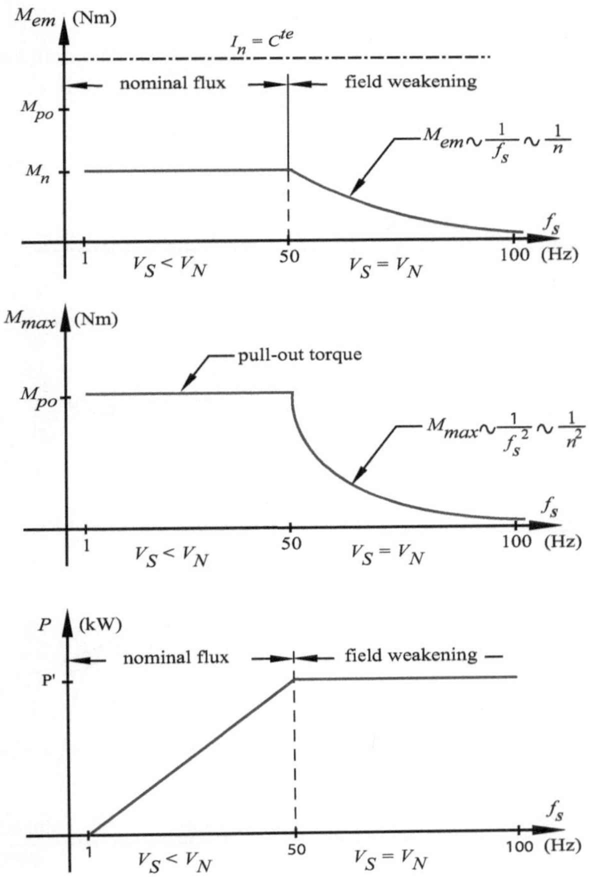

Fig. 20-16: Relationship between moment, power and supply frequency

6.3 Control range

We define the control range = $\dfrac{\textit{smallest speed by nominal flux}}{\textit{largest speed (in field weakening area)}}$ (20-11)

- **Electromechanical torque**

In the assumption that the nominal motor current I_n is flowing, it follows from (20-6) that the torque M_{em} proportionally changes with Φ.

In the field weakening area $\Phi = k_1 \cdot V_S/f_S = k_1 \cdot \dfrac{constant}{f_S}$, so that the torque: $M_{em} \sim 1/f_S \sim 1/n$.

This is indicated by the hyperbole in fig. 20-17.

- **Maximum torque**

In the field weakening area V_S is constant. From (20-8) it follows that in this area the maximum torque quadratically decreases with increasing frequency f_S.

Above F_7 the motor cannot run with constant power, see fig. 20-17. The range wherein the motor power remains constant (up to f_7) depends upon M_n/M_{po} in the area with nominal flux Φ_n.

Often for a motor $M_n/M_{po} = 0.5$, so that : $f_7 \approx 2 \cdot f_4$.

From f_7 the motor can only deliver a certain percentage of its maximum torque because of the losses. In this area of "higher speed" M_{em} is inversely proportional to f_S^2.

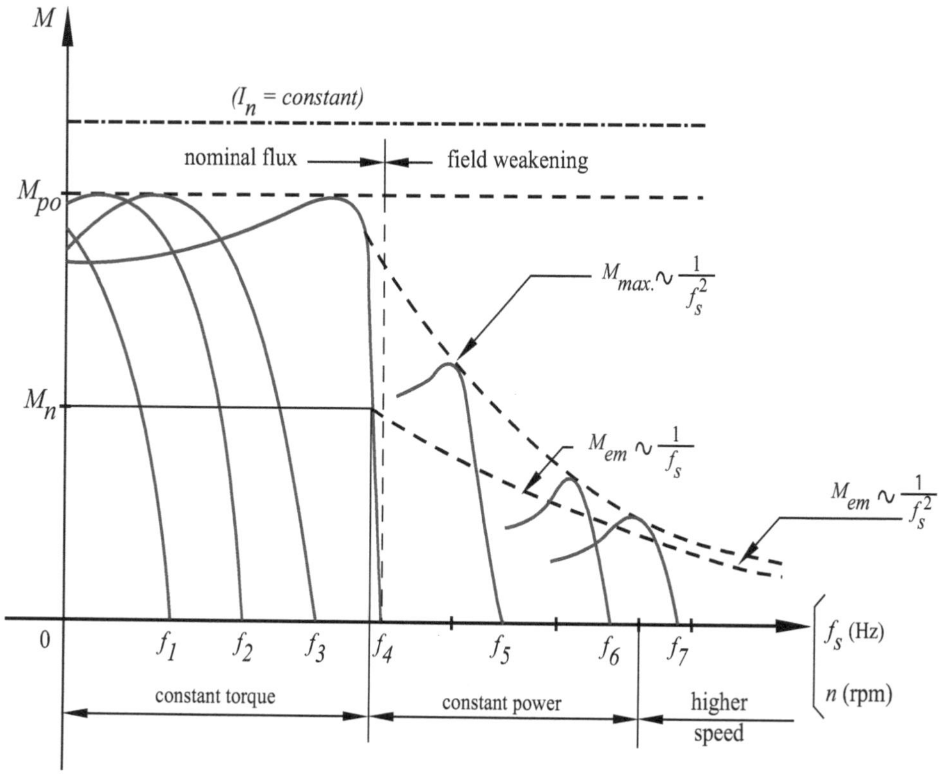

Fig. 20-17: *M-n* curves of a three-phase asynchronous motor

Numeric example 20-1:

Given: motor 400V (Δ) - 50Hz - two poles - slip is negligible
converter: $V_n = 400V$; $f_N = 200Hz$; $f_{min} = 5Hz$

Required: reference speed = 4800 rpm. Determine: control range, *k*-value of motor, border of field weakening, speed up to which maximum torque can be achieved, output frequency of converter at 4800 rpm

Solution:

- with $f_{min} = 5Hz$ the synchronous speed is : $n_S = \frac{60 \cdot 5}{1} = 300$ rpm
- control range $= \frac{300}{4800} = \frac{1}{16}$
- *k*-value motor: $k = \frac{400}{50} = 8$
- border of field weakening lies by $f = \frac{V_N}{k} = \frac{400}{8} = 50Hz$
- speed up to which nominal torque M_n can be achieved:
  $$n_S = \frac{60 \cdot f_S}{p} = 3000 \text{ rpm}$$
- output frequency of converter at 4800 rpm: $f_s = \frac{4800 \cdot 1}{60} = 80Hz$

Photo Siemens:
The Siemens Loher Chemstar motor series is available in efficiency classes IE2 “High Efficiency” for all types of explosion proof classifications EX n, EX e and EX de. Thanks to their chemically resistant paint finish and the electroplated fan cowl, the low voltage asynchronous motors are especially suited to applications in the chemicals and petrochemical sectors as well as the gas and oil industry. The motors are available in 230/400 V, 400/690 V and 500 V; 0.12 to 315kW; 1000/1500/3000 rpm; IP56 to IP67

6.4 Torque reduction

Large control ranges can result in a certain amount of over dimensioning of the motor. This is necessary to be able to supply enough torque over the entire control range.

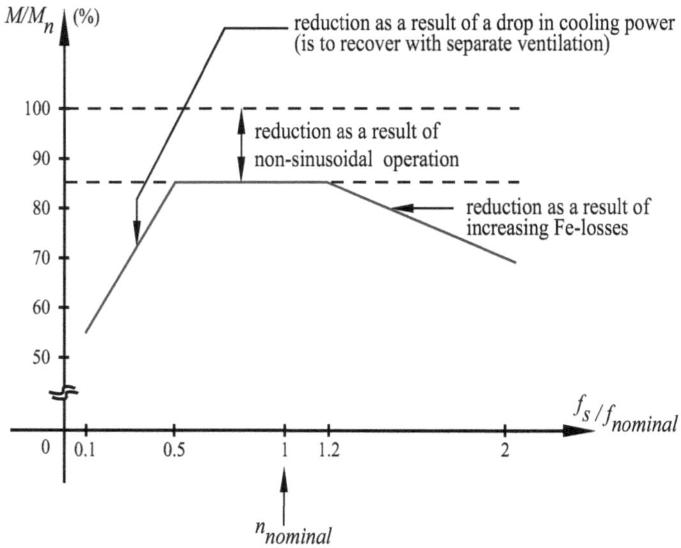

Fig. 20-18: Torque reduction with the use of a standard motor

Since we are using a standard motor the cooling is calculated for the nominal speed (at 50 Hz) of the asynchronous motor. At low speeds as a result the torque that can be produced has to be limited unless independent cooling is possible.
Over the entire range there is also a reduction of 10 to 20 % as a result of the non sinusoidal service (losses due to harmonics!).
In the highest speed range the iron losses increase strongly so that there is a further reduction of the deliverable torque. This is indicated in fig. 20-18. This is referred to as “de-rating” the motor that is fed from a frequency drive.

6.5 Important parameters of a frequency converter

6.5.1 *V/f* - setting

As we saw the *V/f* - ratio determines the flux. Fig. 20-19 provides two typical examples of *V/f*-settings.
In systems with FCC (flux current control) or CFC (current flux control) there is no *V/f* setting required since the flux is automatically adjusted for the motor load.

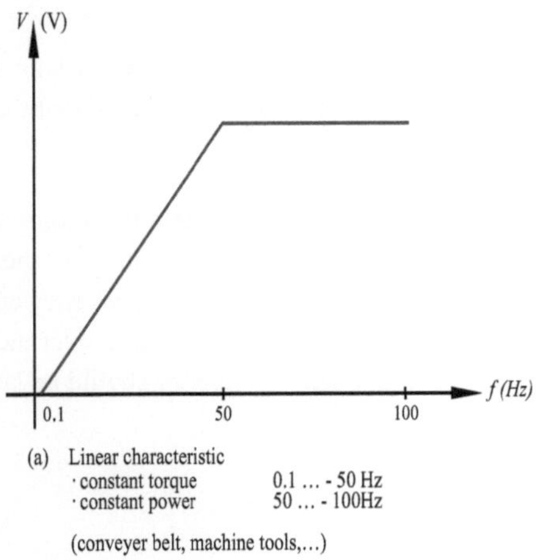

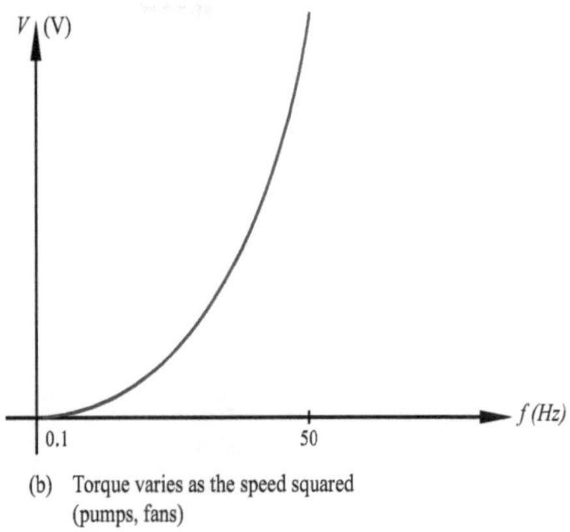

Fig 20-19: *V/f*-settings of frequency converters

6.5.2 Torque gain and IR-compensation

At start-up at low speed we see to it that $V_S/f_S > k$. If we remain below saturation this results in a torque gain. Above that, with low frequencies f_S, the resistive voltage drop $I_S \cdot R_S$ in the motor is important so that $E_S << V_S$. The flux is reduced and the torque decreases. To prevent this, *IR*-compensation is implemented. This *IR*-compensation can also compensate for the voltage drop in the motor cable.

Fig. 20-20 shows the combination of torque gain and *IR*-compensation in the *V/f* curves. The point B is determined by the manufacturer, e.g. 20 to 50 Hz. Point A is a variable parameter and determines the torque gain at start-up of the motor. With another variable parameter, namely the IR-compensation parameter, the angled line rotates around point B as required by the motor.

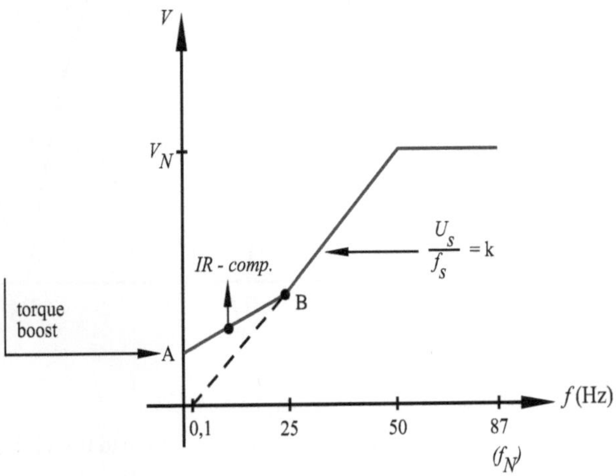

Fig. 20-20: *V/f* compensation and torque gain at start-up

6.5.3 Maximum current

It is possible to limit the output current of a frequency converter just as it is with any electronic motor drive . The output current $I_{out\ progr.} = I_{op}$ is adjustable as a percentage of the maximum possible load current $I_{out\ max.} = I_{om}$ of the converter.

Example $I_{op} = \frac{60}{100} \cdot I_{om}$. With this I_{op} - setting we can safely operate a motor, even when the nominal motor current is much smaller than the nominal output current I_{om} of the frequency converter. If multiple motors are connected in parallel on the output of a frequency converter then an overload relay is required per motor. The relay should be able to function over the entire frequency range. The value I_{om} of the output current of the frequency converter should be larger than the sum of the individual motor currents.

6.5.4 Induction motor acceleration

The acceleration ($\Delta n/\Delta t$) of the motor is limited by the inertia of the motor and load.
Fig. 20-21 shows what this means and what the problems can be. Assume a counter torque M_t and a supply frequency f_1 . The operating point is A. A step increase of the stator frequency to f_2 normally produces *M-n* curve 2 for the motor. Due to the inertia the speed of the motor in the first instant remains unchanged (A → B) and thereafter the speed changes according to curve 2 to the point C: the motor is now operating faster, which is the intention.

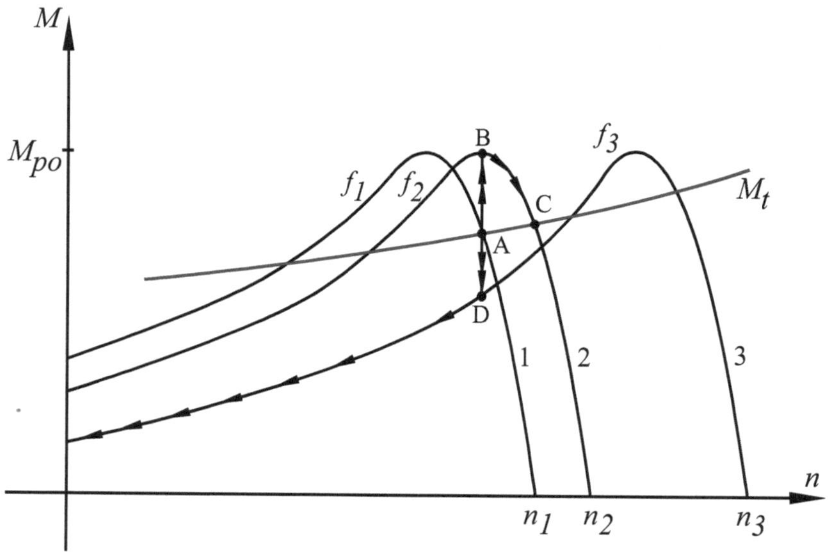

Fig. 20-21: *M-n* curves in the case of frequency change of the supply voltage

With a step change in the frequency from f_1 to f_3 the operating point can travel from A to D due to the inertia (speed unchanged). Then it follows curve 3: the motor slows to a stop since $M < M_t$. It is necessary to adjust the start-up time of the frequency controller as a result of the inertia of the drive train.

Example:
Frequency controller with f_N = 100Hz and variable start-up time between 0 and 30 seconds. This means that with a setting of 20 seconds the output frequency can run from 0 to 100 Hz in 20 seconds or from 0 to 40 Hz in 8 seconds, etc.

6.5.5 Induction motor braking

Consider a motor connected to a frequency-converter with output frequency f_1. The motor speed is given by $n_1 = n_{S1} \cdot (1 - g)$. The frequency of the converter is now reduced to the value f_2. If $- df/dt$ is so large that as a result of the inertia of the load the machines speed is faster than synchronous ($n > n_{S2}$) the machine will operate as a generator: see fig. 20-22. The stator rotating field (n_{S2}) now runs slower than the rotor (n_2) and the machine is operating as a generator.
The induction machine is now converting mechanical energy ($\omega^2 \cdot J_m/2$) to electrical energy which primarily ends up in the DC-link capacitor ($C \cdot V^2/2$) as the diodes of the rectifier do not allow (inverse) current to flow to the power grid. To prevent an excessive voltage increase in the DC-link the run-down time is an adjustable parameter. This parameter is normally expressed as the time to go from f_N to zero.
A practical value for the run-down time is between 0 and 30 seconds.

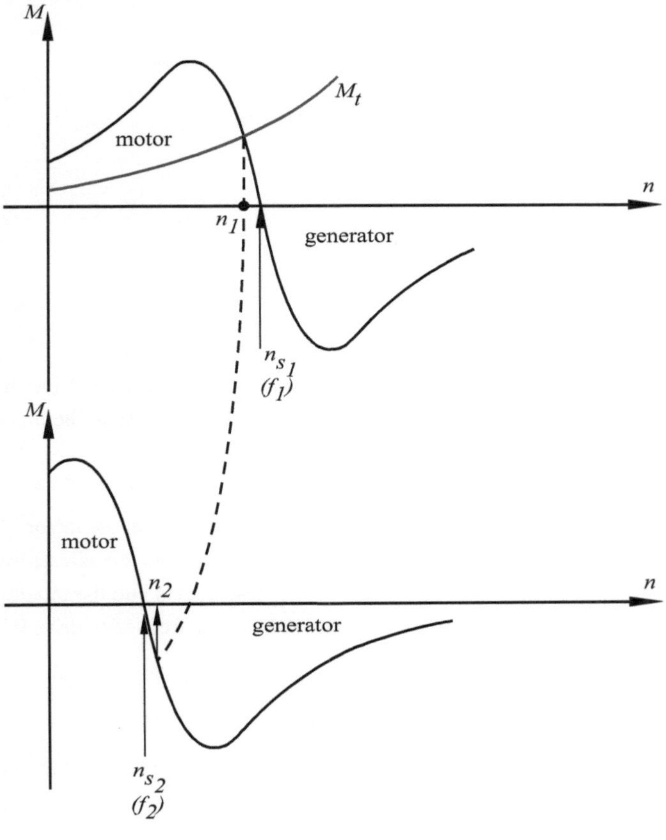

Fig. 20-22: Induction motor braking in generator service

The run down time is in fact a choice whereby the asynchronous machine runs down as a weak generator. The released energy is mostly dissipated as heat in the motor windings and the inverter so that the DC-link capacitor charge hardly changes.
If a faster run down time is required of the machine then a shorter time can be set and the energy be dissipated (externally) in a brake resistor which is switched in for this purpose. The brake resistor R_2 is connected in parallel to the DC-link via IGBT T_1 (see fig. 20-13).
If the DC-link voltage becomes too high (for example 800 V!) despite the brake resistor R_2 then the inverter is automatically switched off.
Fig. 20-23 shows a timing diagram of the adjustable parameters acceleration and run down time. With the majority of frequency drives the angled lines can be adjusted for a more "ergonomic" setting. By rounding off the angles we achieve a more S-shaped acceleration and run down curve.

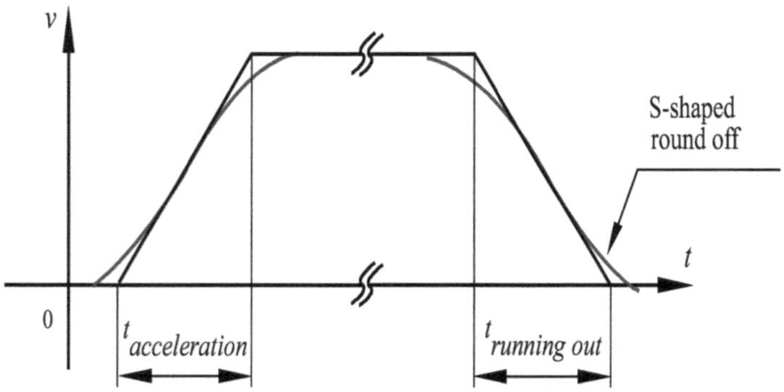

Fig. 20-23: Parameters acceleration and run down time

6.6 General remarks

6.6.1 Parameters

In the previous material we only discussed the main parameters. Frequency drives have a lot more adjustable parameters. For details see the technical information of the manufacturers.

6.6.2 Self tuning

With modern frequency-drives the most important parameters are set in the factory. They can of course be changed by the drive owner. More and more the parameters are adjusted with a self tuning routine. The frequency drive injects AC- and DC- test signals into the motor. From the test results the stator resistance and reactance are determined and the frequency drive will determine its own parameter settings (self tuning).

6.6.3 Direction of rotation of motor

To change the direction of rotation of a motor (in a very simple way) the conduction order of the inverter switches are changed.

6.6.4 Power factor on the supply side

Due to the uncontrolled rectifier on the supply side there will be a power factor of almost unity. Choke coils in the supply can limit the current harmonics. These chokes have other advantages:

- protection of the rectifier bridge against overvoltage
- reduction of overload of any cosφ - compensation battery capacitors that may be present

6.6.5 Braking

The uncontrolled rectifier does not allow energy recuperation to the power grid. For motor braking a temporary braking resistor connected to the DC-link can be switched in. Brake resistors are mainly used to work in quadrants 2 and 4, These resistors are most commonly present with:

- drives with large moment of inertia
- driving loads
- applications with fast cycles

It is also possible to quickly stop a motor using a DC-injection. After the inverter is switched off a DC-current is sent to a stator coil. The stationary stator field quickly brakes the rotating motor.

6.6.6 Traction

If we want to use an induction motor on an existing (DC-) traction net then PWM-control can be used. The PWM-inverter can be supplied direct from the DC grid.

6.6.7 Chokes between the converter and motor

Manufacturers of converters recommend the use of chokes if:

- more than two motors are connected in parallel on the output of a converter
- the supply cable between motor and converter is longer than 50m
- the motor has a poor power factor and low stator inductance

Due to the use of fast operating IGBT's there are a lot of "spikes" present in the output signal. The use of *dv/dt* filters is recommended. If a more sinusoidal output signal is required the so called sine filters may be used.

6.6.8 Earth faults

Earth faults in a motor or motor supply cable are easily detected by measuring the difference (if any) between the input and output current of the DC-link. Modern frequency converters offer protection against earth faults.

6.6.9 Trap circuits

Modern frequency drives can (e.g. after a power grid interruption), drive a rotating motor immediately by automatically producing a suitable output frequency.

6.6.10 Applications

In low voltage drives, frequency converters control motors in 2- and 4-quadrant operation in a power range from under 0.37kW to beyond 1MW. Typical applications include fans and pumps, conveyer belts, textile industry, control valves for flow control, lifting equipment, etc.
Drives for elevators for example have demanding requirements in terms of smooth operation and accuracy of position. Here, drives with outputs of 10kW to 250kW are used for 2- and 4-quadrant operation.
Medium-voltage drives with outputs of 500kW to 5MW are used extensively in heavy industry.
In recent years, multilevel and multi cell topology are increasingly gaining in importance in this market segment. Many IGBT or inverter cells are connected in series, allowing for significantly higher system voltages than the blocking voltage of the power semiconductor. This means that it is possible to use cost-efficient IGBT modules with just 1700V reverse recovery voltage in medium-voltage networks with line voltages of 3.3kV and higher. Moreover, multilevel systems produces low network harmonics, so reducing on filters.

6.6.11 PI(D)-controllers

These days frequency controllers have at least one or more PI or PID controllers on board.
By adding a measurement sensor a closed control loop is easily created.
Example:
We want to control with our frequency converter a fan or a circulating pump.
We need only a sensor to measure the water or air flow. By using the internal PI-controller we can simply realise a flow control.

6.6.12 Communication with the converter

Frequency converters can be controlled in a number of ways:
1 analogue: via a 0-10 V or 4-20 mA signal
2. digital: via a parameterised value or via a digital input
3. serial: mostly an RS485 communication bus (USS protocol) is possible. For point to point communication an RS232 connection may be used
4. via integrated communication (PROFIBUS, PROFINET, CANopen,...)

6.6.13 Technical specifications

Minimal specifications (with an example) are:
- power grid (3x400V/50 Hz)
- maximum power (37 kW)
- output frequency (0.1 to 300 Hz)

In addition to the basis parameters (acceleration, IR-compensation,...) the manufacturer will concentrate on “special” aspects of his product: PI-controller, fault diagnostics, number of binary and analogue inputs and outputs, simplicity of commissioning, integrated PLC functions, ECO (energy saving operation), USB-port, IOP-interface (Intelligent Operator Panel), integrated line filter, etc.

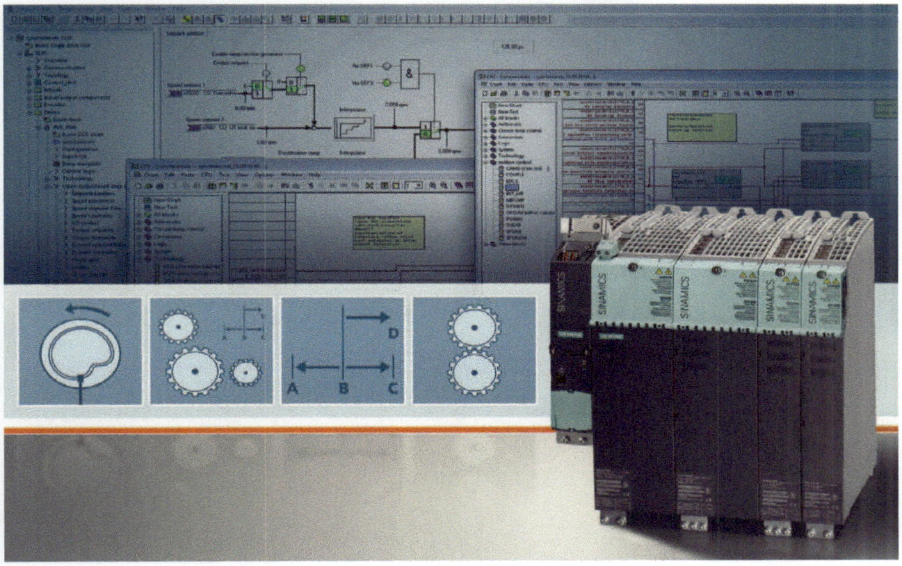

Photo Siemens: Siemens is extending the range of applications for its Sinamics S120 drives family with new, softwarebased Advanced Technology Functions. The specific functions concerned are synchronous operation (1:1 or with gear ratio), camming, and positioning of the synchronous axes

Photo Siemens: The second generation of the Sinamics G120 from Siemens is a modular drive with a higher power density and space-saving frame size. The higher power density is attributable to the new PM240-2 power module. The re-engineered drive range is now available in three voltage versions for connection to 200 V, 400 V and 690 V lines

Photo Siemens: This compact device was developed for universal application in industrial environments. The G120C can be used for driving pumps, compressors, fans, mixers, extruders, conveyor belts and simple handling machines. Powers from 0.55 to 18.5 kW. The frequency controller takes up 30% less room and has 40% more power density than previous types. The Siemens "Safety Integrated" safety technology is included as standard so that the drive can be stopped

Foto Siemens: Crane for loading and unloading container ships

6.7 Four quadrant operation with anti-parallel B_6-bridges

Since we can operate a motor in both directions with a frequency converter, switching in a brake resistor in the DC-link allows four quadrant operation. The brake energy is fed into the resistor. If we want to recuperate this energy into the power grid then the diode rectifier has to be replaced with two anti-parallel B_6-converters. This is drawn in fig. 20-24.

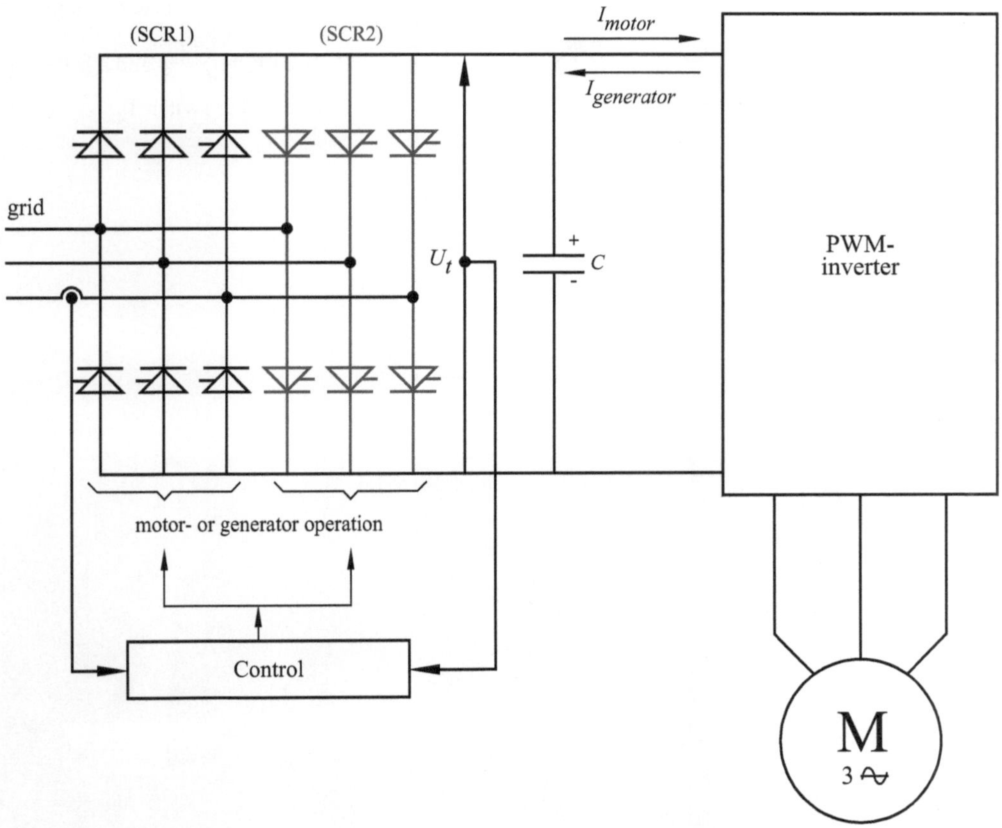

Fig. 20-24: VSI for four quadrant operation with energy recuperation

The B_6-bridge SCR_1 is switched on during motor operation. If we want to brake the machine then the inverter frequency is lowered. At the same time SCR_1 is blocked and SCR_2 is used as an inverter ($90° < \alpha < 150°$) so that energy flows to the power grid. To prevent uncontrollability the first bridge SCR_1 should operate with $\alpha > 30°$ so that we automatically have $\alpha < 150°$ for the second bridge SCR_2 and "firing failure" is avoided.
Analogous with this is the problem that with a weak net, inverter "firing failure" can occur. The result of this will be blowing a fuse designed to protect the thyristors.

6.8 Active in-feed (Siemens)

6.8.1 Raising DC-link voltage -reducing harmonics

One means of producing a sufficiently high output voltage from a PWM-converter is to raising the DC-link voltage. A simple method to increase the DC-link voltage is to use a boost converter as studied in fig. 13-7.

This converter is redrawn in fig. 20-25.

Here $V_t = \frac{V_1}{1 - \delta}$ and δ is the duty cycle of chopper switch T_4.

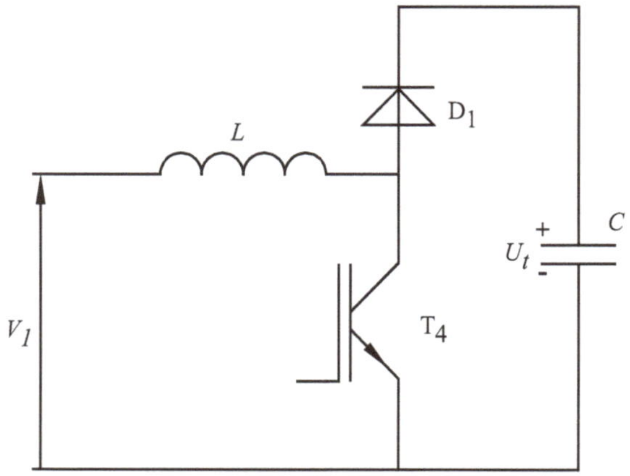

Fig. 20-25: Boost converter

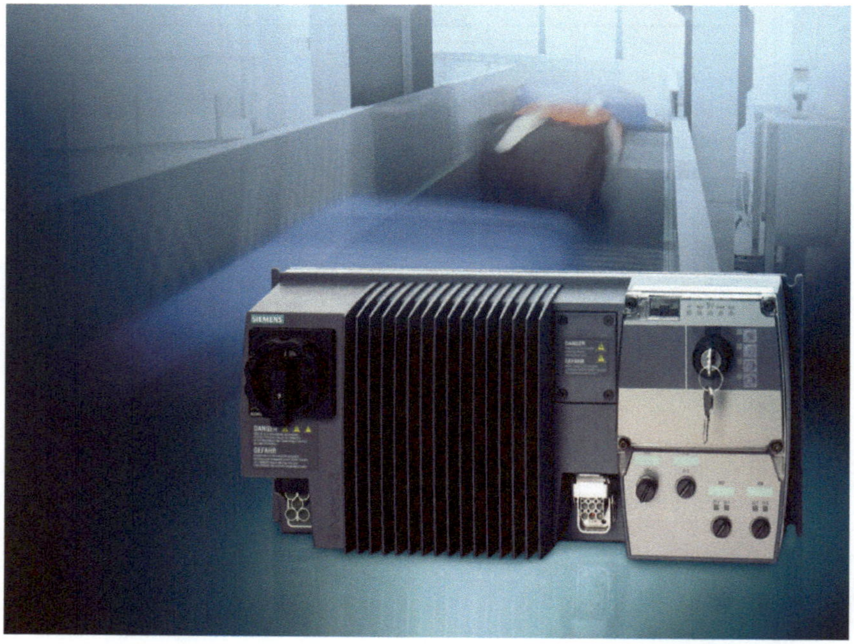

Photo Siemens: This Sinamics G110D frequency converter with IP65 was specifically designed for simple industrial applications for material handling in which a requirement is distributed communication.
The most important applications are airports and logistic distribution centers. Integrated brake control; integrated EMC-filter. Output: 0...650Hz; 0.75 to 7.5kW. Supply: 380...500V , 47...63Hz

6.8.2 Input inverter

If we add an IGBT bridge to the rectifier bridge of fig. 20-13 then we end up with the configuration shown in fig. 20-26.

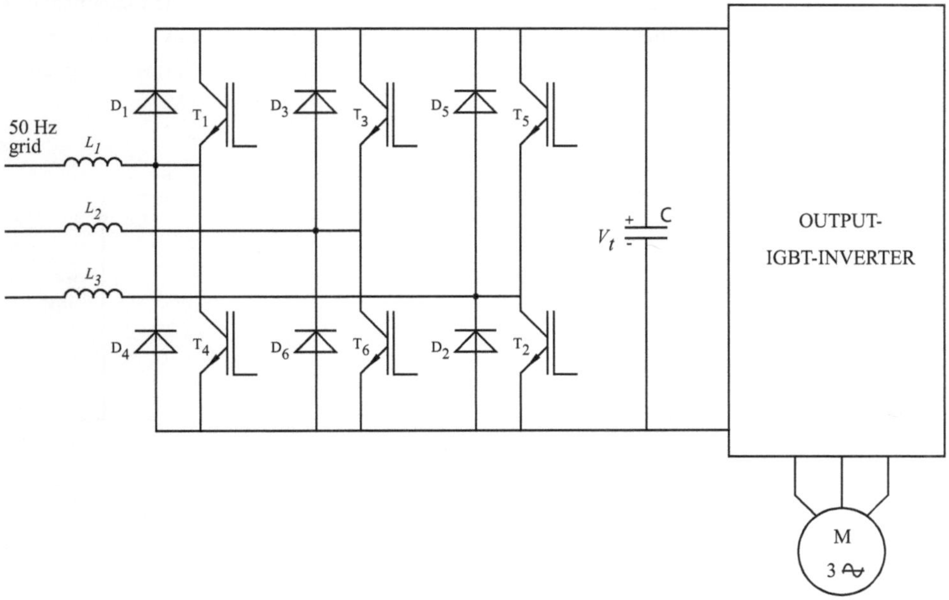

Fig. 20-26: Frequency controller with “active” input

The boost converter of fig. 20-25 apparently is also present in fig. 20-26. During the positive half of every phase voltage there is now a boost converter ($L_1/T_4/D_1/C$; $L_2/T_6/D_3/C$; $L_3/T_2/D_5/C$!). In this way a DC-link voltage of 640 V can be created with a supply of 3 x400V. The chopper frequency is a few kHz.

Since we are working with this elevated DC-link voltage, a practically sinusoidal PWM is used and the motor currents have only a small harmonic content. The motor current is quasi similar to the fundamental. The motor leakage reactance aids in filtering the higher harmonics.

6.8.3 Active input

Assume the motor is braking. The asynchronous machine is operating as an alternator and sends energy back to the DC-link capacitor so that V_t increases. We can now let the input IGBT bridge operate as inverter and the DC voltage from the DC-link is converted to a three-phase voltage which can send current back into the power grid: energy regeneration or recuperation.

Siemens refers to this circuit as an “active in-feed” and provides these frequency converters in the range 16kW-1200kW.

If insufficient energy is fed back to the power grid the DC-link voltage increases. For a 3x400 V grid the normal DC-link voltage with “active in-feed” is 600V. At $V_t \geq 810$V an error message is generated.

6.8.4 Operation "active in-feed"

Assume that the input-IGBT-bridge is operating as an inverter and from the DC-link voltage V_t a three-phase voltage is formed with as RMS value equal to that of the fundamental: V_{inv} .The frequency of this fundamental during nominal operation is the same as the power grid frequency. The equivalent diagram of the input as shown in fig. 20-26 appears as drawn in fig. 20-27.

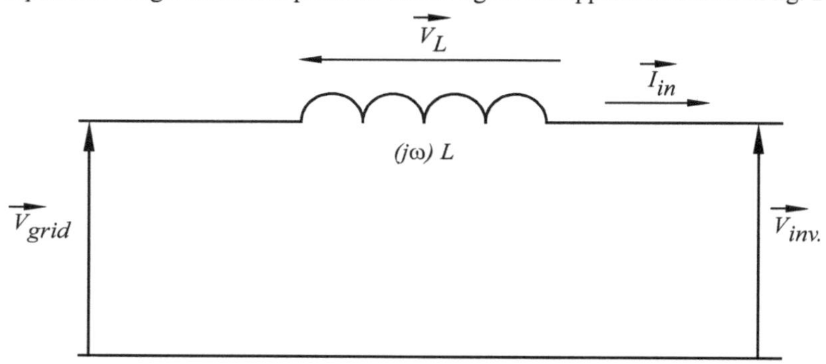

Fig. 20-27: Input circuit of the frequency converter

By controlling the IGBT bridge we can control the following variables:

1. frequency of $\vec{V}_{inv}$
2. value of $\vec{V}_{inv}$

We now look (vectorial) at the influence of controlling $\vec{V}_{inv}$. Here we use the expression $\vec{V}_L = \vec{V}_{grid} - \vec{V}_{inv}$, which follows from fig. 20-27.

1. Assume $\vec{V}_{inv}$ in phase with $\vec{V}_{grid}$. Fig. 20-28a shows the vector diagram. Since $\vec{I}_{in}$ always lags by 90° with respect to $\vec{V}_L$ then $\vec{I}_{in}$ also lags 90° by $\vec{V}_{grid}$.
   Only reactive power flows from the power grid to the DC-link. If $V_{inv} > V_{grid}$ then fig. 20-28b shows that reactive power is exchanged from DC-link to power grid.

2. If the input inverter is controlled in such a way that $\vec{V}_{inv}$ lags $\vec{V}_{grid}$, then we see in fig. 20-28c that the power grid is sending active power to the frequency converter. This finally reaches the motor. If we make $\vec{V}_{inv}$ a little larger (fig. 20-28d) or a little smaller then $\vec{V}_{grid}$ (fig. 20-28e), then we see that the power grid is capacitively or inductively loaded.
3. If we allow $\vec{V}_{inv}$ to lead $\vec{V}_{grid}$, then we see in fig. 20-28 f that active power flows from frequency converter to the power grid. In fig. 20-28g active and reactive power flows from the DC-link to the power grid.

Conclusion: by suitably controlling the input IGBT bridge active or reactive power can flow from or to the power grid. The displacement factor can be unity (fig. 20-28c) which is usely desirable. This converter can also be used to correct the power factor of the grid (fig. 20-28 d and e).

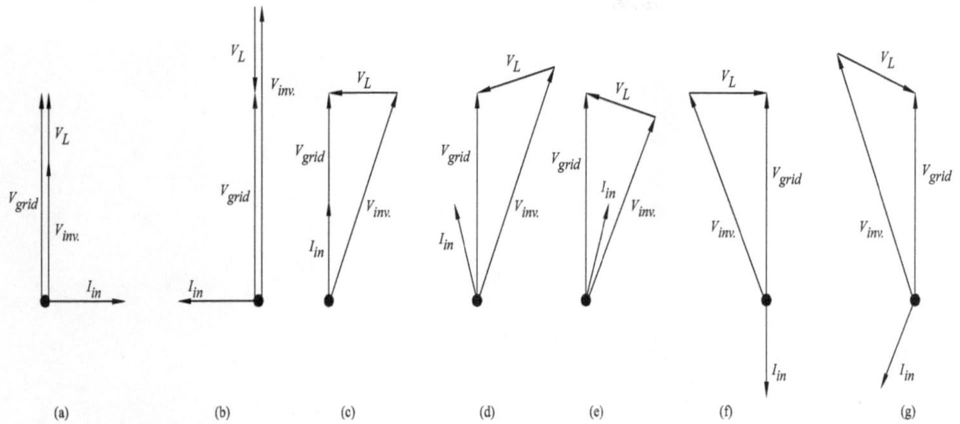

Fig. 20-28: Vector diagrams associated with fig. 20-27

6.8.5 Sinusoidal supply current

Due to the elevated DC-link voltage the (PWM) frequency controller delivers a current with little harmonics. This is identical to delivering output current in motor operation.
In addition to the three line inductors L (fig. 20-26) in the Siemens "Active in-feed" drives a low pass filter is also included in the supply cable. This filter is referred to as the "clean power filter". Here the current is practically sinusoidal.
The use of the boost-converter has another fortunate advantage: fluctuations in power grid voltage are automatically compensated. The boost-converter will maintain a constant DC-link voltage.

6.8.6 Control loop

A. Frequency of $\overrightarrow{V}_{inv}$

How can we cause $\overrightarrow{V}_{inv}$ to lead or lag $\overrightarrow{V}_{grid}$. Very simply by increasing or decreasing the frequency of $\overrightarrow{V}_{inv}$. In nominal service the frequency of $\overrightarrow{V}_{inv}$ is the same as the power grid frequency, but a temporary increase or decrease will cause $\overrightarrow{V}_{inv}$ to lead or lag $\overrightarrow{V}_{grid}$, according to the need. The frequency of $\overrightarrow{V}_{inv}$ determines the phase displacement and therefore the active power that the power grid and inverter must exchange. In practice the set point of the converter frequency is determined by the set point of the motor torque.

B. Amplitude of $\overrightarrow{V}_{inv}$

The value of the inverter voltage V_{inv} determines the magnitude of the exchanged reactive power. The control loop that determines the value of V_{inv} is controlled by the error signal between set point and measured reactive power.

Photo Maxon Motor: Mars Rover with 39 Maxon DC motors. The Mars Rover serves also as cover photo for Maxon Motor

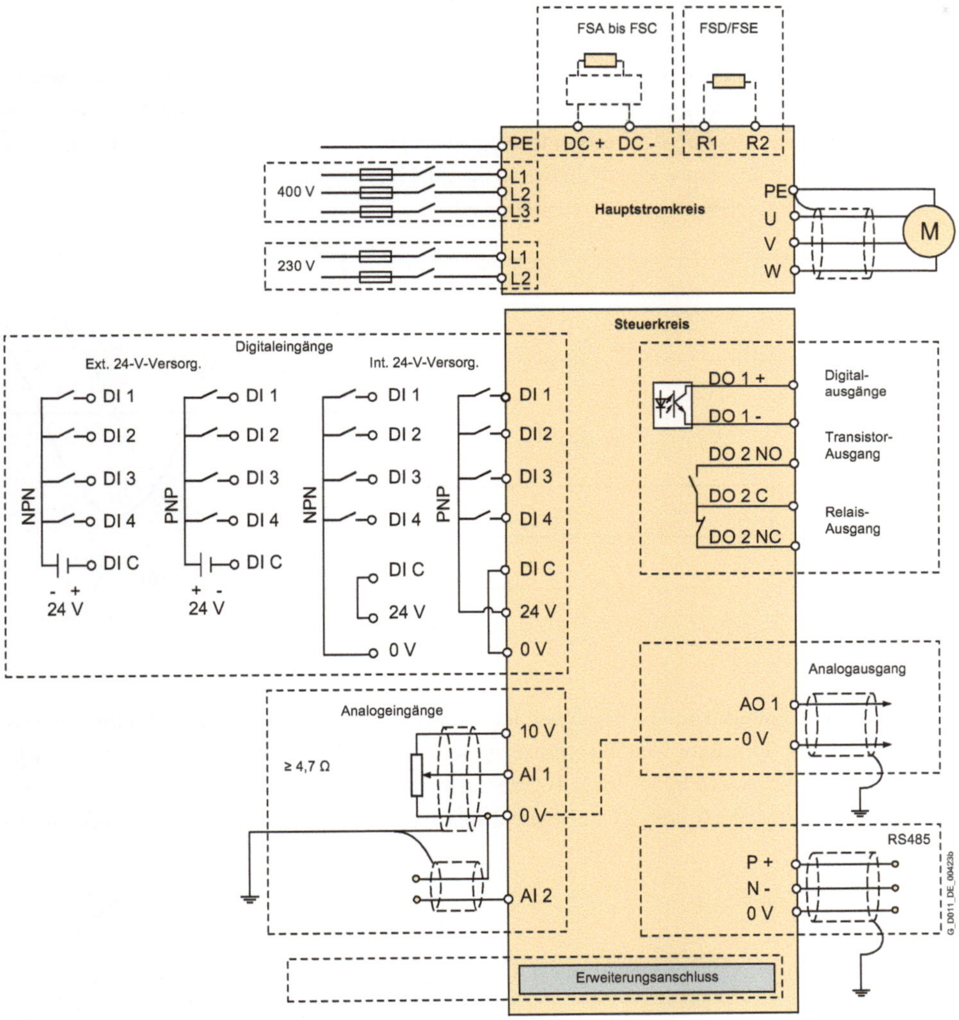

Photo Siemens: Connection diagram SINAMICS V20 .

With five frame sizes, it covers a power range extending from 0.12 kW up to 30 kW.

7 VECTOR CONTROL

7.1 Two phase induction motor in sinusoidal service

In the beginning of electrical distribution two phase networks existed. These power grids had in principle four conductors as shown in fig. 20-29a. Here we had two AC-voltages which were displaced by 90° with respect to each other. This is shown in fig. 20-29b. It is also possible to have a common supply wire as shown in fig. 20-29c.

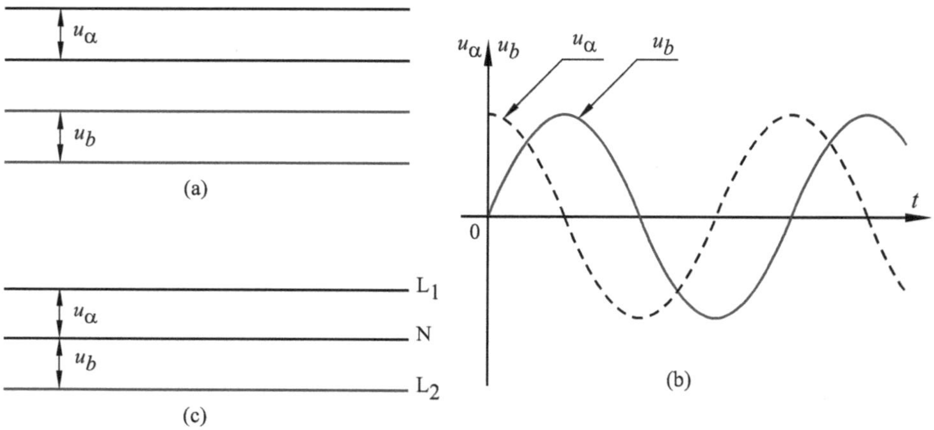

Fig. 20-29: Two phase grid

We could also build a two-phase induction motor in a similar fashion to a three-phase induction motor. The stator winding consists of, for a two pole machine, two coils (α and *β*) which are spatially displaced by 90°. This is graphically represented in fig. 20-30.
Two currents (i_α and i_β) which are displaced by 90° (in time) are now sent through the coils.

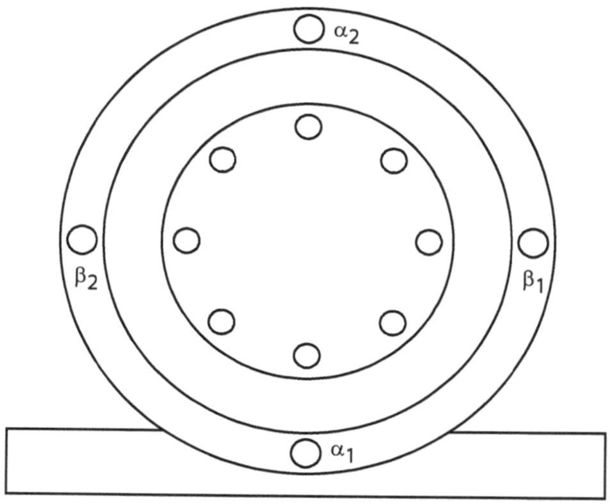

Fig. 20-30: Two-phase asynchronous motor

Both current carrying coils produce a magnetic field. The results of both fields are qualitatively determined in fig. 20-31.

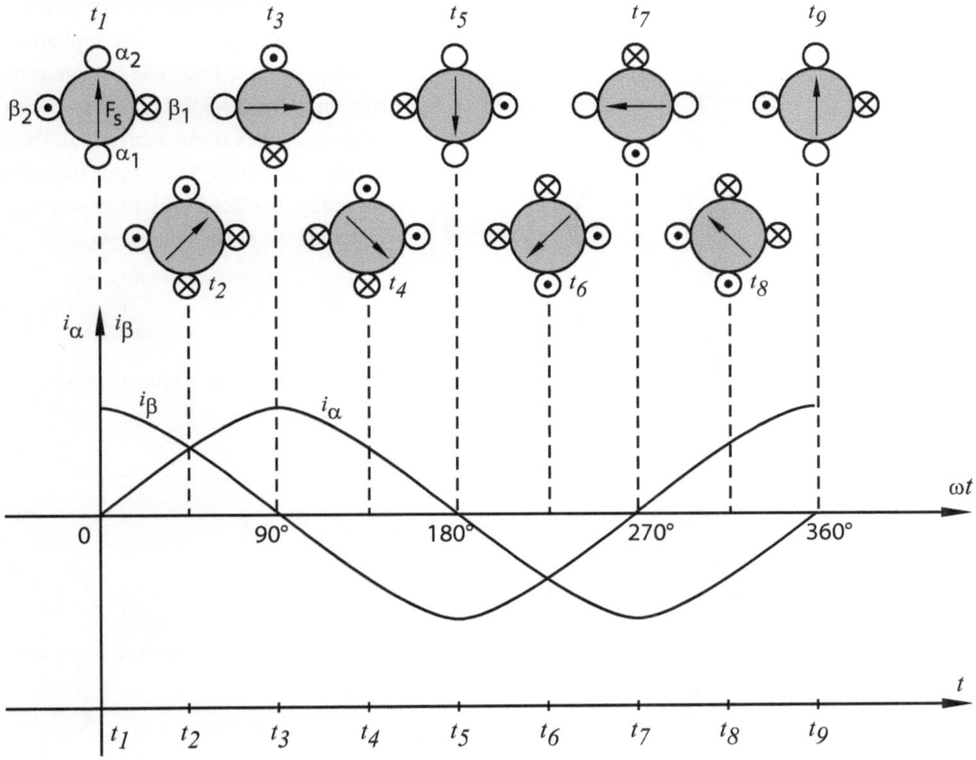

Fig. 20-31: Stator rotating field in the asynchronous motor of fig. 20-30

In the stator iron, just as was the case in a three-phase induction motor, a rotating field Φ_S with speed $n_S = \frac{60 \cdot f_S}{p}$ is created. A cage rotor will be pulled along by this stator rotating field. The rotor also has a rotating field Φ_R that synchronously rotates with the stator flux Φ_S. The rotor rotates with n rpm and manifests a slip $g = \frac{n_S - n}{n_S}$. We recognise the properties of a three-phase machine. The *M-n* curves also look the same.

7.2 Three-phase to two-phase transformation (and vice versa)

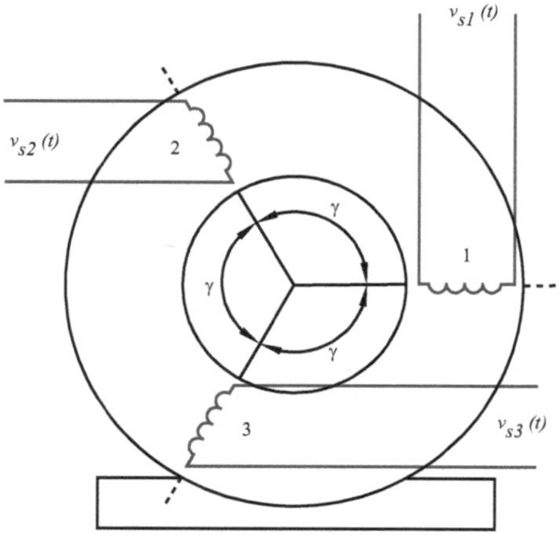

Fig. 20-32: Three-phase system

Assume a three-phase system with voltages as shown in fig. 20-32. The applied voltage is shown in fig. 20-32. The voltage vectors are drawn in fig. 20-33b in the directions (1), (2) and (3) as they occur spatially and as applied to the stator coils. The vector sum of these spatially displaced time functions (v_{S1}, v_{S2}, v_{S3}) is called the stator voltage vector:

$$\vec{v}_S = \vec{v}_{S1} + \vec{v}_{S2} + \vec{v}_{S3} \tag{20-12}$$

We refer to this as a **space vector.**

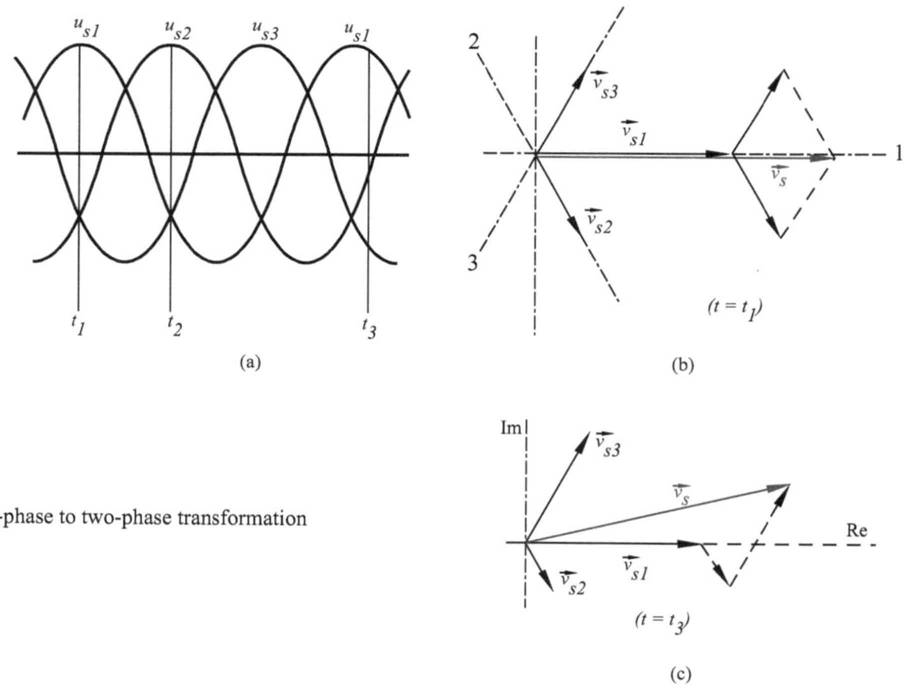

Fig. 20-33: Three-phase to two-phase transformation

In fig. 20-33b and c we have drawn the stator voltage vector at two instants. If we were to do this over an entire period then we would see that $\vec{v}_S$ rotates by 360°. With a sinusoidal progression of the phase voltages, $\vec{v}_S$ is constant and equal to1.5 times the amplitude of one phase voltage. The amplitude of $\vec{v}_S$ can be seen in fig. 20-33b and c.

Here we can conclude that the stator rotating field $\Phi_S = \frac{3}{2} \cdot \Phi_{S1}$.

A similar construction as with the three-phase currents i_{S1}, i_{S2}, i_{S3} provides the stator current vector $\vec{i_S} = \vec{i_{S1}} + \vec{i_{S2}} + \vec{i_{S3}}$. This can be generally expanded in fig. 20-34 and simultaneously we determine the components of $\vec{i_S}$ in a two axis system β, α.

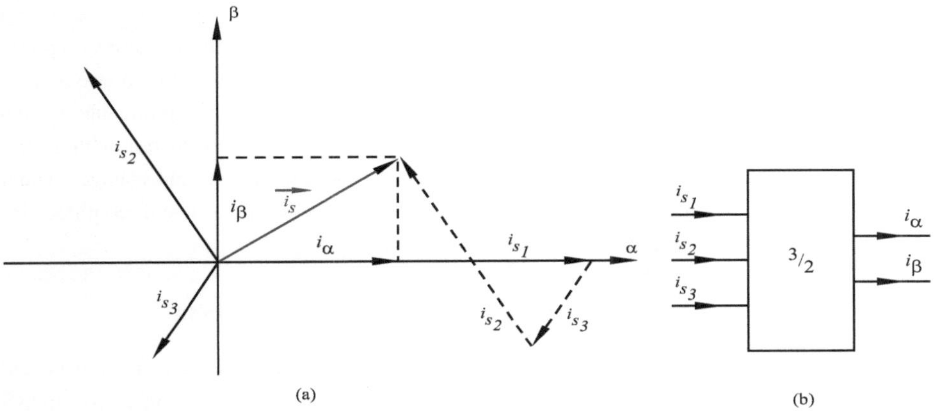

Fig. 20-34: Stator current vector

According to the α-axis: $i_\alpha = i_{S1} - i_{S2} \cdot \sin 30° - i_{S3} \cdot \sin 30° = i_{S1} - \dfrac{i_{S2} + i_{S3}}{2}$

with $i_{S1} + i_{S2} + i_{S3} = 0$ becomes $i_\alpha = \frac{3}{2} \cdot i_{S1}$

According to the β-axis: $i_\beta = 0 + \frac{\sqrt{3}}{2} \cdot i_{S2} - \frac{\sqrt{3}}{2} \cdot i_{S3} = \frac{\sqrt{3}}{2} \cdot (i_{S2} - i_{S3})$

Conclusion:

$$i_\alpha = \frac{3}{2} \cdot i_{S1} \qquad (20\text{-}13)$$

$$i_\beta = \frac{\sqrt{3}}{2} \cdot (i_{S2} - i_{S3}) \qquad (20\text{-}14)$$

The expressions (20-13) and (20-14) represent a three-phase to two-phase transformation in the case of a symmetrical three-phase system. This is known as a Clarke transformation.

In a similar fashion for such a symmetrical three-phase system we can also write expressions for the components of the stator voltage:

$$v_\alpha = \frac{3}{2} \cdot v_{S1} \qquad (20\text{-}15)$$

$$v_\beta = \frac{\sqrt{3}}{2} \cdot (v_{S2} - v_{S3}) \qquad (20\text{-}16)$$

From (20-13) and (20-14) we can, with some simple calculations, determine the inverse transformation: $i_{S1} = \frac{3}{2} \cdot i_\alpha$; $i_{S2} = \frac{1}{\sqrt{3}} \cdot i_\beta - \frac{1}{3} \cdot i_\alpha$; $i_{S3} = -\frac{1}{\sqrt{3}} \cdot i_\beta - \frac{1}{3} \cdot i_\alpha$

This is the inverse Clarke transformation.

A similar transformation is of course possible for the voltages.

7.3 Equivalent two phase motor

We could ask ourselves what the currents i_α and i_β should be so that the two-phase motor in fig. 20-30 would be equivalent (same flux, torque, slip) with a three-phase motor with stator currents i_{S1}, i_{S2} and i_{S3}. The answer lies hidden in expressions (20-13) and (20-14). If we send a current $i_\alpha = \frac{3}{2} \cdot i_{S1}$ through coil α in fig. 20-30 and a current $i_\beta = \frac{\sqrt{3}}{2} \cdot (i_{S2} - i_{S3})$ through coil β, then the motor in fig. 20-30 is exactly equivalent to the three-phase motor of fig. 20-32 (with phase currents i_{S1}, i_{S2} and i_{S3}). This is referred to as an equivalent two-phase motor.
The following study of the equivalent two-phase motor is directly applicable to the three-phase motor. Note that everything we have concluded up to now in connection with three to two phase transformations is equally applicable for non-sinusoidal currents and voltages. The only condition for application of (20-13) and (20-14) is that we have a symmetrical three-phase system. This is of course the case with an asynchronous motor.

Example of a 2- to 3-phase transformation and vice versa

When the switch over was made from two-phase to three-phase networks it was possible to operate with the two-phase devices on a three-phase network by using a Scott-transformer. Fig. 20-35 shows such a transformer. For every coil a specific winding ratio is given.
The associated vector diagrams illustrate the operation.

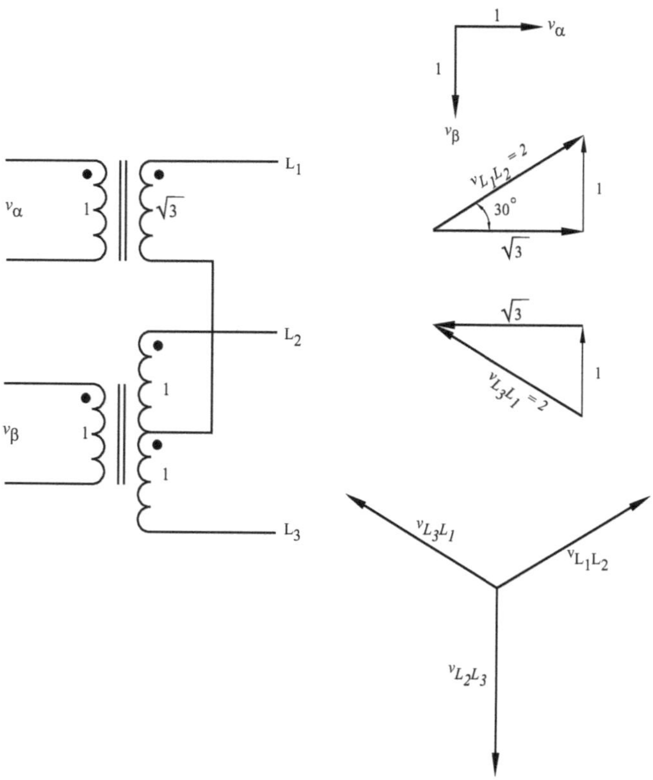

Fig. 20-35: Scott-transformer

Fig. 20-36 shows an equivalent two-phase motor compared to a three-phase motor.

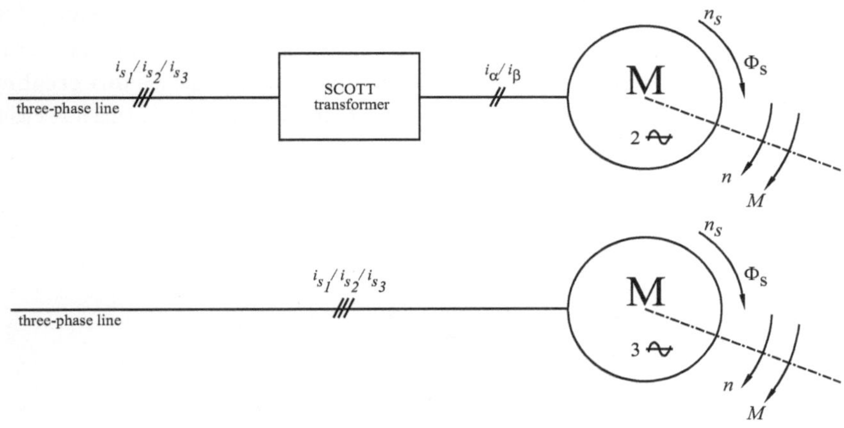

Fig. 20-36: Three-phase asynchronous motor with equivalent two-phase asynchronous motor

7.4 Flux vector and torque vector of an equivalent two phase motor

Fig. 20-37 shows a two-phase motor in which the coils α and β are normally supplied by sinusoidal currents i_α and i_β which are 90° displaced in time with respect to each other. At the instant that $i_\alpha = \hat{i}_\alpha$ (maximum positive) is $i_\beta = 0$.

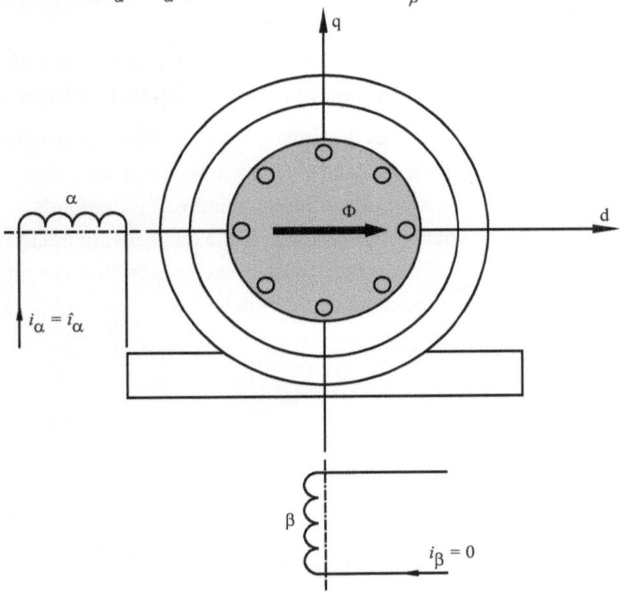

Fig. 20-37: Two phase motor

We now draw a d-q coordinate system. Here d is the axis of the α-coil and q the axis of the β-coil. The notation d and q are abbreviations for **d**irect and **q**uadrature axis, this means that the q axis is perpendicular to the d axis.

7.4.1 Flux component

We replace the coils α and β with two fictive coils D and Q. This is shown in fig. 20-38. A DC-current i_d, just as large as $\hat{i}_\alpha$ was, is sent through D. A flux results as drawn in fig. 20-38. We refer to this flux as Φ_R .The current i_d is called the **flux creating component.** While the DC-current i_d increases from zero to the nominal value the flux changes between zero and the nominal value Φ_R.

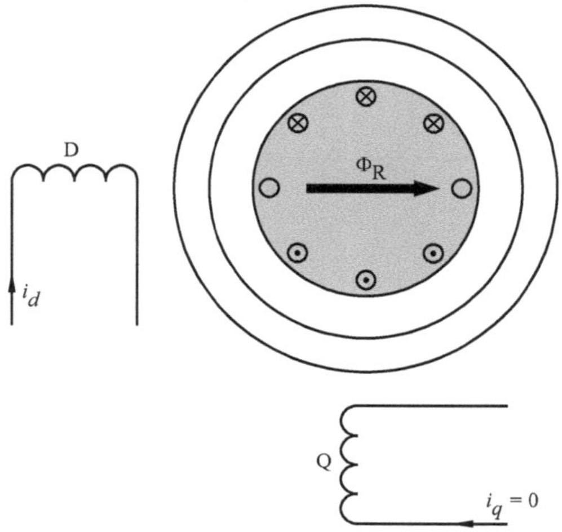

Fig. 20-38: Flux creation in a two-phase motor

During the flux change an induced emf is produced in the rotor conductors and with a short-circuited rotor circuit, currents flow in the direction shown in fig. 20-38. Due to the self inductance of the rotor it takes about five times the time $\tau_R = \frac{L_R}{R_R}$ until the flux reaches its nominal value. At that instant Φ_R is constant and the induced currents in the rotor disappear.

τ_R is called the rotor time constant. This is usually at least a hundred milliseconds.

In the T- equivalent diagram of an asynchronous motor L_0 is the magnetising inductance. The flux Φ_R could be described as $\Phi_R = L_0 \cdot i_d$. This is not always correct since the rotor time constant results in an (exponential) delay, and so the exact expression is:

$$\tau_R \cdot \frac{d\,\Phi_R}{dt} + \Phi_R = L_0 \cdot i_d \qquad (20\text{-}17)$$

Calculation of (20-17) gives: $\Phi_R = L_0 \cdot i_d \cdot [\,1 - e^{-t/\tau_R}\,]$

The flux will thus exponentially rise (or fall). Its **nominal value** is: $\Phi_R = L_0 \cdot i_d$.

7.4.2 Torque creating component

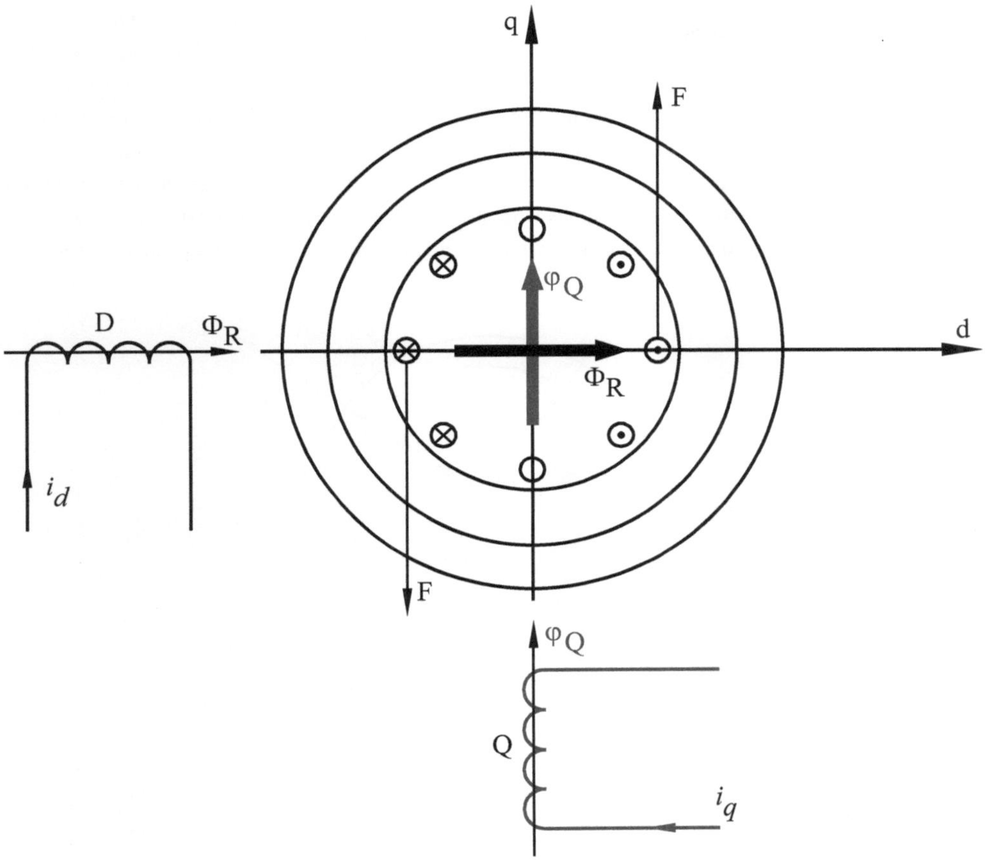

Fig. 20-39: Two phase motor: torque creation

Assume a flux which has reached its nominal value Φ. We now send a DC-current i_q through coil Q. The result is that coil Q creates a flux φ_q. This flux will, in the transition from zero to nominal value, result in induced voltages and currents in the cage rotor. The direction of these currents is indicated in fig. 20-39. The result is that these induced currents together with the flux form a torque. This torque is also drawn in fig. 20-39.

In fact during the transient state of φ_q , the operation of a DC-machine is simulated: namely the operation of a current carrying conductor (rotor current) on a flux (Φ_R). The current in the rotor conductors is the result of a changing i_q .

We therefore refer to i_q as the **torque creating component**.

7.4.3 Continuous motor rotation

Once the transient phenomenon of φ_q is passed φ_q is constant and the induced rotor currents in fig. 20-39 and also the torque disappear. To maintain this torque we use a trick: we rotate the coil Q around the rotor (fig. 20-40) with a speed n_S. This constant, rotating flux φ_q, is experienced by the rotor conductors as a changing flux. The result is that now currents continue to be induced in the rotor conductors. Let us now simultaneously rotate coil D (perpendicular to Q) with Q, then Φ_R remains and therefore also the torque on the rotor. The rotor is pulled along in the same manner as the three-phase motor. It rotates with speed n rpm, which is not much slower than the coils D and Q. If the rotor has the same speed as Q then the induction currents would naturally cease, just as with the three-phase motor.

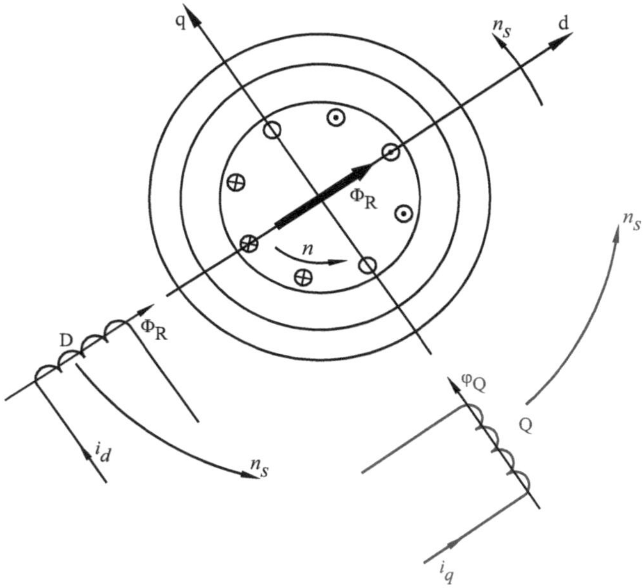

Fig. 20-40: Two phase motor with fictive rotating coils

The slip can also be defined as: $g = \frac{n_S - n}{n_S}$. The larger the counter torque M_t on the rotor, the larger the slip and the greater the induced current in the rotor will be.

Conclusion:

- the rotor flux rotates at n_S rpm
- the rotor rotates at n rpm
- the slip g increases as:
  - the counter torque M_t and the current i_q increase,
  - the rotor resistance increases or in other word the rotor time constant τ_R decreases,
  - the flux Φ_R (according to the d axis) is weaker.

Instead of the speed n_S we could also indicate the angular velocity $\omega_S = \frac{2 \cdot \pi \cdot n_S}{60}$ (rad/s).

In fig. 20-41 we have drawn the rotating d-q coordinate system at a certain instant with an angle λ given by:

$$\lambda(t) = \omega_S \cdot t \tag{20-18}$$

For this we have assumed that at $t = 0$ the d-q coordinate system corresponds to the α-β coordinate system.

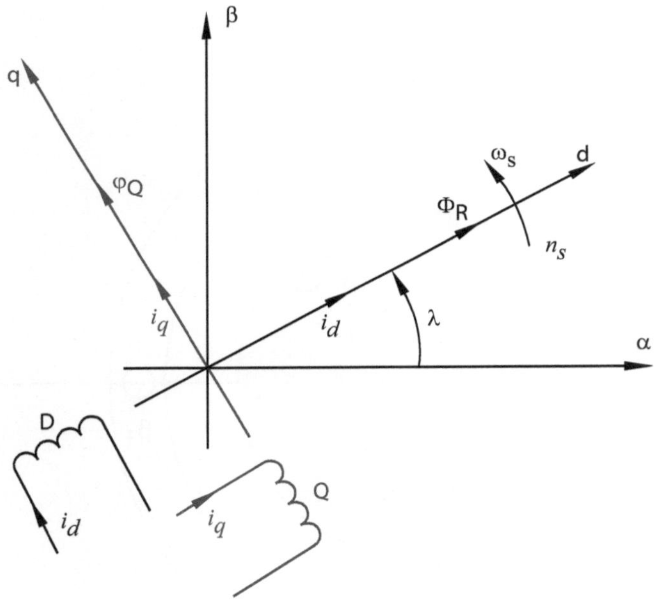

Fig. 20-41: Current and flux vector in fig. 20-40

From fig 20-39 it follows that the torque of an equivalent two-phase motor is given by the product of the flux Φ_R with the current in the rotor conductors. This current is proportional with the flux φ_q and therefore with the current i_q .

For the electromechanical torque of the motor we can write: $M_{em} = k_3 \cdot i_q \cdot \Phi_R$ (20-19)

Here k_3 is the machine constant.

Expression (20-19) is identical with that of a DC-machine : $M_{em} = k_2 \cdot I_a \cdot \Phi$ (19-2)

By controlling the current i_q in our equivalent two-phase motor we are able to control the torque just as we did with the armature current I_a of the DC-machine.

7.5 Coordinate transformation

7.5.1 Transformation formulas

The flux creating and torque creating components of the fictive coils D and Q are in reality obtained from two variable (e.g. sinusoidal) currents i_α and i_β in the real coils of our two-phase motor. These coils are fixed on the stator. We draw the α-β coordinate system that is related to the stator in such a way that the coil α produces a flux according to the α-axis and coil β according to the β-axis. (fig. 20-42).

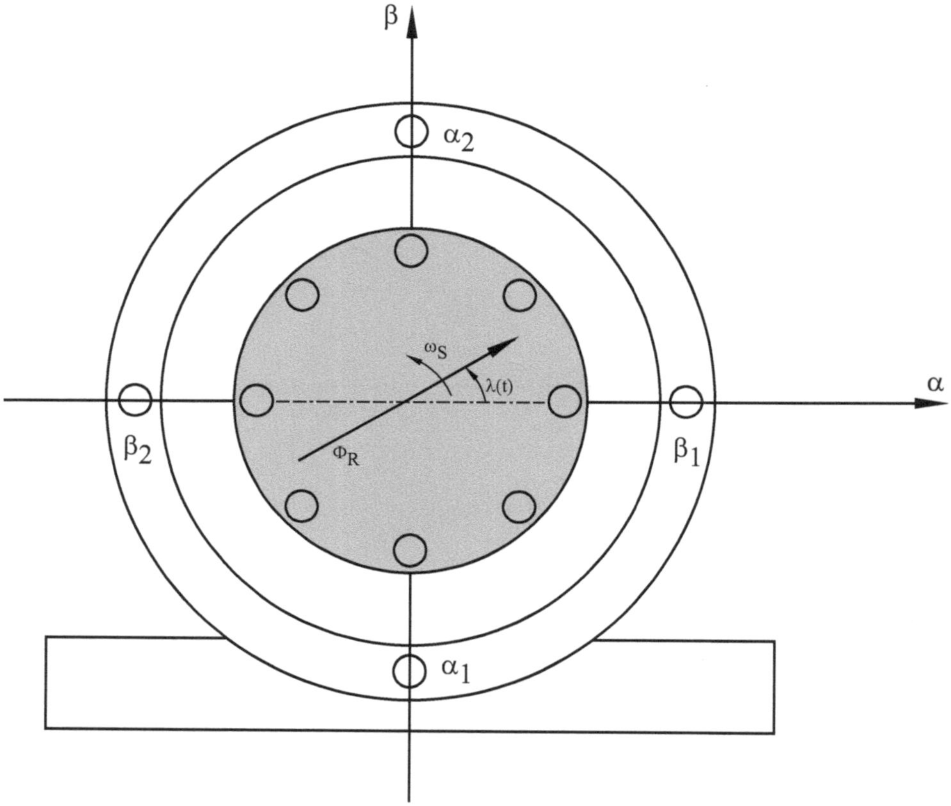

Fig. 20-42: Equivalent two-phase stator coordinate system

We know from fig. 20-31 that the rotating field in the stator has an angular velocity ω_S. As with the three-phase motor the rotor is pulled along with a speed $n < n_S$.
The rotor has a rotor flux Φ_R which also rotates at ω_S rad/s, similar to the three-phase motor.
If we now allow the fictive coils D and Q (and therefore the d-q coordinate system) to rotate with the rotor flux Φ_R as in fig. 20-41, then we can draw fig. 20-43.

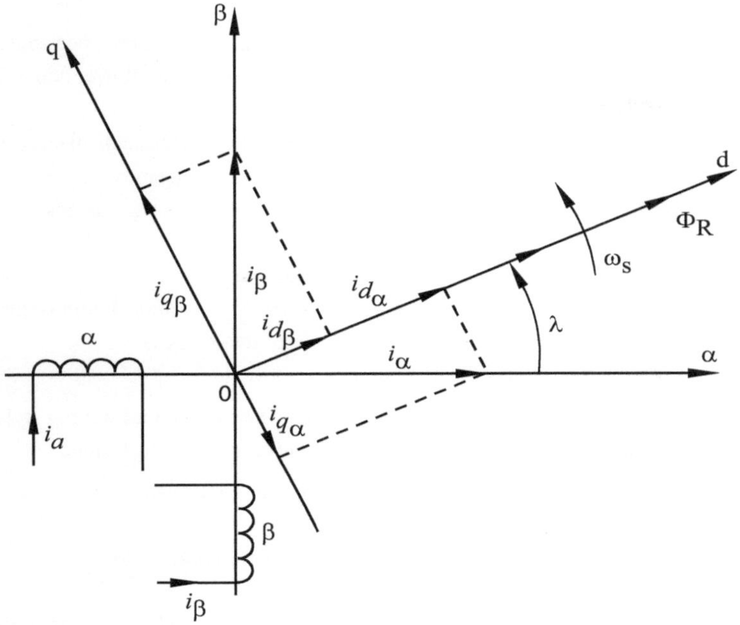

Fig. 20-43: d-q coordinate system, linked to the rotor flux

The current i_α produces the components in:
- d-direction: $i_{d\alpha} = i_\alpha \cdot \cos\lambda$
- q-direction: $i_{q\alpha} = -\, i_\alpha \cdot \sin\lambda$

The current i_β produces the components in:
- d-direction: $i_{d\beta} = i_\beta \cdot \sin\lambda$
- q-direction: $i_{q\beta} = i_\beta \cdot \cos\lambda$

The currents i_α and i_β from the stator coordinate system (α, *β*) provide currents i_d and i_q in a coordinate system that rotates with the rotor flux Φ_R :
- $i_d = i_{d\alpha} + i_{d\beta}$
- $i_q = i_{q\alpha} + i_{q\beta}$

or:

$$i_d = i_\alpha \cdot \cos\lambda + i_\beta \cdot \sin\lambda \tag{20-20}$$

$$i_q = i_\beta \cdot \cos\lambda - i_\alpha \cdot \sin\lambda \tag{20-21}$$

Conclusion: The expressions (20-13) and (20-14) give for a three-phase induction motor the currents i_α and i_β as transformed components of the three-phase stator current (i_{S1}, i_{S2}, i_{S3}). This equivalent system α-*β* is fixed to the stator. We call it the α-*β* stator coordinate system. The coordinate system d-q with its components i_d and i_q are called the field coordinate system since they are fixed to the (rotor) flux.
The expressions (20-20) and (20-21) describe the transformation from stator coordinate system to (rotor) field coordinate system. This is known as the Park transformation.

7.5.2 Notation for various coordinate systems

Consider figure 20-44. At the instant $t = 0$ the rotor flux Φ_R is for example horizontal. This flux rotates at ω_S rad/s, this is synchronous speed. The rotor itself rotates slower and at time t has for example an angle delay of δ.
If we look at the rotor flux with respect to the rotor shaft, then Φ_R has a displacement angle β.
With respect to the stator reference axis the rotor flux has a displacement angle $\alpha = \beta + \delta$.
Whether we take the stator shaft or the rotor flux as reference, nothing changes as far as the magnitude of $\overrightarrow{\Phi_R}$ is concerned.
We can use various notations for the rotor flux, for example:
$(\overrightarrow{\Phi_R})_{\text{rotor}}$ = rotorflux in a rotor coordinates system and $(\overrightarrow{\Phi_R})_{\text{stator}}$ = rotor flux in stator coordinates
The notation which we have chosen for the remainder of this study is:

- $\overrightarrow{\Phi_R}$ = rotor flux in a rotor coordinate system
- $\overrightarrow{\Phi_R} \cdot e^{j\delta}$ = rotorflux in a stator coordinate system. The exponential writing style only means that an angle δ is added to the angle of the flux in the rotor coordinate system.

In this way we find:

- $\overrightarrow{\Phi_R}\,;\ \overrightarrow{V_R}\,;\ \overrightarrow{I_R}$: vectors in rotor coordinates
- $\overrightarrow{\Phi_S}\,;\ \overrightarrow{V_S}\,;\ \overrightarrow{I_S}$: vectors in stator coordinates
- $\overrightarrow{\Phi_R} \cdot e^{j\delta}\,;\ \overrightarrow{V_R} \cdot e^{j\delta}\,;\ \overrightarrow{I_R} \cdot e^{j\delta}$: rotor values in stator coordinates
- $\overrightarrow{\Phi_S} \cdot e^{-j\delta}\,;\ \overrightarrow{V_S} \cdot e^{-j\delta}\,;\ \overrightarrow{I_S} \cdot e^{-j\delta}$: stator values in rotor coordinates

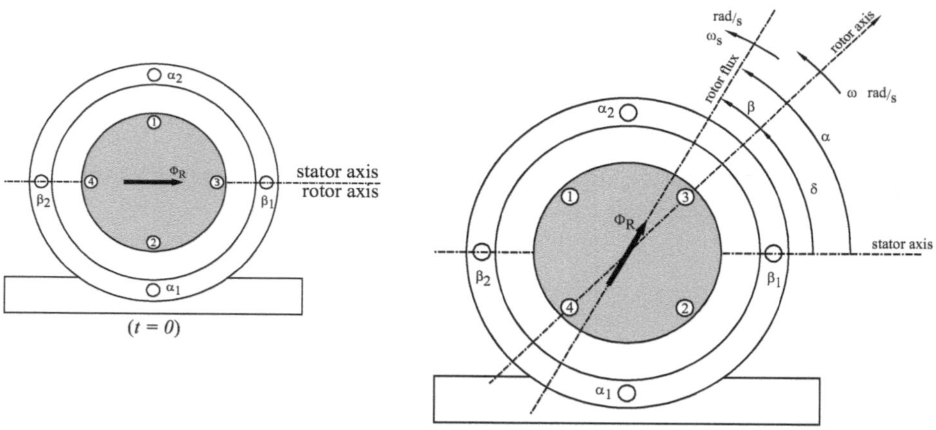

Fig. 20-44: Rotor flux, in both rotor and stator coordinate systems

During the study of the equivalent two-phase motor in which the d-axis was fixed to the rotor we took $\overrightarrow{\Phi_R} \cdot e^{j\delta}$ as the rotor flux, since in that study we looked from the perspective of the stator and the rotor flux should be described in stator coordinates. We redraw a part of fig. 20-43 in fig. 20-45 with the exact notation for the rotor flux: $\overrightarrow{\Phi_R} \cdot e^{j\delta}$. At the same time we have shown the stator current vector $\overrightarrow{i_S} = \overrightarrow{i_d} + \overrightarrow{i_q}$.

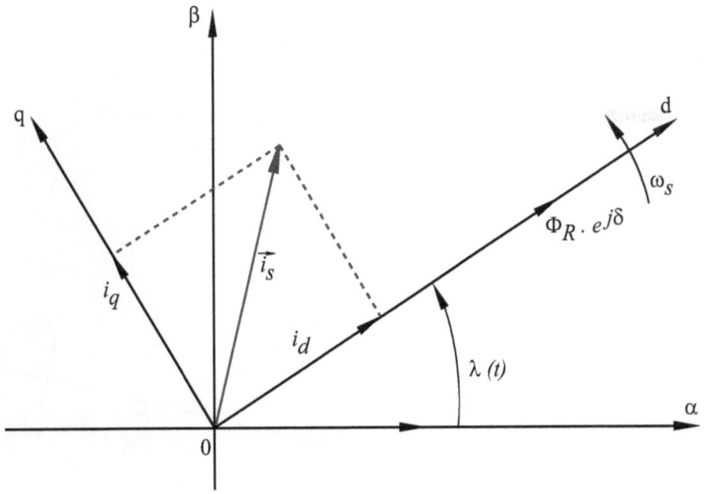

Fig. 20-45: Rotor flux in stator coordinates

In fig. 20-4 we see that $\vec{I_S} + \vec{I_R} = \vec{I_\mu}$. The rotorflux $\vec{\Phi_R}$ is formed (see fig. 20-4!) from the current $\vec{i_S}$ in the coil L_0, together with the current $\vec{i_R}$ in the coil L_R. If we now work in stator coordinates then we need to add a factor $e^{j\delta}$, so that:

$$\vec{i_S} \cdot L_0 + \vec{i_R} \cdot e^{j\delta} \cdot L_R = \vec{\Phi_R} \cdot e^{j\delta} \quad \text{or:} \quad \vec{i_S} + \frac{L_R}{L_0} \cdot \vec{i_R} \cdot e^{j\delta} = \frac{\vec{\Phi_R}}{L_0} \cdot e^{j\delta} \qquad (20\text{-}22)$$

We refer to $\frac{\vec{\Phi_R}}{L_0} \cdot e^{j\delta} = \vec{i_\mu}$ as the magnetizing current, on the stator side, that produces the rotor flux.

Expression (20-22) results in fig. 20-46:

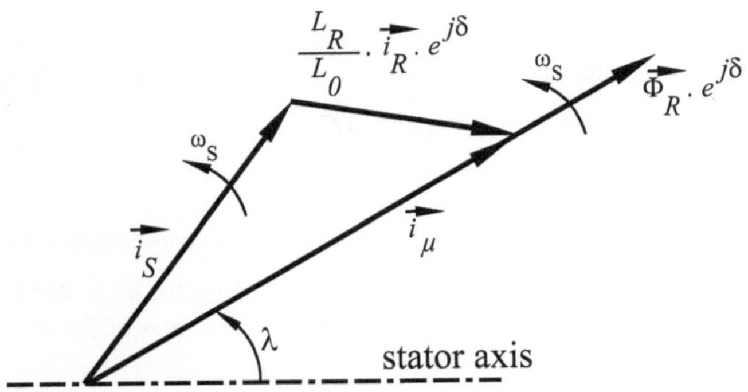

Fig. 20-46: Relationship between the space vectors $\vec{i_S}$, $\vec{i_R}$, $\vec{i_\mu}$ and $\vec{\Phi_R}$ in a stator coordinates system

7.6 Field orientated control

7.6.1 Field orientation-field coordinates

Combining the vector diagrams of fig. 20-45 and 20-46 leads to fig. 20-47.

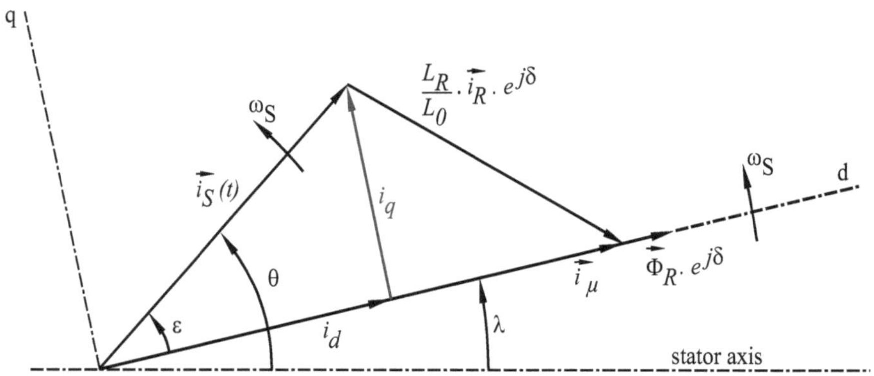

Fig. 20-47: Field orientation

Explanation of the symbols in fig. 20-47:

- $\lambda(t)$ = displacement angle of the rotor flux at instant t with respect to the stator reference axis (α-axis)
- $\delta(t)$ = angle between stator and rotor axis
- $\overrightarrow{i_\mu}$ = fictive magnetising current which, on the stator side, is responsible for the rotor flux
- $\overrightarrow{\Phi_R} \cdot e^{j\delta}$ = rotor flux in stator coordinates
- $\overrightarrow{i_S}\ (t)$ = stator current vector with components i_α and i_β in stator coordinates and i_d and i_q in field coordinates (rotating with the rotor flux)
- L_0 = magnetising inductance of the motor
- L_R = rotor self inductance
- $\overrightarrow{i_R}$ = rotor current vector = $\overrightarrow{i_{R1}} + \overrightarrow{i_{R2}} + \overrightarrow{i_{R3}}$
- $\overrightarrow{i_R} \cdot e^{j\delta}$ = rotor current in stator coordinates.

Since $\overrightarrow{i_S(t)}$ and $\overrightarrow{\Phi_R} \cdot e^{j\delta}$ have the same speed ω_S the angle ε is a time independent angle.

The stator current $\overrightarrow{i_S}$ is fixed orientated with respect to the (rotor) flux, dr. ir. Blaschke referred to this as field orientation. The currents i_d and i_q are the field coordinates of the stator current in a coordinate system that rotates with the rotor flux.

In addition to the familiar formulas (20-19) and (20-17) we will develop equation (20-45) under nr. 8. This describes the relationship between the angular velocity of the rotor and stator rotating field (ω_S) and the angular velocity ω ($= \frac{2.\pi.n}{60}$) of the rotor:

$$\omega_S - \omega = \frac{L_0 \cdot i_q}{\Phi_R \cdot \tau_R} \qquad (20\text{-}45)$$

The formulas (20-19), (20-17) and (20-45) form the **mathematical model** of an asynchronous motor in field coordinates:

$$M_{em} = k_3 \cdot i_q \cdot \Phi_R \qquad (20\text{-}19)$$

$$\tau_R \cdot \frac{d\,\Phi_R}{dt} + \Phi_R = L_0 \cdot i_d \qquad (20\text{-}17)$$

$$\omega_S - \omega = \frac{L_0 \cdot i_q}{\Phi_R \cdot \tau_R} \qquad (20\text{-}45)$$

7.6.2 Field oriented control

From what we have previously studied we can create a block diagram that will allow us to mechanically understand how the torque and (or) speed of a three-phase squirrel cage motor can be controlled. We need to determine the currents i_d and i_q in order to control the torque and flux. To determine these imaginary currents (mathematically) we need to know the amplitude and location of the rotor flux at every instant. For this there are two common methods:

- **direct calculation** from voltage and current measurements on the stator side. In addition a number of motor parameters are required.
- **indirect calculation** from the slip. For this a slip measurement is required (encoder or tachogenerator on the motor shaft) and the rotor time constant needs to be known.

Photo Heidenhain: Collection of encoders

7.6.3 Direct Field orientation

Fig. 20-48 is a block diagram illustrating direct field orientation.

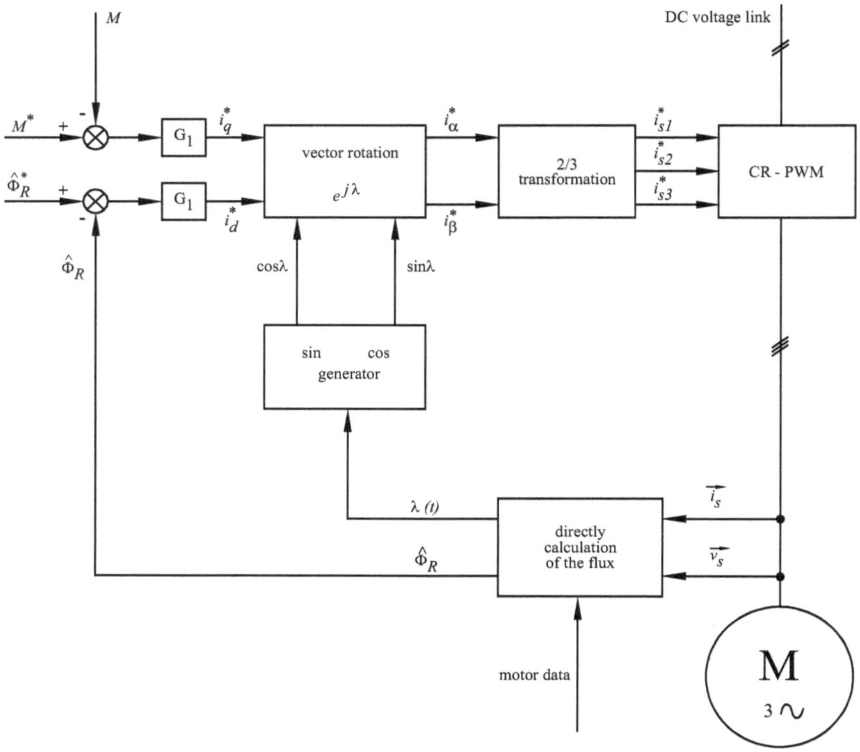

Fig. 20-48: Block diagram direct field orientation

The formulas (20-46) and (20-47), derived under nr. 7.8.5 can be used here:

$$\vec{\Phi}_S = \int_0^t \left[\vec{v}_S(t) - R_S \cdot \vec{i}_S(t)\right] .dt \quad (20\text{-}46)$$

$$\vec{\Phi}_R \cdot e^{j\delta} = \frac{L_R}{L_0} \cdot (\vec{\Phi}_S - \sigma \cdot L_S \cdot \vec{i}_S) \quad (20\text{-}47)$$

Herein:
- L_R = rotor self induction
- L_0 = magnetizing inductance
- L_S = stator stator self inductance
- $\sigma = 1 - \dfrac{L_0^2}{L_R \cdot L_S}$ = total scattering coefficient of the induction motor

If the motor parameters L_R, L_S, L_0, σ and R_S are known then from a measurement of $\vec{v}_S$ and $\vec{i}_S$, the flux $\vec{\Phi}_R \cdot e^{j\delta}$ can be directly calculated with the aid of expressions (20-46) and (20-47).
Flux orientation is in principle possible without the need of an encoder or tachogenerator. It is necessary though to compensate for the thermal effects on R_S.
Note that for demanding applications with vector control an encoder or tachogenerator is required.

7.6.4 Indirect field orientation

A. Determining the position of $\vec{\Phi}_R$

With the aid of (20-45) ω_S can be determined. Integration leads to an angle $\lambda(t)$ as shown in fig. 20-49.

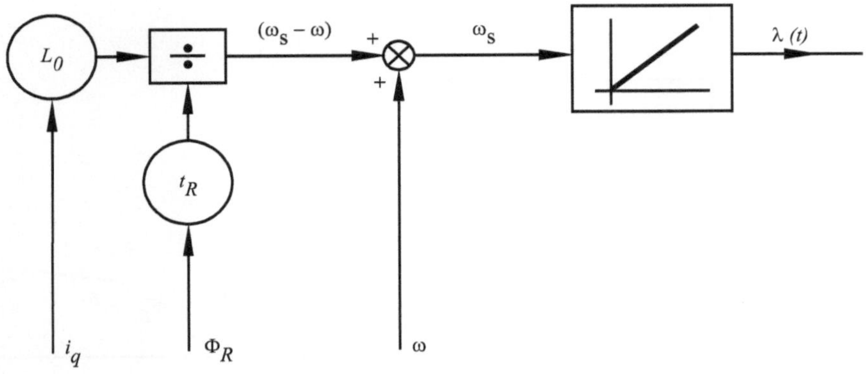

Fig. 20-49: Determining $\lambda(t)$

B. Stator current transformation and vector rotation block

Determining $\vec{i_S}(t)$ is achieved using the Clarke-transformation, see fig. 20-50.

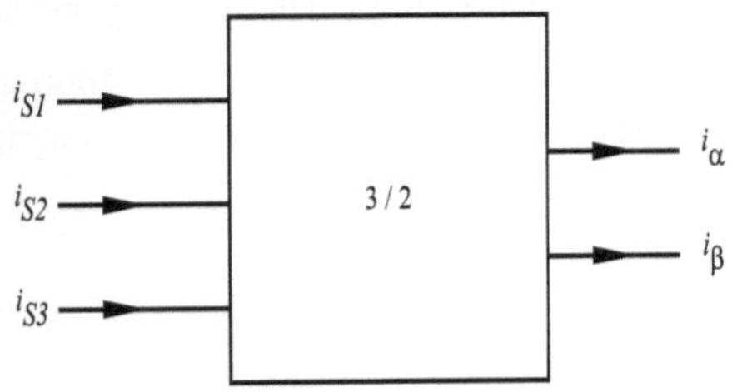

Fig. 20-50: 3/2-transformation

From (20-13) and (20-14) the components i_α and i_β of $\vec{i_S(t)}$ follow:

- $i_\alpha = \frac{3}{2} \cdot i_{S1}(t)$
- $i_\beta = \frac{\sqrt{3}}{2} \cdot [\, i_{S2}(t) - i_{S3}(t) \,]$

With $i_{S1} + i_{S2} + i_{S3} = 0$ then $i_{S1} + i_{S2} = -\, i_{S3}$ and:

- $i_\alpha = \frac{3}{2} \cdot i_{S1}$
- $i_\beta = \frac{\sqrt{3}}{2} \cdot (\, 2 \cdot i_{S2} + i_{S1} \,)$

It is sufficient to have only two of the three stator currents to perform a 3/2 transformation.

In order to transfer between the stator coordinates system α-*β* and the field coordinates system d-q in fig. 20-51 the Clarke-transformation is used with the following expressions (20-20) and (20-21).

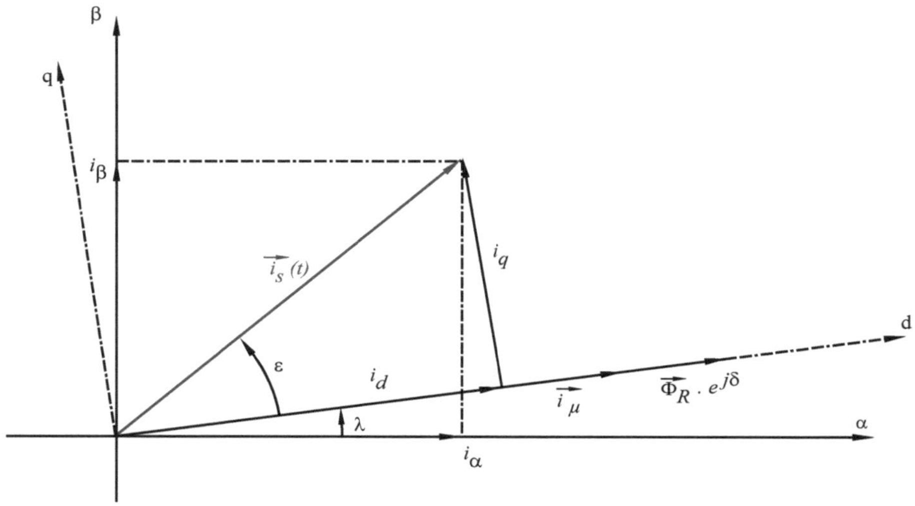

Fig. 20-51: Determining i_d and i_q

The expressions (20-20) and (20-21) are calculated in a module as shown in fig. 20-52.
Such a module is known as a vector rotation block. The vectors i_α and i_β are rotated through an angle λ . A vector in the α-*β* system has to be multiplied with $e^{-j\lambda}$ during the transition to the d-q system, hence the symbolic representation of the vector rotator to the right in fig. 20-52.

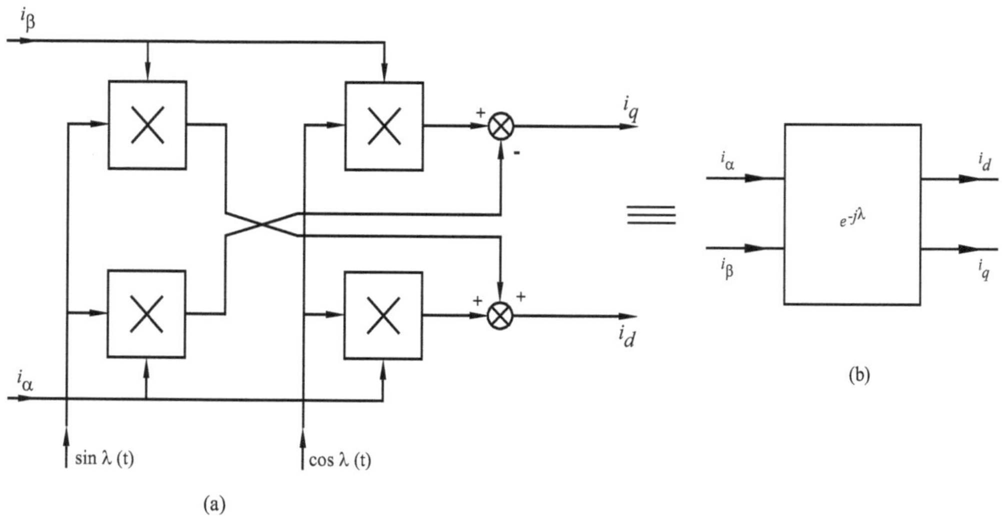

Fig. 20-52: Vector rotation block (Park-transformation)

C. Block diagram of indirect field orientation

The mathematical model (20-19, 20-17, 20-45) allows the exact amplitude and frequency of the injected stator current to be determined. It is now necessary to know the rotor angular velocity ω (or the slip) for which a tachogenerator or angular position meter is required. In addition the rotor time constant plays an important role in the accuracy of this system since τ_R is dependent on temperature and air-gap flux. Fig. 20-53 shows a block diagram for calculating Φ_R and λ from stator currents and angular velocity (or position) of the rotor.

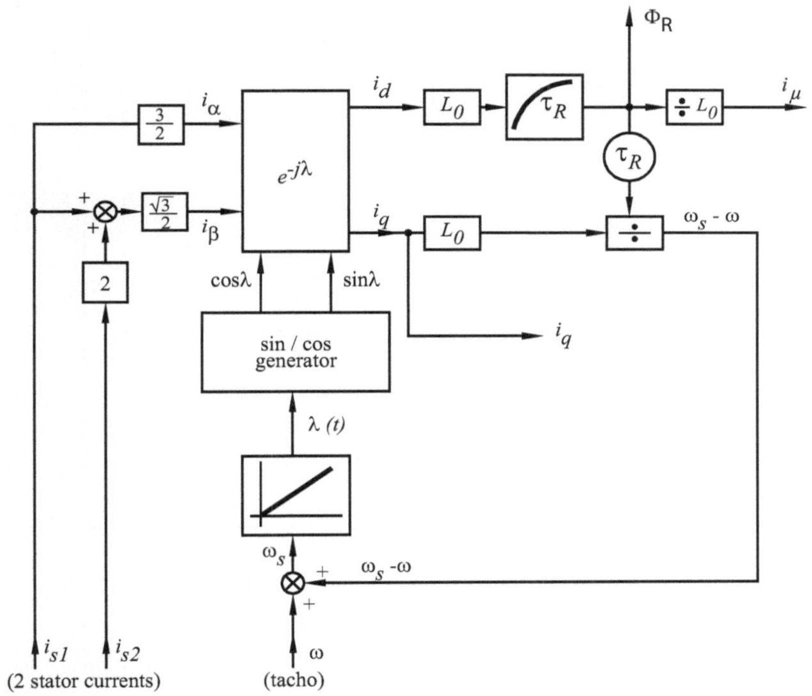

Fig. 20-53: Calculation module to determine Φ_R , i_q and i_μ from ω and $\vec{i_S}$

From (20-46) and (20-47) it follows that: $\dfrac{d(\vec{\Phi_R}\cdot e^{j\delta})}{dt} = \dfrac{L_R}{L_0\cdot}\left[\vec{v_S}(t) - R_S\cdot\vec{i_S} - \sigma\cdot L_S\cdot\dfrac{d\vec{i_S}}{dt}\right]$

$$\frac{d(\vec{\Phi_R}\cdot e^{j\delta})}{dt} = R_S\cdot\frac{L_R}{L_0}\cdot\left[\frac{\vec{v_S}(t)}{R_S} - \vec{i_S} - \sigma\cdot\tau_S\cdot\frac{d\vec{i_S}}{dt}\right] \qquad (20\text{-}48)$$

Equation (20-48) is provided in stator coordinates and allows the amplitude and frequency of the applied stator voltage to be determined.

Here $\tau_S = \dfrac{L_S}{R_S}$ = stator time constant.

Fig. 20-54 shows finally a block diagram with associated PI-controllers for field orientated speed control of an induction motor with imposed stator currents.

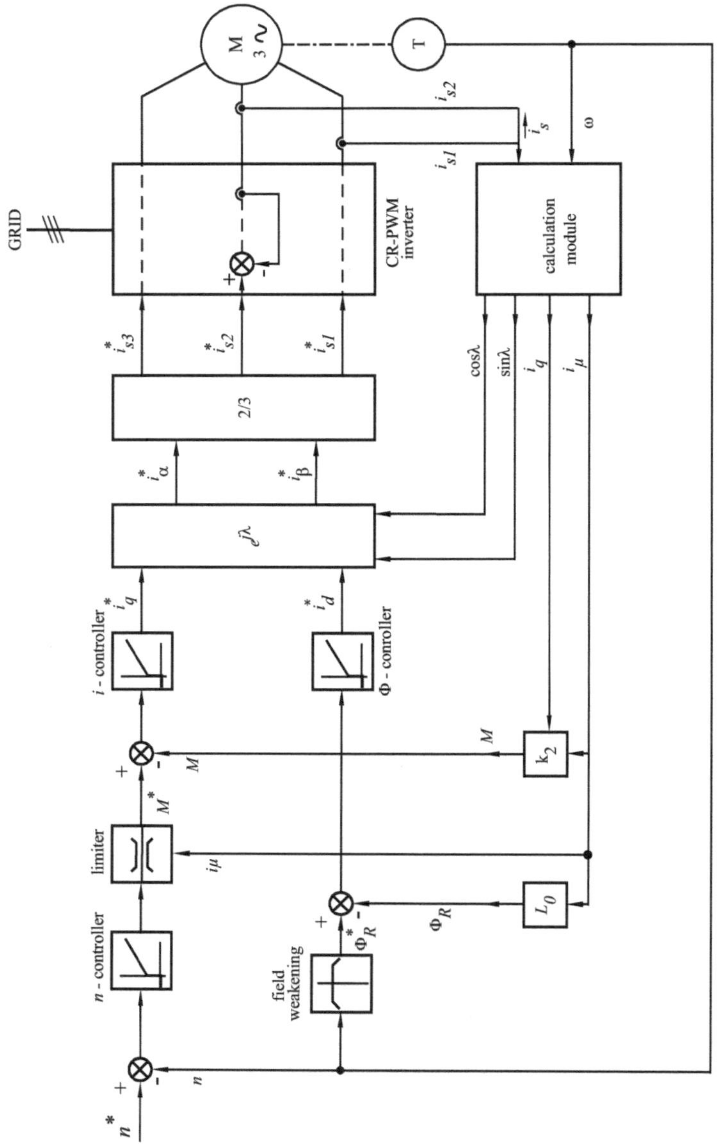

Fig. 20-54: Speed control of induction motor in field coordinates (indirect method)

7.6.5 Applications with closed loop control

Frequency controllers with vector control are used in:
cranes, elevators, lifting equipment, extruders, paper machines, rolling mills in the production of steel, CNC-machines, cable layers (ship based), drums etc.

7.7 Space vector

7.7.1 Stator current vector

Consider the three-phase motor of fig. 20-32. At a random instant the phase currents can have a value as shown in fig. 20-55.

N_S is the number of windings per stator phase. In the case of symmetry the three instantaneous currents can be described by $i_{S1} + i_{S2} + i_{S3} = 0$. These currents are not of necessity sinusoidal. Unless otherwise indicated the expressions that follow in our study are applicable for every waveform of the three-phase currents i_{S1}, i_{S2} and i_{S3}.

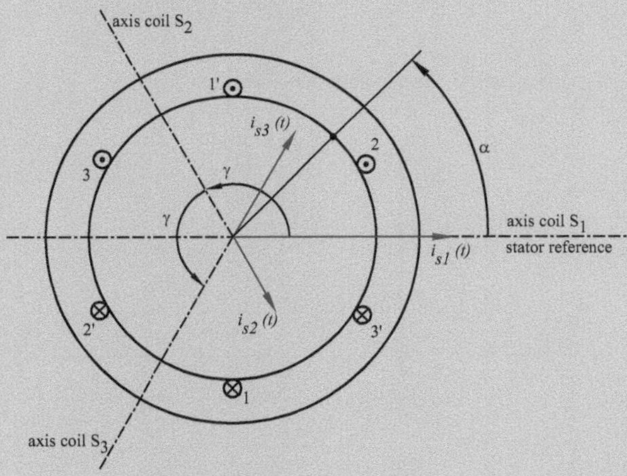

Fig. 20-55: Stator currents at a random instant

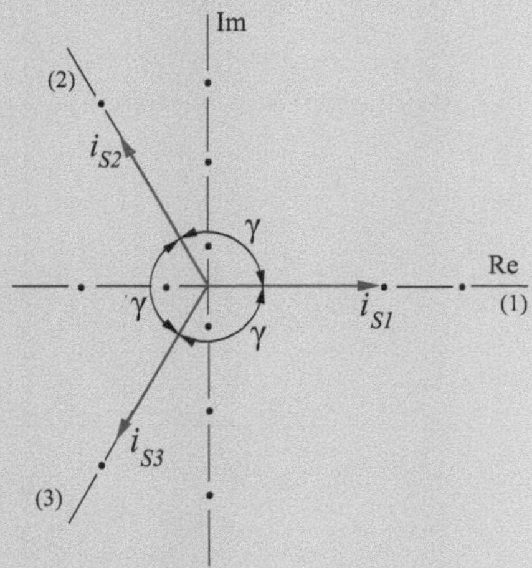

Fig. 20-56

The vector sum of the three stator currents is the stator vector $\overrightarrow{i_S}(t) = \overrightarrow{i_{S1}}(t) + \overrightarrow{i_{S2}}(t) + \overrightarrow{i_{S3}}(t)$.
If we take the axis of coil 1 as a reference and draw the three axis system as in fig. 20-56, then the three stator currents can be described as:

$\overrightarrow{i_{S1}}(t) = i_{S1}$; $\overrightarrow{i_{S2}}(t) = i_{S2}(t) \, . \, e^{j.\gamma}$ and $\overrightarrow{i_{S3}}(t) = i_{S3}(t) \, . \, e^{2.j.\gamma}$.

The stator current vector may also be written as: $\overrightarrow{i_S}(t) = i_{S1}(t) + i_{S2}(t) \, . \, e^{j.\gamma} + i_{S3}(t) \, . \, e^{j.2.\gamma}$. (20-23)

Note that the currents in fig. 20-56 are drawn principally. In reality the three phase currents may not all be positive.

Fig. 20-57 shows the construction of the stator current vector $\vec{i_S}\ (t)$ at the instant that i_{S1} is positive, while i_{S2} and i_{S3} are negative.

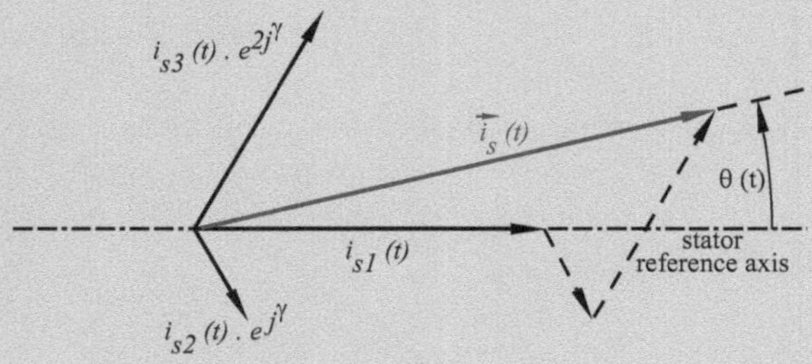

Fig. 20-57: Construction of stator current as a function of the three instantaneous stator currents

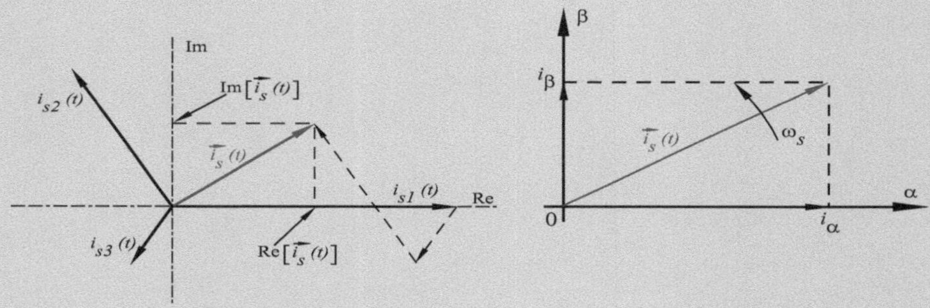

(a) (b)

Fig. 20-58: Stator current vector in the complex plane

Fig. 20-58a shows current vectors in the complex plane. For this figure we can write:

$$\vec{i_S}\ (t) = i_{S1}(t) + i_{S2}(t) \cdot e^{j.\gamma} + i_{S3}(t) \cdot e^{2.j.\gamma}$$

$$\mathbf{Re}\left[\ \vec{i_S}\ (t)\right] = i_{S1} - i_{S2} \cdot \sin 30° - i_{S3} \cdot \sin 30° = i_{S1} - \frac{i_{S2}}{2} - \frac{i_{S3}}{2}$$

With $i_{S1} + i_{S2} + i_{S3} = 0$ this becomes:

$$\mathbf{Re}\left[\ \vec{i_S}\ (t)\right] = \frac{3}{2} \cdot i_{S1}$$

$$\mathbf{Im}\left[\ \vec{i_S}\ (t)\right] = j \cdot \frac{\sqrt{3}}{2} \cdot i_{S2} - j \cdot \frac{\sqrt{3}}{2} \cdot i_{S3} = j \cdot \frac{\sqrt{3}}{2} \cdot [\ i_{S2} - i_{S3}\ ]$$

The vector $i_S(t)$ therefore may be written in complex form as :

$$\vec{i_S}\ (t) = \frac{3}{2} \cdot i_{S1}(t) + j \cdot \frac{\sqrt{3}}{2} \cdot [\ i_{S2}(t) - i_{S3}(t)\ ]$$

The stator current vector is here written as a function of the instantaneous values of the three stator currents.

The complex plane of fig. 20-58a can also be drawn as an β-α coordinate system as in fig. 20-58b. The currents i_α and i_β together produce $\overrightarrow{i_S} = i_\alpha + j \,.\, i_\beta$. This is identical to the stator current vector $\overrightarrow{i_S} = i_{S1} + i_{S2} \,.\, e^{j.\gamma} + i_{S3} \cdot e^{j.2\gamma}$.

In fig. 20-57 we can also write:

$$\overrightarrow{i_S}(t) = i_S(t) \cdot e^{j.\theta} \quad (20\text{-}24)$$

The stator current vector is dependent on time (t) and space (θ), and is referred to as a **space vector**. In the special case where the stator currents have a sinusoidal form as function of time and forms part of a symmetric three-phase system, then the end of $\overrightarrow{i_S}(t)$ rotates in a circular path with constant angular velocity.

7.7.2 Rotor current vector

As we know from the operation of an induction motor the stator rotating field induces emfs in the rotor windings. These emfs result in lagging rotor currents as a result of the rotor inductance. Due to the operation of the rotor currents on the stator rotating field the rotor is as it were pulled by this rotating field. The rotor rotates somewhat slower than the synchronous speed of the rotating field. Assume that at a certain instant t the rotor is at an angle δ as shown in fig. 20-59.
The rotor current vector which replaces the three rotor currents may be written as:

$$\overrightarrow{i_R}(t) = i_{R1}(t) + i_{R2}(t) \cdot e^{j.\gamma} + i_{R3}(t) \cdot e^{j.2\gamma}$$

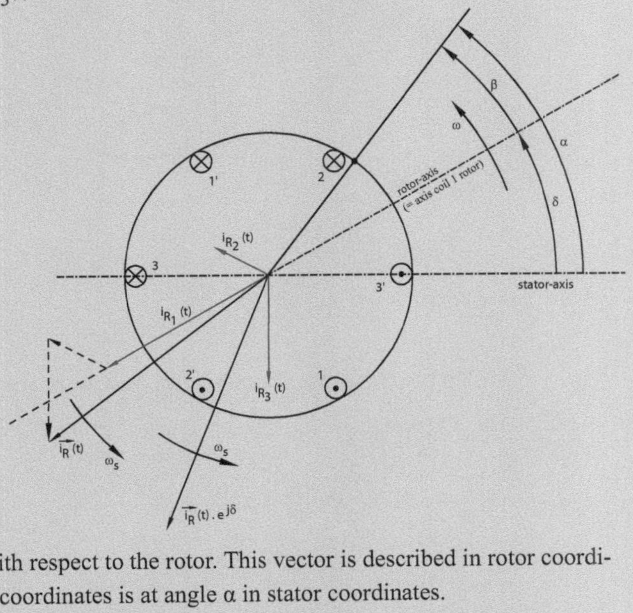

Fig. 20-59: Space vector of the rotorcurrent at instant t

The rotor current vector rotates with respect to the rotor. This vector is described in rotor coordinates. A vector at angle β in rotor coordinates is at angle α in stator coordinates.
From $\alpha = \beta + \delta$ it follows that for the rotor coordinates an angle δ needs to added in order to calculate in stator coordinates, we need to multiply by $e^{j\delta}$.
We could just as easily determine everything in rotor coordinates. In this case $\overrightarrow{i_S}(t)$ needs to be multiplied with $e^{-j\delta}$, in other words stator coordinates are rotated backwards with an angle δ.

7.7.3 Space vectors in the machine

Fig. 20-60 shows a summary of the figures 20-55, 20-57 and 20-59.
Stator and rotor vectors rotate with synchronous speed ω_S.
The rotor itself rotates with $\omega = \frac{d\delta}{dt}$ rad/s.

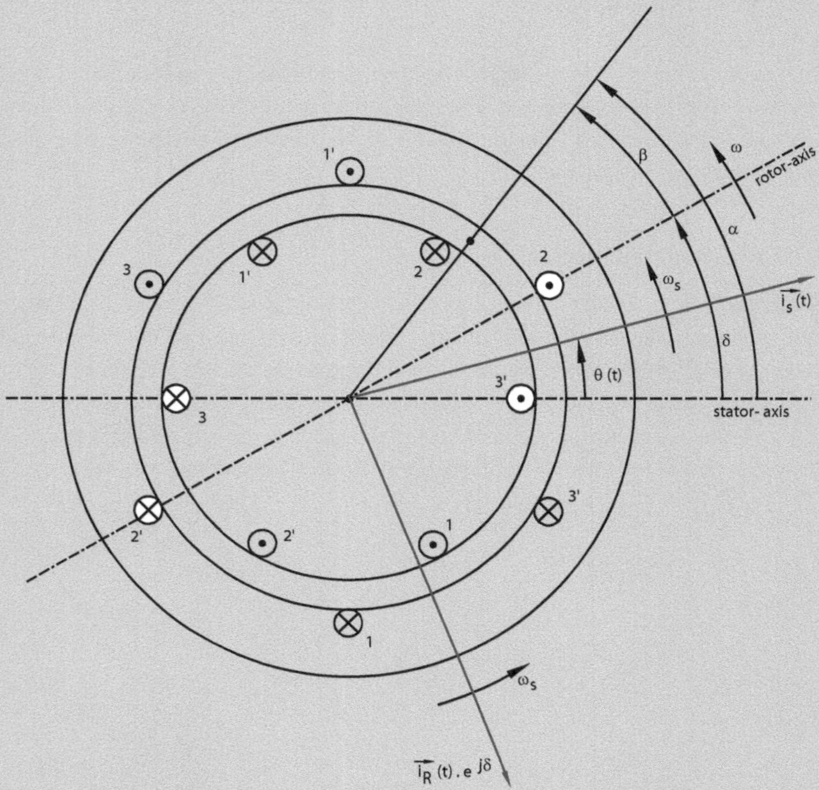

Fig. 20-60: Space vectors $\vec{i}_S\,(t)$ and $\vec{i}_R\,(t)$

7.7.4 Mathematical model. T-equivalent induction motor

Fig. 20-61 shows an equivalent diagram for an induction motor.

We can write

- $v_{S1} = R_S \cdot i_{S1} + \frac{d\Phi_{S1}}{dt}$
- $v_{S2} = R_S \cdot i_{S2} + \frac{d\Phi_{S2}}{dt}$
- $v_{S3} = R_S \cdot i_{S3} + \frac{d\Phi_{S3}}{dt}$
- $\vec{v}_S(t) = v_{S1} + v_{S2} \cdot e^{j\gamma} + v_{S3} \cdot e^{j2\gamma}$
- $\vec{i}_S\,(t) = i_{S1} + i_{S2} \cdot e^{j\gamma} + i_{S3} \cdot e^{j2\gamma}$
- $\vec{\Phi}_S(t) = \Phi_{S1}(t) + \Phi_{S2}(t) \cdot e^{j\gamma} + \Phi_{S3}(t) \cdot e^{j2\gamma}$

Evaluating these six equations leads to: $\vec{v}_S\,(t) = R_S \cdot \vec{i}_S\,(t) + \frac{d\vec{\Phi}_S}{dt}$ (20-25)

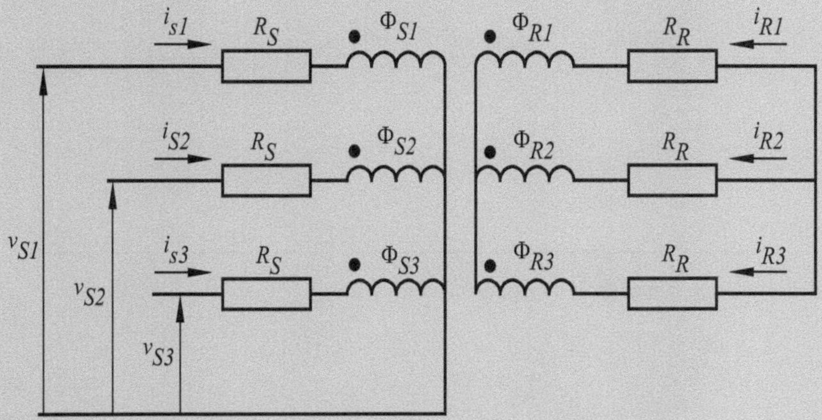

Fig 20-61: Equivalent diagram of the induction motor

In similar fashion to the static transformer we can also draw a T-equivalent model for the induction motor but now including space vectors. Working in the stator coordinates system, the rotor values as previously demonstrated need to be multiplied by $e^{j\delta}$. Fig. 20-62 shows this T-equivalent model.

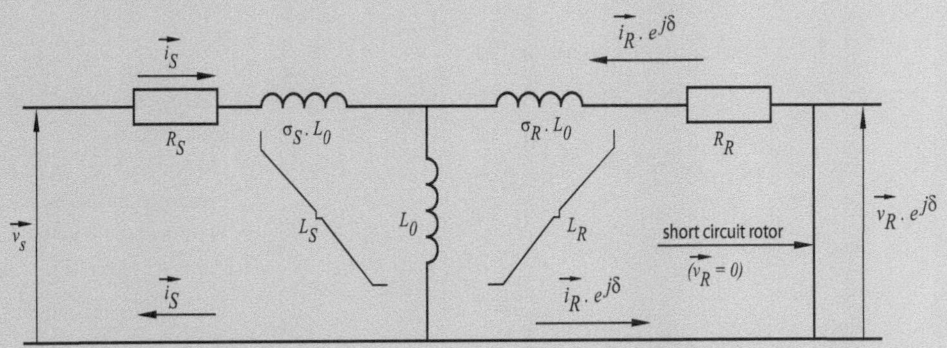

Fig. 20-62: T-equivalent model of a three-phase induction motor in stator coordinates

In fig. 20-62 we can write:

In stator coordinates:

- $$\vec{v}_S(t) = R_S \cdot \vec{i}_S(t) + \frac{d\vec{\Phi}_S}{dt} \qquad \text{(20-25)}$$
- $$\vec{\Phi}_S(t) = \vec{i}_S(t) \cdot \sigma_S . L_0 + \vec{i}_S(t) \cdot L_0 + L_0 \cdot \vec{i}_R(t) . e^{j\delta}$$
- or: $$\vec{\Phi}_S(t) = \vec{i}_S(t) \cdot L_S + L_0 \cdot \vec{i}_R(t) \cdot e^{j\delta} \qquad \text{(20-26)}$$
- $$\vec{\Phi}_R(t) \cdot e^{j\delta} = L_0 \cdot \vec{i}_S(t) + L_R \cdot \vec{i}_R(t) \cdot e^{j\delta} \qquad \text{(20-27)}$$

We can also represent the T-equivalent model of fig. 20-62 in rotor coordinates as in fig. 20.63.

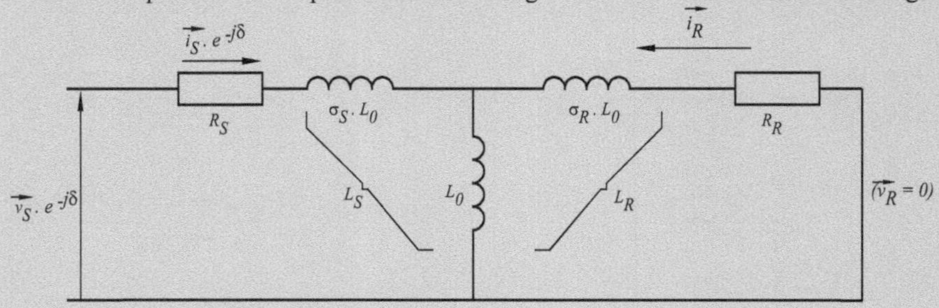

Fig. 20.63: T-equivalent of induction motor in a rotor coordinates system

- $\vec{\Phi_S}(t) \cdot e^{-j\delta} = \vec{i_S}(t) \cdot e^{-j\delta} \cdot L_S + \vec{i_R}(t) \cdot L_0$ (20-28)
- $\vec{\Phi_R}(t) = \vec{i_S}(t) \cdot e^{-j\delta} \cdot L_0 + \vec{i_R}(t) \cdot L_R$ (20-29)
- $\vec{v_R}(t) = \vec{i_R}(t) \cdot R_R + \dfrac{d\vec{\Phi_R}(t)}{dt}$ (20-30)

Comparing (20-26) with (20-28), we note that they are in fact identical expressions. Note that $\vec{\Phi_S}$ the stator flux is in stator coordinates, and $\vec{\Phi_S} \cdot e^{-j\delta}$ the stator flux in rotor coordinates.

7.7.5 Electromechanical torque

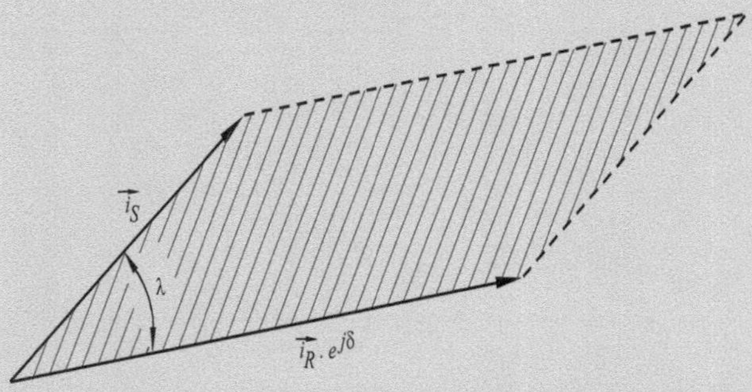

The electromechanical torque of an induction motor is proportional to the vector product of stator and rotor current vectors. Both vectors need to be represented in the same coordinates system. Fig. 20-64 shows the vectors in a stator coordinates system.

Fig. 20-64: Electromechanical torque of induction motor

The vector product of $\vec{i_S}$ with $\vec{i_R} \cdot e^{j\delta}$ is the area of the parallelogram in fig. 20-64 and is described by $i_S \cdot i_R\ e^{j\delta} \cdot \sin(\pi - \lambda) = i_S \cdot i_R \cdot e^{j\delta} \cdot \sin\lambda$

Therefore: $M_{em} = k_1 \cdot i_S \cdot i_R \cdot e^{j\delta} \cdot \sin\lambda$

By neglecting the internal friction torque this last expression practically describes the torque on the shaft of the motor, so that:

$$M \approx M_{em} = k_1 \cdot i_S \cdot i_R \cdot e^{j\delta} \cdot \sin\lambda \qquad (20\text{-}31)$$

Here k_1 is a machine constant.

For the remaining study it is easy to consider the triangle formed by $\vec{i_S}$ and $\vec{i_R} \cdot e^{j\delta}$ (fig. 20-65). The area in fig. 20-64 is double the area of this triangle. The torque is therefore directly proportional to the area of the triangle in fig. 20-65.

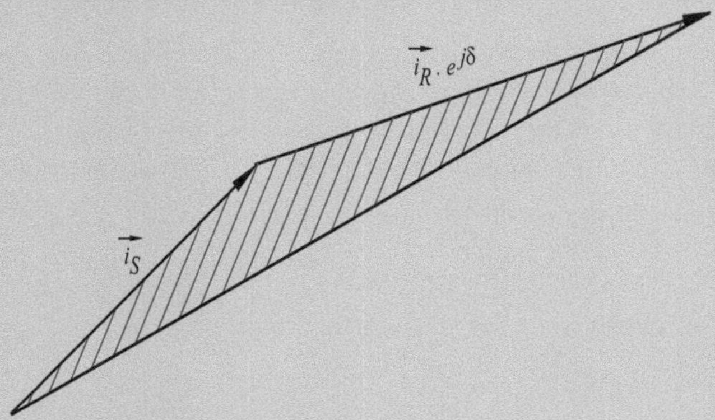

Fig. 20-65: Torque triangle of induction machine

7.7.6 Conclusion

The equations in table 20-2 form the mathematical model of the squirrel cage induction motor. Here $\vec{v_S}(t)$ and $\vec{\Phi_S}(t)$ are described in stator coordinates while $\vec{v_R}(t)$ and $\vec{\Phi_R}(t)$ are presented in rotor coordinates.

Table 20-2

$$\vec{\Phi_S}(t) = L_S \cdot \vec{i_S}(t) + L_0 \cdot \vec{i_R}(t) \cdot e^{j\delta} \quad (20\text{-}26)$$

$$\vec{v_S}(t) = R_S \cdot \vec{i_S}(t) + \frac{d\vec{\Phi_S}}{dt} \quad (20\text{-}25)$$

$$or:\ \vec{v_S}(t) = R_S \cdot \vec{i_S}(t) + L_S \cdot \frac{d\vec{i_S}}{dt} + L_0 \cdot \frac{d}{dt}(\vec{i_R} \cdot e^{j\delta}) \quad (20\text{-}32)$$

$$\vec{\Phi_R}(t) = L_R \cdot \vec{i_R}(t) + L_0 \cdot \vec{i_S}(t) \cdot e^{-j\delta} \quad (20\text{-}29)$$

$$\vec{v_R}(t) = R_R \cdot \vec{i_R}(t) + \frac{d\vec{\Phi_R}}{dt} = 0 \quad (20\text{-}30)$$

$$or:\ \vec{v_R}(t) = R_R \cdot \vec{i_R}(t) + L_R \cdot \frac{d\vec{i_R}}{dt} + L_0 \cdot \frac{d}{dt}(\vec{i_S} \cdot e^{-j\delta}) = 0 \quad (20\text{-}33)$$

$$M \approx M_{em} = k_1 \cdot i_S \cdot i_R \cdot e^{j\delta} \cdot \sin\lambda \quad (20\text{-}31)$$

$M = M_t + M_v$ (20-34) and $M_v = J_m \cdot \frac{d\omega}{dt}$ (20-35) are also applicable here.

7.8 Mathematical model in field coordinates

7.8.1 Fictive magnetising current

In a squirrel cage motor currents and voltages can only easily be measured on the stator side. On the other hand the formulas used in the study of field orientation are quite simple as long as the interaction is considered to be between stator current vector and rotor flux.
In addition by choosing the rotor flux as reference we get a "full" decoupling between flux and current. In the rotor coordinates system we found: $\vec{\Phi}_R = L_R \cdot \vec{i}_R + L_0 \cdot \vec{i}_S \cdot e^{-j\delta}$ (20-29)

We describe the rotor flux in stator coordinates: $\vec{\Phi}_R \cdot e^{j\delta} = L_R \cdot \vec{i}_R \cdot e^{j\delta} + L_0 \cdot \vec{i}_S$

Fig. 20-66 shows the vector diagram that accompanies these expressions.

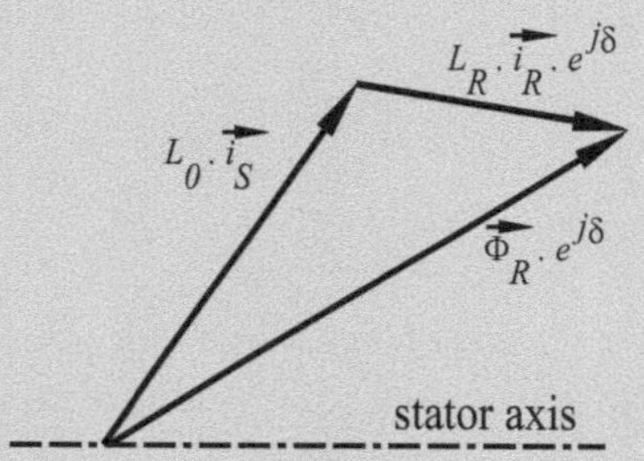

Fig. 20-66: Rotor flux diagram

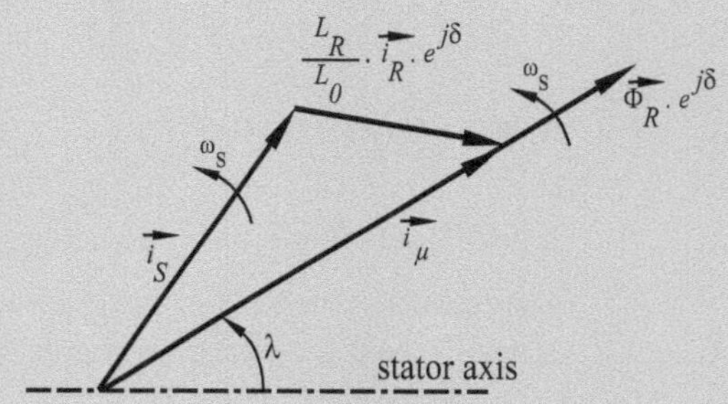

Fig. 20-67: Magnetising current diagram

Fig. 20-67 is easily derived from fig. 20-66. Here $\vec{i}_\mu = \frac{\vec{\Phi}_R \cdot e^{j\delta}}{L_0}$ is a fictive magnetising current which, on the stator side, is responsible for the rotor flux.

From fig. 20-67 it follows that: $\vec{i}_\mu = \vec{i}_S + \frac{L_R}{L_0} \cdot \vec{i}_R \cdot e^{j\delta}$ (20-36)

So that: $\vec{i}_R = \frac{L_0}{L_R} \cdot (\vec{i}_\mu - \vec{i}_S) \cdot e^{-j\delta}$

The rotorflux
- in statorcoordinates: $\vec{\Phi}_R \cdot e^{j\delta} = L_0 \cdot \vec{i}_\mu$
- and in rotor coordinates: $\vec{\Phi}_R = L_0 \cdot \vec{i}_\mu \cdot e^{-j\delta}$

We replace $\vec{i}_R$ and $\vec{\Phi}_R$ in (20-30), see table 20-2 (p.20-63):

$$0 = R_R \cdot \vec{i}_R + \frac{d\vec{\Phi}_R}{dt} = R_R \cdot \frac{L_0}{L_R} \cdot (\vec{i}_\mu - \vec{i}_S) \cdot e^{-j\delta} + \frac{d}{dt}(L_0 \cdot \vec{i}_\mu \cdot e^{-j\delta})$$

$$R_R \cdot \frac{L_0}{L_R} \cdot (\vec{i}_\mu - \vec{i}_S) \cdot e^{-j\delta} + L_0 \cdot \frac{d\vec{i}_\mu}{dt} \cdot e^{-j\delta} - L_0 \cdot j \cdot \frac{d\delta}{dt} \cdot e^{-j\delta} \cdot \vec{i}_\mu = 0$$

$$R_R \cdot (\vec{i}_\mu - \vec{i}_S) + L_R \cdot \frac{d\vec{i}_\mu}{dt} - j \cdot L_R \cdot \frac{d\delta}{dt} \cdot \vec{i}_\mu = 0$$

With $\frac{d\delta}{dt} = \omega$ and $\frac{L_R}{R_R} = \tau_R$ we find: $\tau_R \cdot \frac{d\vec{i}_\mu}{dt} + (1 - j\,\omega \cdot \tau_R) \cdot \vec{i}_\mu = \vec{i}_S$ (20-37)

Fig. 20-68 shows a number of the familiar space vectors in the stator coordinates system.
The magnetising current is described by: $\vec{i_\mu}(t) = i_\mu(t) \cdot e^{j\lambda}$ (20-38)

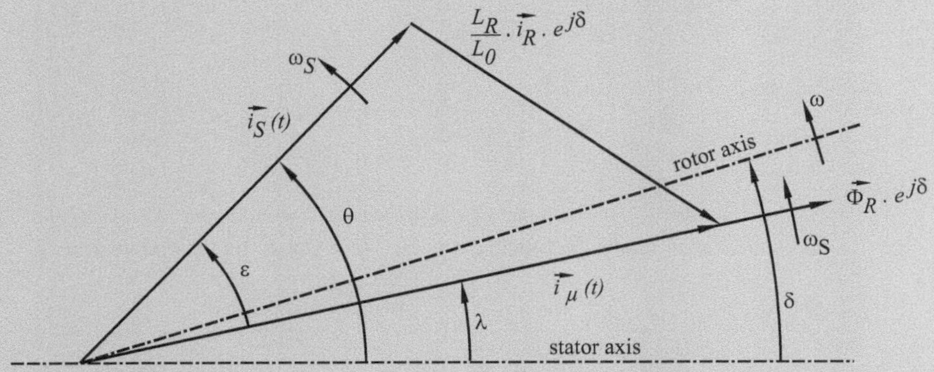

Fig. 20-68: Space vectors $\vec{i_S}$ and $\vec{\Phi_R}$ in a stator coordinates system

7.8.2 Field orientation - field coordinates

We redraw fig. 20-68 in fig. 20-69 and create a new coordinates system (d-q) based on the rotor flux.

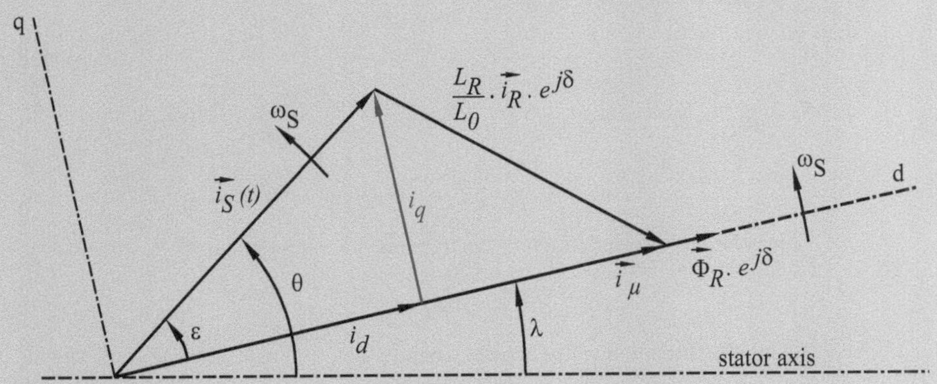

Fig. 20-69: Field orientation

Since $\vec{i_S}(t)$ and $\vec{\Phi_R} \cdot e^{j\delta}$ rotate with the same angular velocity ω_S the angle ε is a time independent angle. The stator current $\vec{i_S}$ is fixed with respect to the (rotor) flux, dr.ir. Blaschke called this **field orientation**.

In fig. 20-69 we can determine the stator current $\vec{i_S}$ and its components i_d and i_q in the newly chosen system d-q: $\vec{i_S}(t) = i_d + j \cdot i_q$

We now have determined the **field coordinates** i_d and i_q of the stator current $\vec{i_S}$.

The indexes d and q refer to the **d**irect and **q**uadrature axis.

7.8.3 Moment of work torque

As shown in fig. 20-65 the work torque in fig. 20-69 is proportional to the area of the triangle formed by $\overrightarrow{i_S}$, $\frac{L_R}{L_0} \cdot \overrightarrow{i_R} \cdot e^{j\delta}$ and $\overrightarrow{i_\mu}$. This area can also be described in fig. 20-69 as:

$1/2 \cdot i_q \cdot i_\mu$ so that: $M_{em} = k_2 \cdot i_\mu \cdot i_q$ (20-39)

With $\Phi_r = L_0 \cdot i_\mu$ is $M_{em} = k_3 \cdot i_q \cdot \Phi_R$ (20-40)

Another method is to assume that M_{em} is proportional with the product of the rotor flux with the component of $\overrightarrow{i_S}$ that is perpendicular to this rotor flux. In fig. 20-69 this leads to :

$M_{em} = k_3 \cdot i_S \cdot \sin \varepsilon \cdot \Phi_R$, so that: $M_{em} = k_3 \cdot i_q \cdot \Phi_R$ (20-40)

Calculation leads to : $k_3 = \frac{3 \cdot p}{2} \cdot \frac{L_0}{L_R}$ with p = number of pole pairs per machine.

7.8.4 Mathematical model in field coordinates

With $\overrightarrow{i_\mu} = i_\mu \cdot e^{j\lambda}$ (20-37) becomes: $\tau_R \cdot \frac{d}{dt}(i_\mu \cdot e^{j\lambda}) + (1 - j\omega \cdot \tau_R) \cdot i_\mu \cdot e^{j\lambda} = \overrightarrow{i_S}$

$$\tau_R \cdot e^{j\lambda} \cdot \frac{d\overrightarrow{i_\mu}}{dt} + \tau_R \cdot i_\mu \cdot j \cdot \frac{d\lambda}{dt} \cdot e^{j\lambda} + (1 - j\,\omega \cdot \tau_R) \cdot i_\mu \cdot e^{j\lambda} = \overrightarrow{i_S}$$

$$\tau_R \cdot \frac{d\overrightarrow{i_\mu}}{dt} + \tau_R \cdot i_\mu \cdot j \cdot \omega_S + (1 - j\,\omega \cdot \tau_R) \cdot i_\mu = \overrightarrow{i_S} \cdot e^{-j\lambda}$$

From (20-24): $\overrightarrow{i_S}(t) = i_S \cdot e^{j\theta}$ it follows: $\overrightarrow{i_S} \cdot e^{-j\lambda} = i_S \cdot e^{j(\theta - \lambda)} = i_S \cdot e^{j\varepsilon}$ (20-41)

This may be expressed in the complex d-q coordinates system as: $\overrightarrow{i_S} \cdot e^{-j\lambda} = i_S \cdot e^{j\varepsilon} = i_d + j.i_q$

so that: $\tau_R \cdot \frac{d\overrightarrow{i_\mu}}{dt} + \tau_R \cdot i_\mu \cdot j \cdot \omega_S + (1 - j\omega \cdot \tau_R) \cdot i_\mu = i_d + j \cdot i_q$

Taking the real and imaginary parts separately:

- $\tau_R \cdot \frac{d\overrightarrow{i_\mu}}{dt} + i_\mu = i_d$ (20-42)
- $(\omega_S - \omega) \cdot i_\mu \cdot \tau_R = i_q$ (20-43)

Multiplying (20-42) with L_0: $\tau_R \cdot \frac{d\overrightarrow{\Phi_R}}{dt} + \Phi_R = L_0 \cdot i_d$ (20-44)

Rewriting of (20-44) using Laplace notations: $\tau_R \cdot s \cdot \Phi_R = L_0 \cdot I_d$

$$\frac{\Phi_R}{I_d} = L_0 \cdot \frac{1}{1 + s \cdot \tau_R}$$

This is the transfer function of a first order system with time constant $\tau_R = \frac{L_R}{R_R}$ (fig. 20-70)
The flux Φ_R is determined by i_d!

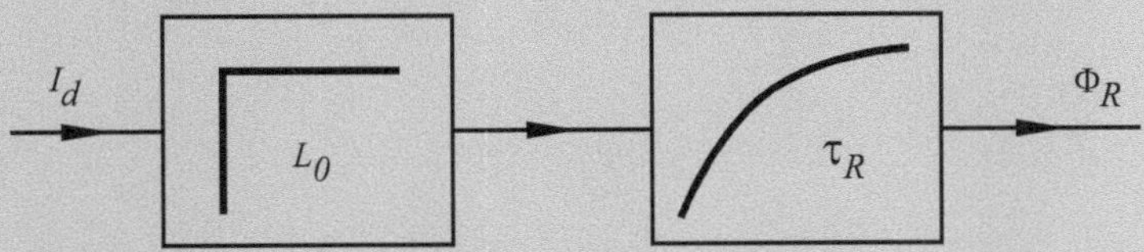

Fig. 20-70: Transfer function of rotor flux

From (20-43) the slip follows: $g = \frac{\omega_S - \omega}{\omega_S} = \frac{i_q}{i_\mu \cdot \tau_R \cdot \omega_S}$

With $i_\mu = \frac{\Phi_R}{L_0}$ then $g = \frac{L_0}{\Phi_R \cdot \tau_R} \cdot \frac{i_q}{\omega_S}$

From (20.43) we can derive:

$$\omega_S - \omega = \frac{L_0 \cdot i_q}{\Phi_R \cdot \tau_R} \tag{20-45}$$

The mathematical model of a three-phase squirrel cage induction motor in field coordinates (reference coordinates as field coordinates rotates with rotor flux) is given by:

$$M_{em} = k_3 \cdot i_q \cdot \Phi_R \tag{20-40}$$

$$\tau_R \cdot \frac{d\Phi_R}{dt} + \Phi_R = L_0 \cdot i_d \tag{20-44}$$

$$\omega_S - \omega = \frac{L_0 \cdot i_q}{\Phi_R \cdot \tau_R} \tag{20-45}$$

If we now determine the stator current components at every instant in a coordinates system that rotates with the rotor flux then the torque of the induction motor may be controlled exactly as with a DC-motor. In fact the expression $M_{em} = k_3 \cdot i_q \cdot \Phi_R$ (20-40) is comparable with the formula for the moment of work torque of a DC-machine: $M_{em} = k_2 \cdot I_a \cdot \Phi$ (19-2).
With an induction motor Φ_R can be kept constant or controlled via i_d as (20-44) suggests.
The speed n of the induction motor can be controlled with (20-45) by influencing ω_S and i_q.
In fact with field orientation via mathematical transformation the way is open to control the torque of an induction motor in the same way as with a DC-motor.

7.8.5 Direct calculation of the flux

From (20-25) follows $d\vec{\Phi}_S = \left[\vec{v}_S(t) - R_S \cdot \vec{i}_S(t)\right]dt \;\rightarrow\;$

$$\vec{\Phi}_S = \int_0^t \left[\vec{v}_S(t) - R_S \cdot \vec{i}_S(t)\right] \cdot dt \qquad (20\text{-}46)$$

We can describe the rotor flux in stator coordinates as:

$$\vec{\Phi}_R \cdot e^{j\delta} = L_0\,\vec{i}_\mu = L_0 \cdot \left[\vec{i}_S + \frac{L_R}{L_0}\vec{i}_R . e^{j\delta}\right] = \frac{L_0}{L_S} \cdot \left[L_S\,\vec{i}_S + \frac{L_R L_S}{L_0}\,\vec{i}_R \cdot e^{j\delta}\right]$$

$$= \frac{L_0}{L_S} \cdot \left[L_S \cdot \vec{i}_S + \frac{L_0 \cdot (1+\sigma_R).L_0 \cdot (1+\sigma_S)}{L_0} \cdot \vec{i}_R \cdot e^{j\delta}\right]$$

With $\sigma = 1 - \dfrac{1}{(1+\sigma_R)(1+\sigma_S)} = 1 - \dfrac{L_0^2}{L_R \cdot L_S}$ = total spreading coefficient of the induction motor is:

$\dfrac{1}{1-\sigma} = (1+\sigma_R).(1+\sigma_S)$, so that:

$$\vec{\Phi}_R \cdot e^{j\delta} = \frac{L_0}{L_S}\left[L_S\,\vec{i}_S + \frac{L_0}{(1-\sigma)} \cdot \vec{i}_R \cdot e^{j\delta}\right]$$

$$= \frac{L_0}{(1-\sigma).L_S}\left[L_S\,\vec{i}_S \cdot (1-\sigma) + L_0 \cdot \vec{i}_R \cdot e^{j\delta}\right]$$

$$= \frac{1}{(1-\sigma).(1+\sigma_S)}\left[L_S \cdot \vec{i}_S + L_0 \cdot \vec{i}_R . e^{j\delta} - \sigma . L_S \cdot \vec{i}_S\right] = (1+\sigma_R) \cdot \left[\vec{\Phi}_S - \sigma . L_S \cdot \vec{i}_S\right]$$

$$\vec{\Phi}_R \cdot e^{j\delta} = \frac{L_R}{L_0} \cdot (\vec{\Phi}_S - \sigma \cdot L_S \cdot \vec{i}_S) \qquad = \Phi_{R(\alpha)} + \Phi_{R(\beta)} \qquad (20\text{-}47)$$

Photo Siemens: Servomotor with planetary gearbox. Suitable for 400-480V. Speed: 34 to 825 rpm. Power 0.3 to 57kW. Torque: 2 to 3400Nm. Resolver, incremental encoder (sin/cos;1Vpp) EnDat absolute encoder. These motors are used for positioning in production machines, filling systems, conveyer belts, robots etc.

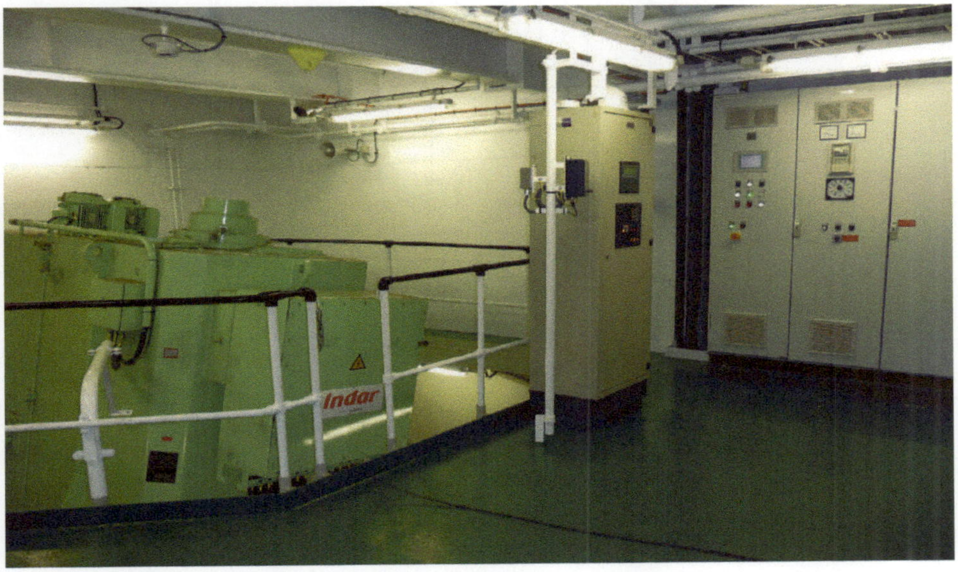

Photo Bakker Sliedrecht: Ship drive train on board the “Seven Oceans”. This is a pipe layer belonging to the company Subsea. The upper photo shows the switching cabinet. The lower photo shows a part of the drive train. The cabinet provides the power for Thruster 2.

Motor details: Indar ACP-50-L/6; 2200kW; 3x600V; 2385A nom; 800rpm nom; 880 rpm max.

DC-link voltage of inverter: 930 V DC. Every DSU delivers 1400 A DC, total DC-current is 2800A

7.9 Direct torque control (DTC)

The firm ABB brought DTC to the market in 1996. Prof I. Takahashi published the theory already in 1985. Where the field orientation of dr. ir. Blaschke references to the rotor flux vector with DTC it is referenced to the stator flux vector. With DTC there is no PWM converter. The inverter is directly controlled. Per cycle of 25μs after a calculation of the motor torque and stator flux a suitable switch pattern for the inverter is determined. The accuracy of the DTC depends on the accuracy of the stored motor model. According to ABB the step response of the torque is better than 2 ms.

7.9.1 Vector inverter with a voltage DC-link

A. Voltage vectors

We consider in fig. 20-71 the switching matrix of a three-phase bridge inverter which is controlling an asynchronous motor.

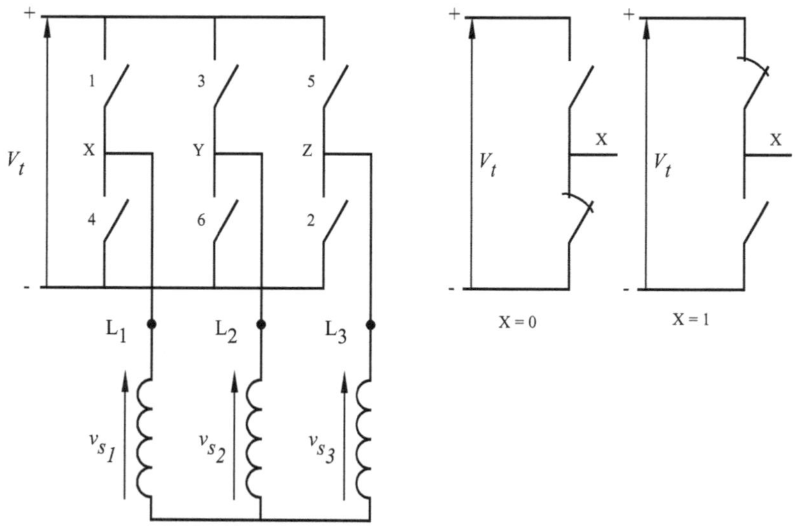

Fig. 20-71: Switching matrix three-phase bridge inverter

Per half branch of bridge (X,Y or Z) the position of the switches can be indicated with binary code: X = 0 (lower switch closed)
X = 1 (upper switch closed)

It is clear that per half branch of the bridge:

1) both switches may not be closed simultaneously, otherwise there is a short circuit across V_t.
2) one of the two switches must be closed, otherwise the corresponding stator coil is current free and would not play a role in the build up of the motor flux and motor torque.

The state of the switching matrix can be expressed with a three bit binary word XYZ.
If for example we have programmed a six step sequence, then table 20-3 shows the possible combinations. For every XYZ-combination we can determine the phase voltages v_{S1}, v_{S2} and v_{S3}.

From (20-15) and (20-16) the components v_α and v_β of the stator voltage vector V follow. This V is on the right hand side of table 20-3 and will be calculated in the following numeric example.

Numeric example:

Question:
Calculate the position of the six vectors drawn to the right of table 2-3.

Solution:
We solve for one vector. The other five vectors are calculated in an identical manner.
Example: Vector V = 5; XYZ = 101; closed switches: 1/6/5; $v_X = V_t\,;\ v_Y = 0;\ v_Z = V_t$

line voltages: $v_{L1L2} = v_X - v_Y = V_t\,;\ v_{L2L3} = -\,V_t\,;\ v_{L3L1} = 0$

phase voltage (see fig. 20-73): $v_{S1} = \frac{V_t}{3}\,;\ v_{S2} = -\,\frac{2}{3}\cdot V_t\,;\ v_{S3} = \frac{V_t}{3}$

v_α- and v_β - components: $v_\alpha = \frac{3}{2}\cdot v_{S1} = \frac{V_t}{2}$

$$v_\beta = \frac{\sqrt{3}}{2}\,(v_{S2} - v_{S3}) = -\,\frac{\sqrt{3}}{2}\cdot V_t$$

vector 5 can now be determined:
- amplitude $= \sqrt{v_\alpha^2 + v_\beta^2} = V_t$
- phase $= \arctan\frac{v_\beta}{v_\alpha} = -\,60°$

Calculation will reveal that all the V-vectors have the same length V_t.
Since the length of the vectors are calculated rather than using a digital word XYZ we could indicate the direction (V = 0, 1, 2 ... 5).

Table 20-3

X	Y	Z	V	
1	0	0	0	active situations
1	1	0	1	
0	1	0	2	
0	1	1	3	
0	0	1	4	
1	0	1	5	
0	0	0	6	BREAK vector = 0 (origin)
1	1	1	7	

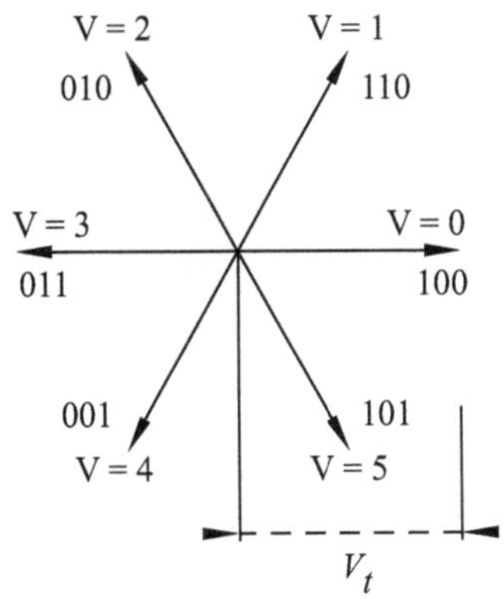

Table 20-3: Table of vectors associated with fig. 20-71

B. Direct control of the inverter switches - Stator flux vector

Assume inverter position 7 (= pause) as shown in fig. 20-72, and accompanied by a motor at stand still without flux.

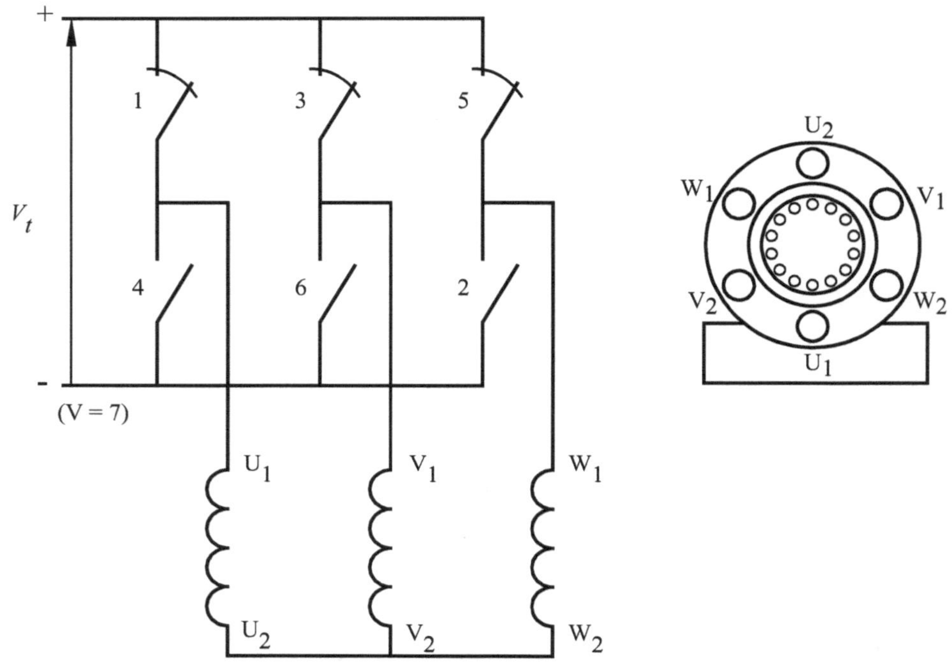

Fig. 20-72: Inverter position at pause (V = 7)

Inverter switch 3 is opened and switch 6 is closed (fig. 20-73). The voltage vector is now V = 5 (XYZ = 101). A magnetic field is built up as shown in fig. 20-73. The flux grows as it were from zero to Φ_5 at instant t.

The next active switch position of the inverter bridge is V = 0 (XYZ = 100), this means that switch 2 has taken over from switch 5 with respect to fig. 20-73. This is shown in fig. 20-74.

The current in coil U_1U_2 is doubled in value, the current in coil V_1V_2 is halved and the current in coil W_1W_2 has reversed direction. The stator flux has rotated 60° to the left.

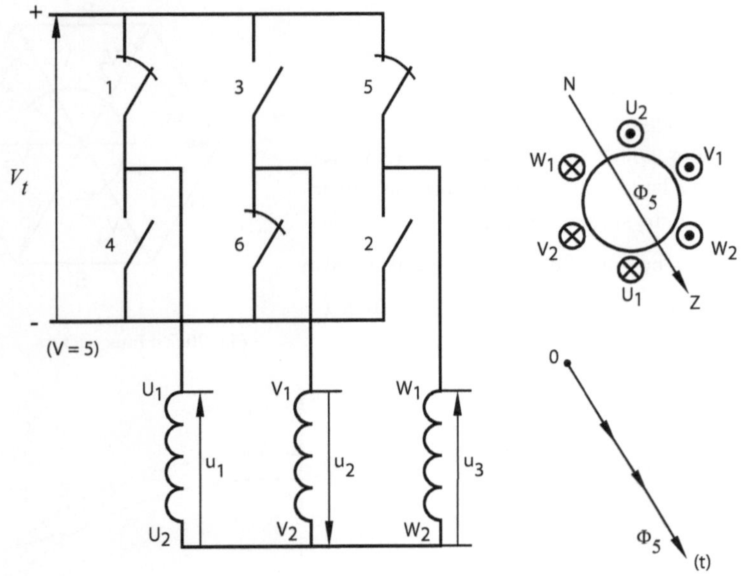

Fig. 20-73: Inverter position V = 5

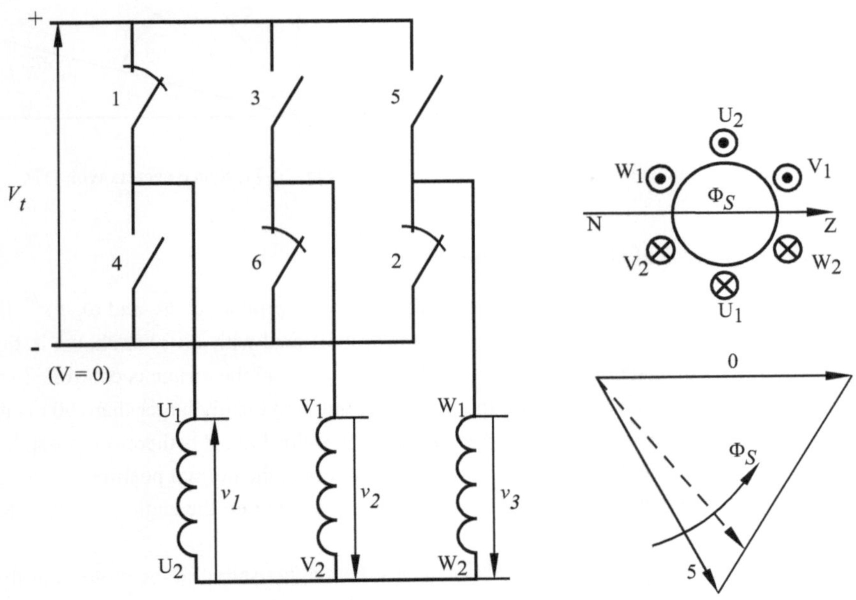

Fig. 20-74: Inverter bridge in switch position V = 0

If now consecutively V = 1, 2, 3... the flux and flux vector acquires the form shown in fig. 20-75. The hexagon does not show an ideal circular form. We can approximate a circular form by for example continually changing the switching order (fig. 20-76). The stator voltage vector and stator flux vector now assume a large number of positions, no longer comparable to the six step sequence.

Together with the stator flux $\overrightarrow{\Phi_S}$ we have also drawn the space vector $\overrightarrow{i_S}$ and $\overrightarrow{\Phi_R} \cdot e^{j\delta}$ in fig. 20-76 in a stator coordinates system α - β .

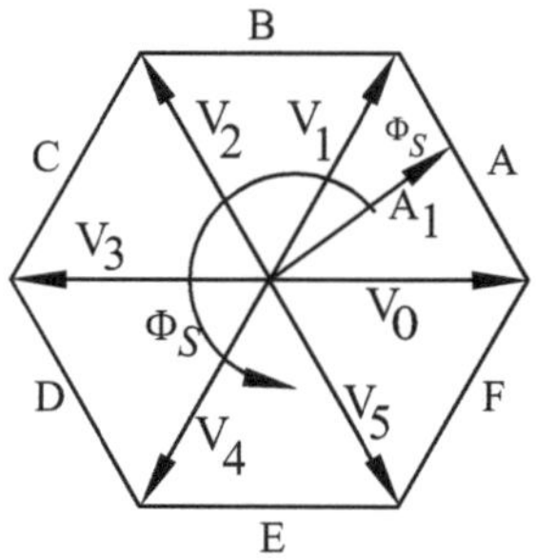

Fig. 20-75: Flux vectors

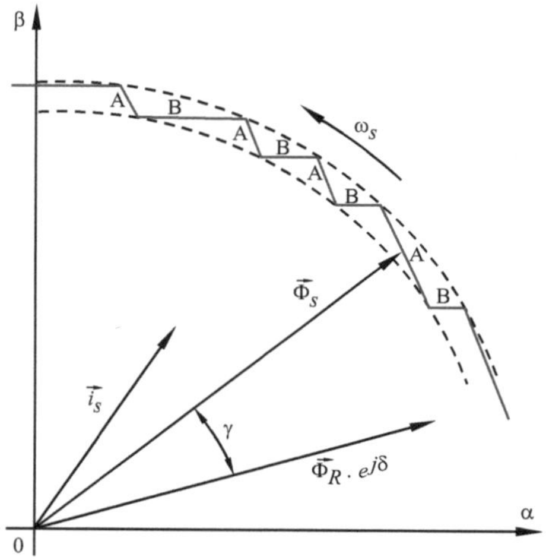

Fig. 20-76: Space vectors of induction motor

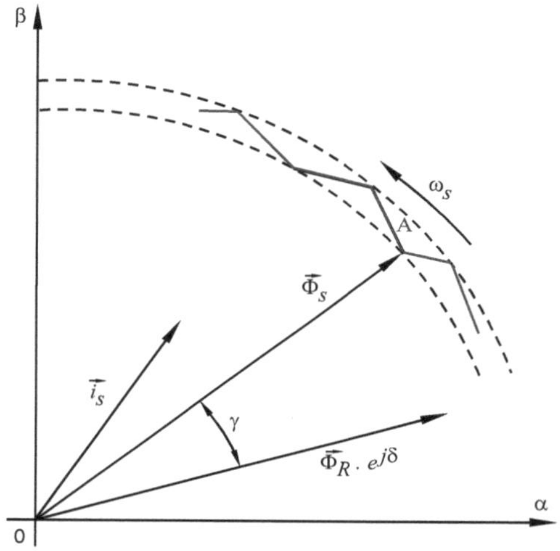

Fig. 20-77: Space vectors with DTC

7.9.2 Control of stator flux and torque

The electromagnetic torque of the motor is the vector product of $\overrightarrow{\Phi_S}$ and $\overrightarrow{\Phi_R} \cdot e^{j\delta}$. If both fluxes form an angle γ , then this vector product is proportional with $\sin(\gamma)$ as shown in fig. 20-64.
The value of the stator flux is normally held constant and the torque is controlled with the aid of γ.
Since the rotor time constant of an induction motor is typically larger than100 ms the rotor flux is quite stable and changes slowly, compared to the stator flux. It is therefore possible to control the torque by rotating the stator flux vector with respect to the normal position. In fig. 20-76 we see that by quickly changing the switch positions of the inverter, the angle γ and therefore the torque is controlled.

The radial change in stator flux also changes the magnetisation of the motor. For this reason care is taken that both the required torque and required stator flux (magnetisation) are achieved. Therefore a hysteresis loop is implemented within which the stator flux and torque must remain.

In the situation of fig. 20-77 the stator flux should increase as it is at it's minimum value. We can allow $\overrightarrow{\Phi_S}$ to increase by letting the inverter in fig. 20-71 switch to the inverter position V_0, V_1 or V_5 (see flux vector $\overrightarrow{\Phi_S} = A_1$ in fig. 20-75). In addition the desired value of the torque should be looked at! If we need to increase the torque in fig. 20-77 then we need to increase γ. With the stator flux vector somewhere between V = 0 and V = 1 (fig. 20-75) we can increase γ with the inverter switching pattern V_1, V_2 or V_3. The condition required by the stator flux vector (V_0, V_1 or V_5) and the torque (V_1, V_2 or V_3) determine that ultimately we switch to V_1 in fig. 20-77, and not as in the principle drawing of fig. 20-76 switching from A to B.
The DTC also contains an ASIC which stores a table that indicates the required inverter switch positions. It is first determined in which 60° sector the stator flux is. For each of these sectors a certain part of the table is used.
Fig. 20-78 shows the acceptable hysteresis of the torque.

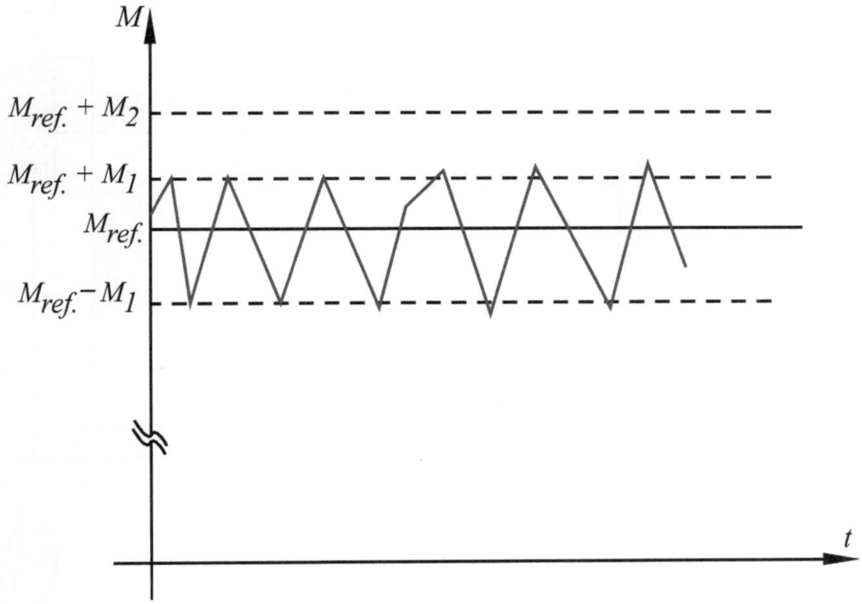

Fig. 20-78: Acceptable hysteresis of the torque

If the calculated motor torque (via motor model, see later) deviates by more than M_1 from the reference torque a suitable stator voltage vector is chosen to adjust in order to return it to within $M_{ref} \pm M_1$. In principle the amplitude of $\overrightarrow{V_S}$ is kept practically constant. At very low frequencies a voltage vector in a tangential direction has a large influence on the torque.
If M is above $M_{ref} + M_2$ it is permitted to change the stator voltage vector to reduce the torque.
In addition by adjusting the hysteresis parameters for torque and flux it is possible to adjust the switching frequency from for example 1.5 to 3.5 kHz.

7.9.3 Torque control loop

Fig. 20-79 shows the heart of a DTC, namely the torque controller. The motor model allows the motor shaft-torque, stator flux and speed to be determined. For this a number of motor parameters are required as well as the stator current vector and the stator voltage vector. The stator current vector can be calculated from a measurement of the stator currents. The stator voltage vector is determined from the measurement of the DC-link voltage and knowledge of the switching position of the inverter bridge.

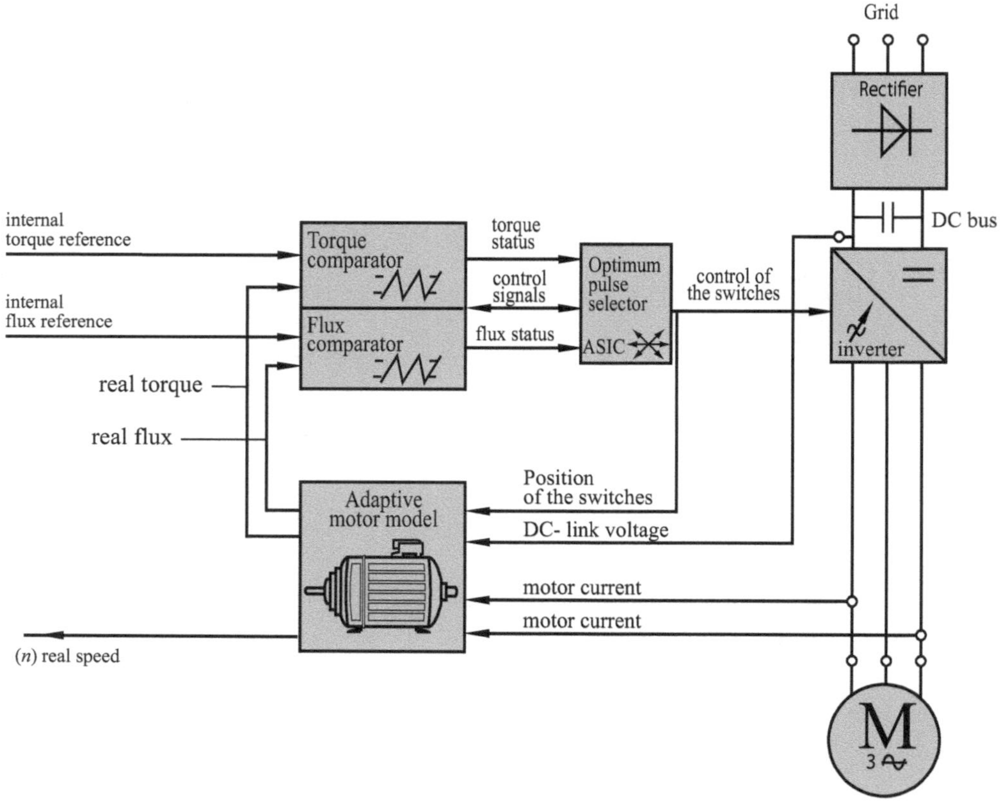

Fig. 20-79: Block diagram DTC torque controller

An external torque reference can be used for the torque controller, or the torque reference can be internally provided by the output of the speed controller, as was the case with the DC-machine. Limiting the torque reference can maintain the torque below the pull out torque. The actual value of torque and flux are compared with reference values via two hysteresis controllers and used to control an ASIC. This ASIC contains the optimum switching logic for controlling the inverter. The feedback information of the inverter state is used together with the DC-link voltage to calculate the actual stator voltage vector. In this process delay time and voltage drops across the inverter switches are taken into account.

7.9.4 Motor model

Accuracy of the motor model is essential for correct operation of the DTC. This motor model is based on the identification of the motor parameters and measuring the stator current of the motor. The most important task of the motor model is to make an exact estimate of the stator flux and to do this 40,000 times a second. Every consecutive control cycle takes 25μs. Every 25μs the stator flux is calculated with the aid of (20-46): $\overrightarrow{\Phi_S} = \int (\overrightarrow{v_S} - \overrightarrow{i_S} . R_S) \cdot dt$

The influence of temperature on the stator resistance R_S is derived from a thermodynamic model of the motor. In addition the estimate of the stator flux is checked with expression (20-26):

$\overrightarrow{\Phi_S} = L_S \cdot \overrightarrow{i_S} + L_0 \cdot \overrightarrow{i_R} \cdot e^{j\delta}$

This check is especially needed at low speed .The electromechanical torque is determined as the vector product of $\overrightarrow{i_S}$ and $\overrightarrow{i_R} \cdot e^{j\delta}$.

With the aid of the motor model the angular velocity ω of the motor and ω_S of the stator flux are calculated.

The angular velocity ω_S follows from the derivative of the angle of the rotor flux vector.

This angle follows from (20-47)

$\overrightarrow{\Phi_R} \cdot e^{j\delta} = \frac{L_R}{L_0} \cdot (\overrightarrow{\Phi_S} - \sigma \cdot L_S \cdot \overrightarrow{i_S}) = \Phi_R(\alpha) + j \cdot \Phi_R(\beta)$ Indeed from this: $\lambda = \arctan \frac{\Phi_R(\beta)}{\Phi_R(\alpha)}$

The angular velocity of the stator flux is then: $\omega_S = \frac{d\lambda}{dt} = \frac{\lambda(t_2) - \lambda(t_1)}{\Delta t}$. Here Δt = 1ms. This allows for calculation of frequencies up to 400 Hz.

The angular velocity of the motor follows from an expression like :

$$\omega = \omega_S - \frac{2 \cdot R_R \cdot M_{em}}{3 \cdot p \cdot \Phi_R^{\ 2}}$$

The motor model is derived from the data input from the motor name plate, together with a test run (auto-tuning) during the commissioning of the drive.

As indicated on p.20.52 the motor parameters L_R, L_S, L_0, σ and R_S are required together with $\overrightarrow{v_S}$ and $\overrightarrow{i_S}$, in order to calculate the flux $\overrightarrow{\Phi_R}$.

7.9.5 Speed control loop

The heart of the DTC as shown in fig. 20-79 results in good torque control and determination of the motor values.
For speed control though a speed controller is still required.
Fig. 20-80 shows a block diagram of a speed controller. The flux reference controller (see 7.9.6 which is the next section) is also shown in fig. 20-80.
The speed controller uses a PID algorithm. The output of this controller is the torque reference.
In addition there is also an acceleration compensation included in the speed controller. During commissioning the mechanical time constant of the drive is determined.
The acceleration compensation will automatically take the mechanical time constant into account.

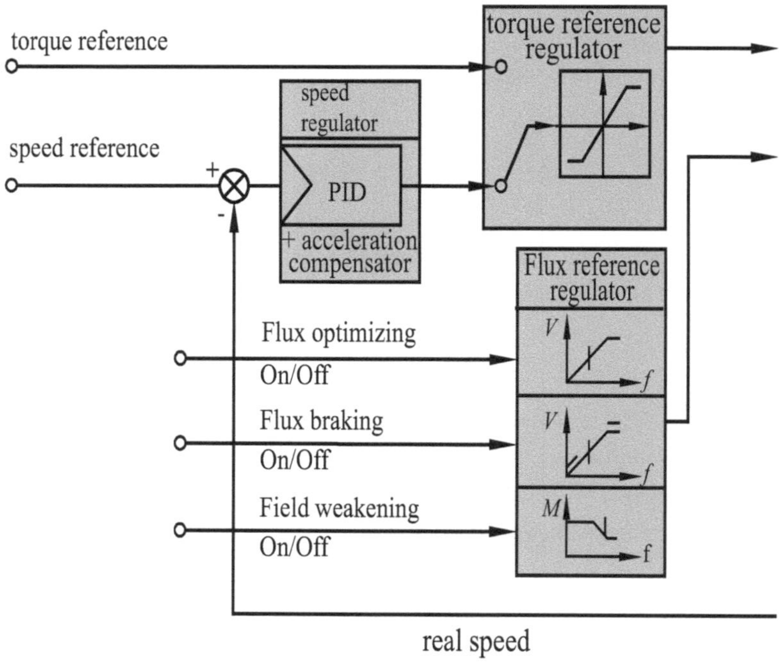

Fig. 20-80: Block diagram speed controller and flux reference controller

7.9.6 Flux reference control

The absolute value of the stator flux can serve as a reference for the DTC-block.
As the possibility exists to adjust this flux reference the converter properties can be adjusted, for example during field weakening or flux optimisation.

7.9.7 DTC Block diagram

Fig. 20-81 provides a complete block diagram. It is a combination of fig. 20-79 and 20-80.

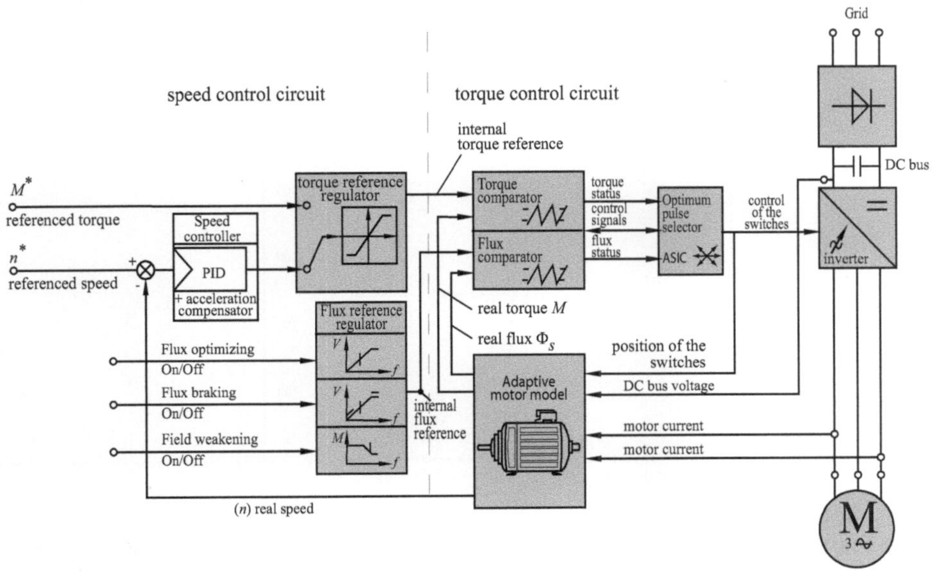

Fig. 20-81: DTC block diagram

7.9.8 Special DTC functions

A starting properties

The motor can be started practically immediately in every electromagnetic state. In the case of a rotating motor the synchronous speed is detected in less than 2ms and the DTC synchronises immediately.

Even when the rotor flux has not completely disappeared an immediate start is possible.

B. Flux braking

The motor can have the brake applied during normal operation by braking the flux. The braking power is the same as DC-braking, but with the DTC the inverter is controlled normally.

C. Flux optimisation

The DTC-model can automatically calculate the optimum magnetisation of the motor, depending on motor load. Flux optimisation raises the efficiency of the drive.

7.9.9 Performance of a DTC-drive

The following data was presented by ABB to demonstrate the performance of the DTC.
In fig. 20-82 the speed and torque performance are presented at a speed of 30 rpm without tachogenerator. The torque data is also provided. Note a step response of 2 ms.
Fig. 20-83 shows the same response but now with a tachogenerator. The difference with fig. 20-82 appears minimal.

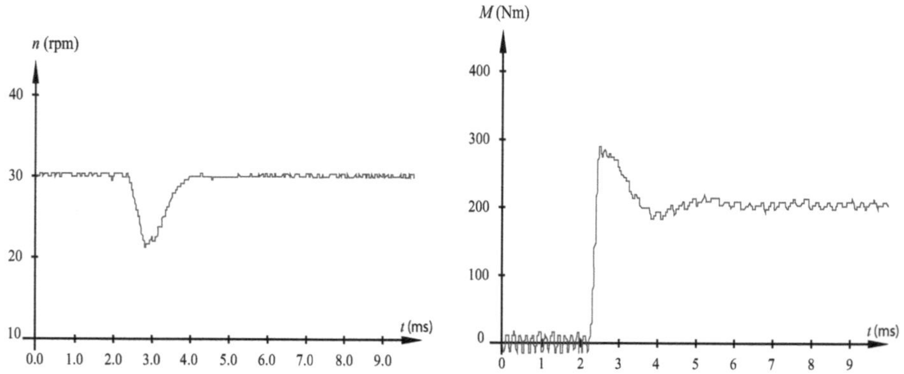

Fig. 20-82: Speed control without tachogenerator. Speed set point 30 rpm (1Hz). Load step = nominal torque. Inverter ACS601-0050-3. Motor: 37 kW, 1475 rpm

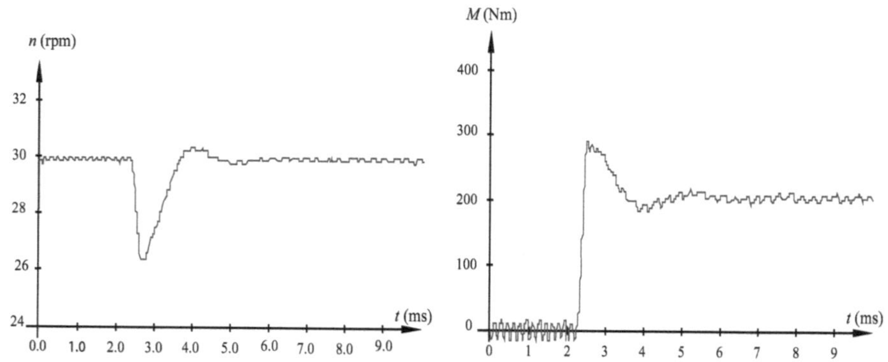

Fig. 20-83: The same data as in fig. 20-82, now with a tachogenerator

8. MICRO ELECTRONICS IN POWER ELECTRONICS

The minimum requirements for the control of an asynchronous motor are:

1. V/f setting
2. Adjustable:
   set-point speed / acceleration / deceleration / torque gain at low speed / IR-compensation / soft start (= time before the frequency begins to rise from the minimum value) / set point torque (in the case of vector control)

3. Additional possibilities:
   reversal of direction / overvoltage -,undervoltage-, earth fault- and overcurrent protection / zero voltage detection /start-stop / brake resistor / DC-brake / trap circuit / short circuit protection / thermal overload of motor ($I^2.t$), can usually replace a thermal relay.

These requirements can only easily be achieved with the aid of micro-electronics.

These days the motor control logic is completely digital. The first three-phase PWM-control IC was presented by Philips in 1979 at an IEEE-conference in London. It was the HEF4752. This is an LSI-circuit using LOCMOS-technology. The HEF4752 generates three signals with phase displacements of 120°. The average value of these signals changes sinusoidally in time in a range between 0 and 200 Hz. They are PWM signals. There after the developers switched to micro-controllers and digital signal processors (DSP). In the first decade (1980-1990) they were mostly 8 bit micro-controllers, followed by 16 bit controllers.

The first DSP from Texas instruments appeared in 1990 and from 1993 the first controller boards based on DSP's became available, for amongst other things applications with vector control of three-phase motors.

The second generation with a DSP power of 20 MIPS appeared on the market in 1997.

The latest generation of DSP chips is optimised for mathematical calculations and control processing. These DSPs are therefore called digital signal controllers (DSC). Such a controller based on a DSP is the TMS320F28x-series from Texas instruments. This series contains cores of between 60 to 150 MIPS, by contrast about 10 MIPS is required to implement a complete FOC- structure (Field Orientated Control) of an AC-induction motor. With 10 MIPS the entire control loop, this includes all calculations and transformations, can be completed in less than 35μs. DSP compilers are so advanced these days that the majority of mathematical operations are immediately written in a

C-environment and no longer in assembly language.

Fig. 20-84 shows a block diagram for the DSP-control of an induction motor.

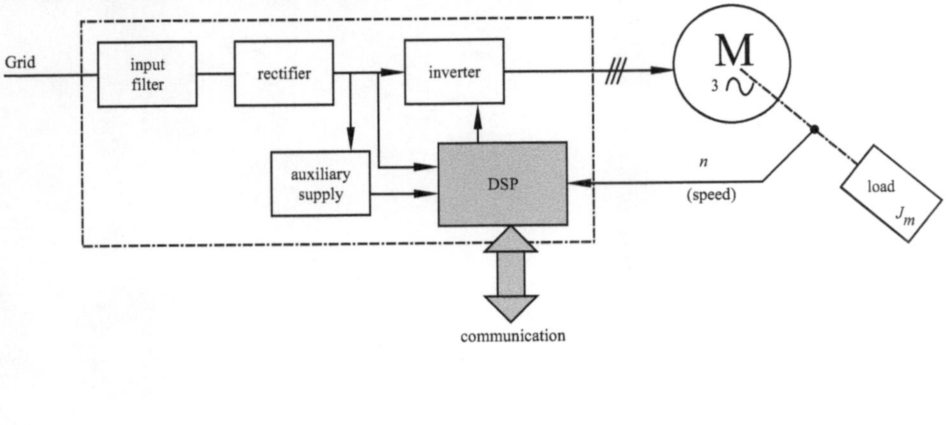

Fig. 20-84: DSP- control of a three-phase induction motor

9. SOFT-STARTERS

The problems associated with starting induction motors are well known

- large starting current
- reduced torque

The following table provides an overview of conventional starting methods.

Here I_n and M_n are respectively the nominal motor current and the nominal motor torque.

Table 20-4

Method	Application	Starting current	Starting torque
direct start	- small motors - start loaded	4 to 8 x I_n	0.5 to 1.5 x I_n
star-delta	- start at no load or small load - motor with six terminals	1.7 to 2.6 x I_n	0.5 x M_n
rotor resistances	- large power motor	4 x I_n	0.5 to 0.75 x M_n
autotransformer	- large power motor	1.7 to 4 x I_n	0.4 to 0.8 x M_n

Principle of a softstarter

With the aid of a three-phase AC controller the motor voltage is progressively raised. Since the currents can be hundreds of amperes, anti parallel SCRs are used rather than triacs (fig. 20-85). The ramp of the voltage can be controlled by an accelerated ramp function (and) or dependent on an adjustable current limit.

Photo Siemens: Sirius soft-starters for circuits from 3.6 to 1214 A. Voltages from 200 to 690V.
Motor powers from 1.1 to 710 kW. The heaviest implementation (Sirius 3RW44) for powers of 15 to 710kW (in line) and 1200kW with internal star-delta have the option of PROFIBUS/PROFINET-communication

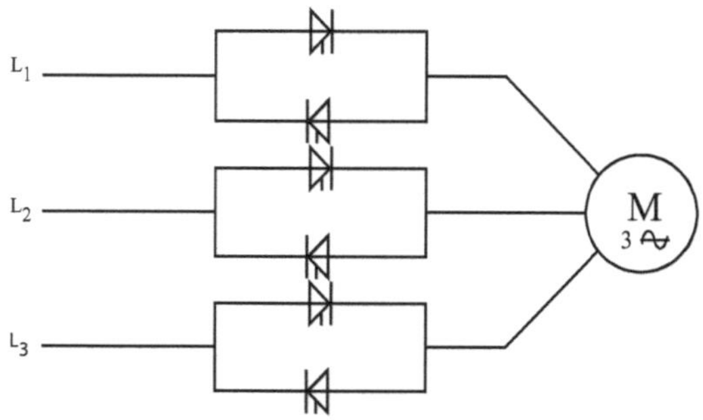

Fig. 20-85: Three-phase AC-controller

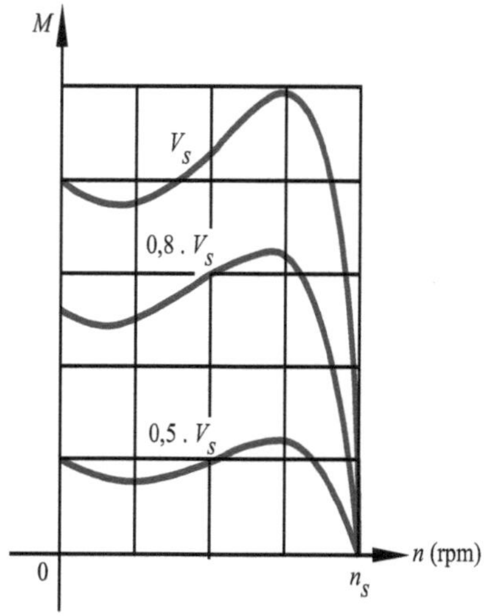

Fig. 20-86: *M-n* curves with the phase voltage as parameter

From the study of the asynchronous motor we know that torque is proportional with $\left[\frac{V_S}{f_S}\right]^2$. With fixed frequency f_S the torque will proportionally change with $V_S^{\,2}$, see fig. 20-86.

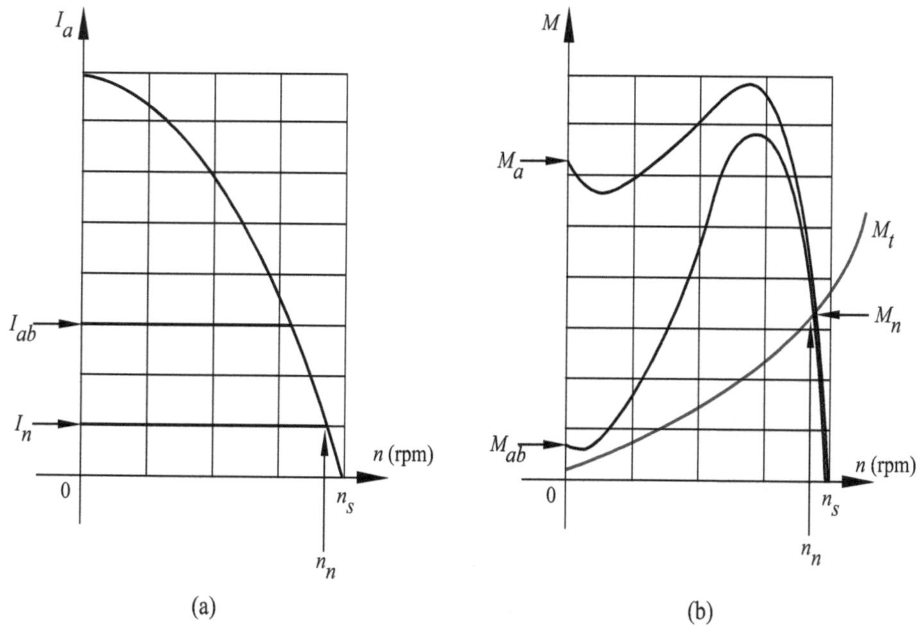

Fig. 20-87: Current progression and *M-n* curve with limited start current

Limiting the starting current I_a to I_{ab} results in reduced starting torque: M_a becomes M_{ab}. This reduction in torque is approximately proportional to the square of the ratio of I_a and I_{ab} , so that:

$$\frac{M_a}{M_{ab}} \approx \left[\frac{I_a}{I_{ab}}\right]^2 \qquad (20\text{-}49)$$

I_n and M_n are respectively called the nominal current and torque of the motor. Assume a motor with $M_a = 2.M_n$ and $I_a = 8.I_n$. Limiting the start current to $I_{ab} = 3.I_n$, produces:

$$I_{ab} = 3 \,.\, I_n = \frac{3}{8} \,.\, I_a \,. \quad \text{Further: } M_{ab} = M_a \,.\, \left[\frac{I_{ab}}{I_a}\right]^2 = 2 \,.\, M_n \,.\, \left[\frac{3}{8}\right]^2 = 0.28 \text{ x } M_n$$

Fig. 20-87a shows the currents and fig. 20-87b shows the torques of the example just considered.
During the study of the loaded AC-controller we considered three situations:

$\alpha < \Phi$

$\alpha = \Phi$

$\alpha > \Phi$

Here:
- α = thyristor firing delay
- Φ = phase angle of the load

It is only at $\alpha > \Phi$ that the circuit operates as an AC-controller.
At $\alpha \leq \Phi$ the thyristor configuration operates as a switch.
An asynchronous motor has as a property that the $\cos\varphi$ changes with motor load but also and especially during motor starting. With some motors the $\cos\varphi$ changes very quickly at the end of start up and with the application of a soft starter can result in large currents and instability because of using the α-control. Indeed as the motor starts the voltage is increased by reducing α and the resulting rapidly changeable Φ can cause a switch over from $\alpha < \Phi$ to $\alpha \geq \Phi$ and vice versa. The thyristor configuration jumps from AC-controller to switch mode and back, causing current pulses and eventual instability. Soft-starters solve this problem with a so called γ - control. With this method the firing delay (γ) of the thyristors is controlled with respect to the zero cross over of the current so that the relationship α >/< Φ plays no role.
Fig. 20-88 shows a γ - control of an inductive load.

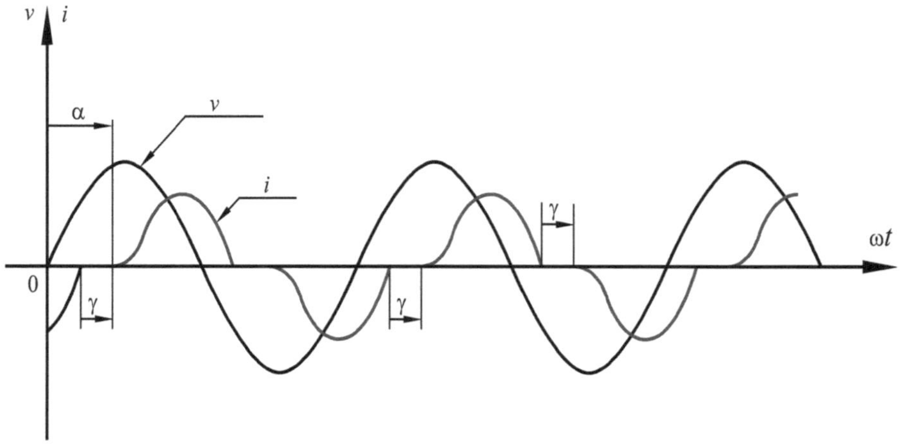

Fig. 20-88: γ-control of an inductive load

To operate as an AC-controller it should be the case that $\gamma > 0$ always maintaining a voltage across the SCR's. The output voltage of a soft-starter is in practice 10 to 20 V less than the input voltage.

Remarks

- A softstarter should not be used between power grid and supply transformer with motor.
- It is also not allowed to connect power factor correction capacitors to the terminals of motors with soft-starters.
- Once the motor has started the soft-starter can be bridged (short circuited) so that the motor is then connected directly to the power grid.

10. INDIRECT FREQUENCY CONVERTER WITH CURRENT SOURCE INVERTER (CSI)

10.1 Current source inverter

The frequency converter with voltage source inverter will impose a voltage on the motor. Depending on the load the motor current will regulate itself. With an inverter of the current source type a constant current is imposed on the motor. Fig. 20-89 shows the switching matrix with associated switch currents and line currents. With the current source inverter one switch in the upper branch of the bridge is closed together with one switch in the lower branch of the bridge. The constant current which is transferred in a suitable manner from one stator coil to the other results in a constant flux that rotates. This rotating flux induces a sinusoiodal emf in the stator windings. During commutation of the switches the quickly changing current in the leakage reactances will produce voltage pulses so that the stator voltage might appear as shown in fig. 20-90.

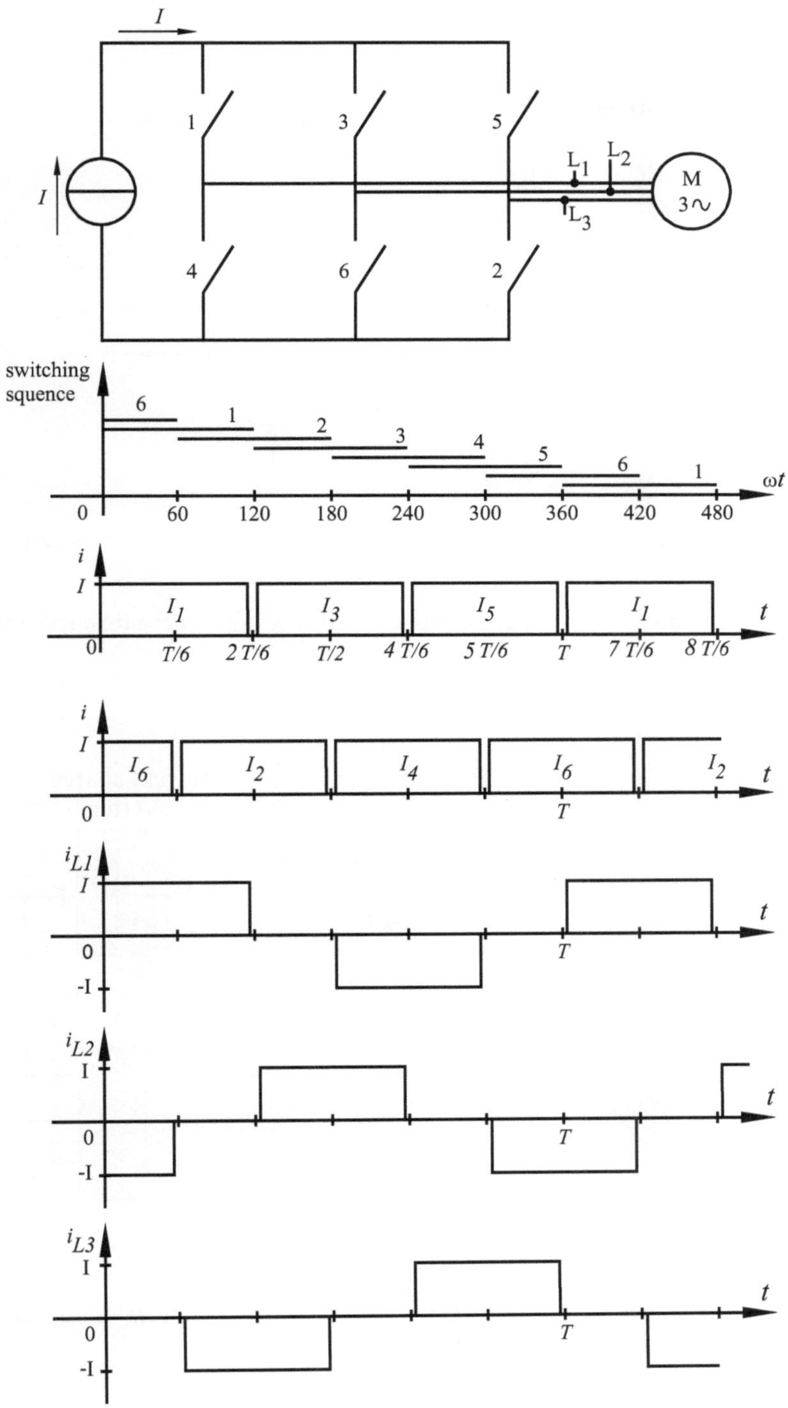

Fig. 20-89: Current source inverter for supplying an induction motor

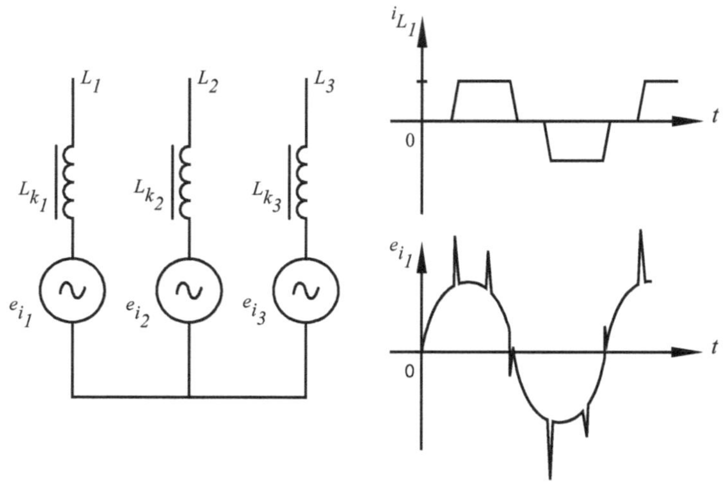

Fig. 20-90: Induced stator voltages as a result of rotating field and voltage pulses as a result of commutation

10.2 Principle configuration of motor control using a current source inverter

Fig. 20-91 shows the block diagram of a C.S.I.

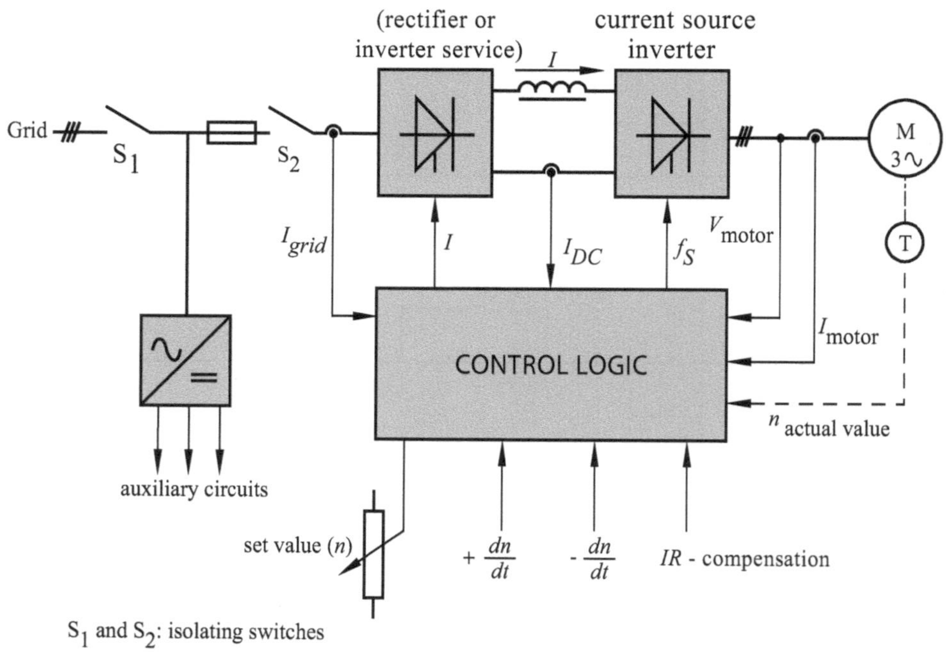

Fig. 20-91: Block diagram current source inverter control of asynchronous motor

The power bridges can look like the one shown in fig. 20-95.

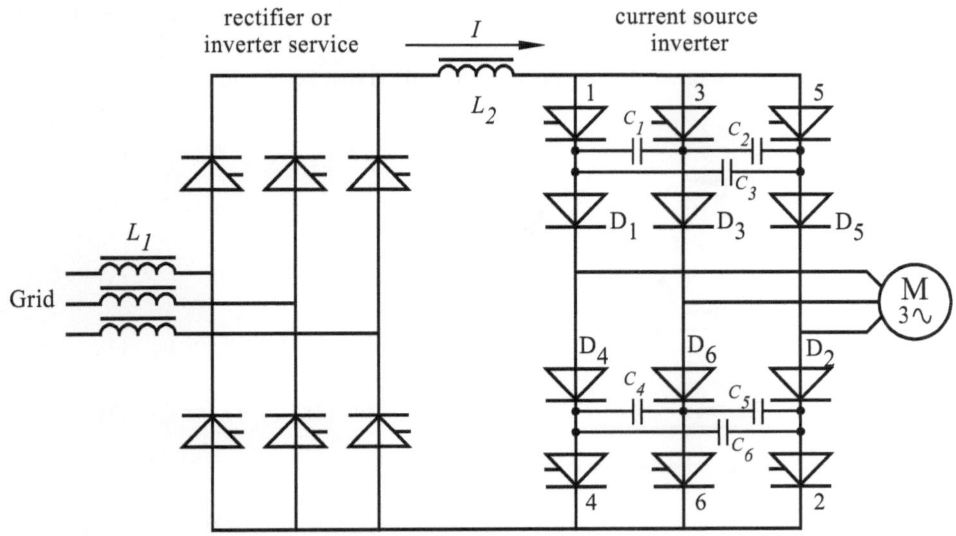

Fig. 20-92: Principle configuration current source inverter of an asynchronous motor

Operation:
The machines rotating field moves with constant speed so that the induced phase voltages have sinusoidal form. The stator current lags the voltage, with the angle depending on the load of the motor.
As the machine load decreases the phase displacement increases and at synchronous speed $\varphi = 90°$ and $\cos\varphi = 0$. At this synchronous speed the machine draws no active power since only the magnetising current I_μ needs to be supplied and this I_μ is lagging 90° behind the induced phase voltage. The DC-link voltage at no load is minimal. If the frequency of the inverter drops quicker than the speed of the motor (due to mechanical inertia of the drive) then it appears to the asynchronous machine as an over synchronised drive in generator mode. The lagging current increases with respect to the synchronous drive, in other words the current i lags more than 90° behind the induced voltage and increases the stronger the generator operation is. With extremely strong generator operation the induced phase voltage in the machine is almost 180° displaced with respect to the phase current. Since the current in the thyristors switches remains flowing in the same direction, generator operation means that the DC-link voltage reverses polarity with respect to motor operation. This corresponds with an energy flux which flows from asynchronous machine to the DC-link. The polarity reversal of the DC-link voltage is possible by controlling the input rectifier as an inverter ($90° < \alpha < 150°$).

10.3 Phase current with delta connection

If the motor is connected in star, then the (imposed) line current blocks of fig. 20-89 are also present in the stator windings of the machine.
In case the motor windings are connected in delta (fig. 20-93) then the instantaneous currents are described by:

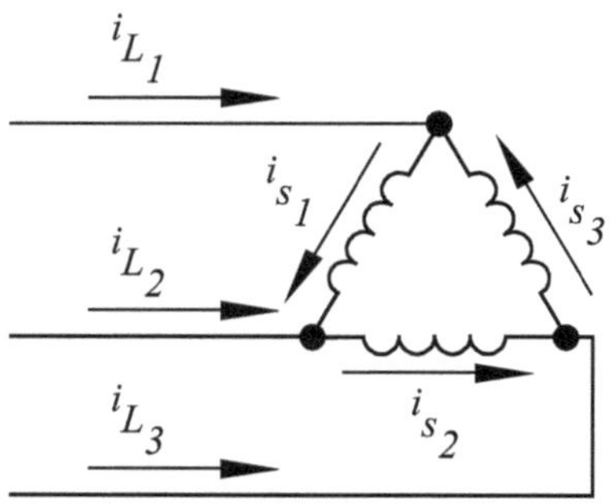

$$i_{L1} + i_{S3} - i_{S1} = 0 \quad (1)$$

$$i_{L2} + i_{S1} - i_{S2} = 0 \quad (2)$$

$$i_{S1} + i_{S2} + i_{S3} = 0 \quad (3)$$

From (1): $i_{S3} = i_{S1} - i_{L1}$ (4)

From (2): $i_{S2} = i_{S1} + i_{L2}$ (5)

Fig. 20-93: Motor in delta with current source inverter

Replacing (4) and (5) in (3) gives: $i_{S1} + i_{S1} + i_{L2} + i_{S1} - i_{L1} = 0$

or :

$$i_{S1} = \frac{i_{L1} - i_{L2}}{3} \quad (20\text{-}50)$$

similarly:

$$i_{S2} = \frac{i_{L2} - i_{L3}}{3} \quad (20\text{-}51)$$

$$i_{S3} = \frac{i_{L3} - i_{L1}}{3} \quad (20\text{-}52)$$

This has been graphically represented in fig. 20-94. The line currents I_{L1}, I_{L2} and I_{L3} are reproduced from fig. 20-89.

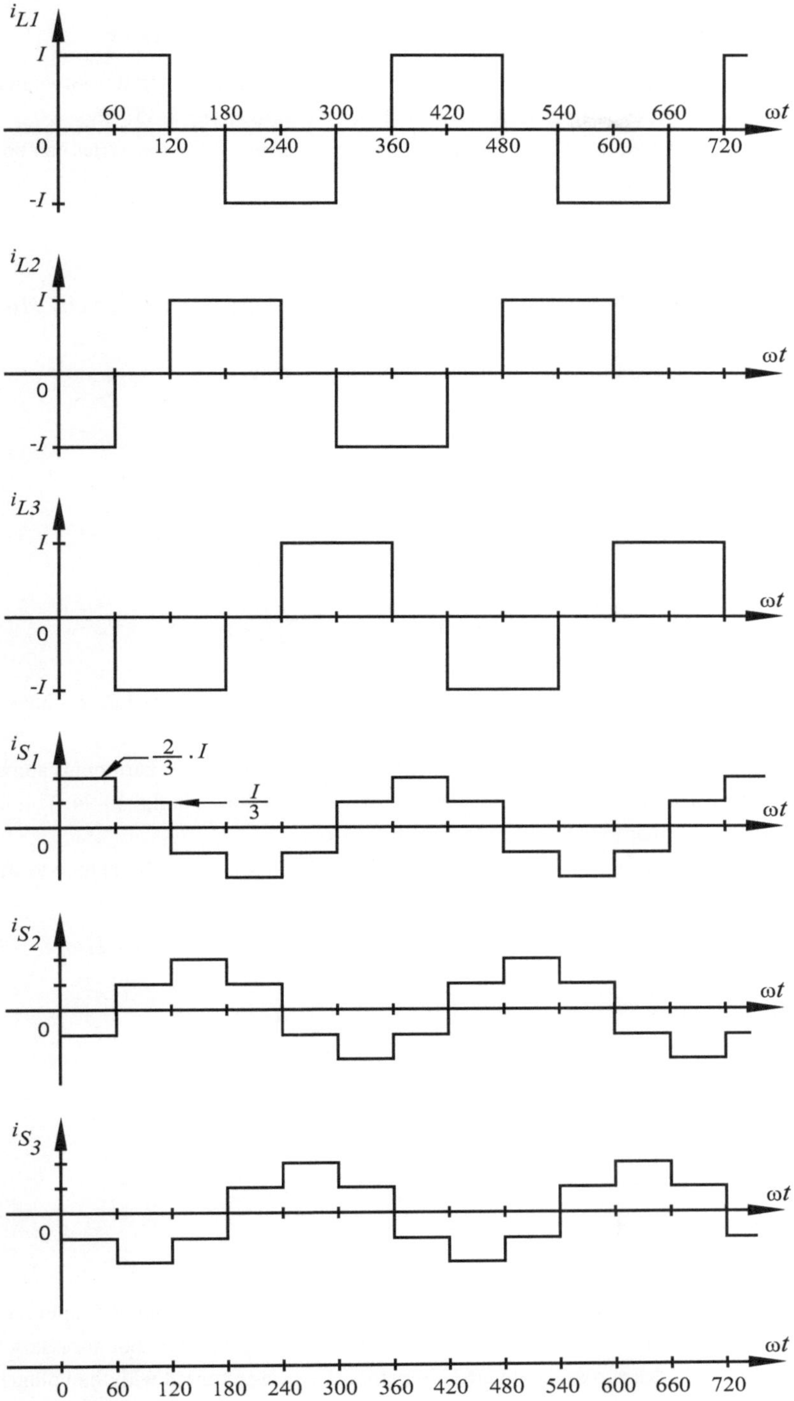

Fig. 20-94: Line currents and phase currents in a delta connected motor, connected to a CSI

10.4 Harmonics on the output of the current source inverter

The line currents which flow from current source inverter to motor are shown in the upper part of fig. 20-94. These currents have the same block shape as in fig. 8-35.
In order to determine the harmonics in the output current, a Fourier series can be expanded and similar to expression (8-28) we then find:

Series expansion:

$$i_L(t) = \frac{2 \cdot \sqrt{3}}{\pi} \cdot I \cdot \left[\cos \omega_S t - \frac{1}{5} \cdot \cos 5\omega_S t + \frac{1}{7} \cdot \cos 7\omega_S t - \frac{1}{11} \cdot \cos 11\omega_S t + \ldots \right] \quad (20\text{-}53)$$

Here I is the amplitude of the block shaped line current.

Fundamental:

- Amplitude: $\hat{i}_1 = \frac{2 . \sqrt{3}}{\pi} \cdot I$
- RMS value: $I_{eff(1)} = \frac{\sqrt{6}}{\pi} \cdot I$

Line current:

- RMS value: $I_{Leff.} = \sqrt{\frac{2}{3}} \cdot I$

Harmonics:

- Amplitude: $\hat{i}_{L(k)} = \frac{2 \cdot \sqrt{3}}{k \cdot \pi} \cdot I$

In the case where the motor winding is connected in star the phase currents have the same form as the line currents and the expressions above are valid.
If the motor is connected in delta the expressions for current and harmonics above remain valid for the line current. The phase currents have a form as drawn in fig. 20-94. This step shaped pattern is similar to the voltage form in fig. 14-6b.
We have already expanded this Fourier series and can use result (14-1) here to write for the phase currents:

$$i_S(t) = \frac{2}{\pi} \cdot I \cdot \left[\cos \omega_S t + \frac{1}{5} \cdot \cos 5\omega_S t - \frac{1}{7} \cdot \cos 7\omega_S t - \frac{1}{11} \cdot \cos 11\omega_S t \; + + - - \ldots \right] \quad (14\text{-}1)$$

Fundamental:

- Amplitude: $\hat{i}_{S(1)} = \frac{2}{\pi} \cdot I$
- RMS value: $I_{S\,eff(1)} = \frac{\sqrt{2}}{\pi} \cdot I$

Harmonics:

- Amplitude: $\hat{i}_{S(k)} = \frac{2}{k \cdot \pi} \cdot I$

10.5 In General

1. *No free wheel diodes*
   Regardless of motor or generator service a current flows through the input rectifier and output inverter and always in the same direction. It is therefore not necessary to place free wheel diodes across the inverter thyristors. This was required with the voltage source inverter. An input converter with six thyristors is enough for a current source inverter (with energy regeneration to the power grid).

2. *Induced machine voltage*
RMS value, frequency and phase displacement of the phase voltage with respect to the phase current are a good criteria for evaluating the operating condition of an asynchronous machine including its load.
3. *Constant current source*
The coil L_2 (fig. 20-92) together with L_1 forms an energy buffer. As a result I may be considered as good as constant. In addition L_2 results in a decoupling between the controlled input rectifier and the inverter. Since no capacitors are present in the DC-link circuit, they cannot be overcharged as with the voltage source inverter.
4. *Commutation of current source inverter*
The capacitors C_1 to C_6 are commutation capacitors. The blocking diodes D_1 to D_6 decouple the asynchronous machine from the commutation capacitors. The commutation capacitors result in extinguishing a conducting SCR when another thyristor is fired in the same branch. For this purpose a blocking voltage is imposed across the conducting thyristor. If the diodes D_1 to D_6 were not present, then the capacitors would loose a portion of their charge across the machine windings during the periods between commutation. A current source inverter has simple commutation methods (no auxuliary thyristors required).
The commutation capacitors should be well matched to motor current and motor reactance.
5. *Four quadrant service*
By altering the firing order of the thyristors of the current source inverter it is an easy matter to change the direction of rotation of the motor. From the description of the operation of a CSI and its simple layout it is clearly easy to operate in four quadrants.
6. *Single motor service*
Due to the constant current (source) this configuration is typically only suited to driving a single motor.
7. *Control behaviour*
Assume stable operation. At a certain instant a step increase in speed is required. As a result the control loop will increase the inverter frequency f_S. Since in the first case the induced motor terminal voltage V_S remains the same then V_S/f_S decreases with resulting flux Φ decrease and therefore a reduced induced E_{S1}. The voltage controller which forms a part of the control logic reacts to this by raising the current set-point and the current controller (which is also part of the control logic) will maximise the input converter output ($\alpha \rightarrow 0°$).
As a result the output voltage of this controlled rectifier is increased, resulting in an increased value of I of the constant current source. The motor torque M increases and the motor accelerates until a new balance at the new set-point is reached.
8. *Starting current*
Since the motor current is controlled via the controlled rectifier on the input it is therefore also possible to limit the starting current.
9. *Applications*
Control of fans, pumps, compressors, mixers, etc...
Typical control range: 1:10 to 1:20. Power: above 1MW.
Frequency range: normally from 5 to 100Hz. Under 5 Hz a form of PWM can be used to prevent pulsating operation.

10.6 Comparison of DC-link voltage and current

Table 20-5

VOLTAGE DC-LINK	CURRENT DC-LINK
• can operate with no load	• cannot operate without load (output voltage current source too high!)
• can drive multiple motors simultaneously	• primarily single motor drive
• with energy regeneration to the power grid anti-parallel bridges required on the input, or an IGBT bridge (see active in-feed from Siemens)	• four quadrant operation with energy regenera tion to the power grid possible with minimum number of power switches (input bridge with six thyristors and inverter with six thyristors.)
• converter is independent of the motor	• inverter needs to be adjusted for the motor (the values of the commutation capacitors are dependent on the motor reactances)
• PWM: uncontrolled power grid rectifier: low electrical pollution levels and power factor = 1	• electrical pollution and poor power factor on the net side
• machine can be operated above synchronous speed	• machine can be operated above synchronous speed
• PWM: converter can be supplied from an existing DC-supply	
• PWM: - fast switches required (IGBT's) - almost sinusoidal motor currents without additional output filters	

21 ELECTRONIC CONTROL of: Appliance- / SR- / Synchronous 3-phase- / & Induction servo-Motors

CONTENTS

For industrial drives three-phase asynchronous motors, independently excited motors and PM-motors are predominantly used. Controlling these motors was the subject of study of the last few chapters. A few motor types remain which are used industrially and "domestically" especially with speed control.
Industrially in a limited number of situations a synchronous AC-motor is used and also the "odd man out": the switched reluctance motor. In addition there is also the universal motor and the single phase induction motor which are especially common in domestic appliances such as drills, mixers, extraction fans,...
In this chapter the electronic control of such motors will be briefly considered.

1. APPLIANCE MOTORS

1.1 Universal motor

1.1.1 General information

Small motors are often of the universal type. Their power varies between tens of watts and a hundred watts. A universal motor can operate from an AC or DC supply. The majority of these motors are constructed to have optimum characteristics at 50Hz. The difference in construction with the DC-motor is amongst other things the fact that the rotor is laminated as well as the stator. Since the AC-supply reverses the current direction in both armature and field winding simultaneously, these series motors can operate on AC or DC. Fig. 21-1 shows the typical curves, together with the principal configuration of electronic control. Large starting torque, variable speed and small dimensions are advantages of the universal motor with respect to the single phase induction motor.

Power Electronics: DRIVE TECHNOLOGY and MOTION CONTROL
http://dx.doi.org/10.1016/B978-0-12-814641-5.50006-8,

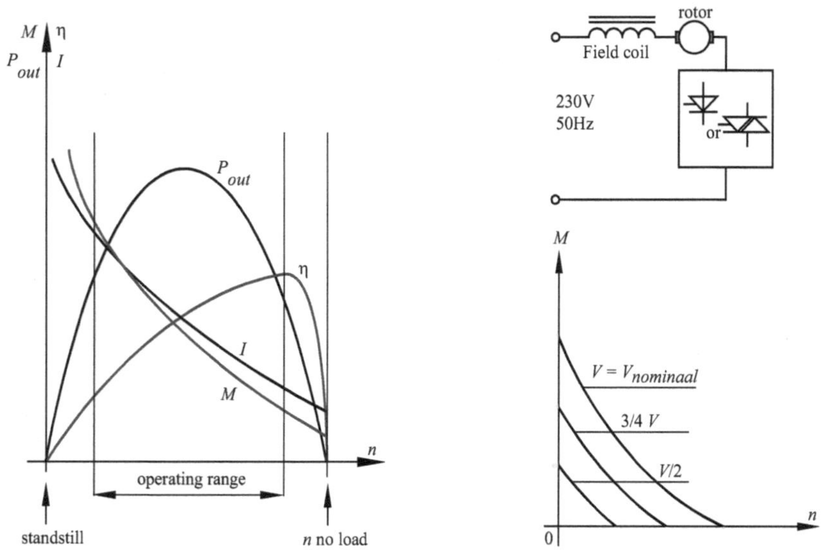

Fig. 21-1: Universal motor: curves, block diagram of the control

1.1.2 Half wave control

Fig. 21-2 shows a typical circuit with voltage feedback. During the time that the SCR is blocking there is an emf in the armature. Indeed the remanent magnetism Φ_0 produces: $E_0 = k \, . \, n \, . \, \Phi_0$. This counter emf is proportional to the speed and is therefore also an indication of the speed change. A speed change for example can occur under the influence of a varying mechanical load.

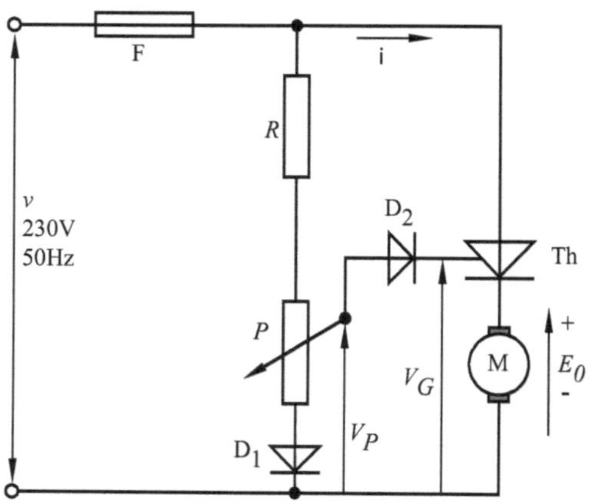

Fig. 21-2: Universal motor control with voltage feedback

During the positive half period the part of the voltage across the tapping point of P is compared to the counter emf E_0 .
If $V_p > E_0 + 2$ volt, then the SCR is fired on. This 2V is the voltage drop across the diode D_2 and the gate-cathode space of Th.
The smallest conduction angle β is 90°.
At motor stand still there is no counter emf E_0 , the SCR fires quicker and as the speed increases the counter emf becomes larger and the conduction angle β smaller.
The balance and therefore the speed are adjustable with the potmeter P.

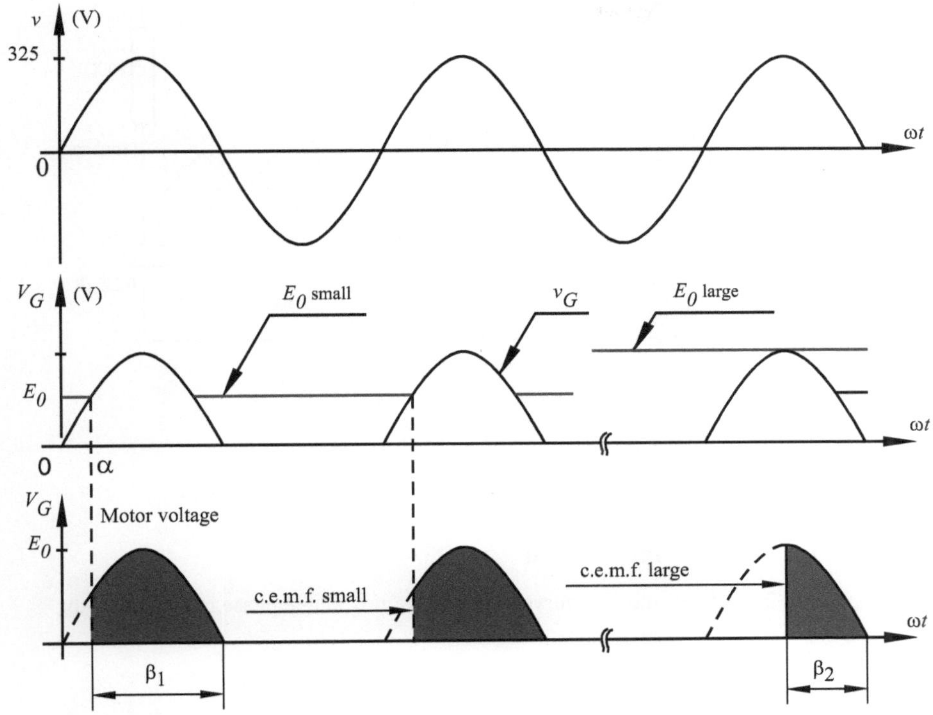

Fig. 21-3: Voltage waveforms of the configuration of fig. 21-2

If the mechanical load of the motor increases (decreases) then the speed increases (decreases) and the counter emf decreases (increases) so that the firing angle α reduces (increases), as a result of which the speed change is counteracted.

If a universal motor is used at low speed with a heavy mechanical load, then the motor can stall resulting in a large current through the SCR.
For the protection of the SCR account needs to taken of I_{FSM}!
The circuit shown is usable for 230 V universal motors up to1 kW.

1.1.3 Double sided control

If we want control with full power and small conduction angles then a circuit with double time constants is required. Fig. 21-45 shows such a configuration.

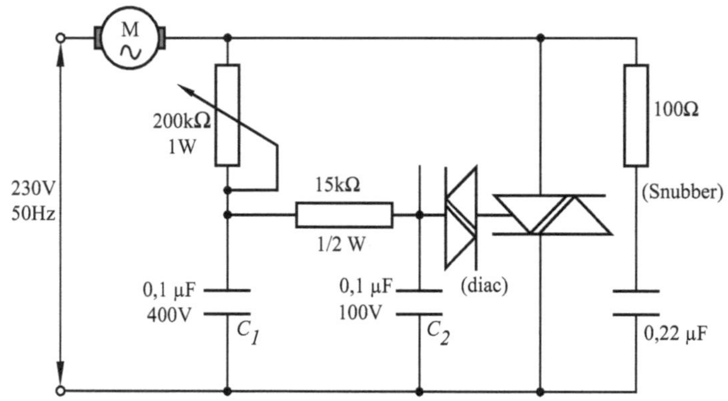

Fig. 21-4: Double sided control of a universal motor

1.2 Single phase induction motor

1.2.1 General information

Fig. 21-5 shows the *M-n* curves of a self starting single phase induction motor.

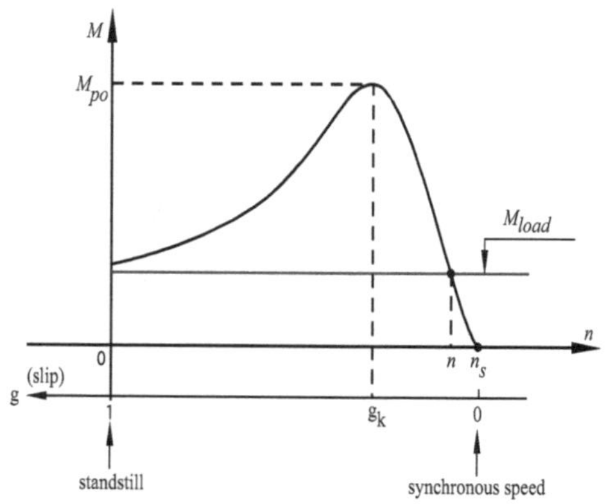

Fig. 21-5: *M-n* curves of an asynchronous motor

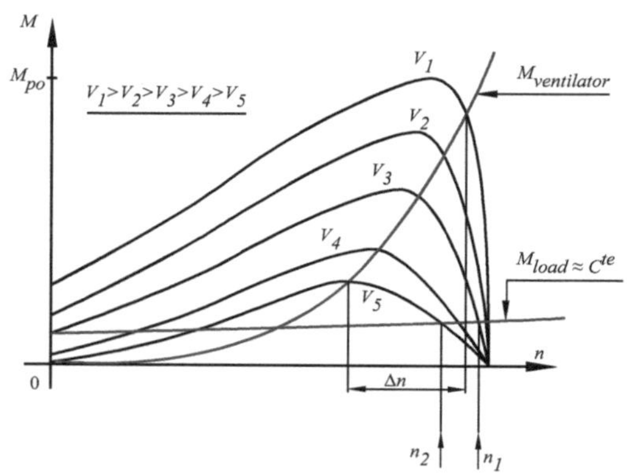

21-6: Curves of an asynchronous motor with variable load

$$n_S = \frac{60 \cdot f}{p}$$ = synchronous speed rpm; slip% : $$g = 100\,\frac{n_S - n}{n_S}$$

p = number of pole pairs
n = rotor speed (rpm)

1.2.2 Control of single phase induction motor

We know that a good control system for an induction motor simultaneously changes the voltage and frequency. This especially finds application in the *V/f* control of a three-phase asynchronous motor. Small single phase motors often have a fan or pump as load. These are loads with a quadratic characteristic. For this a variable voltage with constant frequency is sufficient. In fig. 21-6 we redraw the *M-n* curves of a self starting single phase induction motor together with the *M-n* curves of a pair of loads. Note that in the case of constant load (as with a machine tool) only a limited speed change is possible. A fan can handle a much larger control range.
The motor voltage can be made variable using phase control of a series connected triac. Only the main winding will be controlled (fig. 21-7) otherwise the phase displacement between i_1 and i_2 would not be maintained. Fig. 21-8 shows an example of a circuit for controlling a fan motor. At extremely low speed the motor can stall since the *M-n* curve in that area is extremely flat.

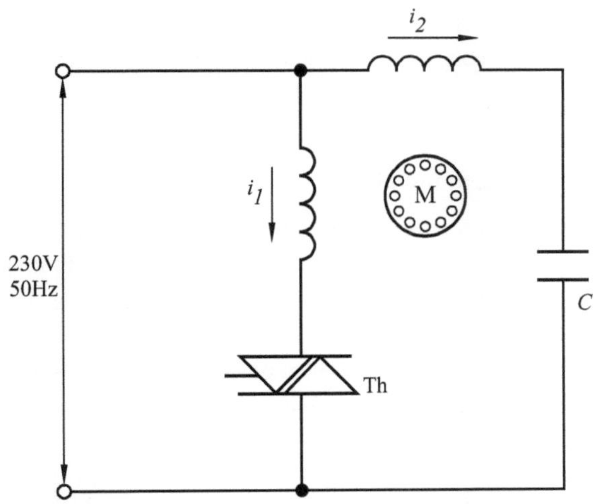

Fig. 21-7: Principle control circuit

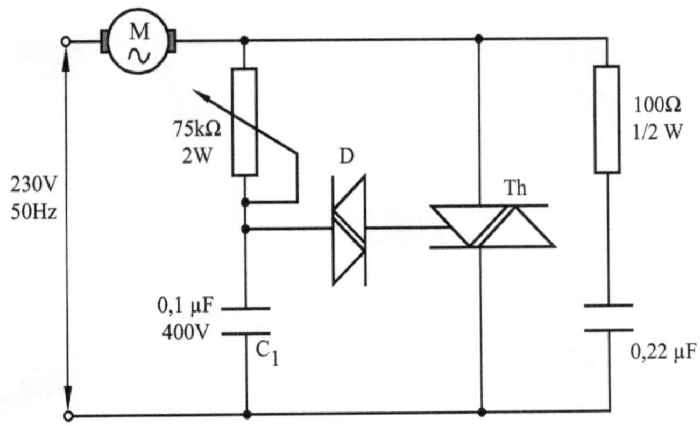

Fig. 21-8: Speed control of an asynchronous motor with “fan”load

2. SWITCHED RELUCTANCE MOTOR

2.1 History

In 1842 Robert Davidson attempted to let a battery powered train ride between Edinburgh and Glasgow. This locomotive was driven by a reluctance machine. This test failed for two reasons:

1. lack of mechanical expertise to build such an apparatus. The technology of mechanical construction came to fruition several decades later.
2. lack of the required electrical switching technology. It took another 150 years before useful "electronic" switching technology became available.

During this period three-phase motors and DC commutating machines were developed and the reluctance motor was almost forgotten. Due to the availability of good and cheap semiconductors a return was made to using the interesting potential of the reluctance motor.
The SRM (Switched Reluctance motor) combines the advantages of AC and DC machines, namely simple construction, good efficiency and the controllability of the DC machine together with the low maintenance costs of an AC-machine.

2.2 Operating principle

A laminated stator with salient poles together with a laminated rotor forms a reluctance motor. The rotor is not wound and also has salient teeth. This rotor has an extremely small moment of inertia (e.g. 0.0083 kgm^2 for a 4kW motor). Due to the minimal moment of inertia reluctance motors can achieve high speeds.
A common version is with 8 salient poles and six rotor teeth. Other combinations are 4/2, 6/4 and 12/10. We consider an example with eight stator poles and six rotor teeth as in fig. 21-9.

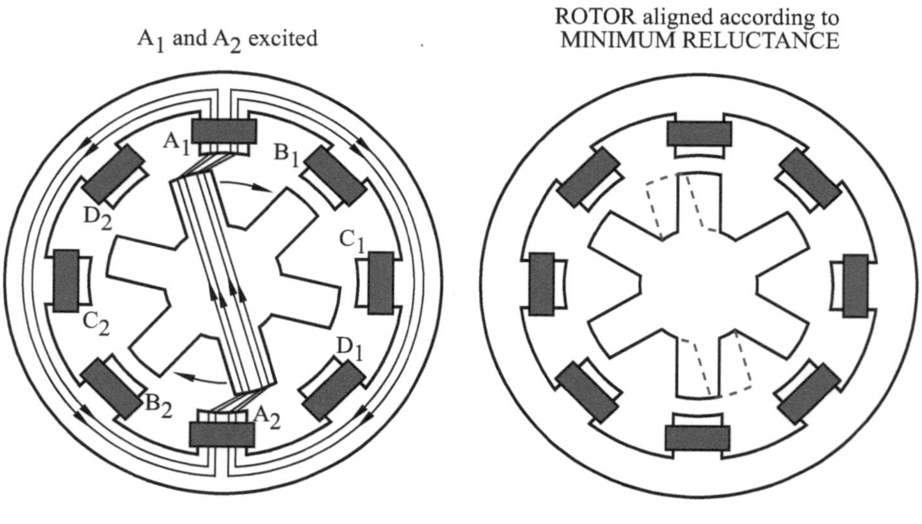

Fig. 21-9: Principle configuration of a reluctance motor

By powering the appropriate stator coils at the right moment the rotor will continuously line itself up according to the least magnetic resistance (reluctance) and so a smooth movement is obtained. The coils on two opposing poles are connected in series or parallel. The current through the coils is only interrupted when the rotor is lined up. For the operation of this motor a rotor position detector is required. Since the poles are continually switched we speak of a switched reluctance motor (SRM).
Fig. 21-10 shows an example of a cross section of an SRM. The housing can be identical to that of an asynchronous motor.
There is no heat dissipation in the rotor coils as was the case with other motors. The simple stator coils are easily insulated from the poles. In fact an SRM has a simple robust construction.

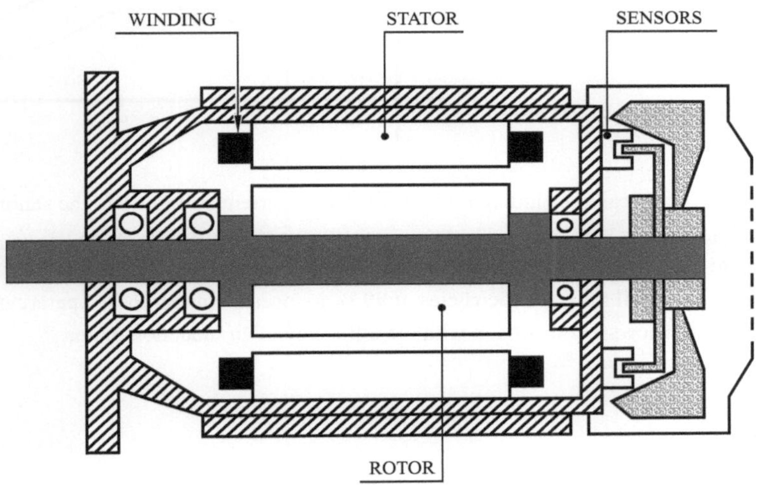

Fig. 21-10: Cross section of an SRM

Advantages SRM:

- cheap motor
- very high speeds possible
- high efficiency
- large overload capacity
- high starting torque
- good reliability
- good dynamic behaviour
- works well in harsh environment (damp, dust, temperature explosive atmospheres...)

Disadvantages:

- the motor requires special control modules to operate
- large ripple in the torque due to the discontinuous character of the phase control, the more rotor poles, the smaller the torque ripple but the greater the motor losses
- noise (as a result of the torque ripple)

2.3 *M-n* curve

Considering the supply voltage as constant, then the current rises between zero and $n_{nom.}$.
The power will rise linearly as with a DC-motor. This corresponds with a practically constant torque. Once the nominal current is reached the power remains constant and the *M-n* curve decreases ($M = \frac{P}{\omega}$).

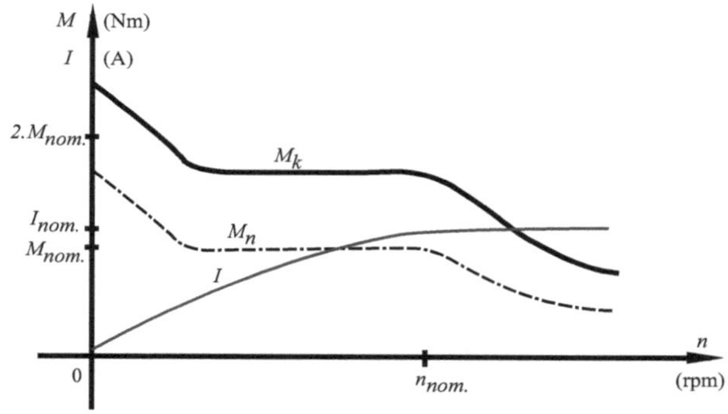

Fig. 21-11: Torque and current path of an SRM

Since in many cases including rotor at stand still, current still flows in the stator, this motor has a holding torque.
Due to the simple electrical construction (single coil on the stator) such motors have a high efficiency at full torque. Efficiencies of 90 % are possible under these operating conditions.
Fig. 21.12 shows a comparison with the efficiency of an induction motor.

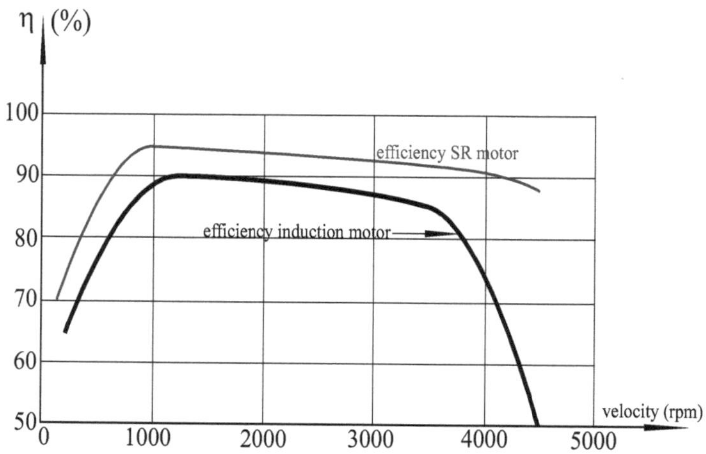

Fig. 21-12 (PSI Control Mechatronics): Comparison of efficiency between an induction motor and an SRM which serves as the main drive of a Picanol loom. The SRM has a nominal power of 11kW and maximum torque of 540 Nm and a base speed of 1000 rpm.

We find SRM's on the market from 10 W to above 140 kW. The speed goes from less than 1 rpm to 10,000 rpm.

Fig. 21-13 shows the *M-n* curves of a 132-size SR-motor. This drive from SRD (Switched Reluctance Drives - UK) is indicated as a 30kW-motor at 3000 rpm. This corresponds to a nominal torque of 95.5 Nm. Reluctance motors have a large starting torque that easily can reach 150% of nominal torque.

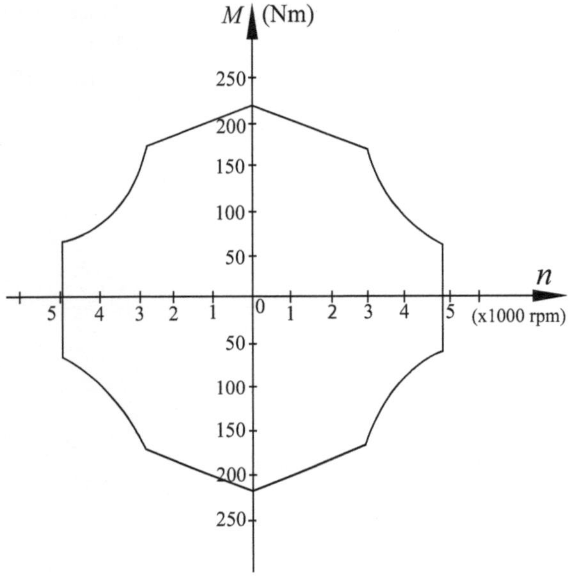

Fig. 21-13: *M-n* curves of a 4 quadrant drive with a 132-size SR-motor from SRD

The motor has an independent fan. Fig. 21-14a finally shows the coupled flux as function of the current for different positions between the aligned and non-aligned rotor positions. These curves contain a lot of information for the designer of a SR motor and are often referred to as the finger-print of the SR-motor. In fig. 21-41b the *M-n* curves of the same motor is shown. Note that the maximums of the SR motor are determined by the combination motor, control and cooling. The main drive of a Picanol-loom is conceived in such a way that it's overload capacity is more than ten times nominal torque.

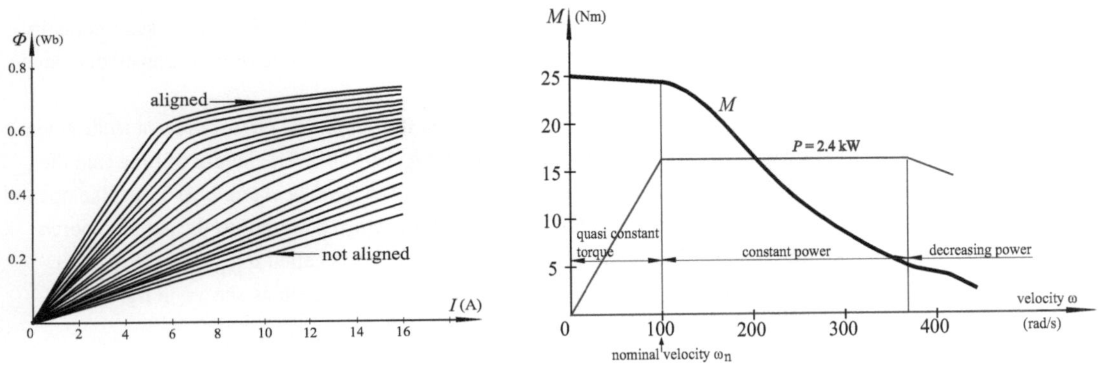

Fig. 21-14: PSI Control mechatronics: Fingerprint (coupled flux) and *M-n* curve of an SR-motor of 1500 W. Nominal, maximum torque 25 Nm and base speed 1000 rpm.

2.4 Power circuit

Fig. 21-15 shows one half of a H-bridge. This circuit is probably the most often used topology for the power circuit of an SRM. IGBT's and MOSFET's are the main choices as switches. The control is implemented per stator phase (two coils in series or parallel).
The difference with the full H-bridge of fig. 19-49 is that now only the transistors T_1 and T_4 and the diodes D_2 and D_3 remain as shown in fig. 21-15.

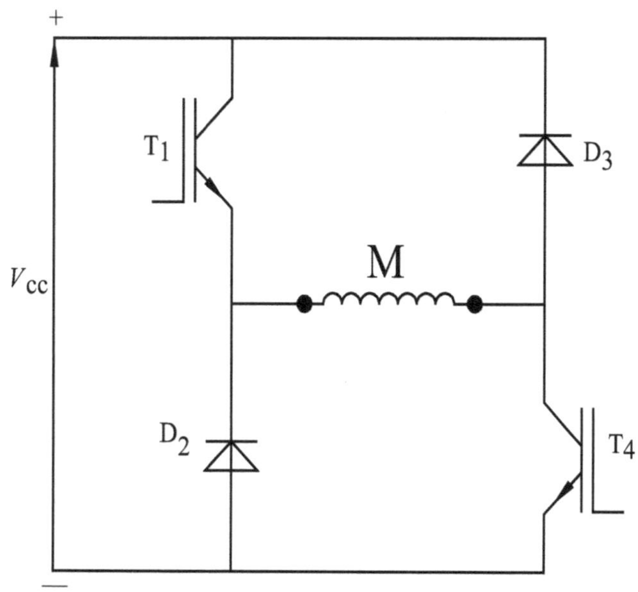

Fig. 21-15: Half H-bridge for the control of one motor phase M

The operation of this bridge is identical to the unipolar operation of the full bridge as was explained with the aid of fig. 19-52.
If we want to energise the stator phase, then T_1 and T_2 are switched "ON". This is schematically redrawn in fig. 21-16a. This is called the ON state. The conduction time of both transistors can be expressed as an angle β_{ON}.
If T_4 is now opened, then the stored magnetic energy in the stator coil causes current to flow via D_3 and T_1 as shown in fig. 21-16b (compare this with fig. 19-52b!). The length of time that the situation in fig. 21-16b lasts is expressed as an angle β_{FA} , the freewheel angle. If T_1 is also opened (fig. 21-16c) then the remaining magnetic energy can also return a current to the supply source (compare with fig. 19-52c). The angle created by this situation is called β_{OFF}.
Thereafter T_1 and T_4 are again switched on and the cycle begins again as shown in fig. 21-16a.

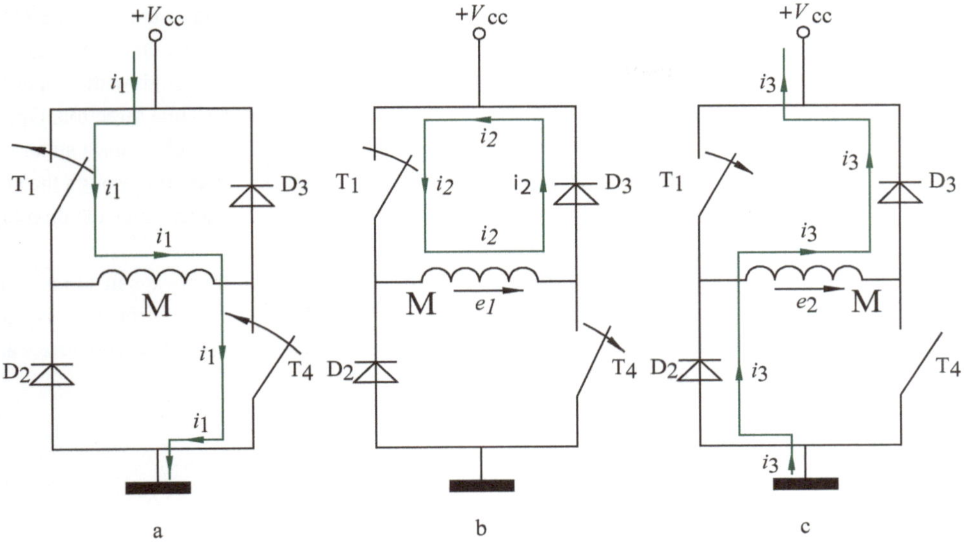

Fig. 21-16: Switch states of the half H-bridge of fig. 21-15

Photo Switched Reluctance Drives (Emerson Electric Co)

The photo shows a simple, compact and robust construction of a switched reluctance motor with "D90 frame size"

Fig. 21-17 shows a typical three-phase SR-converter. Since the stator coils of an SRM are totally insulated from each other a fault in one stator coil will have no effect on other "legs" of the converter. The motor can remain operating even with a faulty coil, although with copper losses. Note also that the current through a phase winding always flows in the same direction. Opposing poles are provided with windings which are in parallel or in series and each forms a stator coil.
The polarity of the motor torque is independent of the current direction through the stator coil. It is the position of the rotor teeth with respect to the stator that determines the direction of rotation.
In fact the diodes operate as freewheel diodes. The transistors are always in series with a coil in which the current always flows in the same direction. The problem of protection is a lot simpler compared to the three-phase inverter of an induction motor. These SRM- converters are more robust than classic three-phase inverters.

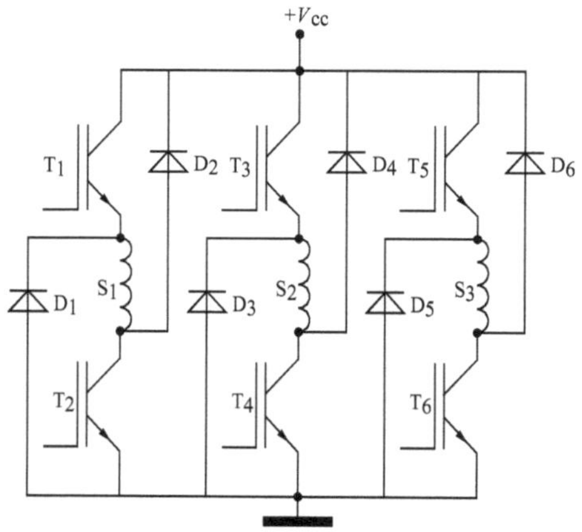

Fig. 21-17: Typical three-phase inverter used to supply an SRM

2.5 Speed-, position- and torque control

The topology of a closed control loop is comparable to that of a DC-motor. With an SRM in addition feedback of the rotor position is required for commutation of the inverter.
Fig. 21-18 shows the closed loop for a position controller.

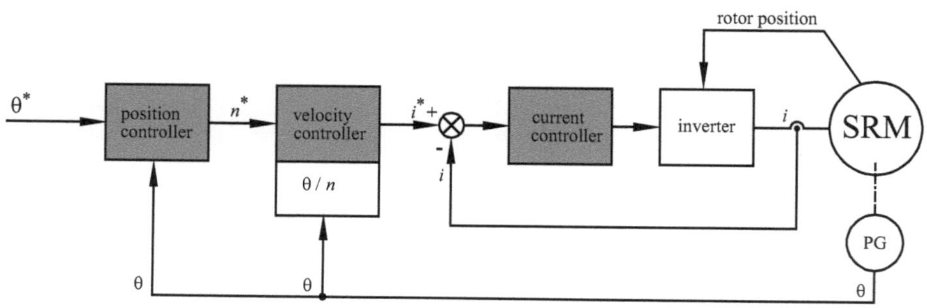

Fig. 21-18: Position control using an SR-motor

The current control is here also the inner loop that we learned about with the control of a DC-motor. The current controller is also controlled by the output of the speed controller. The speed set-point is provided by the output from the position card. The actual speed is derived from the angular position provided by the PG (pulse generator = encoder) on the shaft of the SRM.

Photo (PSI Control mechatronics): SR-motor with a nominal power of 150W, a maximum torque of 12 Nm and a base speed of 1500 rpm

The torque control often occurs with an open loop circuit (fig. 21-19). Look-up tables are often used to convert the requested torque into a set of control parameters (value of the current, angle β_{ON}, freewheel angle β_{FA} and out-angle β_{OFF}).
The angle β_{ON} determines when the stator phase is active within the electric cycle. From that instant the current loop is responsible for reaching the set-point value. The free travel angle β_{FA} indicates when there is no more energy drawn from the supply. The out-angle β_{OFF} determines the continuation of the demagnetising of the stator coil.

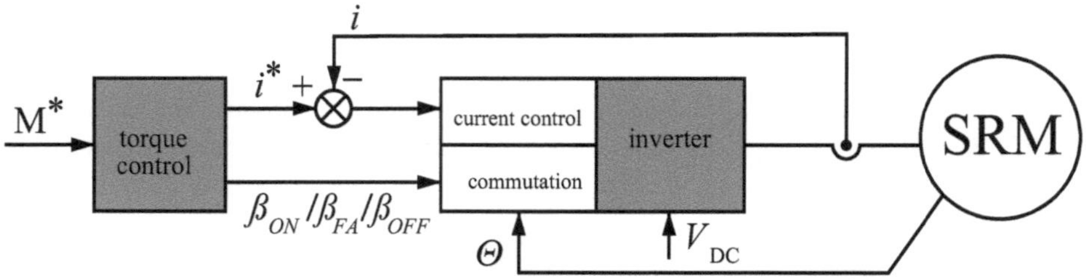

Fig. 21-19: Control of the torque of an SRM

Controlling the commutation of the inverter involves turn-on, the discharge and switch off of the stator coils. This method of torque control requires that the control parameters be very accurately determined for every operating point in the M-n plane. It is also so that more than one control set can produce the same average torque. In addition the SRM has a non-linear character.
PSI Control mechatronics developed a platform so that for a certain cost function, composed of the weighted sum of behaviour functions (torque ripple, efficiency, noise,...) the most favourable set of control parameters is found for the desired operating point in the *M-n* plane. Such optimisation platforms result in an increased flexibility and a large reduction in the development cycle and time-to-market.

2.6 Applications

- automotive
- applications with high speed (domestic applications : washing machines, fans, vacuum cleaners,...)
- compressors
- golf buggies
- pallet movers
- conveyor belts
- wrapping material at a constant pulling force
- textile machines (spinning mills,...)
- a typical application is the main drive of a loom from Picanol (manufacturer of textile machinery, Belgium, see photo on p. 21.13).

Photo Heidenhain: Encoders with Profibus connection

3. SYNCHRONOUS RELUCTANCE MOTOR

3.1 Construction

The synchronous reluctance motor (syncRM) has a three phase stator winding with sinusoidal winding distribution similar to the induction motor (IM) and permanent magnet synchronous motor (PMSM). In the past these motors were optimised for torque and framesize but today the emphasis is on efficiency.

Synchronous reluctance motors have an equal number of stator and rotor poles. The projections on the rotor are arranged to introduce internal "flux barriers". These are holes which direct the magnetic flux along a so-called direct axis.

Typical are 4-pole machines (nominal speed 1500 rev/min).

Fig.21-20 shows the principle. We designate the low magnetic reluctance path as the direct axis d of the rotor and the high magnetic reluctance path is the quadrature axis q (fig. 21-21).

Both axis form an angle of 90° (electrical!).

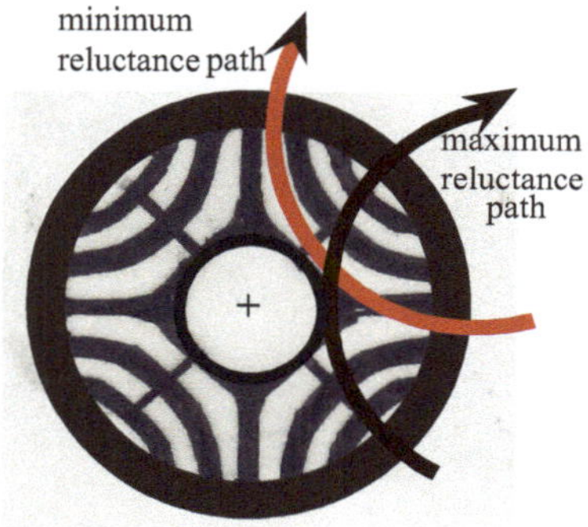

Fig. 21-20: Reluctance paths in rotor of syncRM

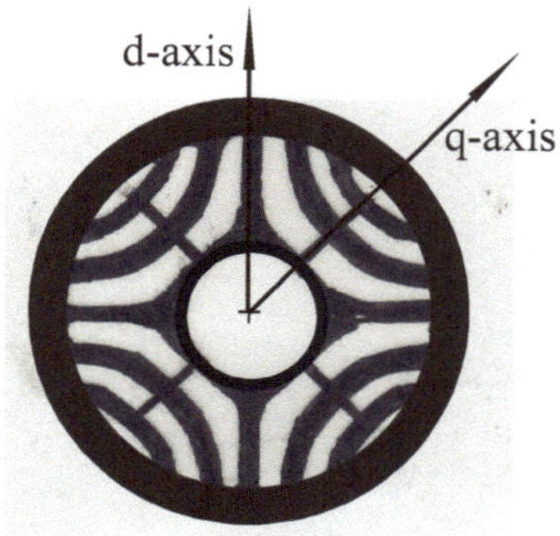

Fig. 21-21: d-q coordinates rotorfield syncRM

Photo Siemens: Cross section rotor VSD4000

The syncRM has a simple and robust rotor construction. The rotor is a stack of punched laminated steel sheets. Due to the lack of a cage, the rotor inertia is low and the rotor losses are small what contributes to a good efficiency level of the syncRM.

This achieves values of 90% at full load what is equivalent to the IE4 efficiency standard.

3.2 Properties

The syncRM presents also an excellent efficiency performance for partial load, so that this motor is extremely suitable for applications who are frequently partially loaded such as fans and pumps.

Due to the need for magnetization of the punched rotor sheets the syncRM needs a relative high supply current, resulting in a low power factor.

As the rotor is operating at synchronous speed and with the absence of current-conductors in the rotor, rotor losses are minimal compared to an asynchronous induction motor.

Reluctance motors can be overloaded up to 20% of their nominal load and this even for longer periods.

Synchronous reluctance motors are comparable to PMSMs in terms of performance and efficiency.

ABB announces that they will add low cost ferrite magnets in the flux air gaps of the rotor. These magnets add a constant flux, resulting in a further reduction of losses by approximately 20%.

3.3 Vector control of a synchronous reluctance motor

The field orientated control (FOC) as studied in chapter 20 concerning an induction motor (fig.20-54) is also applicable for a syncRM (fig. 21-22) and for a PMSM (fig. 22-43).

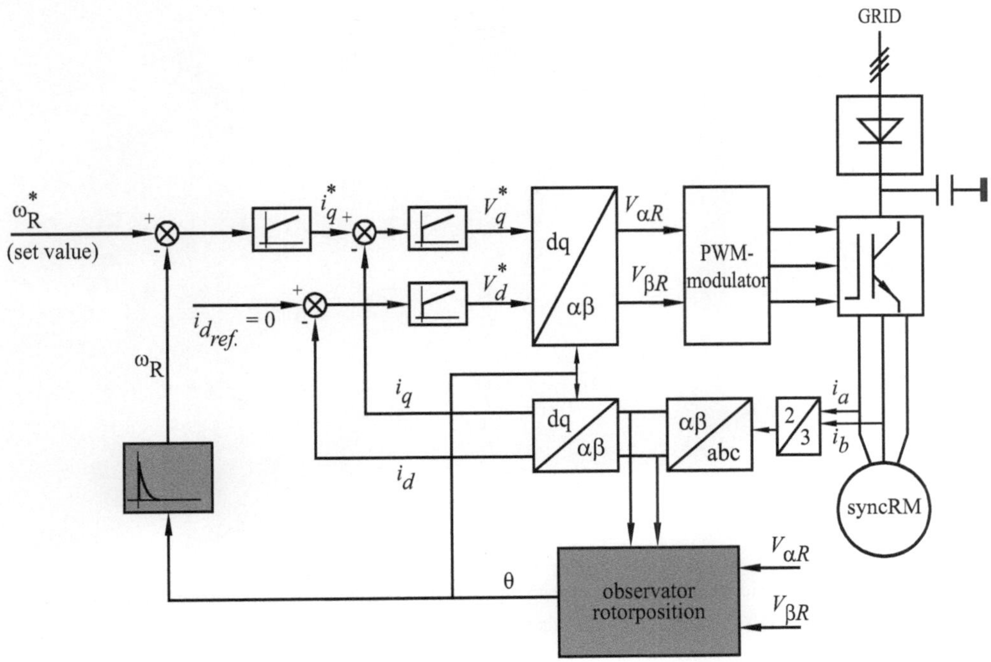

Fig. 21-22: Vector control of a synchronous reluctance motor

In the schematic shown, we assume no position sensor since a syncRM is a synchronous machine and the rotor position is easier to determine than for an induction motor. The "observer" is a mathematical model that requires the parameters and structure of the entire control system.

Remark that the composition of the rotorflux for a synchronous reluctance motor differs strongly from the rotorflux in dq-coördinates for an induction motor.
The reluctance torque is generated by a difference between the inductance L_d of the rotor (in the d-axis direction) and the inductance L_q in the q-axis direction. The difference between L_d and L_q results in a torque ripple and the saliency factor $\frac{L_d}{L_q}$ determines the power factor. A larger saliency factor gives an improved power factor.

Manufacturers of vector control equipment include control algorithms for IM, PMSM and syncRM in their frequency converters. In case of a syncRM the control algorithms strive mostly to result in a maximum torque per ampere (MTPA).

Photo Siemens: The Simotics reluctance motors with an aluminum or gray cast-iron enclosure cover an output range from 5.5 to 30 kW, achieving high levels of efficiency in full and partial load range

3.4 Torque and power characteristics

Fig. 21-23 and 21-24 show torque and power characteristics of a synchronous reluctance motor.

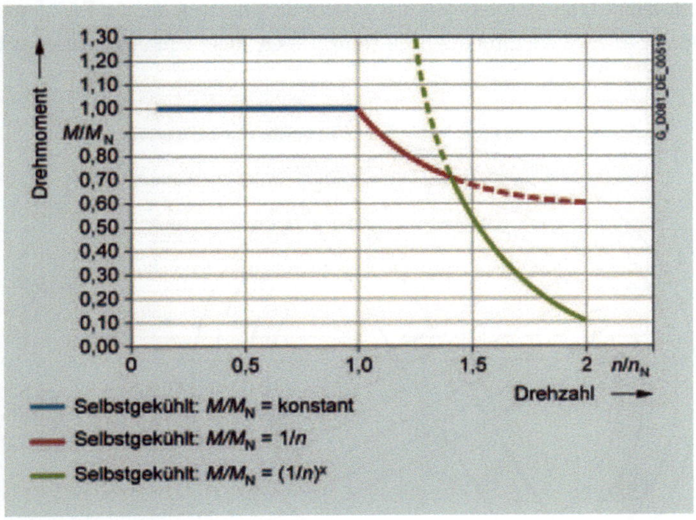

Fig. 21-23: *M-n* characteristic VSD4000 Siemens

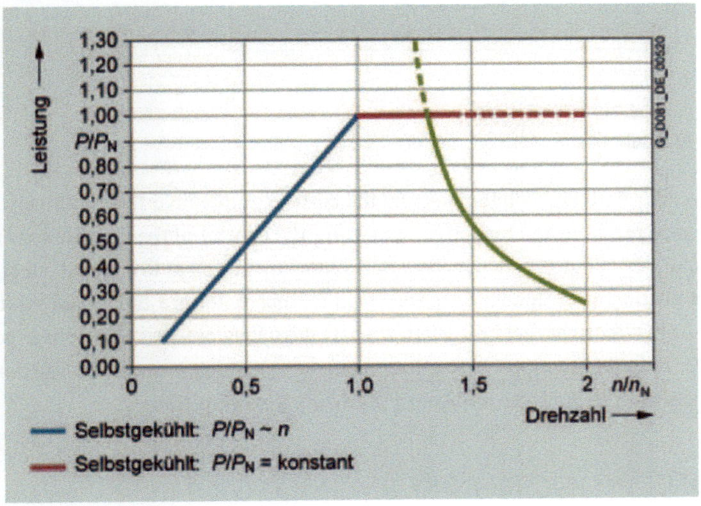

Fig. 21-24: *P- n* characteristic VSD4000 Siemens

The characteristics in fig. 21-23 and 21-24 show curves similar to these of DC- and induction motors.

4. SYNCHRONOUS AC-MOTOR

Fig. 21-25 shows a cross section of a two pole synchronous machine. The stator windings are sinusoidally distributed and three-phase just as with the induction motor. For simplicity only the central conductor of each winding has been shown. A DC field winding is responsible for the rotor field.

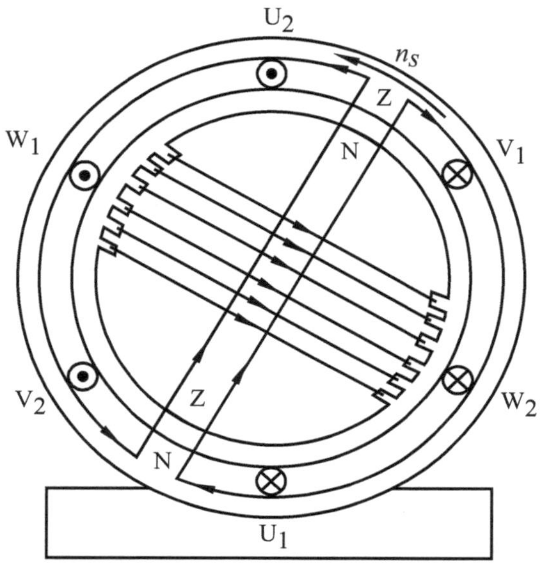

Fig. 21-25: Cross section of a two pole synchronous AC-motor

The PWM inverter with DC voltage link and the current source inverter are already studied with the asynchronous machine and may also be used for the control of the synchronous machine.For speed control of very large power synchronous motors operating at low speed (steel mills, cement ovens) a combination of cycloconverter and synchronous motor is a functional solution. Due to the presence of two anti-parallel bridges (fig. 10-1) the cycloconverter can deliver active and reactive power in both directions so that four quadrant operation is possible. The disadvantage of this type of converter is the large number of thyristors it requires (36).

5. ASYNCHRONOUS INDUCTION SERVOMOTOR

About twenty years ago the first three-phase induction servomotors with low power levels appeared on the market. Now large power types are available (for example the 1PH8 from Siemens: 2.8 to 1340 kW). In applications where the most important property of the main drive is a continuous soft operation (paper machines, textile machines, rolling mills, winders, lifting machines,...) these motors can offer a cost effective alternative for synchronous motors.
The photo on p.16.65 includes an asynchronous servomotor.

22 Electrical Positioning Systems

CONTENTS

The basis of modern positioning systems was laid with the firing control systems which were developed during the second world war, the successor of which are the modern rocket systems. With the current technology electrical positioning systems are much used, for example in:

Instrumentation:	plotters, recorders, printers, scanners, disc drives,...
Industrial applications:	robots, NC-machines, conveyor belts, flow control, actuators,...

Power Electronics: DRIVE TECHNOLOGY and MOTION CONTROL
http://dx.doi.org/10.1016/B978-0-12-814641-5.50007-X,

1. SERVOMECHANISMS

1.1 Definition

A servomechanism complies with the following definition:

- is a feedback system
- has large energy gain
- the controlled output is a mechanical position or a derived time function of this position (speed, acceleration).

If the output is a position then we refer to a positioning system.

1.2 Performance of a positioning system

We distinguish two groups of applications:

a. Point to point positioning
The requirements which are applicable are the speed to go from point A to point B together with the accuracy of the positioning and the dynamic behaviour.

b. Trajectory control
In addition to the requirements above accurately and fluently following a predetermined trajectory is important. Fluent means a speed change of maximum 0.1% during one revolution of the motor shaft.

1.3 Requirements of a servo-system

A servo-system should meet the following requirements

- large torque at stand still
- high impulse torque so that a large acceleration and a fast response is possible
- large speed control range
- good controllability at (very) low speed
- low torque ripple
- high accuracy.

1.4 Servomotor

Important properties of a servomotor are:

- high torque
- low inertia
- linear *M-n* curves.

Electric servomotors have a high nominal speed and a low nominal torque so that a (gearwheel) reduction is required to match the motor to the load. A reduction N will reduce the output to N times slower rotation and simultaneously the available torque will be N times larger.
If a linear movement is required then a transfer mechanism (e.g. a worm-wheel) is required between rotating motor shaft and linear movement of the load.
These days in a number of situations a linear motor is used.

Remarks

The name “servomechanism” is often reserved in the literature for a positioning system.

For high-dynamic positioning and torque control electrical servo-systems used only DC-servomotors or permanent magnet synchronous motors.

Due to the technological developments in the previous years the flux vector control of an asynchronous motor has dramatically improved. The result is that this vector control is now capable in many applications of competing with traditional servo-systems. The border between drive system and servo-system is becoming more vague. Certain company’s such as Yaskawa use a “radar” card on which the differences in performance between servo-systems and flux vector control (VC) are indicated. This is shown in fig. 22-1.

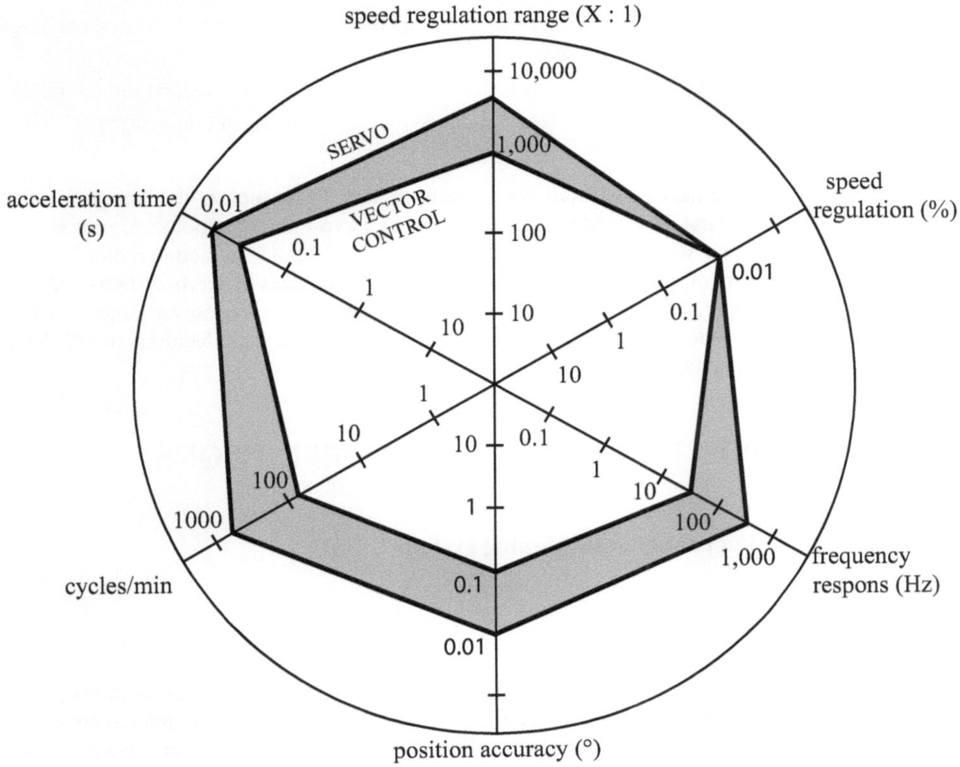

Fig. 22-1: Points of difference servo-system with VC-system
Source : Yaskawa and Control Engineering

Manufacturers of sophisticated VC- systems currently build their “drives” in a “motion controller”. The VC-drive functions as an amplifier, controlled by an external motion controller. The VC-drive is then exclusively configured as a torque controller. When the cycle time is 2 seconds or less then a VC is a viable option as a positioning system. The “drive” can operate with vector- functionality for the highest torque accuracy or with servo-functionality for the best dynamic behaviour. The major difference remains the motor used.

SERVOMOTOR	“CLASSIC” INDUCTION MOTOR
- small moment of inertia: gives a powerful, fast response - smaller than an induction motor for the same speed - normally not available in powers above a few kW. Becomes extremely expensive for larger powers - normally available up to IP65 - only short time rated overload - motor current used to form torque (flux produced with permanent magnets!)	- cheap motor, available up to IP 54 - quickly available (off the shelf) up to large powers - for closed loop control : sensor required (include in cost calculation!) - motor current can be split into a torque forming and flux forming component: can cause delays - overshoot and oscillations larger than with servomotors - can tolerate long duration overload above nominal current

Up until now we have considered synchronous motors, linear motors and “torque”- motors as servomotors. These days gradually more and more servo-induction motors are entering the positioning world with better servo properties than the “classic” induction motor.
For the majority of industrial applications the extra advantages of servo-systems with respect to VC-systems are not necessary. Servo-systems dominate in the low power range. For larger powers (250 kW to 4500 kW) the VC-systems dominate. The manufacturers of VC-drives are steadily approaching the lower powers (to 0.37 kW).

2. ELECTRICAL POSITIONING SYSTEMS. DEFINITIONS

2.1 Block diagram of a positioning system

Standard motors are:

- stepper motors
- servomotors (DC-motor, AC-motor)

Servomotors and stepper motors occur in instrumentation and control applications.
The stepper motor can also be used in open loop circuits and has the additional benefit that they are easily controlled from a micro-controller.An electric positioning system normally is implemented as one of the configurations shown in fig. 22-2. In fig. 22-2a a stepper motor is shown, while in fig. 22-2b a system with servomotor is represented.

Advantages of the stepper-solution:

- few components
- no motor maintenance
- low cost

Advantages of the solution with servomotor:

- high speed
- large torque
- closed loop control

An AC-motor used as servomotor requires no maintenance, in contrast to a classic DC-servomotor. As far as fig. 22-2b is concerned the following remark can be made. Control theory has shown that for a stable positioning system, position and speed feedback are required. Previously an angular position sensor (position) and a tachogenerator (speed) were placed on the output shaft of the motor. These days the angular position signal is differentiated to produce a speed signal. In this way a tachogenerator is not necessary.

As mentioned previously a positioning system is normally constructed with an internal speed controller surrounded by an external feedback for the positioning. In addition during the study of single quadrant operation of a DC-motor it was shown that a speed controller is comprised of an internal current feedback surrounded by an external speed control loop. To keep fig. 22-2 b simple, the internal current loop is not shown. The complete internal speed controller is also not shown. In fig. 22.7 and 22.12 the feedback loops are shown. Even in digital positioning systems the motor remains an analogue type so that a D/A converter is required. In the case where a resolver is used as angular position sensor as in fig. 22-2b the analogue resolver signal needs to be converted to binary code. This takes place in an RDC (resolver to digital converter). An accurate positioning system requires for example 11 to 12 bits from this RDC. This corresponds to an angular position measurement of a few arc minutes.

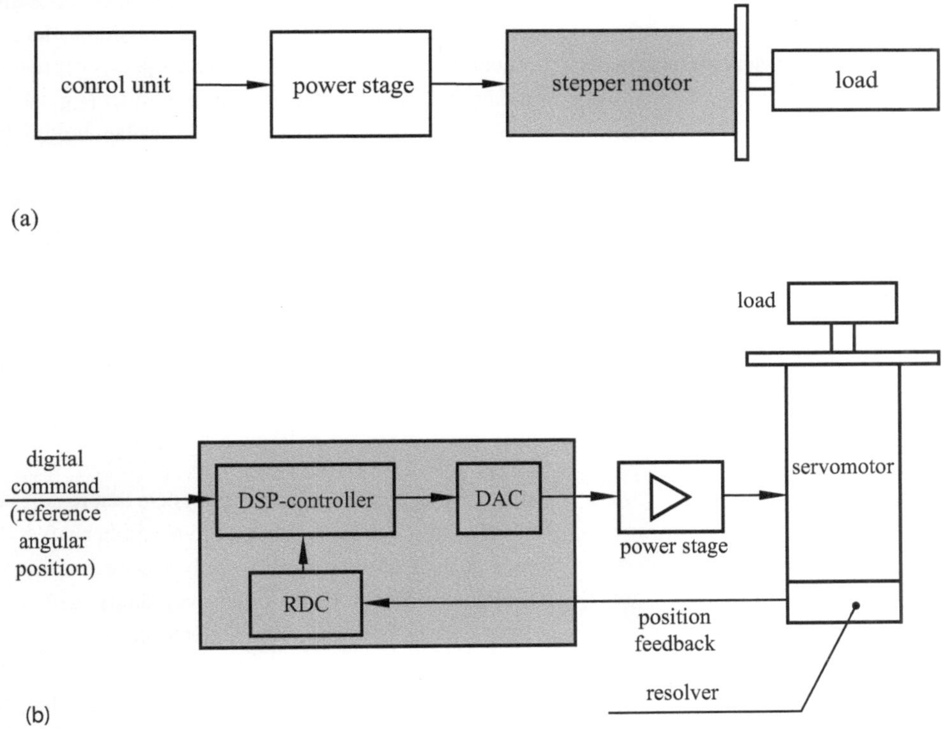

Fig. 22-2: Block diagram of positioning system: (a) with stepper motor; (b) digital controlled servomotor

2.2 Fundamental equations

With an angular displacement θ we find:

$\omega = \frac{d\theta}{dt}$ = angular velocity (rad/s) ; $a = \frac{d\omega}{dt}$ = angular acceleration (rad/s^2)

$\gamma = \frac{da}{dt}$ = jerk (rad/s^3)

Written in another form: $\theta = \int \omega \cdot dt$; $\omega = \int a \cdot dt$; $a = \int \gamma \cdot dt$

The jerk is a measure of the fluency or "soft" rotation of the shaft. Fluent movement is often required, think about an elevator.

2.3 Definitions

- **Accuracy:** Is expressed as the maximum error in achieving the desired position. Is provided in linear units (e.g. mm) or in angular units (e.g. arc min utes).
- **Resolution:** Is the smallest angular movement of the motor that can be programmed. The resolution of the entire system is dependent on additional factors such as the mechanical play in the transmission system.
- **Servo-resolution:** Is determined by the number of encoder pulses per revolution together with the multiplication factor of the controller (e.g. x1, x2, x4).
  An encoder with 500 "stripes"and a controller multiplication factor of 4 has a resolution of 2000 steps /rev or 0.18°.
- **Stepper-resolution:** With a stepper motor the resolution is normally given by the step angle (e.g. 0.9°) and sometimes by the number of steps/rev.

2.4 Terminology

An important part of servo-terminology is already familiar: encoder, resolver, feedback, servo-mechanisms, pulse width modulation, overshoot, commutation,...
A few new terms follow:

Shaft: Every moving part of a machine or system that requires control. Multiple shafts can be coordinated into a multi shaft system.

Multiple shaft system: In a large number of applications only a loose coupling is required between the different shafts. In this way single shaft positioning systems can be used with simple references between the different drive systems.
In machine tools for example it may be necessary that multiple shafts are synchronised to work together. In addition the "state" of the drives must be controlled at every instant.

Electronic gearbox: A method whereby a mechanism is simulated. A closed loop control system functions as electrical "slave". The slave follows another shaft (master) with a variable ratio.
Different configurations are possible:
1. fixed latching between the shafts
2. latching only when the slave has achieved the speed of the master
3. synchronisation between slave and master switched off during a specific time...

Indexer: Electronic unit that converts signals coming from a host computer, a PLC or an operator panel, to step and rotation pulses for the control of a stepper motor.

Tuning: Optimising configuration of a PID-filter in the servo-controller.

Servo-controller: Calculates the difference between the set-point position and the measured position, determines the position error (POS/ERROR) and produces a signal to correct this POS/ERROR with a servomotor.

3. POSITION CONTROL WITH A DC-SERVOMOTOR

3.1 DC-servomotor

In a DC-servomotor a magnetic field is produced using permanent magnets on the stator. The motor should have good commutation at a large speed range. To produce a rotor with the smallest possible inertia primarily two types are used: the staff armature and the disc armature. With the staff armature a long thin rotor is used while with the disc armature an implementation with pressured current circuits is used. By way of an example in fig. 22.3 a tachogenerator and angular encoder is included as was common in the past. These days only an encoder is used.

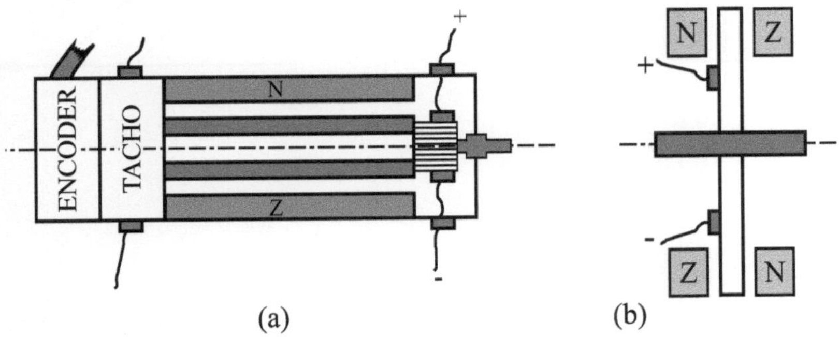

Fig. 22-3: DC-servo motors with a) staff armature and b) disc armature

3.1.1 *M-ω* curves of a DC-servomotor

The *M-n* curves of a PM-motor were shown in fig. 19-47. It is clear that, with the exception of a scale factor, the *M-ω* curves have an identical form. To arrive at more general expressions the relationship between M and ω is determined.

(19-2) $\rightarrow$ $M = K_M \cdot I_a$

(19-1) $\rightarrow$ $E = K_G \cdot \omega \rightarrow I_a = \frac{M}{K_M} = \frac{V_a - K_G \cdot \omega}{R_i} \rightarrow V_a = \frac{R_i}{K_M} \cdot M + K_G \cdot \omega$

(19-4) $\rightarrow$ $I_a = \frac{V_a - E}{R_i}$

At no load $M = 0$ and the no load speed $\omega_N = \frac{V_a}{K_G}$. The stall torque is $M_S = K_M \cdot \frac{V_A}{R_i}$.
The path of M can be written as:

$$M = (V_a - K_G \cdot \omega) \cdot \frac{K_M}{R_i} \tag{22-1}$$

or: $M = M_S - K \cdot \omega$. This is shown in fig. 22-4a.
For different values of the armature voltage V_a, the *M-ω* curve will shift parallel, as shown in fig. 22-4b.

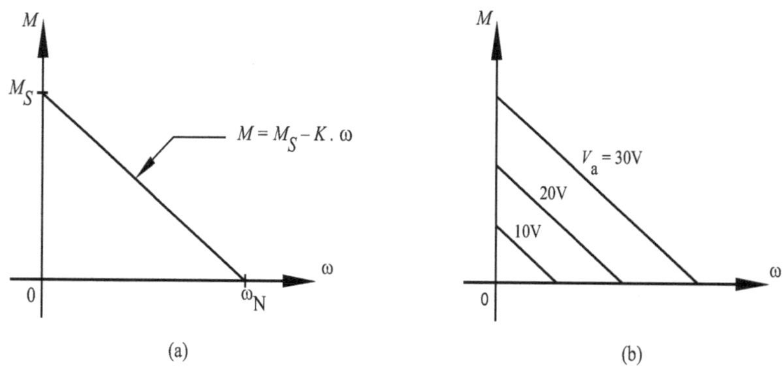

Fig. 22-4: *M-ω* curves of a DC-servomotor

3.1.2 Transfer function of the motor

Assume a servomotor with permanent magnets, so that the flux Φ is constant.
If we call M_{tm} the sum of the counter torques in the motor, this is described by :

$$M_{tm} = M_{fric.} + D \cdot \omega$$

With: $M_{fric.}$ = the constant friction torque in the motor

$D \cdot \omega$ = the sum of the viscous friction torques and all other torques which are proportional to speed.

With the load torque M_b the total counter torque on the shaft of the motor is:

$$M_t = M_{tm} + M_b = M_{fric.} + M_b + D \cdot \omega \tag{22-2}$$

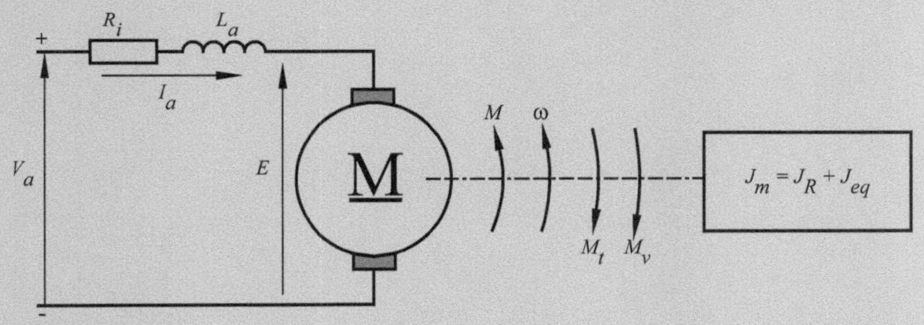

Fig. 22-5: Equivalent of a loaded motor

From (19-3), (19-7), (19-6), (19-2), (19-1) and (22-2) the required relationships to determine the transfer function $\frac{\omega}{V_a}$ follow.

$$V_a = (R_i + s \cdot L_a) \cdot I_a + K_G \cdot \omega \tag{22-3}$$

$$M = K_M \cdot I_a = J_m \cdot s \cdot \omega + M_{fric.} + M_b + D \cdot \omega \tag{22-4}$$

Here:

J_m = total moment of inertia = $J_R + J_{eq}$
J_R = moment of inertia rotor + angular encoder + eventual tachogenerator
J_{eq} = reflected moment of inertia of the load to the motor shaft.

If we just consider the motor as one building block in the entire system then for the transfer function $\frac{\omega}{V_a}$ the torques M_{fric} and M_b play no role so that:

$$M = K_M \cdot I_a = \omega \cdot (D + s \cdot J_m) \; ; \quad I_a = \omega \cdot \frac{(D + s \cdot J_m)}{K_M} \tag{22-5}$$

Replacing (22-5) in (22-3):

$$V_a = \omega \cdot \left[\frac{(D + s \cdot J_m) \cdot (R_i + s \cdot L_a)}{K_M} + K_G \right]$$

From this the transfer function follows:

$$H_m = \frac{\omega}{V_a} = \frac{K_M}{(D + s \cdot J_m) \cdot (R_i + s \cdot L_a) + K_G \cdot K_M} \tag{22-6}$$

For all practical motors: $D \cdot L_a << J_m \cdot R_i$ and $D \cdot R_i << K_G \cdot K_M$
so that a good approximation is:

$$H_m = \frac{\omega}{V_a} = \frac{K_M}{s^2 \cdot L_a \cdot J_m + s \cdot J_m \cdot R_i + K_G \cdot K_M} = \frac{1}{K_G} \cdot \frac{\frac{K_G \cdot K_M}{L_a \cdot J_m}}{s^2 + s \cdot \frac{R_i}{L_a} + \frac{K_G \cdot K_M}{L_a \cdot J_m}}$$

Set :

- $\tau_a = \frac{L_a}{R_i}$ = electrical time constant of the motor
- $\tau_m = \frac{J_m \cdot R_i}{K_G \cdot K_M}$ = mechanical time constant of the motor

then this becomes:

$$H_m = \frac{1}{K_G} \cdot \frac{\frac{1}{\tau_m \cdot \tau_a}}{s^2 + s \cdot 1/\tau_a + 1/\tau_m \cdot \tau_a} \tag{22-7}$$

This is the transfer function of a second order system, written in general terms as:

$$\frac{1}{K_G} \cdot \frac{\omega_0^2}{s^2 + 2 \cdot \beta \cdot \omega_0 \cdot s + \omega_0^2}$$

With:

- natural resonance pulsation $\omega_0 = \frac{1}{\sqrt{\tau_m \cdot \tau_a}}$
- damping factor $\beta = \frac{1}{2} \cdot \sqrt{\frac{\tau_m}{\tau_a}}$
- poles $s_{1,2} = -\omega_0 \cdot \beta \pm \omega_0 \cdot \sqrt{\beta^2 - 1}$
- pole zero gain $K_{pz} = \frac{\omega_0^2}{K_G} = \frac{1}{K_G \cdot \tau_m \cdot \tau_a}$

We now calculate the system poles of the transfer function (22-7):

$$s_{1,2} = -\frac{1}{\sqrt{\tau_m \cdot \tau_a}} \cdot \frac{1}{2} \cdot \sqrt{\frac{\tau_m}{\tau_a}} \pm \frac{1}{\sqrt{\tau_m \cdot \tau_a}} \cdot \sqrt{\frac{1}{4} \cdot \frac{\tau_m}{\tau_a} - 1}$$

$$= -\frac{1}{2 \cdot \tau_a} \pm \sqrt{\frac{1}{4 \cdot \tau_a^2} - \frac{1}{\tau_m \cdot \tau_a}}$$

$$s_{1,2} = -\frac{1}{2 \cdot \tau_a} \pm \frac{1}{2 \cdot \tau_a} \sqrt{1 - \frac{4 \cdot \tau_a}{\tau_m}} \tag{22-8}$$

The transfer function may be written as:

$$H_m = \frac{1}{K_G \cdot \tau_m \cdot \tau_a} \cdot \frac{1}{(s-s_1)\cdot(s-s_2)} \qquad (22\text{-}9)$$

With the exception of small servomotors often: $\tau_m > 4.\tau_a$.
We find then that $\beta > 1$ so that the poles are negative and real.
Fig. 22-6a shows the pz-plot of the transfer function.

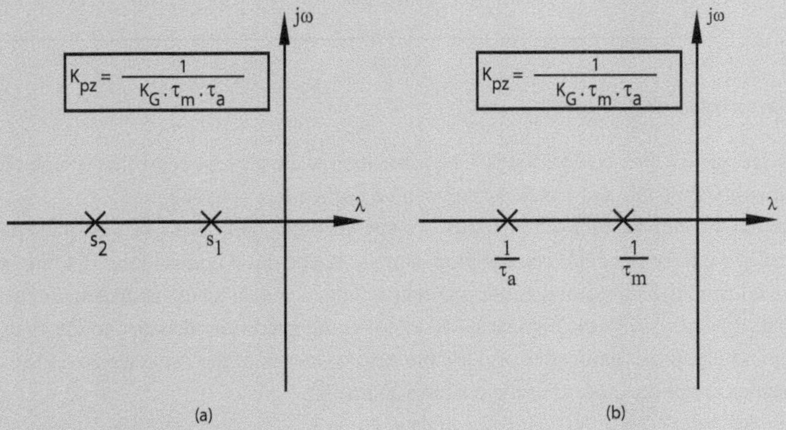

Fig. 22-6: pz-plot of a loaded DC-servomotor

Remark

In exceptional cases $4 \cdot \tau_a > \tau_m$ and the poles are then negative and complex.
If on the other hand $\tau_m >> 4 \cdot \tau_a$ the approximation $\sqrt{1-x} \approx 1 - \frac{x}{2}$ (for $x << 1$!) is applied so that: $\sqrt{1 - \frac{4 \cdot \tau_a}{\tau_m}} \approx 1 - \frac{2 \cdot \tau_a}{\tau_m}$

The poles in (22-8) in this case can be described as

- $s_1 \approx -\frac{1}{2 \cdot \tau_a} + \frac{1}{2 \cdot \tau_a}\left(1 - \frac{2 \cdot \tau_a}{\tau_m}\right) = -\frac{1}{\tau_m}$
- $s_2 \approx -\frac{1}{2 \cdot \tau_a} - \frac{1}{2 \cdot \tau_a}\left(1 - \frac{2 \cdot \tau_a}{\tau_m}\right) = -\frac{1}{\tau_a}$

The transfer function (22-9) can be written as:

$$H_m = \frac{\omega}{V_a} = K_{pz} \cdot \frac{1}{(s-s_1)\cdot(s-s_2)} = \frac{1}{K_G \cdot \tau_m \cdot \tau_a} \cdot \frac{1}{(s+\frac{1}{\tau_a})\cdot(s+\frac{1}{\tau_m})} \qquad (22\text{-}10)$$

The pole-zero plot is then shown in fig. 22-6b.

3.2 Block diagram servo drive

Fig. 22-7 shows the control block diagram of an electrical servo-system.

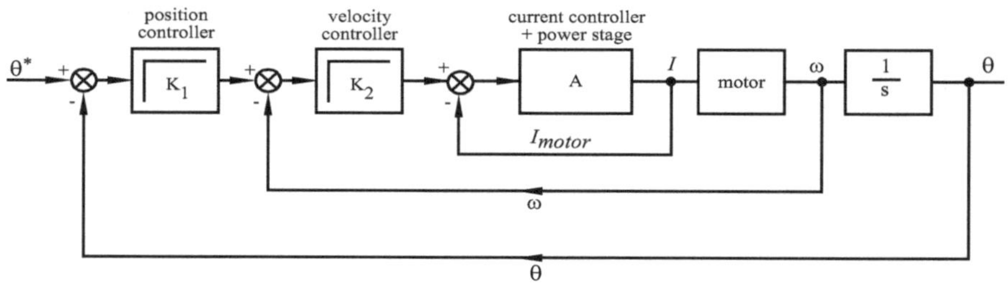

Fig. 22-7: Block diagram of an electrical servo-system

We recognise three control loops. The inner loop is the current controller which controls the torque and simultaneously the motor current can be limited to a safe value.

Around the current control loop there is a speed control loop just like a normal speed controller. Finally we have the outer control loop, namely the position control loop. As long as we are far removed from the set-point position the speed controller will have maximum output, in other words we try (usually) with the highest speed to cover the maximum distance to the position set-point. Depending on the application and the moment of inertia of the drive we cover the remaining distance under the control of the position controller.

3.3 Speed profile of the positioning

The speed profile of the positioning system is the plot of the motor angular velocity as function of time. The speed profile can be optimised as a function of various parameters (minimum peak current, minimum top speed, etc...). In general the heat dissipation in the armature is the limiting factor. Calculations show a parabolic speed profile (fig. 22-8) is the most advantageous from the viewpoint of heat dissipation in the armature. In fig. 22-8 t_p is the time to arrive at the position θ.

A trapezium shaped speed profile as shown in fig. 22-8b results in 12.5% more energy dissipation in the armature than the parabolic path and 21% less dissipation than the triangular profile (fig. 22-8c). From calculations it follows that the most advantageous trapezium shaped speed profile looks like that shown in fig. 22-8b, in other words using one third of the time for acceleration normal speed and deceleration.

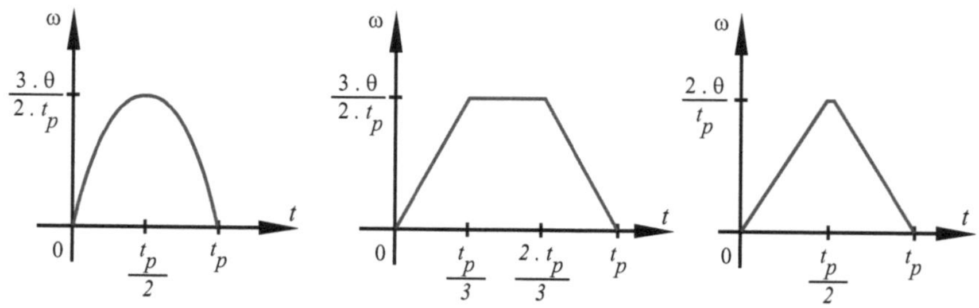

Fig. 22-8: Different speed profiles of a positioning system

Fig. 22-9 shows a trapezium shaped speed profile with the covered distance in the case of a linear positioning system. The angular deviation of the motor shaft is also shown.

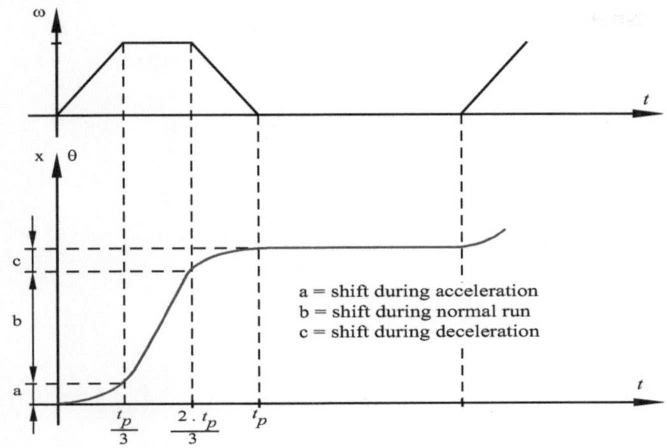

Fig.22-9: Covered distance in the case of a trapezium shaped speed profile

In fig. 22-10 the acceleration and jerk are also shown and this for two different speed profiles. On the left of the figure a constant acceleration/deceleration profile and to the right a sinusoidal acceleration change is shown. In this last situation the jerk (derivative) and the angular velocity (integral) are also sinusoidal functions.

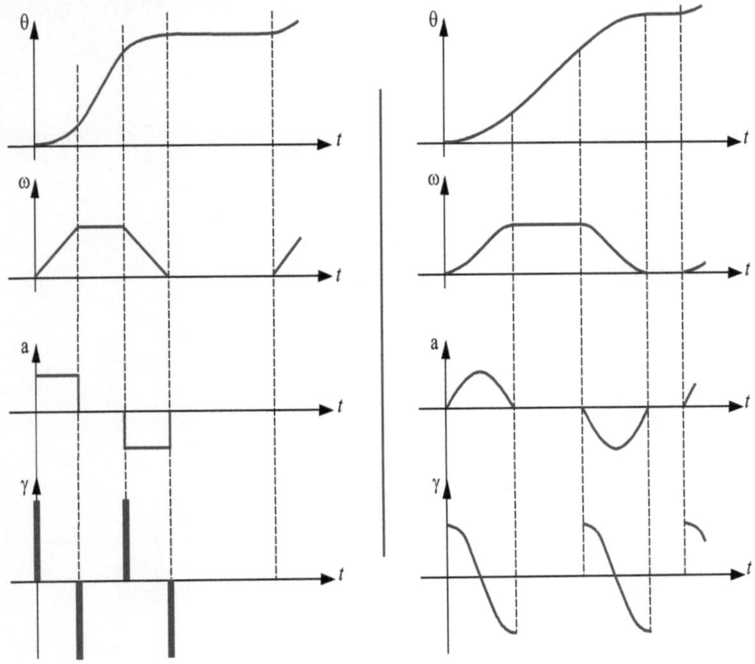

Fig. 22-10: Covered distance acceleration and jerk of two different speed profiles

3.4 Calculation with a trapezium shaped speed profile

3.4.1 Motor acceleration

An important factor for a servomotor is its acceleration. To determine this we write:

(19-2) → → → $M = K_M \cdot I_a$

[(19-6) and (19-7)] → → → $M = M_t + J_m \cdot \ddot{\theta}$, with $\ddot{\theta} = \frac{d\omega}{dt}$

M_t = sum of all friction torques and torques caused by forces
J_m = sum of all moments of inertia

Normally M is negligibly small with respect to $J_m \cdot \ddot{\theta}$,

so that: $M \approx J_m \cdot \ddot{\theta} = K_M \cdot I_a$

The acceleration a $(= \ddot{\theta} = \frac{d\omega}{dt})$ can be written as:

$$a = \frac{K_M}{J_R + J_{eq}} \cdot I_a \qquad (22\text{-}11)$$

a = rotor acceleration (rad/s²)
K_M = torque constant of motor (Nm/A)
I_a = armature current (A)
J_R = rotor moment of inertia (kgm²) , including angular encoder and tachogenerator
J_{eq} = moment of inertia of load (kgm²), reflected to the shaft of the motor.

From the acceleration a we can determine the time $\Delta t = t_2 - t_1$ (fig. 22-11) to go from stand still to a motor speed ω_{SC}: $\frac{\omega_{SC}\,(rad/s)}{a\,(rad/s^2)} = t_2 - t_1$

The acceleration is determined by K_M, J_R, J_{eq} and the setting of I_a^* of the internal control loop (fig. 22-12).

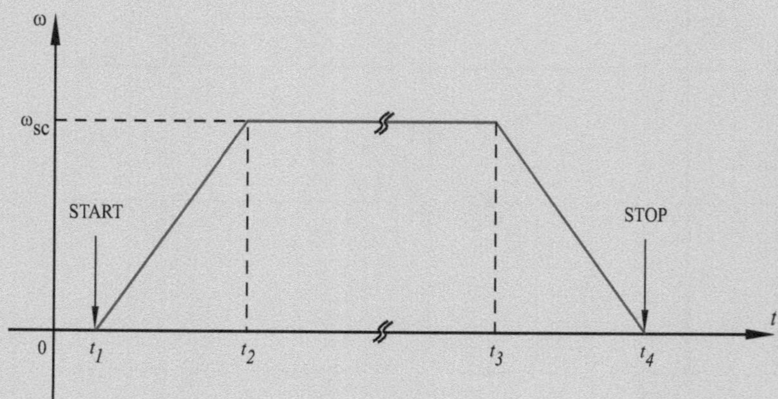

Fig. 22-11: Trapezium shaped speed profile

Photo Maxon motor Benelux: Maxon compact drive intelligence and high performance in a small space

This intelligent positioning drive consist of an EPOS controller, encoder and brushless DC-motor in a modern space saving aluminum housing. A compact drive with high performance.
The new MCD EPOS P is a very dynamic, maintenance free positioning drive with CANopen bus connections and 60 W maximum output power. Programmable with a PC via an RS 232 port using EPOS P programming software conforming to IEC 61131-3. For extra torque the MCD can be extended with planetary gears GP 32 C (1.0-6.0 Nm) and GP 42 C (3.0-15.0 Nm)

Fig. 22-12 shows the principle configuration of a DC-servo-system.

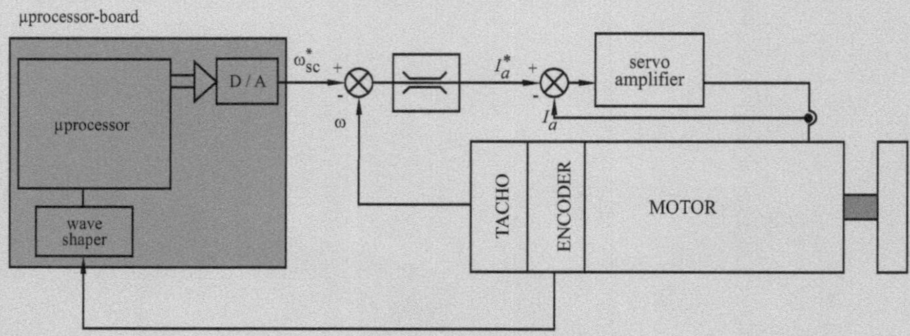

Fig. 22-12: DC-servo-system with the three control loops

3.4.2 Controlled speed

Once the value ω_{SC} is reached the speed controller comes into action (SC = Speed Control). Depending on the setting for ω^*_{SC}, the ω of the motor will be controlled so that $\omega = \omega^*_{SC}$.

3.4.3 Delay

In order to stop at the desired position a time t_3 has to be determined at which the motor should start to decelerate. In reality we determine the number (k) of pulses that the encoder still needs to give until the motor shaft is at the set-point location.

Expression (22-11) allows for calculation of the negative value I_a required to decelerate with a value $\gamma = -\frac{d\omega}{dt}$

Fig. 22-13a shows that $\omega_{SC} = (t_4 - t_3) \cdot \gamma$. From fig. 22-13b it follows that the shaft of the motor is rotated by an angle $\theta = (t_4 - t_3) \cdot \frac{\omega_{SC}}{2}$ between t_3 and t_4.

Elimination of $(t_4 - t_3)$ in both of the previous formulas gives

$$\theta = (t_4 - t_3) \cdot \frac{\omega_{SC}}{2} = \frac{\omega_{SC}}{\gamma} \cdot \frac{\omega_{SC}}{2} = \frac{\omega_{SC}^2}{2 \cdot \gamma}$$

Assume that the speed is controlled at 3000 rpm, then: $\omega_{SC} = \frac{2 \cdot \pi \cdot n}{60} = 314.1593$ rad/s

If for example (according to the micro-controller countdown) we are still $k = 3800$ encoder pulses removed from the set-point position and the encoder generates 2000 pulses/rev, then: $\theta = 11.938$ rad.

The required delay is then: $\gamma = \frac{\omega_{SC}^2}{2 \cdot \theta} = 4133.6925 \text{ rad/s}^2$

With $J_R + J_{eq} = 3.9 \cdot 10^{-5}$ kgm² and $K_M = 5.4 \cdot 10^{-2}$ Nm/A becomes:

$$I_a = \frac{\gamma \cdot (J_R + J_{eq})}{K_M} = -2.985\text{A}$$

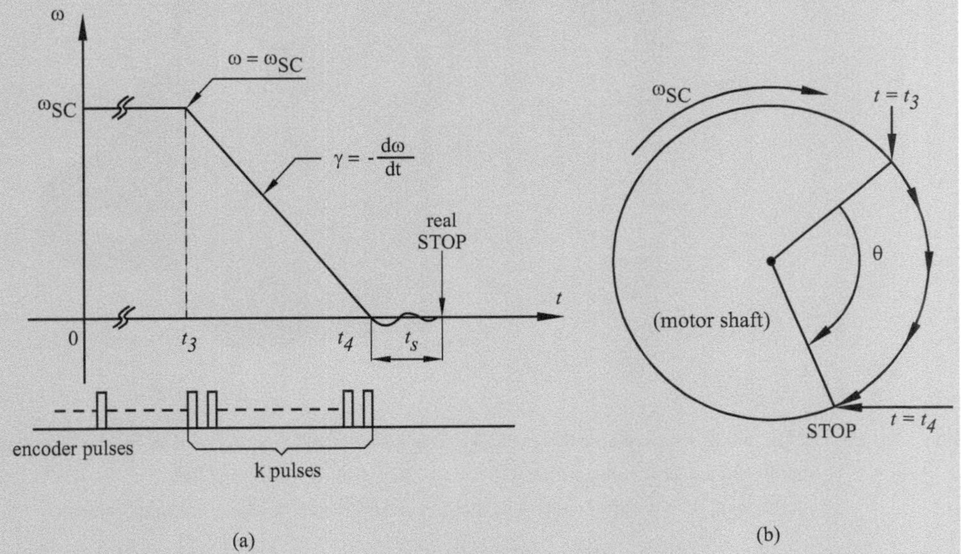

Fig. 22-13: Deceleration of the servomotor to a stop

3.4.4 Stopping

It is clear that the calculation above is dependent on the linear operation during deceleration. This is difficult to know for amongst other reasons because of the digital control of the speed which is always in"steps". Normally a positioning system is controlled in such a way that the rotor with only a few oscillations can be brought to a halt. The time t_S of these oscillations is known as the settling time. Mostly the speed controller is switched off in the region close to the angular position set-point and only the position controller is operating so that when the stop position is achieved a good holding torque is maintained.

Photo Rotero: BLDC-motor with separate drive. Flange sizes 42-60-80-90 mm. Power 5 W to 200 W. Voltage 24 VDC. Supply single phase voltage from 110 V and 230V or three-phase 3x230 V.
Torque up to 68 Nm (after reduction). Protection IP40-IP65. Reducer housing optional

Courtesy of ESO: Very Large Telescope (VLT), Paranal, Chile. In this telescope a large number of Heidenhain angle encoders are used

4 POSITION CONTROL WITH BRUSHLESS DC-MOTOR

4.1 Brushless DC-motor - Determining

In addition to good dynamic control behaviour classic DC-motors have a number of disadvantages such as:

- brushes and collectors require maintenance
- spark production of the brushes making them unsuitable for operation in explosive atmospheres
- open construction prevents operation in damp environments.

By using a rotor with permanent magnets and winding the stator a brushless DC-motor results. Brushless DC-motor is abbreviated to BLDC. Originally motors were only built rated for a few watt, but the use of powerful magnets allows for the construction of more powerful motors. Due to the placing of the windings on the stator this motor has better heat dissipation than the "classic" DC-motor with armature windings. As a result a brushless motor has a larger torque and delivers more power than an equivalent DC-motor of the same size. A BLDC can meet the following specifications: "a motor without brushes with the same *M-n* curves as a classic PM-motor".

4.2 Switched three-phase PM-motor

4.2.1 Construction and operation

The most common type of BLDC has a three-phase winding on the stator. In fig. 22-14 a cross section is shown of a two pole three-phase brushless DC-motor. The difference with the three-phase synchronous motor is that per stator winding there is not a sinusoidal flux distribution but a practically uniform distribution of the conductors. The rotor contains two permanent magnets which each comprise almost 180° of the circumference.

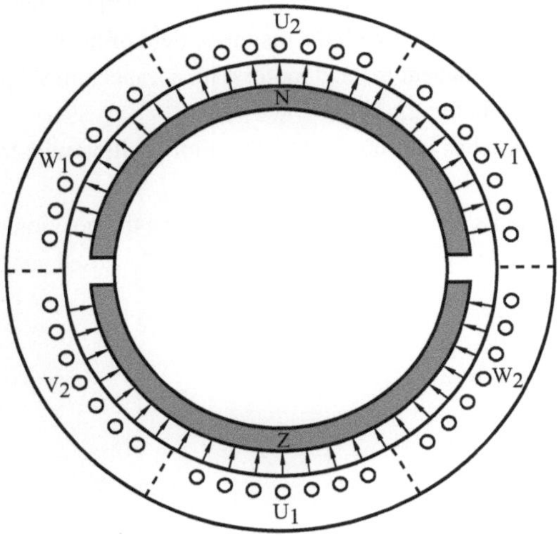

Fig. 22-14: Cross section of switched three-phase PM-motor

An inverter as shown in fig. 22-15 is suitable for supplying the stator coils. It is a three-phase inverter of the 120° type as was studied on p.14.25.
Some manufacturers switch their BLDC-motor is delta. We progress with a connection in star, which is the most common connection.

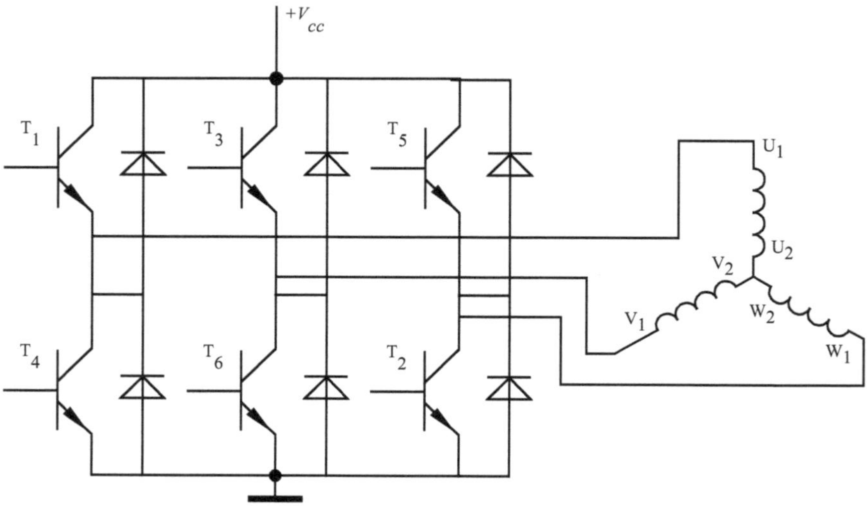

Fig. 22-15: Power bridge of a three-phase brushless motor

The switches operate in pairs, with consecutive numbers. There are always two stator coils switched in series to the DC-voltage, hence the name "switched" PM-motor. In fig. 22-16 the six switches are drawn.
For simplicity the rotor is represented by a bar magnet and only one winding per stator coil is shown. Note that the rotor runs synchronised with the magnetic field of the stator. A BLDC-motor is often defined as a synchronous permanent magnet motor with a trapezium shaped counter emf. This trapezium waveform is shown in fig. 22-18.
The opening and closing of the switches is controlled by the rotor position with the aid of for example hall-sensors.
In this way when the rotor reaches its end position in fig. 22-16a via the position encoder, switch T_6 is opened and T_2 closed. In fig. 22-16b T_1 is opened and T_3 closed, etc...

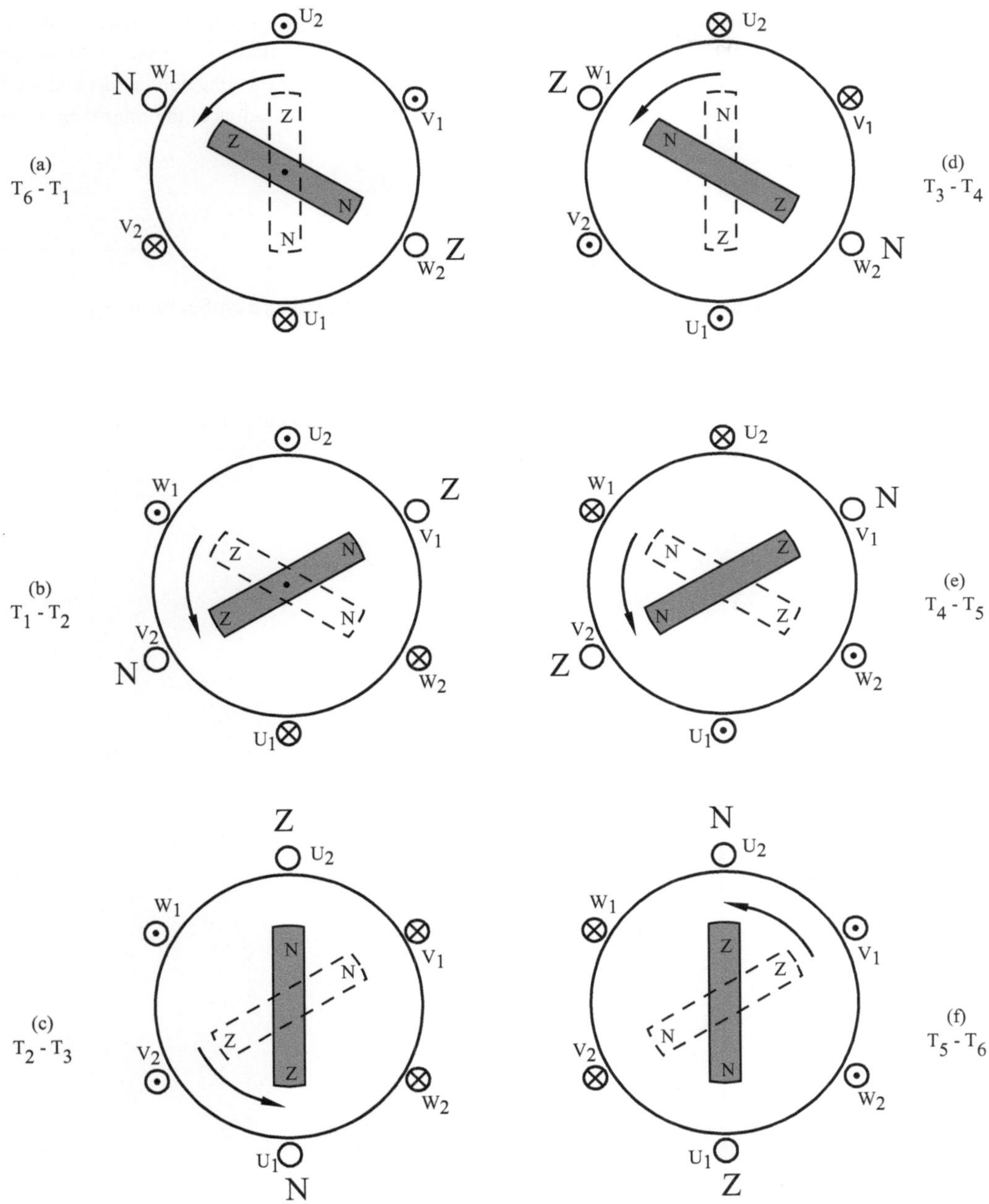

Fig. 22-16: Stator field and rotor position for the six switching sequences of the inverter of fig. 22-15

4.2.2 *M-n* curves

If a current I flows through two series connected stator coils and the magnetic induction of the rotor field is B, then per conductor there is a force $F_1 = B \cdot l \cdot I$. Here l is the axial rotor length. With N_S windings per stator coil there are therefore $4 \cdot N_S$ current carrying conductors and the force on the rotor circumference is then $F = 4 \cdot N_S \cdot B \cdot l \cdot I$. If r is the radius of the rotor then the motor torque is:

$$M_{em} = F \cdot r = 4 \cdot N_S \cdot B \cdot l \cdot r \cdot I = k_1 \cdot I$$

$$M_{em} = k_1 \cdot I \qquad (22\text{-}12)$$

Set the mechanical angular velocity of the rotor as ω_m. The induced emf in the two series connected windings (with $4 \cdot N_S$ conductors) is:

$$E = 4 \cdot N_S \cdot B \cdot l \cdot v = 4 \cdot N_S \cdot B \cdot l \cdot r \cdot \omega_m = k_1 \cdot \omega_m$$

If R is the resistance of the stator coils and V_t is the DC-link voltage, then:

$$E = V_t - R \cdot I = k_1 \cdot \omega_m$$

or:

$$\omega_m = \frac{V_t}{k_1} - \frac{R \cdot I}{k_1} = \frac{V_t}{k_1} - \frac{R \cdot M_{em}}{k_1^2}$$

With $\omega_m = \frac{2 \cdot \pi \cdot n}{60}$ becomes: $n = \frac{V_t}{k_2} - \frac{M_{em}}{k_4}$

or:

$$M_{em} = k_3 \cdot V_t - k_4 \cdot n \qquad (22\text{-}13)$$

The *M-n* curve (fig. 22-17) of the brushless DC-motor is therefore identical to the PM-motor as drawn in fig. 22-4.

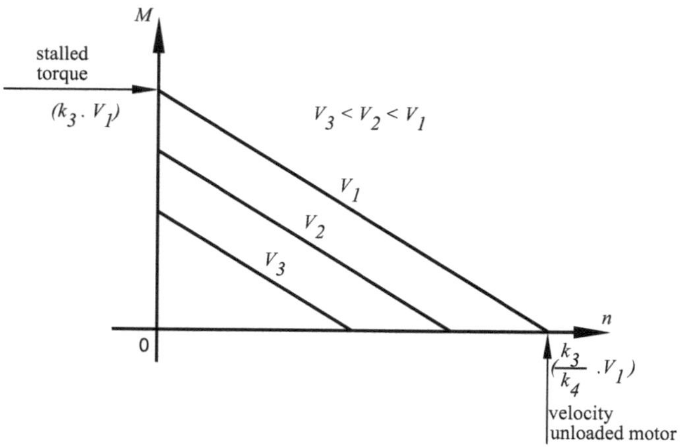

Fig. 22-17: *M-n* curves of a brushless DC-motor

4.2.3 A number of waveforms of a BLDC-motor

Fig. 22-18 shows the waveforms for a configuration such as fig. 22-15, where continually two coils are connected in series.

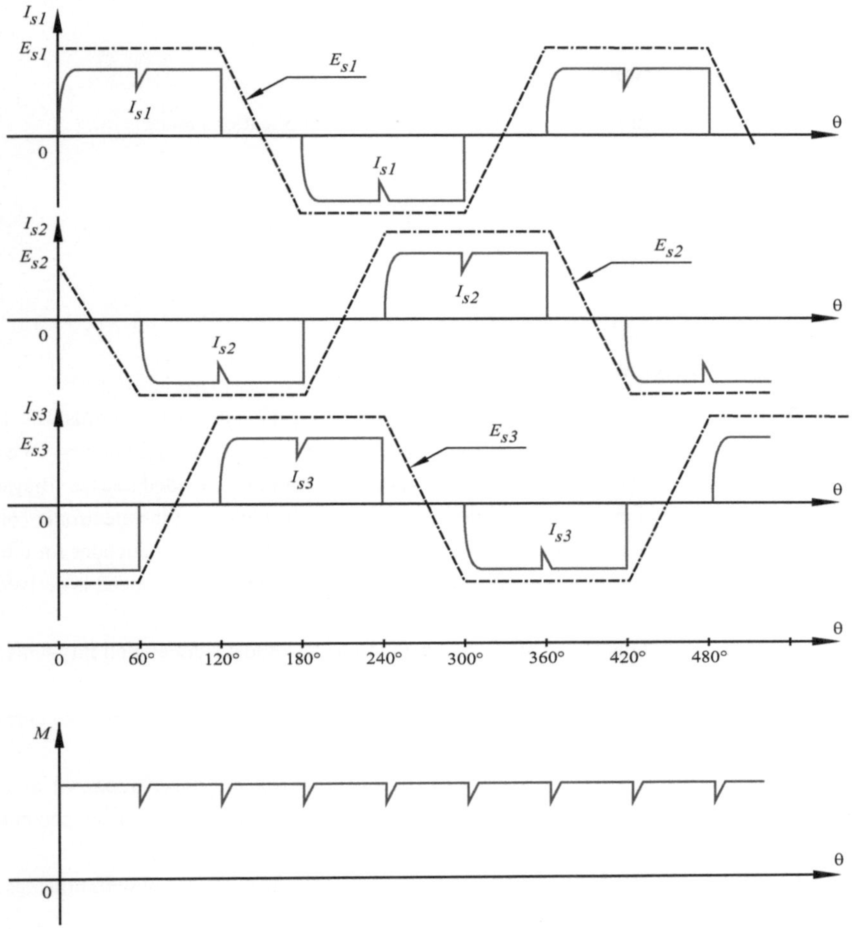

Fig. 22-18: Phase current, phase counter emf and motor torques (with ripple) for the hookup in fig. 22.15

As far as the current measurement is concerned the hook-up as shown in fig. 22.15 is advantageous since:

1. only one current at a time has to be measured, resulting in:
2. only one current sensor is required, for example in the DC-link.

A cheap resistor as current sensor in the lower (ground) conductor has the additional advantage that no galvanic separation is required, since we are measuring a current (as a voltage across the resistor) with respect to ground.

To produce maximum torque a trapezium shaped counter emf is developed (as a result of the motor construction). Since the current is not instantly at nominal value, in the coil every 60° a ripple in the current is produced and therefore in the torque. This torque ripple is a disadvantage of the BLDC-motor.

4.2.4 Applications

BLDC- motors are frequently used. A number of applications being:

- robots
- machine tools
- playing equipment for CD's (CD-ROM, CD-i, etc.)
- bar code readers
- ticket printers etc...

4.3 Position control with a BLDC-motor

Expression (22-12) and (22-13) together with fig. 22-17 indicate that a brushless DC-motor can be controlled in exactly the same way as a "classic" DC-motor with brushes. Due to the addition of an external position control loop the speed controller is extended to a servo-system.
Except for the internal rotor position-detection for the control of the electronic commutation of the stator coils of the motor there is also an externally applied angular encoder for the servo-system. From the signals of this external encoder PG (pulse generator) the speed is derived so that no tachogenerator is required.
The motor torque is obtained by the operation of an almost right angled flux with an almost right angled (stator) current form.
By controlling the voltage the motor speed can be controlled and the internal current control loop allows the torque to be controlled.
Since the current waveform is not exactly square waved, but rather trapezoid we talk about a motor with trapezium shaped current. This in contrast to the sinusoidal shaped current of the AC-servomotor (see later).
Fig. 22-19 shows once more the similarity between position control with brushless DC-motor and with a "classic" DC-motor.

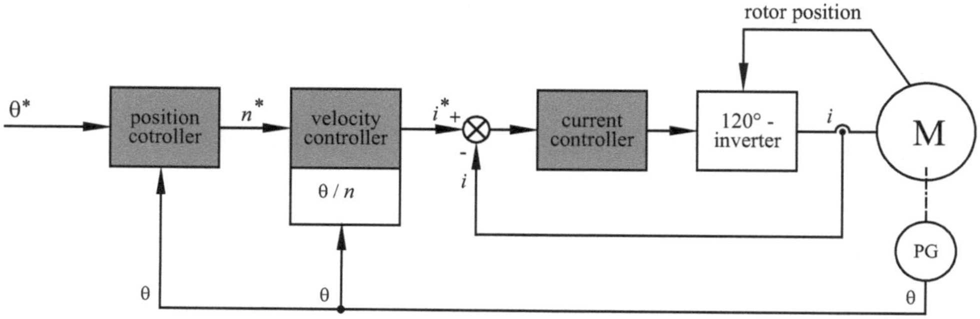

Fig. 22-19: Position control with the aid of a brushless DC-motor

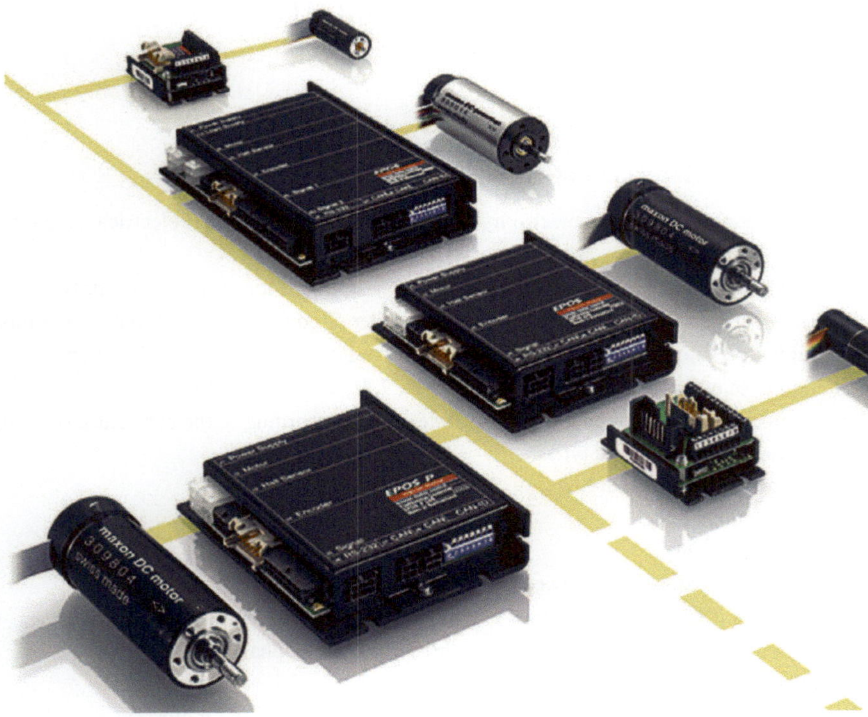

Photo Maxon motor Benelux: EPOS Programmable position controller
EPOS (Easy to use Positioning system)

In addition to all properties of the online controlled EPOS slave version, the EPOS Master version possess a built-in processor and memory for stand-alone use and the independent controlling of up to 127 drive elements via the CANopen bus. Programmable in accordance with the IEC 61131-3 standard.

The EPOS P 24/5 is a freely programmable position controller with integrated amplifier for brushed and brushless DC-motors up to 120 Watt. As a result dynamic precision drives and control systems with minimal dimensions can be realised. Multiple drives and naturally also the EPOS slave versions can be configured and coupled within a network due to the standardised CAN-open interface.

With EPOS P motor controls many operating configurations are possible.

With the brushless EC-motors a sinusoidal current commutation and space vector control (SVC) are implemented. This results in higher efficiency and minimal motor noise.

Extra digital and analogue inputs ensure extra functionality.

The EPOS P is programmed in a very efficient software environment. Multiple program versions (ST, IL, FBD, LD, SFC) available in accord with IEC 61131-3.

Power: 1 to 700 Watt.

Supply: 9-24V DC - 1A / 11-24 V DC - 5A / 11-70 V DC - 10 A

5. POSITION CONTROL WITH STEPPER MOTOR

5.1 Stepper motor

5.1.1 General information-types

A stepper motor (SM) is an electro mechanical converter of a digital electrical signal to a discrete angular position.
The shaft of the stepper motor rotates around a fixed angle for every control pulse.
The number of steps is exactly equal to the number of control pulses so that an SM-motor is useful in a positioning system. This motor can even be used in open control loop.
A number of examples of step angle: 90°; 15°; 7.5°; 3.75°; 1.8°; 0.9°.
The different types of stepper motor are categorised according to the configuration of the rotor.
We distinguish three basic types:

- permanent magnet SM
- reluctance SM
- hybrid SM

A further distinction is made according to circuit and number of stator windings:

- circuit: unipolar or bipolar
- number of windings: two-phase; three-phase; four-phase.

Important advantages of stepper motors are:

- quick to assume a new angular position (quicker than DC-motors!)
- operates accurately in open control loops (i.e. without feedback)
- direct implementation in the case of digital control instructions
- large speed range
- constant speed with stable frequency control
- short starting time and braking time
- reliable operation
- no maintenance
- large holding torque at stand still

Applications are very diverse:

- robots
- coordinate tables
- recorders and plotters
- printers and writing machines
- disc drives
- cash registers
- telex and telefax
- ...

The SM is especially suited to position control systems where short and very fast angular rotations are required.

5.1.2 Two-phase and four-phase stepper motor - unipolar and bipolar

Fig. 22-20 shows an example of a stepper motor. Note the two wound stators and a permanent magnet as rotor. When every stator coil has a centre tap then via a single pole switch a supply terminal is switched and the flux direction of the motor is changed. This is called a unipolar SM. Using stator coils without centre tap then a double pole switch is required (fig. 22-21) to change the current direction in the coil and hence change the direction of the flux. This is called a bipolar SM. The terms **unipolar** and **bipolar** refer to the **change-over** of the **polarity** of the supply voltage in order to change the direction of the stator flux. One **phase** of an SM corresponds with a part of the total winding between supply line and the centre tap point for a unipolar stepper or between two supply lines for a bipolar SM.

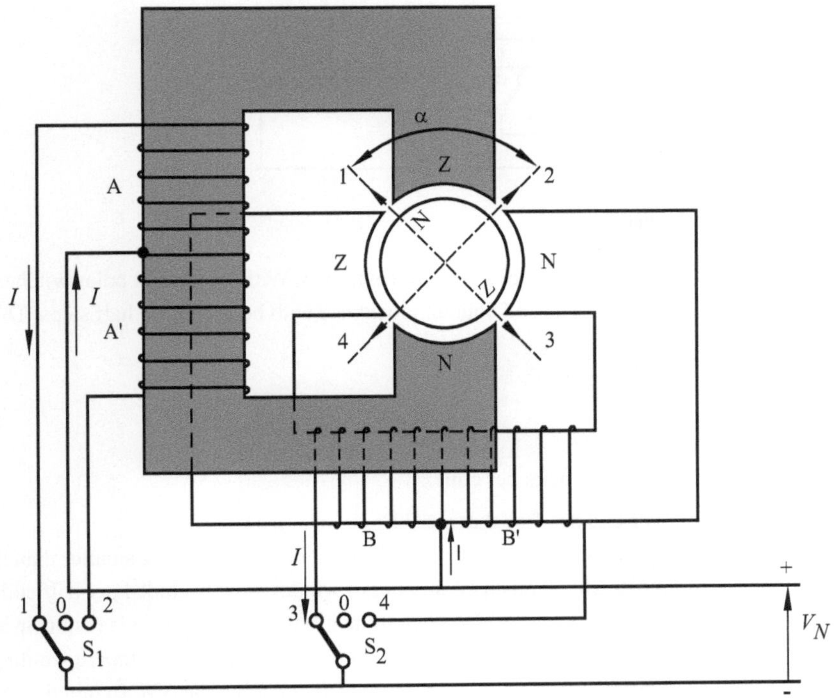

Fig. 22-20: Four-phase unipolar stepper motor

With the indicated position of the switches S_1 and S_2 in fig. 22-20 the rotor north pole is in position 1. If S_2 is set to position 4 coil half B'carries current instead of part B and the stator magnetic poles of coil BB' change names. The permanent magnet rotor will come to position 2. The step angle is 90°.

Due to the combination of positions of S_1 and S_2 it is possible to move the rotor continuously by one step (of 90°). If we have S_2 in position 3 and S_1 in position 2 then the SM would rotate 90° counter-clockwise rather than clockwise.

With S_1 in position 0 and S_2 in position 3 the rotor will make a half step (= 45°) to the left with respect to the output position of fig. 22-20.

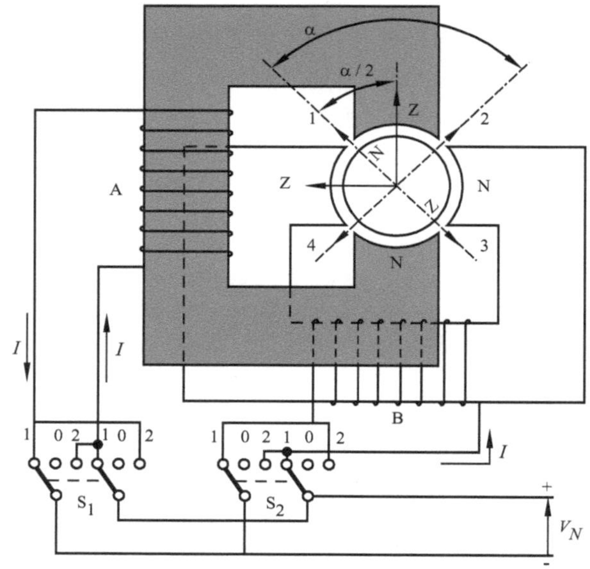

Fig. 22-21: Two-phase bipolar stepper motor

Fig. 22-21 shows a SM with coils without centre tap. With the double pole switches we can cause the motor to step clockwise and counter clockwise with both full and half steps. The step angle is given by:

$$\alpha = \frac{360^\circ}{2 \cdot m \cdot p} \tag{22-14}$$

- m = number of coils
- p = number of pole pairs per coil.

In fig. 22-21 m = 2 and p = 1 so that α = 90°.
Since with a unipolar SM half of every coil is used the torque will be smaller than that of an equally sized bipolar SM. If an equal number of windings are included per half coil of a unipolar SM as are in the coil of a bipolar SM, then thinner wire is required. This results in a larger resistance and lower current so that the unipolar SM has in any case less ampere windings and for the same construction size has a lower output torque than the bipolar SM. In practice the output torque at a lower step rate is about 70% with respect to that of a bipolar SM. At high speed the torque is in practice almost equal.

5.1.3 Stepper motor rotor types

A. Permanent magnet stepper motor

In fig. 22-20 and 22-21 the principle configuration of a permanent magnet SM is shown. The rotor is a cylindrical ceramic permanent magnet which is for example radially magnetised. By suitably switching the coils a rotating magnetic field is produced that pulls the rotor. To obtain a step angle less than 90° a SM with more rotor poles is required.

The number of stator windings also increases. The disadvantage of this solution is that the mass and inertia of the rotor can become unacceptably large.

Due to the force of attraction between the permanent magnet rotor and stator the SM can produce a small torque, even if it is not excited. This is known as detent torque.

Advantages:	**Disadvantages:**
Detent torque	Large rotor inertia
Good damping	Possibilities reduced by decreasing (permanent)
Good accuracy	magnetic strength
Large output power	Large step angle
Low resonant phenomenon	

Application area's: Large torque, low speed.

Remark

A rotating permanent motor induces voltages in the stator windings. As a result of this the step frequency is limited to a certain maximum.

B. SM with variable reluctance

A cylindrical rotor is constructed from soft iron lamella and has radial teeth. Coils are wound around the stator teeth. The rotor has less teeth than the stator. In fig. 22-22 the distance between two consecutive rotor teeth is 45° while for the stator the distance is 30°. Every 15° there is a rotor position with minimum reluctance: this is the lined up position of a pair of rotor teeth and a pair of stator teeth. The three-phase reluctance SM shown here has a step angle of 15°. Note that in this configuration only one of the three coils has been drawn. The other stator teeth are also wound, respectively with the second and third stator coil. A stator coil in this example is wound on four stator teeth.

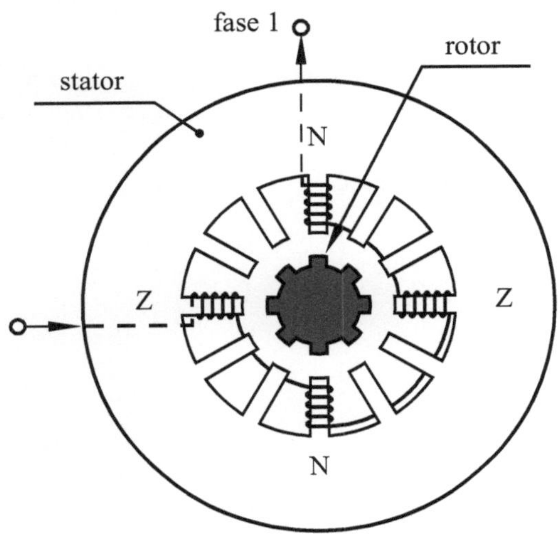

Fig. 22-22: Three-phase reluctance SM

Normally the rotor is built up from multiple rotor disc's (e.g. three) which have the same number of teeth. The stator then consists of three sections. The purpose of this triple implementation is to produce a smaller step angle. The teeth of the stator are lined up but the rotor disc's are shifted with respect to each other depending on the number of phases (by 1/3 tooth distance in the case of three phases).

The phases are consecutively energised: radial in the case that the SM has a simple stator and axial in the case of a multiple stator. The rotor also seeks the position of minimum reluctance for every set of energised windings. Since the rotor has no permanent magnet the diameter can remain small which results in a low moment of inertia. As a result these motors can accelerate quickly. This type of SM normally has a step angle of 5° to 15°.

Advantages:	**Disadvantages:**
Large $\frac{\text{torque}}{\text{inertia}}$ ratio	Resonance can occur at some step speeds
High step speed	Low damping
Low rotor inertia	No detent torque
Small step angle	Low output power

Application area's: High speed, low torque.

C. Hybrid stepper motor

Again the torque is produced using reluctance. With stepper motors of variable reluctance only electromagnetism was used, while with the hybrid SM a combination of electromagnetism (wound rotor teeth) and permanent magnetism (permanent magnet in the rotor) is used.

The cylindrical rotor is axially magnetised. By increasing the number of stator coils the step angle may be decreased. This type of SM can have a small step angle (e.g. 1.8°)

Advantages:	**Disadvantages:**
High resolution	Large rotor inertia
High step speed	Possibility of resonant phenomena at certain step rates
Relatively large detent torque	Average damping
Large output power	
Average to large $\frac{\text{torque}}{\text{moment of inertia}}$ ratio	

Application area's: Large torque, average speed.

Remark
Two-, three- and five-phase stepper motors exist. These days primarily two- and three-phase are used.

5.1.4 Terminology, characteristics

A. Terminology

- **dead band:** the zone between clockwise and counter-clockwise rotation within which the rotor of an energised SM can stand still
- **overshoot:** is the maximum amount by which a step position is exceeded
- **phase:** the part of the winding between supply line and the tapping point (of a unipolar SM) or between two supply lines (in the case of a bipolar stepper-motor)
- **holding torque:** maximum torque that can be applied to the rotor shaft without it starting to rotate continuously with energised motor and zero step control
- **detent torque :** (residual torque) maximum torque of a de-energised motor that can be applied to the shaft without it starting to continuously rotate
- **settling time:** the total time between the application of a step change signal and the time it takes the motor shaft to stabilise at this new step value
- **maximum slew rate:** is the maximum switching pattern for which an unloaded SM can remain synchronous
- **mode:** certain pattern of energisation of the different motor phases
- **pull-in torque:** the maximum load torque with which the SM can start or stop at a certain control frequency and moment of inertia without loosing a step
- **pull-out torque** (stalling torque): the load torque for which the SM looses synchronisation when driving a certain inertia at a set number of steps
- **pull out rate:** the number of steps (= control frequency) at which the SM looses synchronisa tion at a given inertia load
- **step angle:** the angle whereby the shaft of the SM rotates for every control pulse when the consecutive phases are energised one by one. The majority of SM have a step angle between 0.45° and 90°
- **step accuracy:** (step angle tolerance) indicates the maximum error in angular position which can occur. The error is not cumulative and is expressed as a percentage of a step angle
- **stepper motor name:** a SM is named according to: number of stator phases, the switching method, the rotor type. **Example:** two-phase bipolar PM-SM (fig. 22-21).

B. M-n curves

1. Static torque

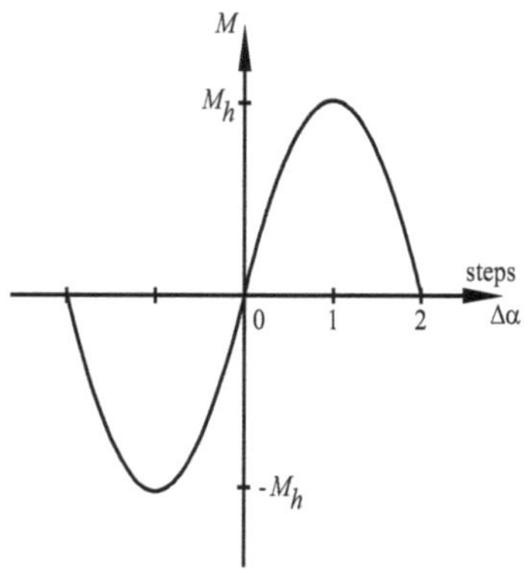

The torque that is required for a stationary SM (zero steps and nominal current through one phase) to move one step is called the holding torque.

Fig.22-23: Static torque of a stepper-motor

2. Dynamic torque

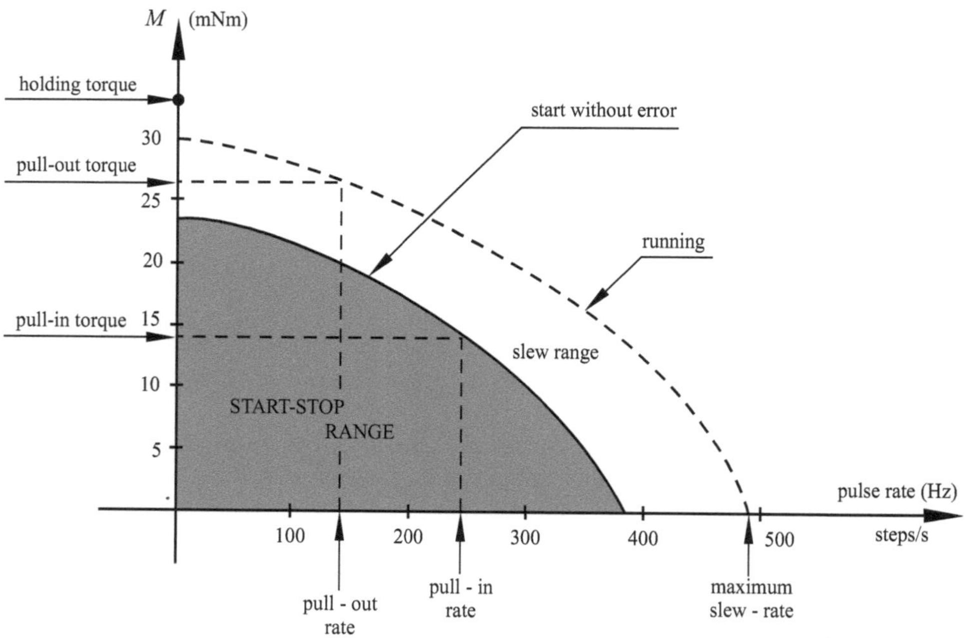

Fig. 22-24: Path of the dynamic torque of a SM

With a higher step frequency the average stator current (and flux) will decrease. The torque decreases as a result.

The earlier definitions (pull-out torque,...) can be found in the curves of fig. 22-24. Both curves are sometimes called “running” and “start without error”. The “start without error” indicates with which torque the SM can start or stop without loosing a step for a constant number of steps. The “running”- curve indicates the available torque when the motor is slowly accelerated to the actual step speed. This is the actual dynamic torque developed by the motor. This running curve is sometimes called the “slew”-rate.
Based on this dynamic torque the motor choice and control method will be made.
If the load consists only of friction torque and the control circuit allows no acceleration, then the “start without error”- curve is used for the motor choice.
In applications where the control circuit allows a variable acceleration the “running”- curve is used to select a suitable motor. A variable acceleration of the “pulse rate” is generally used and is certainly necessary if the load posses a significant inertia.
A practical acceleration is for example 50 Hz/ms.

C. Overshoot

Assume the end of a long series of steps. If the damping is too low then the motor runs through too far (point A in fig. 22-25). The torque is directed to achieving the step position so that the direction of the motor reverses and shoots through to point C. After a number of oscillations the motor achieves equilibrium in the correct position. The largest deviation (AE) with respect to the correct rotor position is called the overshoot.

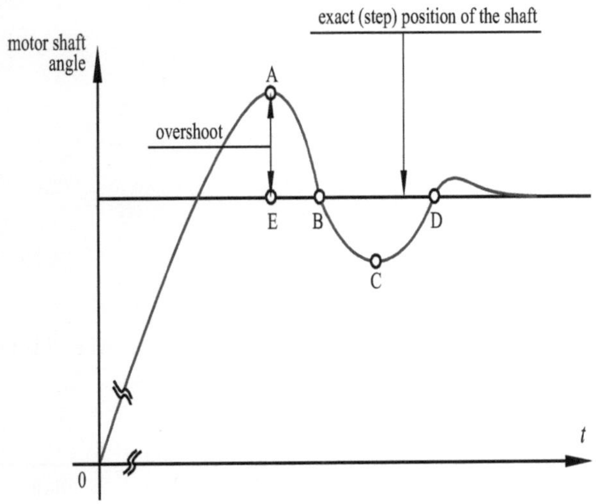

Fig. 22-25: Overshoot

In fig. 22-25 it is clear that during certain time intervals the motor cannot restart or the direction of rotation cannot be changed since the rotor speed is too high or in the wrong direction. In point B it is easier to change direction but maybe it is not possible to operate in the previous direction since the speed is maximum in the opposite direction. Point D is a good point to start again but not so easy to operate in the opposite direction.

D. Resonance

Just as every mechanical system the rotor of the SM has a mechanical resonant frequency. Especially variable reluctance and hybrid stepper motors are susceptible to this. Their resonant frequency normally lies between 10 and 200 Hz. With a stepper impulse frequency in this range close to the resonance frequency or matching it (or with one specific part of it), without sufficient damping, then mechanical resonance can occur and be amplified. The SM becomes unstable in this step range and it is possible that when applying the next step the rotor actually returns to the previous one rather than stepping through.

Permanent magnet stepper motors loose torque around 1000Hz as a result of resonance.

The resonance can be minimised if the load has a large inertia, but this is at the cost of a fast response for a given acceleration.

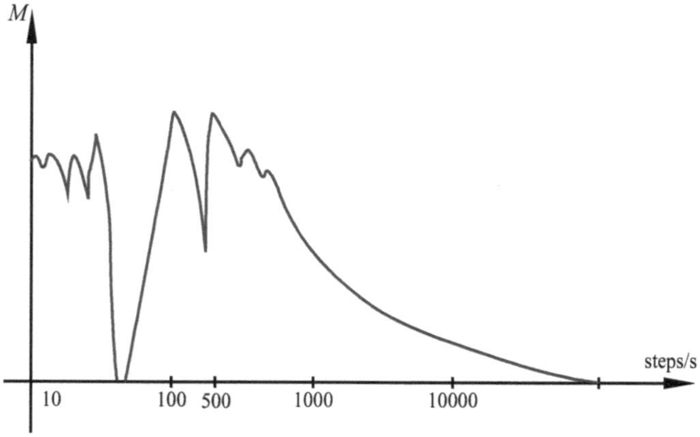

Fig. 22-26: Resonant phenomena of a stepper motor

5.1.5 Control and step modes of stepper motors

A. Full-step drive

The usual control is a so called four step sequence. Fig. 22-27 indicates what this means. By reversing the sequence the direction of rotation of the SM changes.

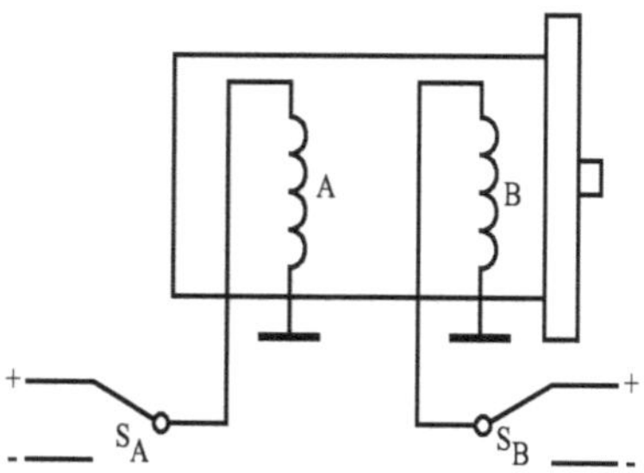

CW rotation

STEP	COIL A	COIL B
1	+	+
2	+	-
3	-	-
4	-	+
1	+	+

CCW rotation

Fig. 22-27: Four step sequence of a two-phase permanent magnet stepper motor

Operating at a constant frequency the input signal is a two-phase (90° displaced !) square wave as shown in fig. 22-28.

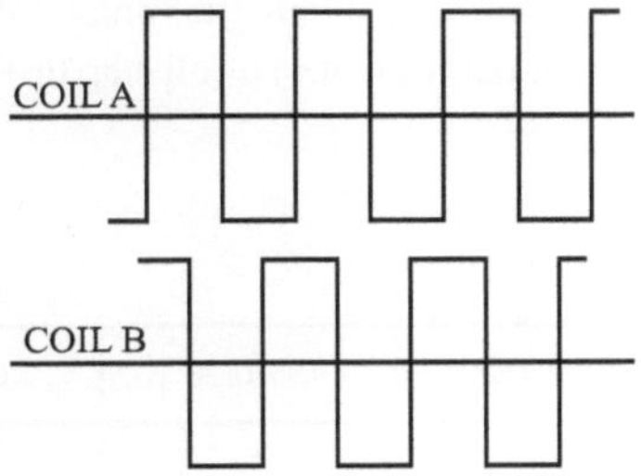

Fig. 22-28: Voltage waveform of a four step sequence with a constant operating frequency

Fig. 22-29 and 22-30 show the control of bipolar and unipolar steppers. With the bipolar SM the transistors are operated in pairs.
For coil A this means T_1-T_4 or T_2-T_3 and for coil B T_5-T_8 or T_6-T_7 .

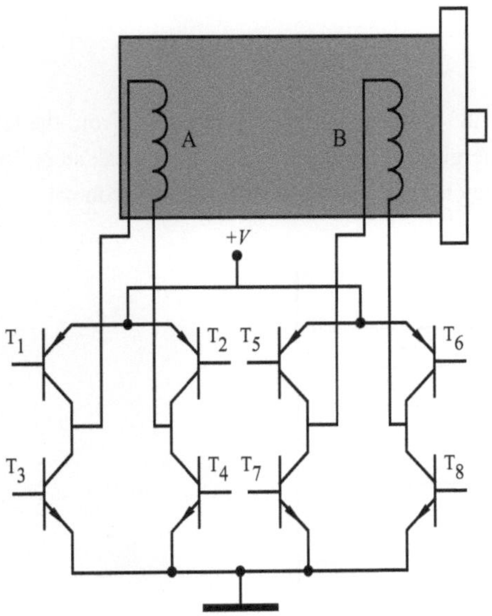

Fig 22-29: Control of a bipolar SM

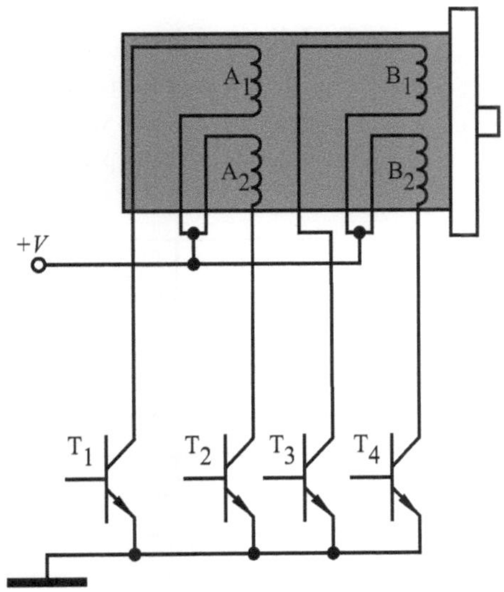

Fig. 22-30: Control of a unipolar SM

In addition to the standard (full-step) four step sequence of fig. 22-27 other control methods are possible:

- wave drive
- half step drive
- bi-level drive
- chopper drive
- micro-stepping

Of these control methods the chopper drive and micro-stepping are the ones of practical interest.

B. Wave drive

If only one winding at a time is energised then we we speak of a wave drive. This is shown in fig. 22-31. Since only one winding is producing flux the static torque (holding torque) and the dynamic torque (running torque) are in practice about half of that of a full step drive.

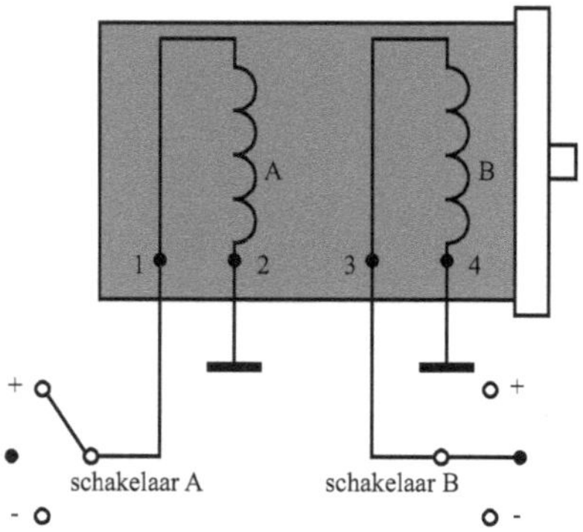

CW rotation ↓ / CCW rotation ↑

STEP	COIL A	COIL B
1	+	OFF
2	OFF	+
3	-	OFF
4	OFF	-
1	+	OFF

Fig. 22-31: Switching sequence of a wave drive

C. Half step drive

Fig. 22-32 shows such a sequence. A 1.8° motor will now step per 0.9°. From the table it follows alternately one or two windings are energised. There is a strong and a weak step. The flux is not the same for every step and the accuracy is not as good as with full step control.

STEP	COIL A	COIL B
1	+	+
2	+	OFF
3	+	-
4	OFF	-
5	-	-
6	-	OFF
7	-	+
8	OFF	+

Fig. 22-32: Half-step sequence (= eight step sequence)

D. Bi-level drive

For this two voltage supplies are required. When acceleration is required the SM is connected to the voltage source which is higher than the nominal phase voltage. When the motor is "running" then it operates with constant load and speed from the lower supply voltage.

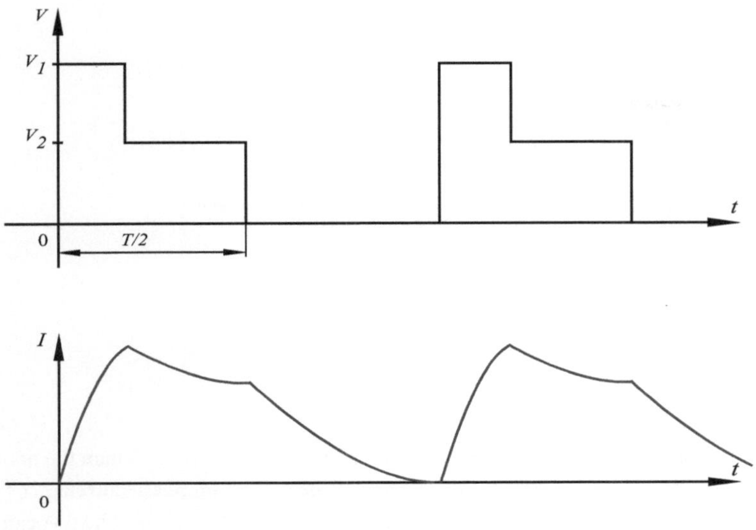

Fig. 22-33: Bi-level drive SM

E. Chopper drive

If the stepper windings are fed from a voltage source then the current form is as shown in fig. 22-34.

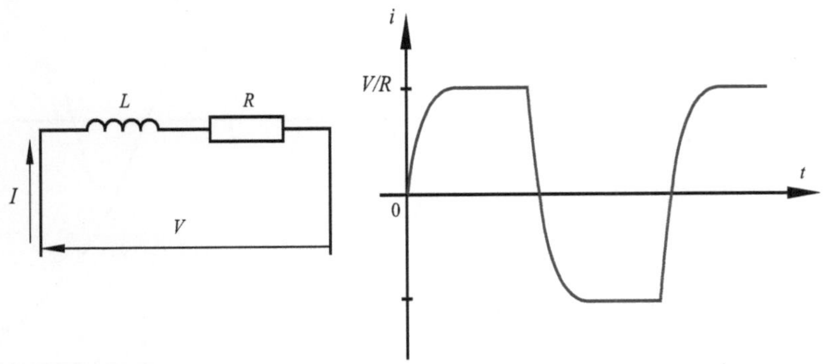

Fig. 22-34: Current form in one phase of the SM

A practical time constant is for example 10 ms ($=\frac{L}{R}$) so that the current reaches it's steady state value after 30 or 50 ms.

With a higher switching frequency the current waveform will look like that shown in fig. 22-35. Note that the current strength remains low and the torque will decrease sharply.

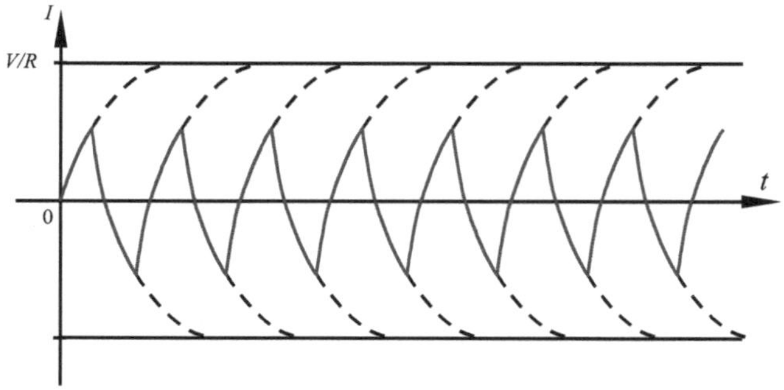

Fig. 22-35: Current form of stepper phase at high control frequency

If we now take a voltage source with a much higher voltage level (m . V) than the nominal supply voltage (V), but at the same time limit the current to the maximum phase current I_F $(=\frac{V}{R})$ of the stepper motor, then we obtain the current wave form shown in fig. 22-36. The SM can once again operate with its nominal torque at a higher control frequency.

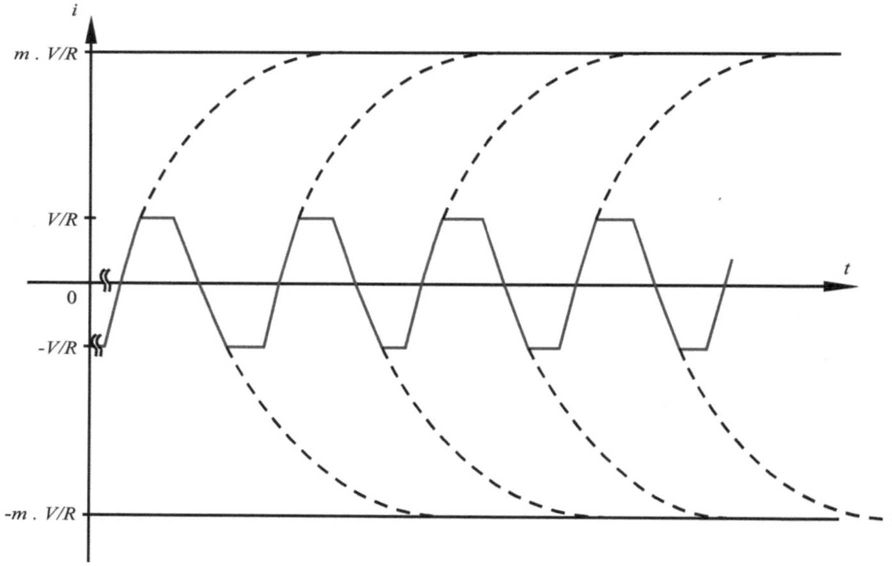

Fig. 22-36: Current waveform of a stepper motor with higher supply voltage and current limiting

In practice a supply voltage that is 10 to 30 times the nominal voltage is used. To limit the phase current $(\frac{V}{R})$ a chopper is used (fig. 22-37).

The phase current is measured using a "sensing"- resistor R_S. Comparing the voltage drop $I_f . R_S$ with a reference voltage V_{ref} switches the chopper "ON" and "OFF".

V_{ref} (e.g. 2V) is often produced in the control-IC. The choice of R_S (e.g. 2Ω) determines the average $I_f (= \frac{V_{ref}}{R_S} = 1\text{A})$.

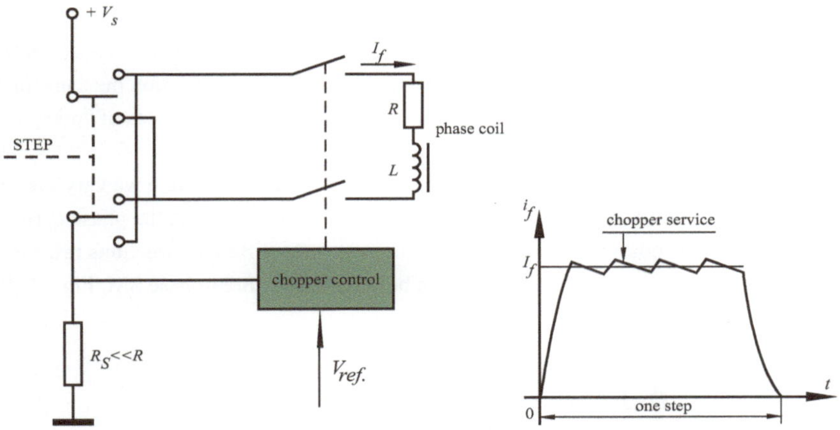

Fig. 22-37: SM with a chopper drive

Photo Rotero: Stepper motors with associated drives

F. Micro-stepping

A typical small step angle of a hybrid stepper is 1.8°(sometimes 0.9°). This corresponds to 200 (or 400) steps per revolution. To obtain a higher precision a reduction mechanism or gearbox can be used. An electronic solution for obtaining higher precision consists of "micro- stepping".
A full step of the SM is divided into smaller steps.
Assume an SM with p pole pairs on the rotor. For the full step drive we vary the phase current between $+I_f$ and zero or between zero and $-I_f$. If we now change the phase current in k small steps then it is possible to orientate the flux in $m = k.p$ different directions resulting in the rotor assuming m possible positions. This can be applied to a three-phase SM. Fig. 22-38a shows the basic cross section of such a stepper.

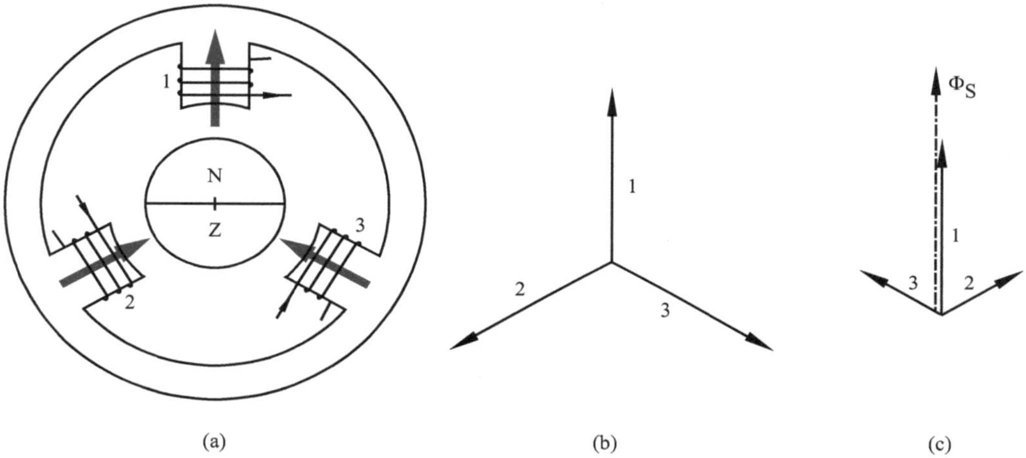

Fig. 22-38: Three-phase SM with the stator flux vector

In fig. 22-38b the three stator flux components are shown. These components are called positive if for example they induce a magnetic south pole on the rotor side of a stator tooth.
With maximum positive and $I_{f2} = I_{f3} = -0.5\ x\ I_{f1}$ we obtain a resulting stator flux as shown in fig. 22-38c.
A practical application is to change the stator current in small steps and therefore the stator flux, and superimpose this on a sinusoidal signal as shown in fig. 22-39a.
With a (p =) 50-pole machine and $k = 40$ we find that $m = k.p = 2000$ steps per revolution. This is shown in fig. 22-39. Fig. 22-39b shows the first 11 states of the stator flux.
The marking 1, 2,...of the stator flux correspond to the indexes 1, 2... of the stator current changes in fig. 22-39a. Position 11 in fig. 22-39b corresponds with fig. 22-38c. This mechanism is known as micro-stepping. In this way practically 10,000 to 25,000 steps per revolution are possible.

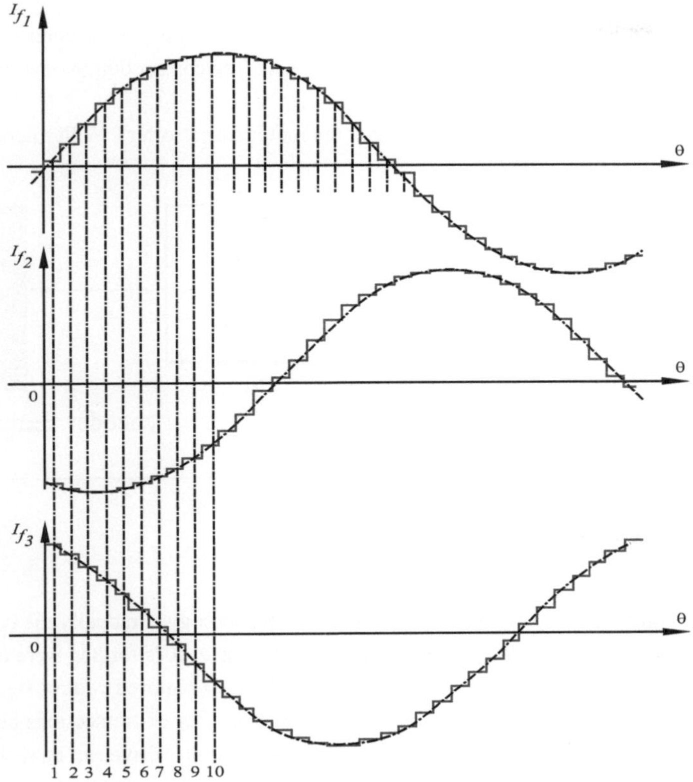

Fig. 22-39a: Stator currents during micro-stepping

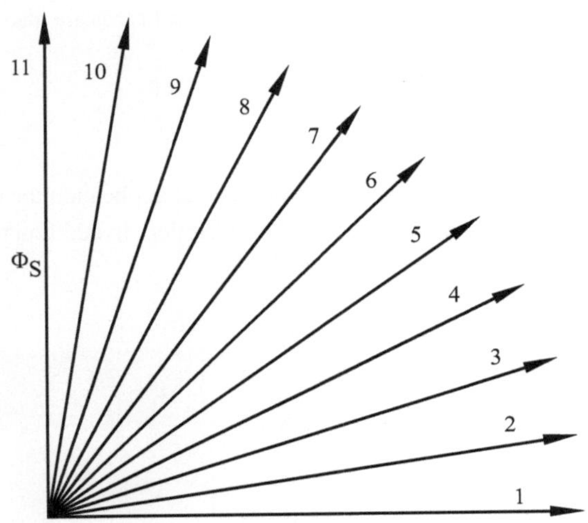

Fig. 22-39b: Stator flux during micro-stepping

Micro-stepping can be implemented with open loop control, but for precise control a closed loop system in which an encoder (or resolver with R/D) detects the correct output position will be chosen. Fig. 22-40 shows the block diagram in this case.
An additional property of micro-stepping is that the motor runs "softer". With micro-stepping high step frequencies are not possible.

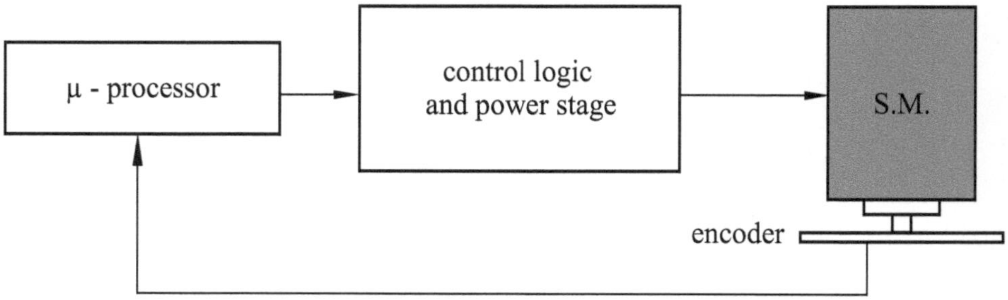

Fig. 22-40: Closed loop control of a stepper motor

G. Control circuits and Power circuits

Originally the control logic was constructed using TTL technology. Gradually the control logic was replaced by specially developed control-IC's. These control-IC's for SM were especially developed as interface between a microprocessor and the SM. All required control signals can be sent from the micro-controller to the IC. With a minimum number of connections between microprocessor and IC and between IC and SM a complete stepper drive can be realised.
If required booster transistors can be added for increased motor current. Also in the case of high supply voltage special transistor stages can form a solution as an interface between IC and SM.
In practice the stepper motor manufacturer provides control and power cards. This card may be controlled from a PC via for example an RS485 connection. Control cards are also available that can be installed in a slot on the PC.

5.2 Position control of a stepper motor

In order to make a suitable choice of stepper motor for a particular application the required torque needs to be determined as well as the required speed and acceleration. In addition to the formulas (17-1) and (17-14) the following formulas are also useful:

1. Drive frequency of motor

In the case with a worm wheel transmission: (see fig. 17-18):

$$p = \frac{360 \cdot v}{l \cdot S} \qquad (22\text{-}15)$$

p = pulse drive frequency (Hz)
v = linear speed of load (mm/s)
l = pitch of worm wheel (mm)
S = step angle (°)

2. Angular position and speed of motor shaft:

After n steps the motor shaft has an angle position:

$$\theta = n \cdot S \quad \text{(degrees)} \tag{22-16}$$

or $$\theta = n \cdot S \cdot \pi/180 \quad \text{(radians)} \tag{22-17}$$

with: S = step angle in degrees

n = number of steps

With a certain pulse frequency p (Hz) we obtain for the motor speed:

$$\dot{\theta} = S \cdot p \,(\,°/\text{s}\,) = S \cdot p \cdot 60 \,(\,°/\text{min}\,) = \frac{S \cdot p \cdot 60}{360} \quad \text{(rpm)}$$

$$\dot{\theta} = \frac{S \cdot p}{6} \quad \text{(rpm)} \tag{22-18}$$

or: $$\dot{\theta} = \frac{S \cdot p \cdot \pi}{180} \quad \text{(rad/s)} \tag{22-19}$$

3. Total motor load

The required motor torque during acceleration is indicated by:

$$M = M_t + J_m \cdot \ddot{\theta} \quad \text{(Nm)} \tag{22-20}$$

Here:

M = motor torque (Nm)

M_t = sum of all friction torques as a result of forces (Nm)
For example a pulley transmission:
M_t = total torque produced by friction + F . R (pulley)

$\ddot{\theta} = \frac{d\omega}{dt}$ = angular acceleration (22-21)

J_m = sum of all moments of inertia (kgm^2) = $J_R + J_{eq}$ (22-22)

J_R = moment of inertia of rotor (kgm^2)

J_{eq} = moment of inertia of load, reflected to the shaft of the motor (kgm^2)

4. Motor output power

$$P = \omega \cdot M = \dot{\theta} \cdot M \tag{22-23}$$

P = output power (W)

M = output torque (Nm)

ω = $\dot{\theta}$ = angular velocity (rad/s)

5. Numeric example:

A load of 150kg has to be moved 20cm in 16 seconds by a ball screw jack. The accuracy of the position must be at least 0.025 mm.
For acceleration and deceleration of the load 0.1s is acceptable.
Determine the required stepper motor.

Screw jack: steel ($\gamma = 7.8$ kg/dm^3)
diameter: 20 mm
pitch: 4mm
length: $L = 50$ cm
friction force load: 15N
efficiency of **ball screw** jack: $\eta = 0.9$

Solution:

- Mass of screw jack:

$$m_2 = \frac{\pi \cdot 20^2 \cdot 500}{4} \cdot 7.8 \cdot 10^{-6} = 1.225 \text{ kg}$$

- Moment of inertia of the load, reflected to shaft of the motor:

$$(17\text{-}10) \rightarrow J_{eq} = \frac{m_1 \cdot l^2}{4 \cdot \pi^2} \cdot 10^{-6} + \frac{m_2 \cdot d^2}{8} \cdot 10^{-6} = \frac{150 \cdot 4^2}{4 \cdot \pi^2} \cdot 10^{-6} + \frac{1.225 \cdot 20^2}{8} \cdot 10^{-6}$$

$$= 12.198 \cdot 10^{-5} \text{ (kgm}^2\text{)}$$

- Friction torque:

$$(17\text{-}11) \rightarrow M_{fric.} = \frac{F \cdot l \cdot 10^{-3}}{2 \cdot \pi \cdot \eta} = \frac{15 \cdot 4 \cdot 10^{-3}}{2 \cdot \pi \cdot 0.9} = 1.06 \cdot 10^{-2} \text{ (Nm)} = M_t$$

- Step angle S?

$l = 4$ mm; angle of rotation α for 0.025 mm linear movement is:

$$\alpha = \frac{0.025}{4} \cdot 360° = 2.25° \text{ ; practical value of step angle S} = 1.8°$$

- Motor speed:

Fig. 22-41 shows speed and distance as a function of time. We assume the same time for acceleration and deceleration: $t_a = t_c = 0.05$ seconds

Total time: $T = 16 = t_a + t_b + t_c = 0.05 + 15.9 + 0.05$

Total distance: $x = a + b + c = 200$ mm

$$x = \frac{v_{max} \cdot t_a}{2} + v_{max} \cdot t_b + \frac{v_{max} \cdot t_c}{2}$$

With $t_a = t_c$ this becomes $x = v_{max} \cdot (t_a + t_b)$ and $v_{max} = \frac{x}{t_a + t_b} = \frac{200}{15.95} = 12.539$ mm/s

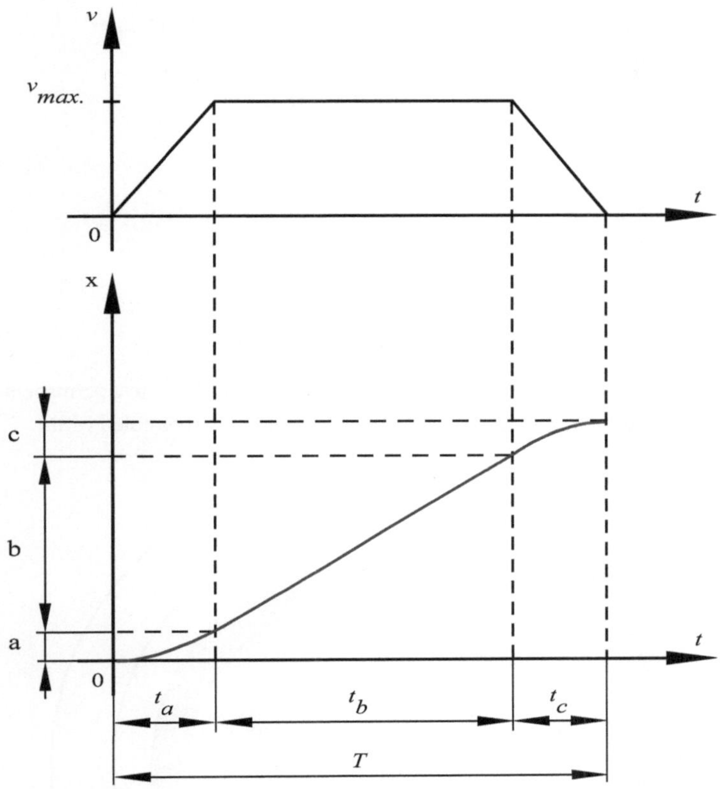

Fig. 22-41: Speed and distance

(22-15) → $p = \frac{360 \cdot v}{l \cdot S} = \frac{360 \cdot 12.539}{4 \cdot 1.8} = 627$ Hz

(22-19) → $\dot{\theta} = S \cdot p \cdot \frac{\pi}{180} = \frac{1.8 \cdot 627 \cdot \pi}{180} = 19.697$ rad/s

- Acceleration ?

$$\ddot{\theta} = \frac{\dot{\theta}}{t} = \frac{19.697}{0.05} = 393.95 \text{ rad/s}^2$$

- Motor torque ?

(22-20) → $M = M_l + J_{eq} \cdot \ddot{\theta}$. Provisionally we fill J_{eq} in, since the motor is not know, and therefore J_R is not known.

$M = 1.06 \cdot 10^{-2} + 12.198 \cdot 10^{-5} \cdot 393.95 = 0.05865$ Nm = 58.65 mNm

We now choose a motor and find for the moment of inertia $J_R = 10^{-5}$ kgm²
The total required motor torque is then:

$M = 0.05865 + 10^{-5} \text{ x } 393.95 = 0.06258 \text{ Nm} = 62.58 \text{ mNm}$

The motor has to be able to provide this torque at the calculated motor speed (627 Hz).

6. POSITION CONTROL WITH AC-SERVOMOTOR

6.1 Permanent magnet synchronous motor (PMSM)

Three-phase synchronous motors of small to medium power often have permanent magnets instead of a wound rotor. Fig. 22-42 shows a cross section of a possible PMSM.

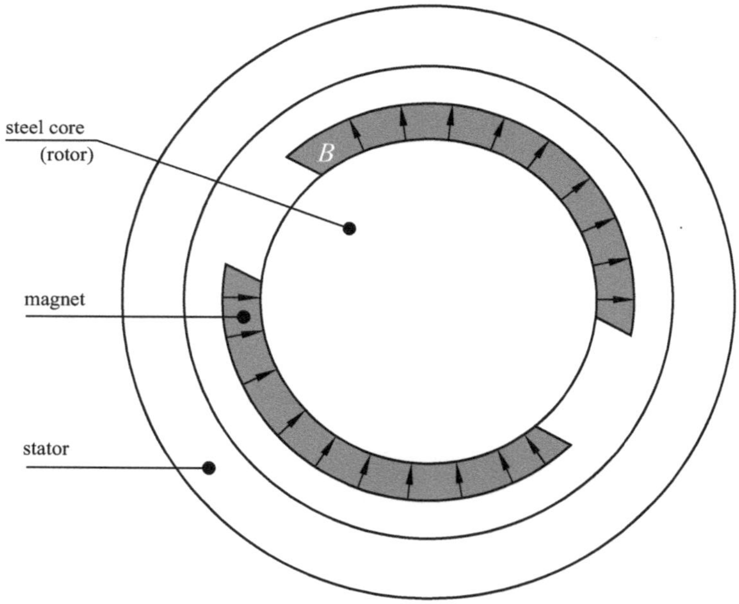

Fig. 22-42: Cross section PMSM

The rotor contains permanent magnets and the stator winding is a three-phase implementation with a sinusoidal winding distribution similar to a classic AC motor.

6.2 Power stage

The stator windings are supplied via a PWM inverter of the 180° type as used to control asynchronous motors.

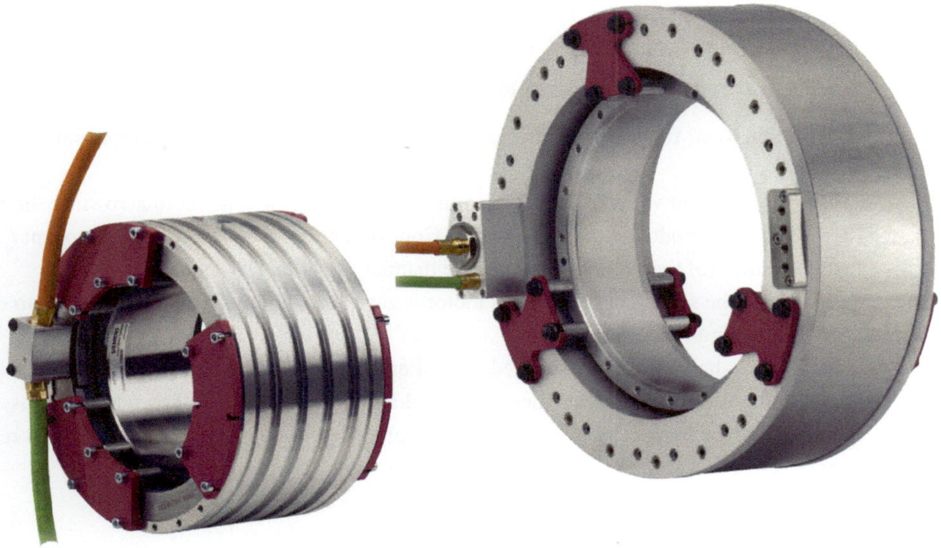

Photo Siemens Drive Technologies: Torque motor 1FW6 from Siemens Drive Technologies. As a direct drive for all rotating shafts with high driving torque and low speed. For example round motors for extremely high torques of machine tools.Water cooling. 109 to 5760Nm – 17 to 61.3kW – 38 to 650 rpm – 400…480V – overload capacity to 2 x Mn

Photo Siemens: Siemens Drive Technologies Division presents the Sinamics G110M, a distributed inverter designed as a motor-integrated inverter for Simogear geared motors. The result is an Integrated Drive System (IDS) that is characterized by its space-saving design, and is easy to install.
The optimal interplay of the Simogear geared motors with the new, motor-integrated inverters provides a high level of efficiency and functionality, particularly for conveyor applications

6.3 Vector control of a PMSM

6.3.1 Internal speed control loop

More and more frequency controllers with vector control are being used for motion control. Two remarks here:

1. the set-point value of the flux forming component i_d will be set to zero since the permanent magnets produce a constant torque. Controlling the torque forming component i_q allows control (of the PMSM), in the same manner as with a DC-motor.
2. since the slip of a synchronous motor is zero, the rotor position will be easier to determine than for an induction motor.

Fig. 22-43 shows the internal speed controller of the vector controlled PMSM. Obviously an external position control loop needs to be added. In the schematic shown here it is assumed that no position or speed sensor is present. The "observer" is a mathematical model that requires the parameters and structure of the entire control system.
The model is supplied with measured data, in this case i_a, i_b, $V_{\alpha R}$, $V_{\beta R}$.
Differentiation of the rotor position θ produces the rotor speed ω_R.

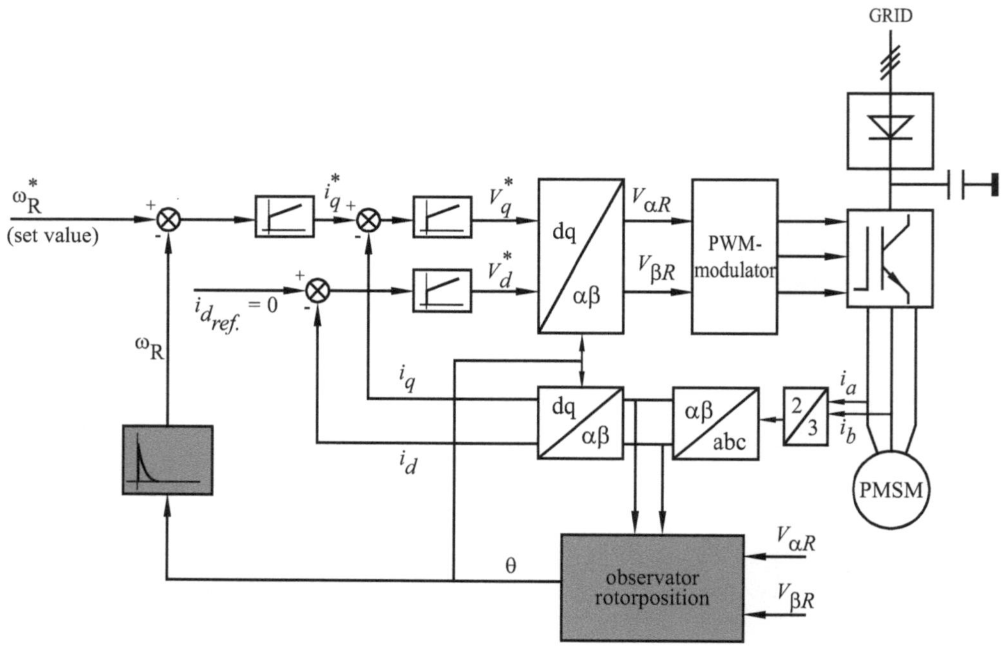

Fig. 22-43: Speed control of a PMSM with the aid of vector control

6.3.2 Power amplifier

We use a classic frequency controller with:

- input filter: EMI and PFC (power factor correction)
- voltage DC-link
- three-phase inverter

This frequency drive can be controlled with a DSP.
Input signals for the DSP:

- the speed- or position signal from a sensor on the shaft of the motor
- the value of the DC-link voltage V_t (via a voltage divider)
- measured value of the motor line currents. An alternative is to measure the DC-link current. From this the motor current can be calculated.

The DSP communicates via a serial line with the external world. In many applications an RS485 is used.

7. POSITION CONTROL WITH LINEAR MOTOR

7.1 "Direct drive"

The demand for shorter production times from the production industry sets increasingly higher demands on feed through times and position accuracy of moving parts in machines. An electrical motor can have two disadvantages for a number of applications:

1. the motor delivers a movement in the form of a rotation
2. without extra equipment the nominal speed of the "classic" AC-motor is less than 3000 rpm with a 50 Hz power grid.

To counteract these disadvantages the following is required:

1. convert the rotation into a translation, for example with a ball jack screw
2. increase the rotation speed for example with a transmission belt.

The mechanical solutions produce additional (dynamic) limits. That is why more and more the choice is made for "direct drives". We distinguish between rotating and translating direct drives.
Rotating: the motor consists of a rotor and stator. The load is directly connected to the rotor. This is called a "torque motor" (photo see p. 22.47).
Translating: the motor consists of a fixed part and a translating part. The load is directly connected to this translating part. This is referred to as a linear motor.

7.2 Linear motors

Different types of linear motors exist:

- induction motor (asynchronous)
- stepper motor
- motor with permanent magnets.

In the remaining study the last type will be considered. Fig. 22-44 shows the construction. A row of permanent magnets is found on the fixed part and windings on the movable part.
The configuration is in fact an unfolded version of the brushless DC-motor (BLDC) which was drawn in fig. 22-14.

Also here current is sent through two of the three phases. Commutation takes place electronically. The moment of switch on is often determined with the aid of Hall generators that are in the movable part (with the windings). The current waveform can be trapezium shaped as with the BLDC-motor or sinusoidal as with the AC-servomotor.
Sinusoidal commutation is used with smooth low vibration movement such as a scanner for example. Sinusoidal commutation produces about 25% less force than trapezium shaped current.
As far as construction is concerned the coils can be wound in an iron core or in an epoxy for example. The epoxy types have a lower efficiency but the construction is lighter, and they are used for applications that require an extremely constant speed. Non-ferrous cores have the advantage that there is no cogging. This is a force variation which occurs as the result of magnetic variation in the winding slots in the iron core.
This cogging-force can amount to 5% of the continuous motor power.

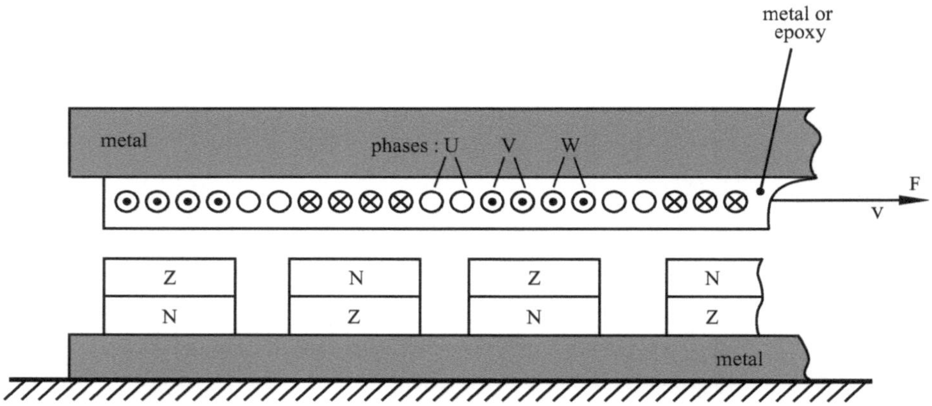

Fig. 22-44: Three-phase linear motor

Force and counter emf of the motor in fig.22-44:

If there are N current carrying conductors with length l with a magnetic pole with induction B, then the force on the conductors is: $F = N \cdot B \cdot l \cdot I = K_M \cdot I$.
By moving the windings with respect to the poles a counter emf is produced in these windings. The magnitude of this counter emf is given by: $E = N \cdot B \cdot l \cdot v = K_G \cdot v$.
Here $K_M = K_G$.

So that: $$F = K_M \cdot I \qquad (22\text{-}24)$$

$$E = K_G \cdot v \qquad (22\text{-}25)$$

7.3 The complete drive

In order to form a complete drive system with a linear motor a number of additional parts are required:

- slide + rail + bearings
- linear position sensor
- current-, speed- and position control loop
- power stage for the motor.

7.3.1 Recording position

As far as the linear encoder is concerned it should be noted that a high resolution is required so that at low speeds enough information is available. In high speed applications this high resolution results in a high input frequency. The input frequency is the number of pulses per second of the encoder. This frequency can be a number of MHz.

Typical position sensors are:

- linear encoder
- laser interferometer
- LVDT

7.3.2 Electronic control

As with other servo-systems the control is digitally implemented. The purpose of the servo-control is amongst others:

1. decoding of the position feedback signal
2. calculation of the desired speed and position profile

7.3.3 Slide - rail - bearing

The slide and bearing need to be available for the construction of the machine in which the motor is to be placed. Normally the part with the permanent magnets is fixed to the machine base. The part with the windings is then fixed to the moving part of the machine. The air-gap between parts is usually in the order of 0.5 to 0.8 mm.

7.4 Advantages and disadvantages of linear motor

7.4.1 Advantages

- large speed range (from 1 μm/s to more than 10m/s)
- large acceleration possible (more than 10.g or $100 m/s^2$). Conventional transmissions reach $10\ m/s^2$!
- low speed ripple (up to 0.01%)
- high level of reliability
- can be used in vacuum and clean rooms
- silent operation
- high resolution (up to 0.6nm with laser interferometer)
- high level of accuracy (0.5 to 3μm with linear encoder and up to 0.1μm with laser-interferometer)
- high level of repeatability (up to 30nm with laser and 0.5μm with linear encoder)

- excellent servo behaviour as a result of the great stiffness (180 to 250N/μm). This is because of the " direct" connection with the load. In fact there are no mechanical connections to reduce the stiffness. The high level of mechanical stiffness (800Nm/μm) results in a large bandwidth (>100Hz) of the servo-system. A large bandwidth results in high speed (>20m/s) and high dynamic accuracy of the system. The settling time is only a few milliseconds
- unlimited stroke: a number of elements can be placed next to each other without diminishing the servo performance. The total length may be tens of meters
- simple assembly and disassembly of the motor
- simple synchronisation possible between different "parallel" linear motors
- large power possible (≥ 10,000N)

7.4.2 Disadvantages

- linear position-measurement systems are expensive
- not suited for driving rapidly changing heavy loads since there is no reduction in the moment of inertia such as with a rotating motor and transmission system
- in the case of a vertical shaft this needs to be balanced (counter weight, spring or air cylinder)
- limited holding force. A position error is required to build up force.

7.5 Applications

- **Semiconductor industry**
  - chip production
  - inspection of wafers
  - writing masks
- **"Smooth" movements**
  - scanners
  - plotters
  - printers
- **Machining**
  - grinding
  - drilling
  - milling...
- **Micro-electronics**
  - construction /inspection of printed circuits
  - SMT (surface mounting technology)
  - pick and place...
- **Packing industry**
  - coordinate measuring machines
  - "visible" inspection
  - test equipment ...

Photo Siemens: 1FN3 linear motors meet the highest requirements of accuracy and dynamic behaviour and are optimised for operation in precision-machines as well as in production units. Water cooled 1FN3-motors are available in two versions, for peak load (200 up to 8100N, max. 836 m/min) or for continuous load (150 up to 10,375N, max. 435 m/min). Supply: 400...480V (control with Sinamics S120 convertor)

8. COMPUTER CONTROLLED "STAND ALONE" POSITION SYSTEM

8.1 Block diagram

A typical servo system is shown in fig. 22-45.

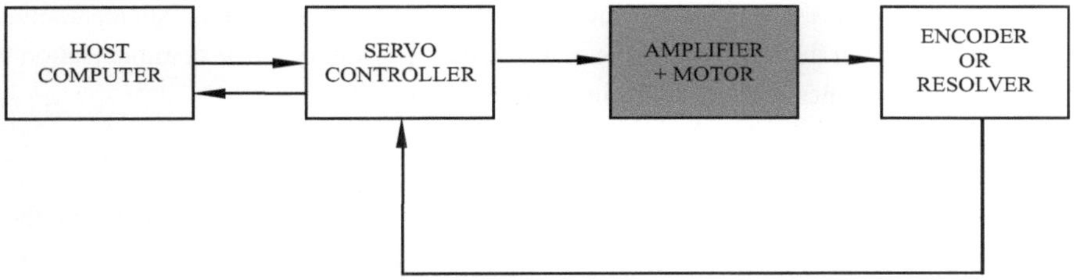

Fig. 22-45: Block diagram of computer controlled "motion control"

8.2 Host computer

The host computer programs the servo-controller with the set-point position. The software is comprised of the digital motor control algorithms. The development "tools" are often accompanied by a user friendly graphical environment. Normally there are control algorithms for: BLDC- motor/ stepper motor / AC-induction motor / DC- and AC- servomotor / variable reluctance motor.

The graphical developments tools have the following minimum "features":

- programming the controller: type and structure
- programming the movement, speed and acceleration profiles
- graphic display of: block diagrams of control structure / movement profile, ... /details of the sensors, limit switches , etc...

Specific modules may include:

- PWM- generation, digital current control, data acquisition,...
- digital motor control algorithms with specific speed and position controllers (PID, state variables, vector control,...)
- reference generator for a specific position profile...

8.3 Servo-controller

The servo-controller calculates the set-point position and generates the error with the actual position. The position error is processed in a PID-filter. The servo controller produces an output signal to eliminate the position error. The servo controller may be external in a separate housing or be a motion controller card plugged into the PC. Standard processors in these servo-controllers can be both fixed and floating point DSP's.

8.4 Amplifier and motor

The output signal of the servo controller needs to be amplified and serves to activate the motor as long as the position set point is not reached.

8.5 Encoder or resolver

Is the position sensor and transforms this position to an electrical signal that is fed back to the servo-controller.

9. INTEGRATED SYSTEM (SIMOTION FROM SIEMENS)

As an example of an integrated positioning system we consider Simotion (Siemens motion control system). Integrated systems have multiple tools. The use of standard modules each with its own intelligence results in not only time saving during development and implementation of a project but in addition there is a multitude of application possibilities. The communication between the different modules occurs via Profibus and Profinet.
In fig. 22-46 we see the Siemens Simotion family.
There are three platforms (Simotion C, D and P).
Siemens has enabled the engineering to become even more efficient by updating the SIMOTION high-end Motion Control System to version 4.4 and integrating it into the TIA Portal.
(TIA = Totally Integrated Automation).

Simotion D is a very compact solution.
Here closed loop control, PLC-functions and motion control are integrated into the module. This is comparable to a certain extent to the "stand-alone" solution of fig. 22-45.
Connection of the Simotion D (fig. 22-47) occurs via the ET200. The ET200 (ET = external terminal) is a decentralized (periphery) module for communication (Profibus or Profinet) via a local network cable to avoid long signal cables. Connection of the Simotion D with Profibus, Profinet or Ethernet results in visualisation and operator control.
HMI (fig. 22-47) means "human machine interface".

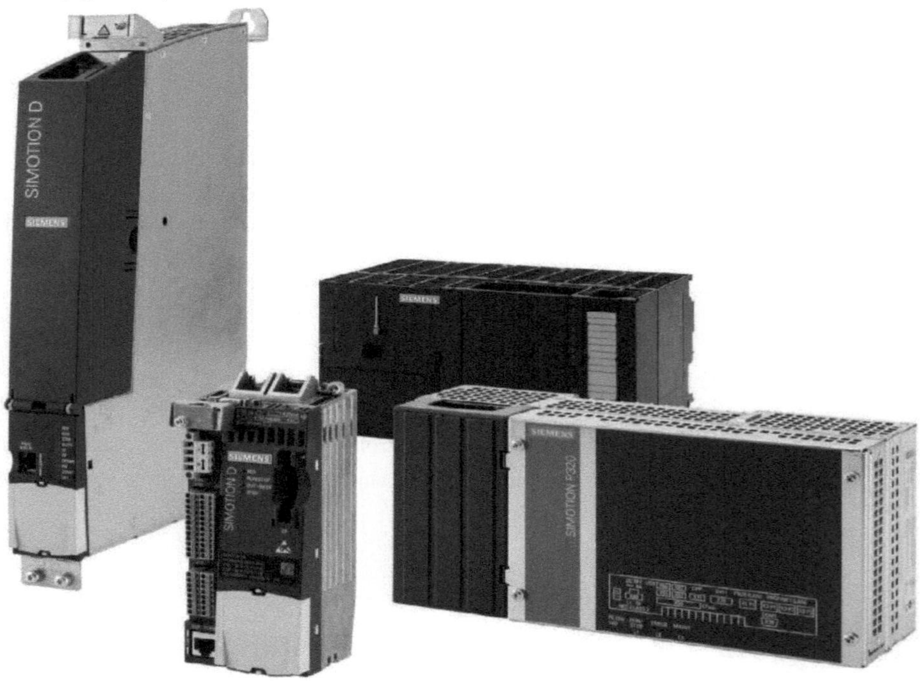

Fig. 22-46: Siemens Simotion family with D410

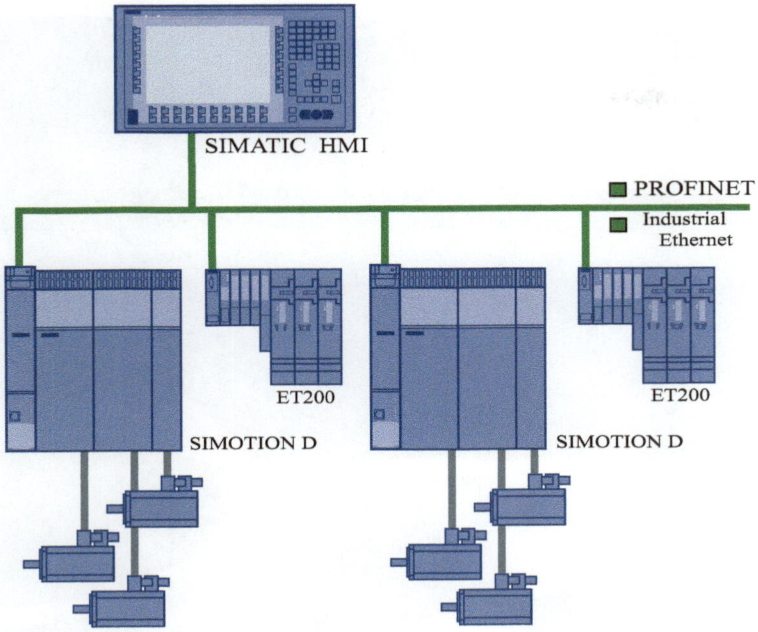

Fig. 22-47: Simotion D topology (drive based)

Machines with multiple shafts are a challenge for every position control system. The reason is that every shaft increases the load on the system and on the bus. With Simotion the machine parts are split into different modules, each controlled by it's own Simotion system. The individual systems are then connected via PROFIBUS or PROFINET. The Simotion D version is highly suited for this type of applications, on the one hand because of it's compact design and on the other hand because of it's fast communication within a multiple shaft system.

Simotion C is a modular solution and has the largest field of application. There are four interfaces on board for analogue and stepper drivers and also multiple digital input and outputs. Due to the segregated architecture, drives and I-O ports can be connected via PROFIBUS or PROFINET.

Simotion P is an (open) system for applications where an open PC-environment is required. To benefit fully from the options of Simotion there is an engineering system (SCOUT) that allows motion control, PLC, technical functions and drives to dealt with in one tool. Configuring, programming, testing and commissioning can all be graphically dealt with from one "work" space. After the drive has been commissioned all data can be requested via the central administration. In addition via LAD (ladder diagram), FBD (function block diagram) and text based ST (structured text) the positioning sequences can be programmed via MCC (motion control chart). More extensive control tasks can be implemented by using DCC (drive control chart).

Photo LEM: Current and Voltage sensors for traction applications

23

e - MOBILITY

CONTENTS

At the start of the twentieth century there were already electric cars in existence. Due to the cheap availability of fossil fuels the electric car was unable to compete with the benzine engine. In addition the action radius of electric cars was small due to the problem of limited energy that could be stored in batteries. A problem that we still have today. Good battery technology is probably the most important condition for the mass breakthrough of electric auto-mobiles.

In addition to road transport there has always been electric traction (train/tram/metro). In addition electric bicycles, scooters and sailing vessels are candidates for electric drives. The complete packet of electrical transport is referred to as electro-mobility or in short e-mobility.

E-mobility deals with speeds from 25 km/h (bicycle) to more than 320 km/h (train) and this with powers of 250 W (bicycle) to more than 9MW (train). Electric cars are somewhere in between with powers of between 40 and 200 kW with maximum speeds between 80 and 200km/h.

In this chapter we look at a number of examples of traction, cars and boats as applications of electronic motor control.

Power Electronics: DRIVE TECHNOLOGY and MOTION CONTROL
http://dx.doi.org/10.1016/B978-0-12-814641-5.50008-1,

1. RENEWABLE ENERGY

Since fossil fuels (oil, gas, coal) are quickly becoming scarce, in the future e-mobility will be supplied by renewable energy. The energy on earth comes for a large part from the sun. The remainder comes from the heat of the earth, tide power and nuclear energy. The sun produces 3.89. 10^{26} watt. The earth is at a distance of 150 billion km from the sun and about 174 petawatt reaches the earth. Right now humanity consumes about 15 terawatt, this is less than one ten thousandth of 174 petawatt! The following table is a memory aid in connection with multiples and subdivisions of units.

Table 23.1

SUBDIVISIONS		MULTIPLES	
yocto =	10^{-24}	kilo =	10^{3}
zepto	10^{-21}	mega	10^{6}
atto	10^{-18}	giga	10^{9}
femto	10^{-15}	tera	10^{12}
pico	10^{-12}	peta	10^{15}
nano	10^{-9}	exa	10^{18}
micro	10^{-6}	zetta	10^{21}
milli	10^{-3}	yotta	10^{24}

Can we satisfy our energy needs with solar panels and windmills? A small calculation shows:

- 174 petawatt of solar energy reaches the earth
- 25 petawatt hits the ground as radiation energy. If we could convert one thousand of one percent to electrical energy then we would have 25 TW, more than the current consumption of the entire population of the planet
- 0.5% of the suns energy is converted to air currents. If we could convert 2% of this with windmills into electrical energy then we would have 174 x 10^{15} x 0.5% x 2% = 17.4 TW.

This is still more than the required 15 TW of the entire world.
We will generate renewable energy in the most favourable locations (solar panels in the Sahara and windmills in the sea and off the coast of Mexico!). Then the problem presents itself of an electrical-network capable of transporting the energy over vast distances.
Such a network is called a super-grid.
As transmission system HVDC (high voltage direct current) presents itself as a candidate.
DC-cables are cheaper than AC-cables and the losses of HVDC are less than 3% per 1000 m.
The disadvantage is the high investment costs of the converters at the beginning and end of the line. This is only viable for large distances (more than 250km by land and 100km offshore) and heavy loads. A super-grid requires international cooperation and enormous investment.

2. ELECTRICAL TRACTION

Electrical traction (tram, train, metro) occurs over a rail network (train) or over a local city network (tram, metro). In addition a number of local tracks exist for example in factories.

The rolling stock is mostly comprised of a motorised part (locomotive), followed by passenger or freight carriages.
Electric locomotives have speeds of up to 230 km/h and powers of more than 9MW.

There are international differences in the supply via the overhead lines and rail. We find 1.5 kV DC, 3kV DC , 15 kV-$16\frac{2}{3}$ Hz, 25 kV-50 Hz, mixed 3kV DC/25kV-50 Hz. In addition there is a system with a DC third rail. This is usually in metro systems.
A number of examples of standard voltages: Germany(15 kV-$16.\frac{2}{3}$ Hz), France (25kV-50 Hz), Belgium (3kV DC and 25 kV-50 Hz) , Netherlands (1.5 kV DC).
Turkey, India and the Congo also work with 25 kV-50 Hz. The new electrification networks in Western Europe 90 % of make use of 25 kV-50Hz. For ease of transport on international lines multiple voltage locomotives are used.
Urban transport operates with DC: 400 V/600V/650 V/750V/900V.
Most common are 600 V and 750 V.

In addition to the “classic” train/tram/metro another term has become popular in recent years namely “light rail”. There is no clear definition for this term since “light rail” is more a collective term for fast trams (inter local), rail vehicles that operate on tram and train tracks and light weight trains (light and cheap trains). There is for example the Docklands Light Railway in London in which you can connect from a metro and ride to the Docklands.
Special metro lines are for example the Thais “Special Skytrain Bangkok” with a maximum speed of 80 km/h and operating on 750 V DC - third rail, or the automatic metro in Nuremberg-Germany which operates without driver.

In addition to speed there is also the factor acceleration /deceleration. It is accepted that the comfort value of acceleration and deceleration is $1 m/s^2$.

With regenerative braking the motor operates as generator. The energy is either returned to the power grid or stored in batteries or supercaps.
Supercaps are capacitors which can store a large amount of energy. Table 23-2 provides an idea of the energy density and lifespan of batteries and supercaps.

Table 23-2

COMPONENT	ENERGY DENSITY	LIFE SPAN
Lead acid battery	30 - 50 Wh/kg	3 - 5 years
Li-ion	90 - 150 Wh/kg	5 - 10 years
Supercap	3 - 5 Wh/kg	unlimited

2.1 The Flexity Outlook Tram (Brussels) by Bombardier Transportation

Fig. 23-1 shows the block diagram of the primary electronics of the Brussels tram. The Flexity Outlook tram by Bombardier Transportation has two models. The "long" tram has a length of 43.2m and the "short" model is 31.8m long.
The tram is constructed with four (six) asynchronous motors depending on the model.
Every motor (photo lower right on p. 23.5) is designed for 3 x 430V-60Hz and has a nominal power of 112 kW. Via pantograph PT, power switch DJ and input filter L_{10}-C, the 650 V supply is applied to the inverter. This inverter is constructed with IGBT's, operates according to the PWM method, has a switching frequency of 2.5 kHz and is driven by a computer board (with microprocessor MC68360, DSP 56203 and FPGA XC4013).
The output frequency of the inverter is 0 to 300 Hz.
At 150 Hz the tram speed is 70 km/h.
During braking energy is regenerated as long as the catenary wire voltage is less than 920V.
Above 920 V it is switched over to dissipation braking via the brake chopper to resistors of 2.6 Ω.

On the lower left of p. 23.5 there is a photo of a traction converter.

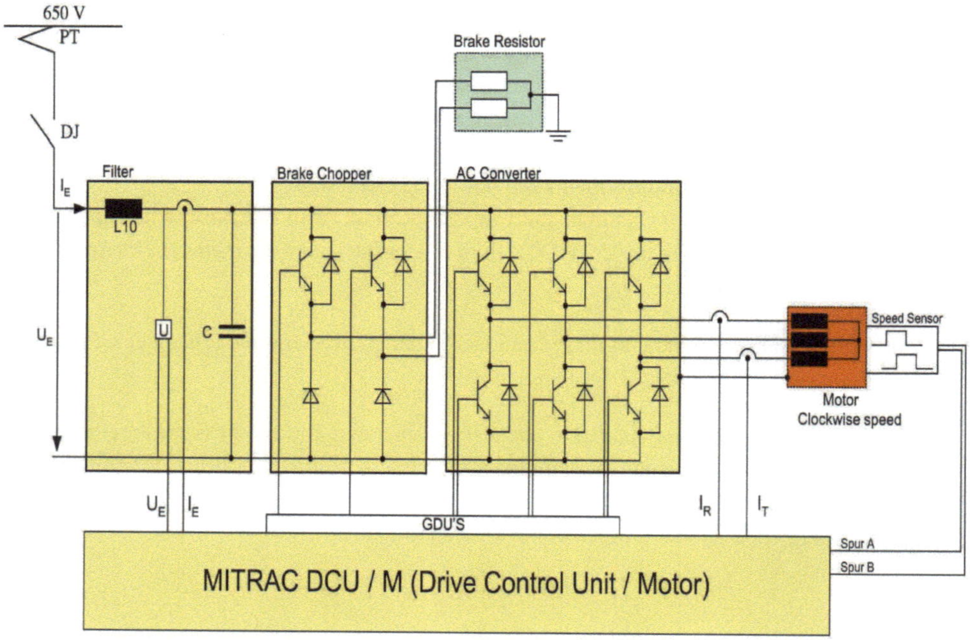

Fig. 23-1: Block diagram of the primary electronics of a Flexity Outlook tram (Bombardier Transportation)

Photo Flexity Outlook: Tram in Brussels (Bombardier transportation)

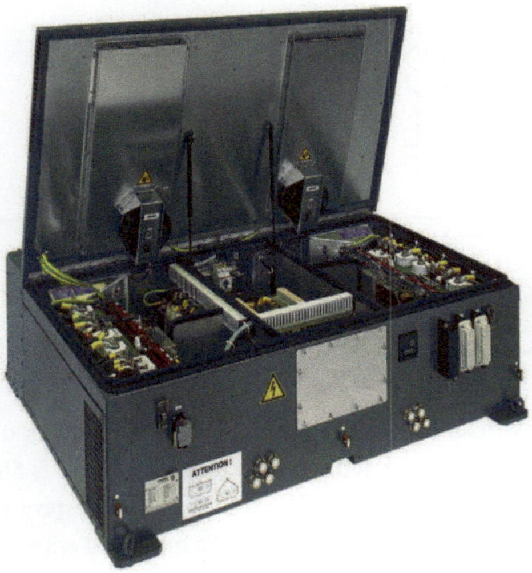

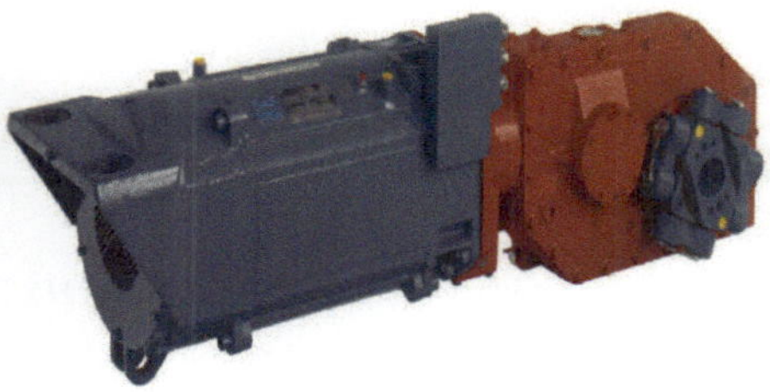

Photo Bombardier transportation:
Traction converter of the tram shown above (Dimensions: b = 1600mm; h = 450 mm; d = 1255mm; Weight 424 kg)

Photo Bombardier transportation:
Motor of the tram shown above. The grey part has dimensions 850 x 385mm

2.2 HST- locomotive Eurostar (ALSTOM)

The Eurostar locomotive from ALSTOM contains four asynchronous motors each 1040 kW. Every boogie has two motors. These motors are constructed for 152 Hz and 1340 V. In addition to the locomotives two motors are also built into the first carriage. Since every train is comprised of two locomotives (one at each end) there are twelve motors available in the drive train. The maximum power of a Eurostar with 18 carriages and 393.4 long is dependent on the supply voltage:

- 15 MW at 25kV-50 Hz (classic lines in France and HST- lines in France and Belgium)
- 7.5 MW at 3kV-DC (classic lines in Belgium)
- 6 MW at 750 V-DC (classic lines between the channel tunnel and London).

These Eurostar locomotive can operate on four different voltages (25 kV-AC, 750 V-DC with third rail, 1500 V DC and 3000V-DC) and serves the connection between Brussels-London via the Chunnel. The French manufacturer calls this a “quadricourant” or a TGV-Q locomotive. Fig. 23-2 shows the input circuit in the case of 3kV-DC overhead wire supply while fig. 23-3 provides the principle schematic for a 25kV-AC supply. As shown in fig. 23-4 the DC-link voltage is 1900 V and the inverters are constructed from GTO-switches. At the output of the inverter a PWM-wave is produced with a maximum frequency of 200 Hz. This 200 Hz corresponds to a maximum train speed of 300 km/h. Dissipative braking occurs with the aid of a chopper.

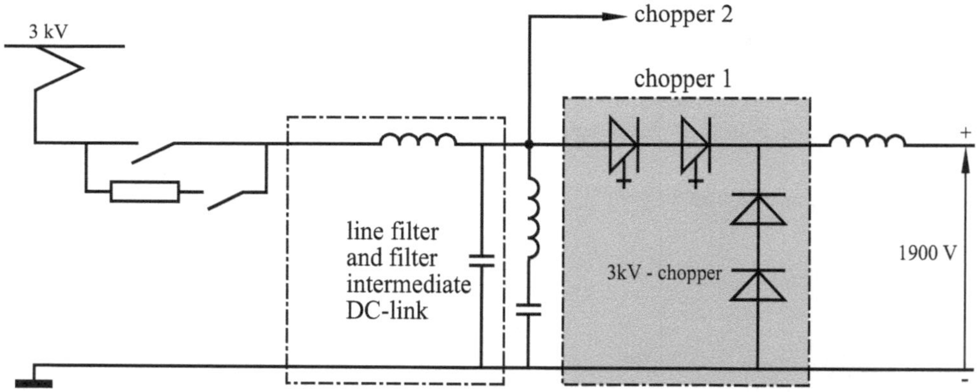

Fig. 23-2: Input circuit (3kV) Eurostar locomotive (ALSTOM)

Remark

Another method to supply the HST is with an AC-voltage 2 x 25 kV. The secondary of the supply transformer in the substation has a centre tap connected to the train tracks. There is 50 kV across the entire secondary. One end is connected to the overhead wire (25 kV with respect to the rails) and the other end is connected to a supply cable that runs next to the overhead wire. Every 5 to 15 km an autotransformer is placed. The entire coil is at 50 kV and the centre tap is connected to the rails. The train is supplied with 25kV (between overhead line and rails) in the section between two autotransformer’s. Between substations and this section the supply is at 50 kV with half current in the locomotive. In the rails between section and substation practically no current flows.
Such a system is in operation between Paris and Lyon and on the new high speed line between Paris and the chunnel (Fréthun).

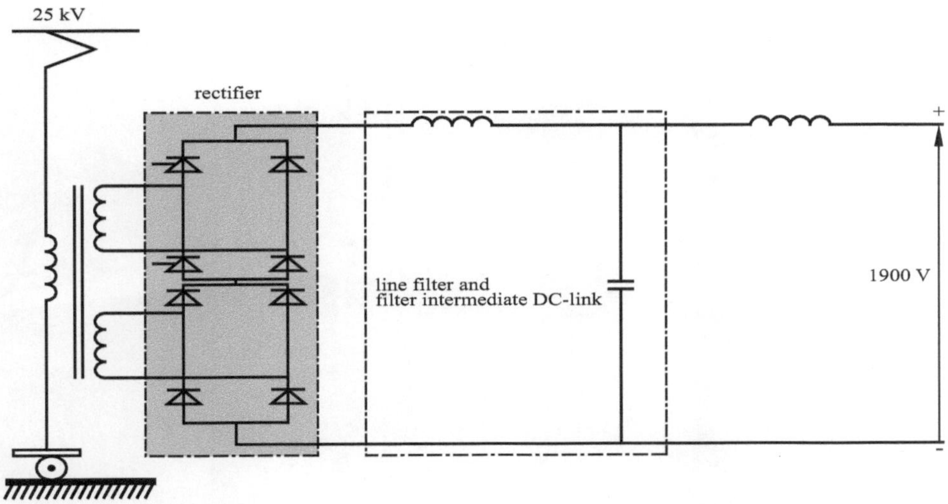

Fig. 23-3: Input circuit (25 kV) Eurostar locomotive (ALSTOM)

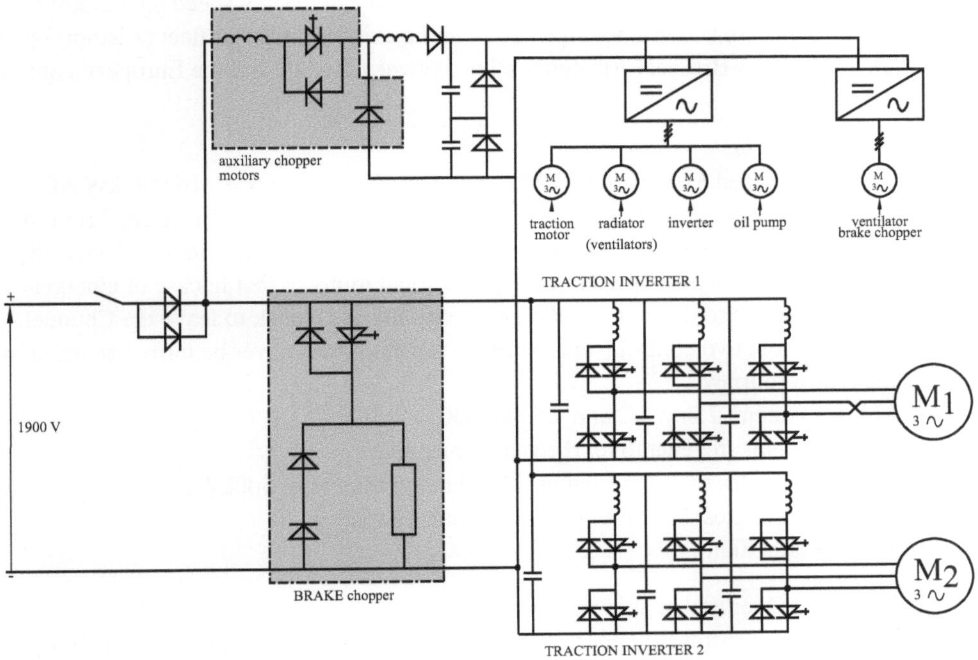

Fig. 23-4: Simplified schematic of the primary electronics of a single motor block of the Eurostar-locomotive

2.3 VELARO D - EUROSTAR e320 from SIEMENS

Photo Siemens: Eurostar HST e320 so called because it can reach a speed op 320 km per hour. These trains are added since April 2018 to the existing Eurostar fleet (Alstom) operating on the London-Brussels-Amsterdam line between the UK and the European continent.

Technical data:

Maximum speed 320 km/h AC (220 km/h DC). Traction power: 16,000 kW AC.
Output frequency inverter is 250Hz at train speed of 320km/h. Basic configuration: the train is comprised of two 8-car units to form a train of 16 cars with two locomotives (one at each end) always coupled in two symmetrical trains so that in case of emergency the train can split and the intact section can continue or go back to leave the Chunnel.
As there exist two control (-locomotives) the train must never be turned at arrival or for the return-journey
Voltage system: 25kV AC and 1.5/3kV DC.
Brake systems: regenerative, rheostatic, pneumatic.
Number of axles 64 (32 driven by 32 induction motors of 500kW).
Number of bogies: 32. Maximum axle load: 17 ton.
Train length 400m;

Fig. 23-5: Overview traction circuit:

- **4QC**: four quadrant chopper: provides the 1800V DC-link voltage for the PWM-inverters. Four quadrant? to allow regenerative braking!
- **V-det**: to detect kind (AC or DC) and value of overhead wire supply
- **ECC**: chopper for eddy current braking on the wheels
- **ACU**: auxiliary power supply (outlets on-board, 60Hz supply cooling fans motors,...)

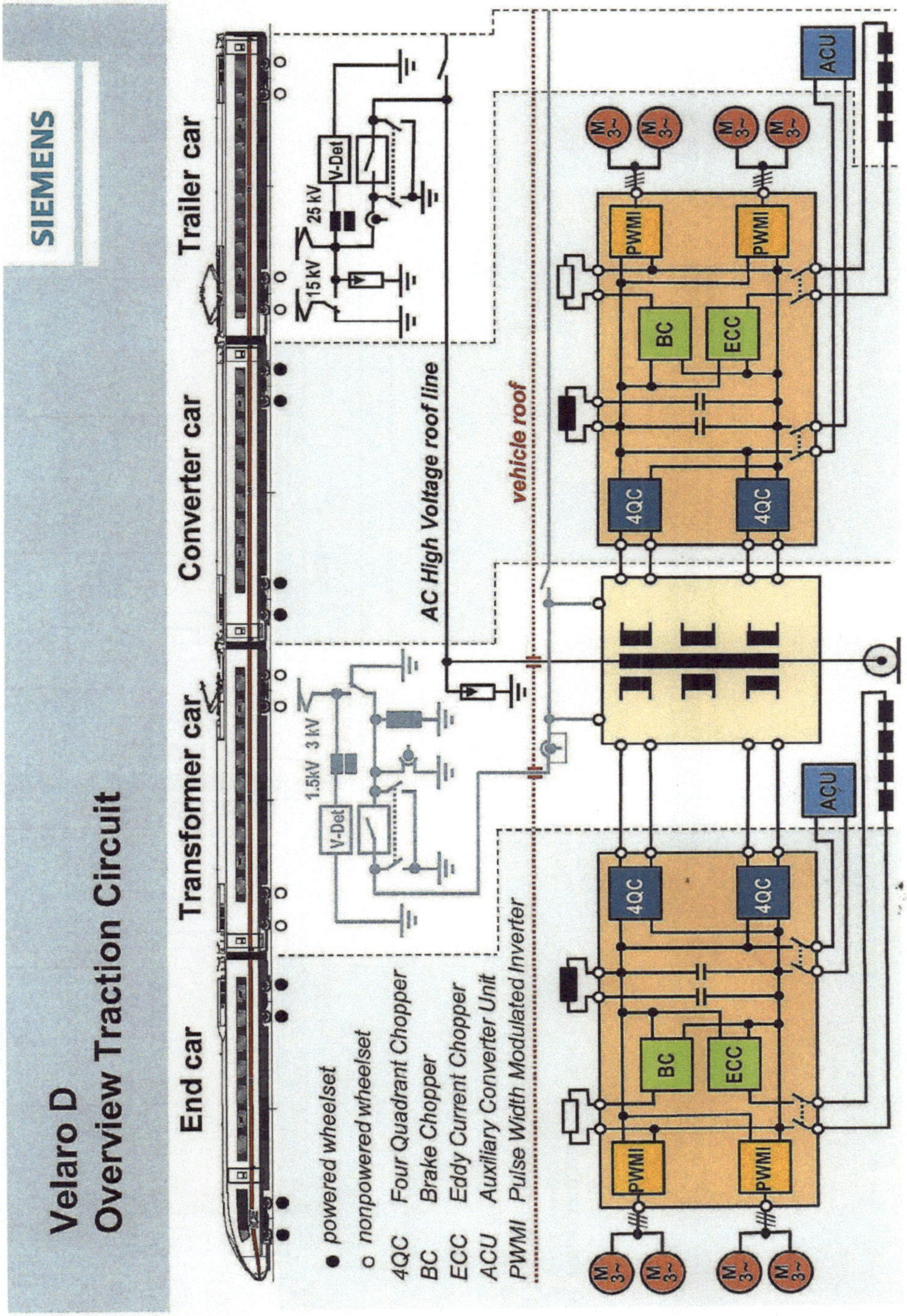

Fig. 23-5: Foto Siemens: e320 Eurostar: Overview traction circuit

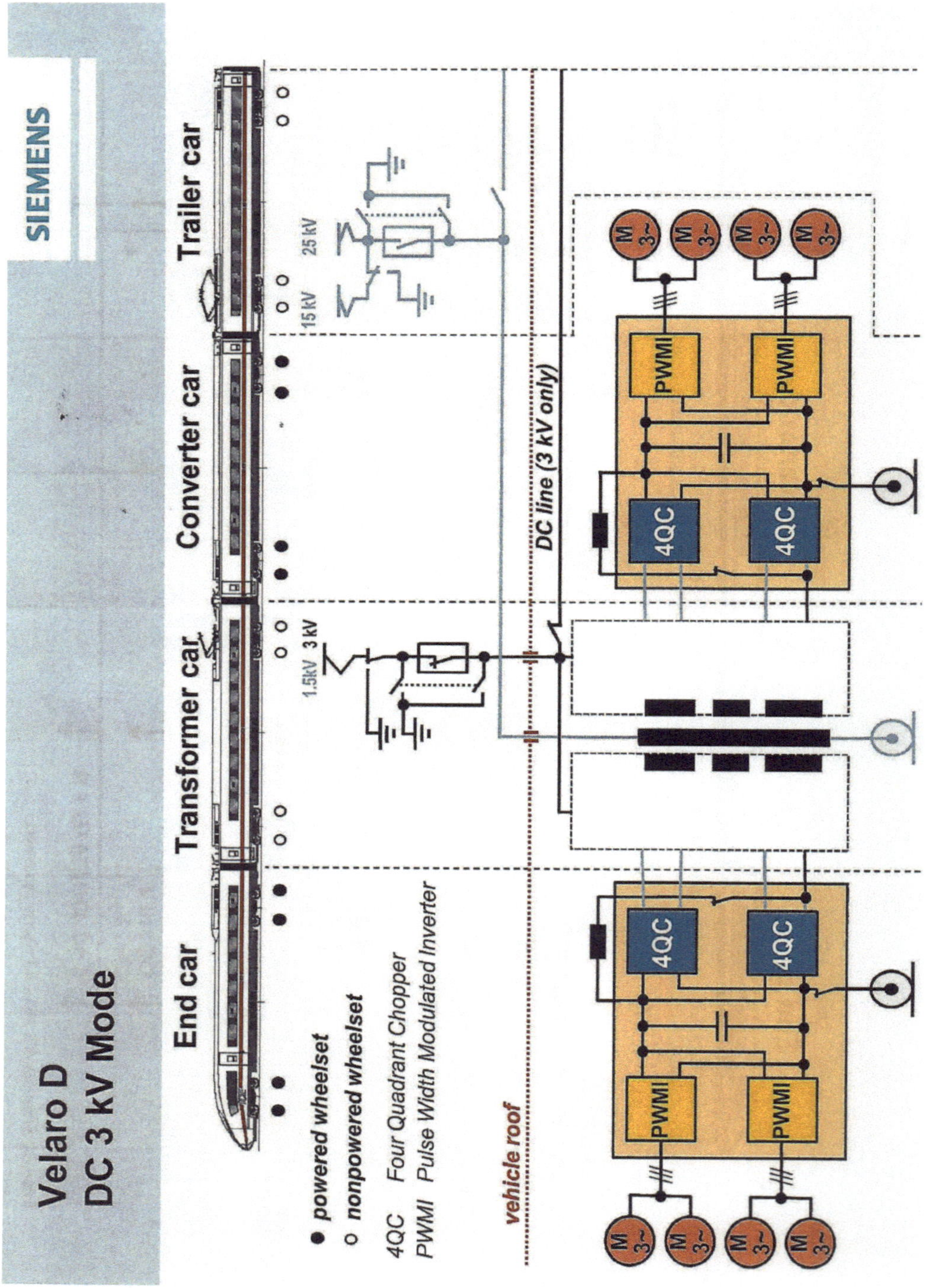

Fig. 23-6: Foto Siemens: e320 Eurostar: DC 3kV mode

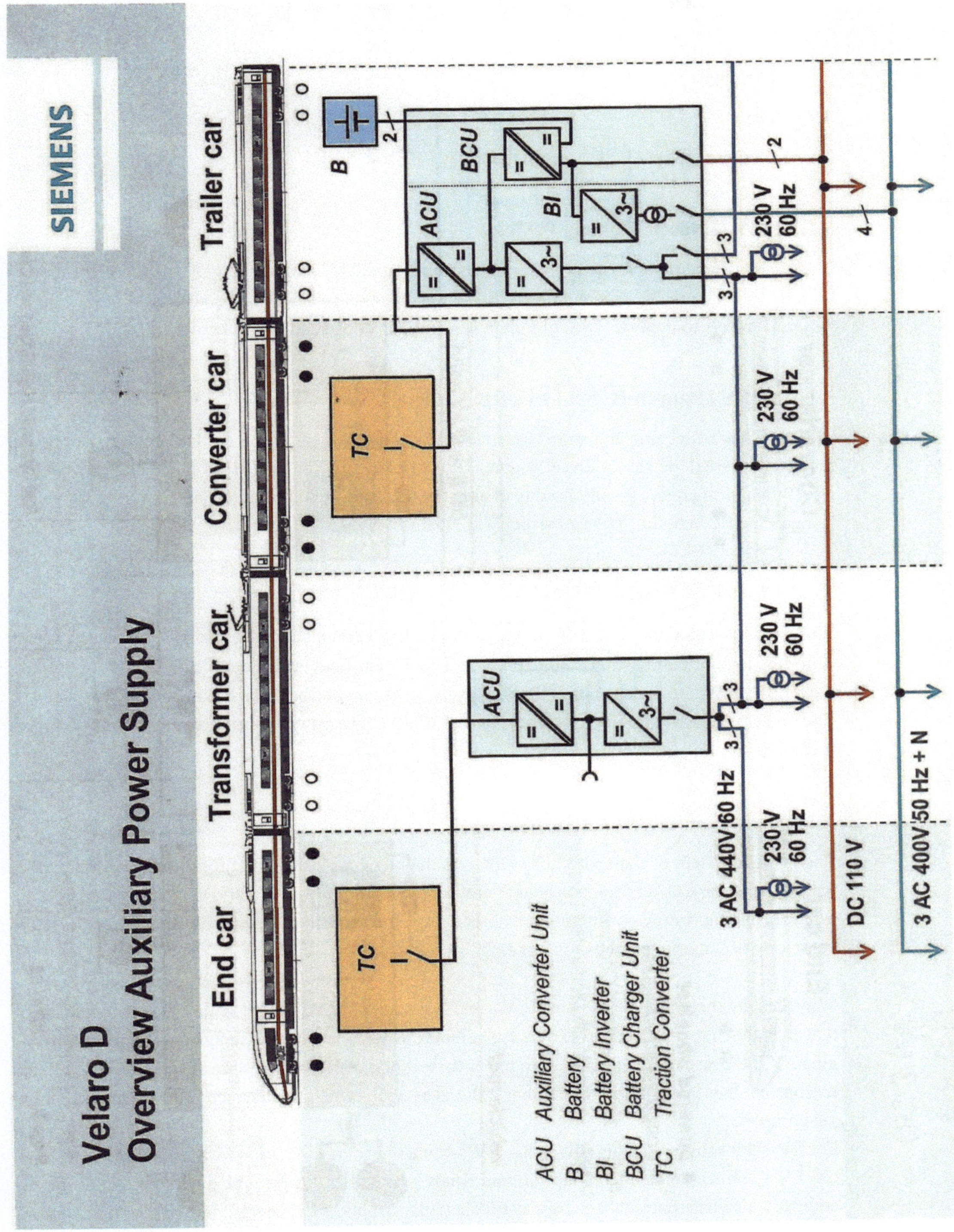

Fig. 23-7: Foto Siemens: e320 Eurostar: Auxiliary power supply

3 ELECTRIC AUTOMOBILE

3.1 Types

There are three types of electric cars: the HEV, PHEV and EV.

3.1.1 HEV (Hybrid Electric Vehicle)

These cars have a relatively large internal combustion engine and a relatively small electro-motor with battery. The HEV is the only electric car without external electrical connection.
"Hybrid" refers to the two types of motor. The Toyota Prius is an example of a HEV.

3.1.2 PHEV (Plug-in Hybrid Electric Vehicle)

This type has a fuel tank and an external electrical connection. There is a relatively large electro-motor and a small petrol- or diesel engine. The petrol engine can also assist in charging the batteries. If the petrol engine is only used to charge the batteries then it is called an ER-EV. (Extended Range Electric Vehicle). The Chevrolet Volt is an ER-EV.

3.1.3 EV (Battery Electric Vehicle)

This is a "fully electric" car. It has an external electrical connection and also a large battery. There is no fuel tank or combustion engine present. The Nissan Leaf is an example of an EV. He has a power of 80kW and a torque of 280Nm. The action radius lies between 85km and 200km.
The power of electric cars lies between 40 and 200kW. The power consumption is about 0.15kWh/km.

3.2 General information-History

The present system of transportation is not sustainable. People are using fossil fuels with finite reserves and the CO_2-emission produces environmental pollution. Electric vehicles will play a role in reducing energy consumption and reducing emissions in contrast with conventional vehicles with internal combustion engines.

Why, despite these advantages are HEV's and EV's not used on a large scale? The most important reason is the high purchase price of these vehicles. A HEV has two drive systems (combustion engine AND electric motor). In addition the batteries for electric vehicles are very expensive. Other reasons are that the number of charging stations as well as the maintenance infrastructures are not yet complete.
For EV's the action radius is still small. EV's come in for consideration as a second car or as a city car. EV's could also be interesting for companies with a client base in the direct vicinity. If the company also has renewable energy available (solar panels, wind energy) then EV's could also be an interesting option. They will definitely be interesting at the time of GRID PARITY. This is the time when electricity from solar panels will be cheaper than electricity from the socket.

In a study conducted by the European commission (DG Climate) the potential market for electric cars and the environmental effects was studied. There was an extensive analysis made of the possible innovations in battery technology. The conclusion of the study was that the share of electric cars on the road would remain limited to a few percent until 2020.
Still, despite everything the market for HEV's and EV's has already started. This is possible because manufacturers with a long term vision on energy efficient HEV's and EV's have started production and simultaneously governments have introduced advantages for proper, energy efficient vehicles.

A special category of PHEV's is the "parallel" HEV, whereby both motors (electric and internal combustion motor) can transfer mechanical energy to the wheels. Another type is ER-EV where the vehicle always operates electrically while the battery charged from the power grid is discharged. Then the internal combustion engine comes into action to charge the batteries and thus extend the action radius. (ER = Extended Range). This is referred to as a "series" power train. In this case the wheels are never driven by the internal combustion engine. The ER-EV has a much larger action radius than a "normal" PHEV.
An important advantage of a PHEV is that the batteries can be charged at night.

A group of HEV- and EV- experts (IA-HEV) operates under the IEA (International Energy Agency) with the intention of implementing cooperation between different countries in the area of HEV- and EV- technology programs. The twelve members of the IEA are Austria, Belgium, Canada, Denmark, Finland, Italy, Netherlands, Sweden, Switzerland, Turkey and the USA.

3.3 Battery

3.3.1 Types of battery

Stanford Ovshinsky is the inventor of the NiMh-battery (Nickel-Metal hydride). These batteries have not only double the energy-storage of lead batteries but also a longer lifespan.

GM purchased the majority share in the Ovshinsky-company and tried to convince the California legislature that EV's were not possible. GM destroyed all electrical cars and sold the shares of the Ovshinsky company to Chevron. After many court cases Toyota could use the NiMh batteries in the Prius. Also the second generation of EV's with these batteries (action-radius 150-200km) came onto the market in 1999, but GM sent all EV's to the scrap yard despite a long list of buyers!
It is clear that a light and cheap battery forms the key to a real breakthrough of electric cars.
Most of today's PHEV's and EV's use lithium-ion batteries, though the exact chemistry often varies from that of consumer electronics batteries. Research and development is ongoing to reduce cost of batteries to less than 100 Euro/kWh, to increase the range EV to 500km and to reduce the charging time to 15 minutes or less.
Nickel-metal hydride batteries have been used succesfully in EV's and are widely used in hybrid electric vehicles. The main challenge with this nickel-metal hydride batteries are their high cost, high self-discharge and heat generation at high temperatures.

Ultracapacitors store energy in a polarized liquid between an electrode and an electrolyte. Energy storage capacity increases as the liquid surface area increases. Ultracapacitors can provide vehicles additional power during acceleration and hill climbing and help recover braking energy. They may also be useful as secondary energy-storage devices in electric drive vehicles.

3.3.2 Battery charging

The charger is present (AC-DC converter) in the car and needs to be connected via charging cables with a charging point at 230V AC. This is what we call a mode-2 charging. Worldwide there are two types of sockets in use. Both comply with the IEC62196 standard.
Type 1 is used in Japan and the USA. With this type a maximum of 32 A may be drawn on one phase. Type 2 was developed in Germany and is in use in all of Europe. This has a threephase socket which uses a maximum current of 63A.

Most cars for the time being have both sockets. The charging cable contains a communication cable to communicate between the loading station and car. Extension cables may not be used.

In the Netherlands for example the charging stations are provided with type 2 sockets and a "mode 3 controller". Mode 3 is a communication protocol with the goal of safely charging a car. The charging station checks (with the communication protocol) first and foremost if the car and charging station are correctly connected via the charging cable. There is also communication with the car to determine the maximum charge. Then the sockets in the car and the charger are locked and the current is switched on. Charging can begin. During charging the car is locked prevented driving away.

Mode 4 is an off-board fast charger with DC. The charger station determines the whole charging process. This station is in direct contact with the car battery. There is no need for an AC-DC converter on-board. The fast off-board DC-charger (mode 4) is the ticket to e-mobility. In 2015 there was a 24% share for this off-board EVC's (electric vehicle chargers) and there is a predicted share of 62% for 2020.

The expectation is that when the majority of car manufacturers have an EV or PHEV on the market, that ten thousand electric cars will come into use in Europe. This expected increase will result in electrical networks having to be upgraded as a result of this significant load increase. This will result in major investments (breaking opening roads, raising power capacity...). To avoid this SMART GRIDS are being considered.

A "smart grid" looks continually to its own operation and reacts immediately, for example by requesting appliances to temporarily cease operation. EV's and HEV's are major consumers but the charging moment is easy to shift since cars on average are parked for 90% of the time. This grid then gives priority to cars that (unexpectedly) need to drive. The rest of the cars can charge later or at night.
Another possibility is wireless charging via induction. The cars are parked above an electric coil in a parking space. The magnetic field of this coil induces voltage in a coil in the car and this voltage charges the battery.
Systems also exist for changing the batteries. In Israel for example a network of such stations exist where in two minutes a battery pack is automatically changed.

3.4 The power train of an electric city bus (with Siemens ELFA © drive technology)

City's are interesting areas to drive electric. The distances are within the action radius of electric vehicles and environmental pollution from diesel and petrol engines is completely avoided.
We now look at an example of a city bus fitted with "ELFA drive technology".
ELFA stands for "Electric Low Floor Axle". This drive technology can be widely used, independent of the energy source. The possibilities are: pure battery service, diesel-electric group, gas-motor or fuel cells.
We consider her only pure battery service. Since a city bus makes many start stop moves this application is especially suited to convert the kinetic brake energy into electrical energy. It may also be stored in ultra capacitors (quickly charged and discharged) or in batteries (slow to charge but energy is available for longer time). The required battery capacity of a battery for a small city bus is at least 150 kWh since all auxiliary energy also needs to be provided.
Fig. 23-8 shows a block diagram of a possible configuration for an ELFA-drive.
The brake energy is used to charge batteries or ultra-caps but if these are fully charged the excess energy is dissipated in brake resistors.
All auxiliary services such as heating, cooling, compression and auxiliary supplies are supplied with energy by this ELFA-system. ELFA is water cooled as a result of which it is extremely compact and silent.
The DICO controller is used for complete energy management. It controls amongst other things the motor speed as a function of the position of the accelerator. Since the DICO (Digital Input Control) controls and manages all energy flows, efficiency is the result.
The efficiency of this system is high and it requires little or no maintenance.

3.4.1 City bus Van Hool (Belgium)

Fig. 23-9 shows the block diagram of an electrical bus from manufacturer Van Hool.
In the traction module an inverter is represented, together with a pair of asynchronous motors, connected to a gearbox. The output of this box transfers torque to the differential.
There is obviously one inverter per motor. In this example two motors were required to produce sufficient torque for a city bus. Asynchronous motors of the type 1PV5138-4WS24 were chosen. These 85 kW-motors produce 200 Nm of torque at 4000 rpm. With a gearwheel ratio of 4.05 a maximum torque of 200 x 2 x 4.05 = 1620 Nm is produced. The maximum speed of the motor is 10,000 rpm.

3.4.2 City bus VDL (The Netherlands)

Fig. 23-10 shows a block diagram of a possible configuration for an ELFA-drive of aVDL city bus. The direct drive is connected with a drive shaft to the differential in the rear axle. In this example one motor (PMSM, type: 1DB2016-2NB06) is required to produce sufficient torque for the bus. This 160kW motor produces 739 Nm of torque at 1500 rpm. The maximum speed of the motor is 3600 rpm. The photo on p.23.17 shows the charging with a "roof mounted pantograph". VDL uses here a mode 4 charger (with CCS communication protocol). Power charging station is 300kW.
It takes about 20 minutes to charge the battery pack for 80%. Battery: 666V, 180kWh.

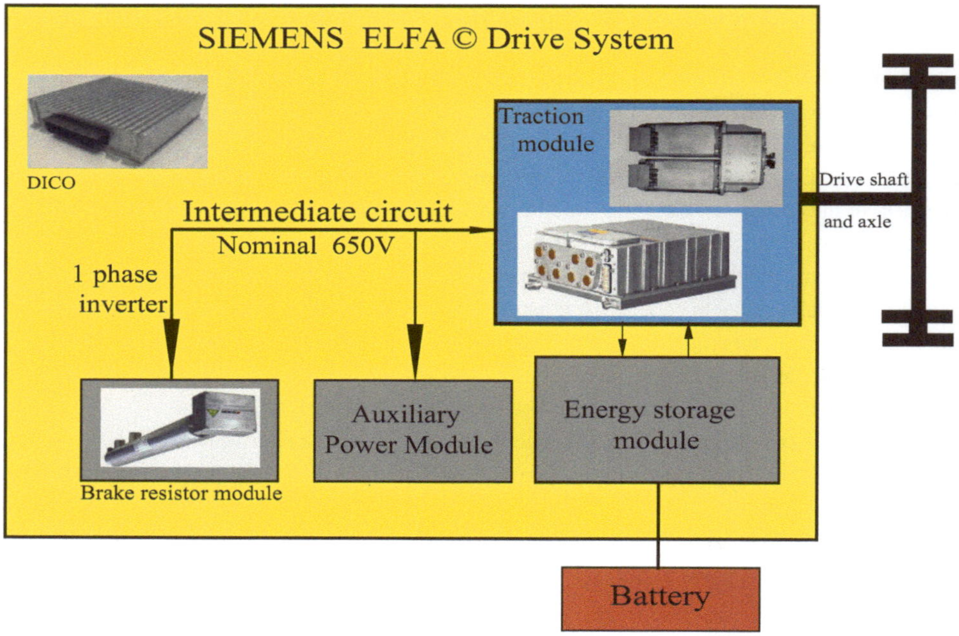

Fig. 23-8: Block diagram ELFA-drive system

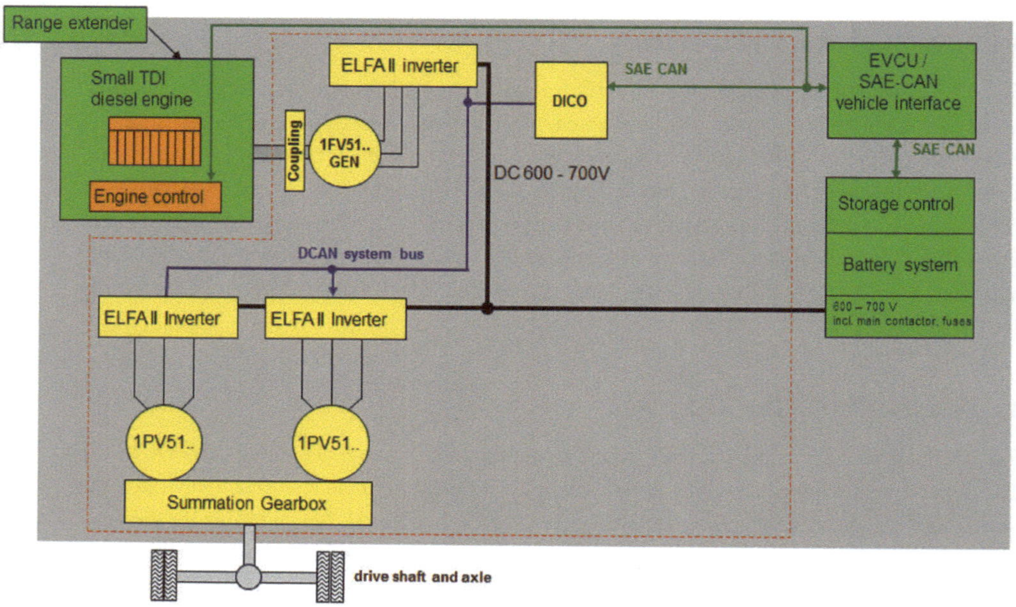

Fig. 23-9 Photo Siemens: Block diagram of the ELFA © drive system (Van Hool city bus)

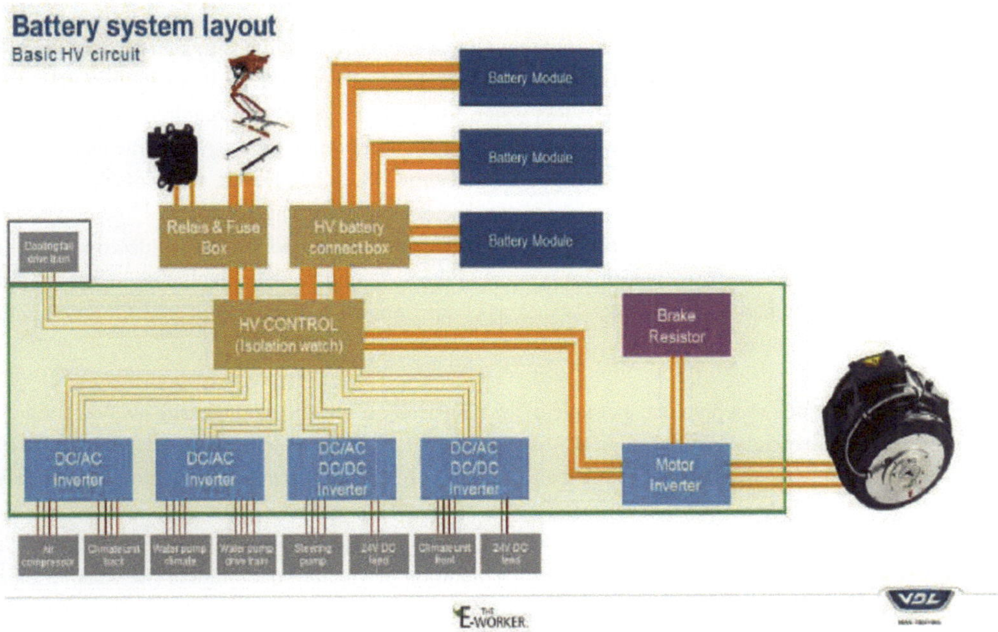

Fig. 23-10: Photo VDL. Energy efficient electrically driven city bus (using Siemens ELFA © technology)

Photo VDL: Roof mounted pantograph charging battery pack electrical bus

4. ELECTRIC BOATS

Advantages of electric boats are there almost silent drives and positive environmental effects. Due to the limited amount of energy that can be stored in batteries it is very important to choose a boat with the right shape. The less water displaced the less power required for propulsion.
In connection with the power the hull shape is an important factor. Above the hull speed waves are created and this means a large loss of energy. Water line length, shape of the boat, shape of the bow, shape of the stern are all important factors that affect the hull speed.
In addition to the hull speed the water displacement is another factor that determines the motor power. The water displacement is equal to the weight of the boat, including motor, batteries, baggage and passengers. Materials such as composites are more advantageous for boat building than wood or steel.

Let's look at a practical example of an electrically driven boat. Dreaming away we travel to the province Salzburg in beautiful Austria. From the European capital Brussels we take the electric train in the direction of Munich and Salzburg. In Salzburg we hire an electric car and drive 50 km northwards to the village of Mattsee. In our hotel we rent an electric bicycle to explore the region. On one of our tours we discover in the Seestrasse 23 a modest boatyard "Steiner Nautic".
We make acquaintance with the fifth generation of Steiner-boat builders (Hermann Steiner V!). We have found a good address because Steiner-Nautic builds electric boats. Interesting because in Austria, from the viewpoint of environmental pollution many lakes are prohibited for combustion engines.

The Steiner company built amongst others the "Seenland" (photo p.23-19) named after the region Mattsee. The boat was launched in 1998. The boat is intended for tourist cruises and can accommodate 50 passengers. The Seenland is 12m long.

The boat, for reliability reasons, has two independent drive systems. Both motors are three-phase asynchronous motors of 20 kW suitable for 100V-50 Hz. Each system is comprised of 12 lead sulphide batteries of 230 Ah in series to produce 144V. A frequency converter converts this to the required three-phase voltage of 100 V-50Hz. The converter power is 15kW. A larger power than 15 kW is not required because this would not result in the boat going faster since its hull speed is reached at 15 kW. With more power waves would be created which would reduce the speed. When both motors are operating at full power there is enough energy in the batteries for about one hour. This is never required during normal cruising where a total of 8kW is used.
In addition every system has its own charging generator of 6kW.

Special feature on the Seenland is that this is the only boat in Austria that travels between two lakes, the Mattsee and the Obertrumer Sea. Both lakes are prohibited for boats because of a 60 cm deep channel and above the channel there is a bridge on an important roadway. The hull has to be extremely flat because of the 60cm channel. In addition the roof has to be as low as possible to make it under the bridge. This is solved by a hydraulic roof that can be lowered during the passage between lakes. To lower the boat at periods of high water (from major rainfall) four water tanks are built into the floor. Full water tanks lower the boat by 5cm.

Photo Steiner-Nautic: The electrically propelled boat "Seenland" by his launch

In addition to the Seenland and their electric boats the Steiner company has built an electric speedboat. This boat can exceed 70 km/h, has 450 kg of Li-ion batteries on board and a 120kW motor. The speed boat can tow a water skier for one and a half hour.

Photo Steiner-Nautic: Speed boat for water skiing

More on the website www.steiner-nautic.at.

5. ELECTRIC BICYCLES

Electric two-wheelers are an important part of the global mobility program. They require minimal space, produce no pollution and no noise.

5.1 Pedelec and E-bike

There are two types of electric bicycles, the "Pedelec" and the "E-bike".
Pedelec (Pedal Electric Cycle) is a bicycle whereby you only can move forward by pedalling. You are supported by a small electric motor. The torque produced by the motor depends on the force exerted by the cyclist. The first Pedelec was presented by the well known Japanese motor brand Yamaha in 1998. The Pedelec will possibly in time replace the standard bicycle but this will take years. On the other hand people are attached to their bicycles and the new Pedelec is a lot more expensive than a bicycle.

The E-bike is an electric bicycle which can move forward without the cyclist pedalling.
The speed is determined by the cyclist with the aid of a button or an "accelerator". In contrast to the Pedelec in the Netherlands and Belgium the E-bike is not considered a bicycle and wearing a helmet is compulsory. The E-bike has to be inspected and approved before being allowed on public roads.

5.2 Motor and battery

Directly connected to a wheel of the bicycle is a hub-motor. This coupling is clearly important for a high efficiency. There are two types of hub-motor, those with and those without gears.
The motors without gears are to be compared to electric motors whereby the shaft is fixed and the housing otates. In the motor with gears a smaller motor is used with a high speed. The speed is reduced, mostly with a planetary gear-system. This type results in smaller dimensions of the whole but with a torque that is larger than the other type.
In many countries the maximum power is limited to 250 W and the pedal support operates up to 25 km/h.

Four types of batteries can be used. The first is the sealed lead acid batteries. In addition there is the classic Nickel-cadmiums (NiCd) which are more expensive but can be charged quicker. Then the Nickel-Metal hydride (Ni-Mh) which can store more electrical energy and are a bit lighter in weight. Finally the lithium-ion (Li-ion) which for a smaller volume has even more capacity but is also more expensive than the other types.

Consider also that the motor of 250W runs on 24 V and draws 10A and you quickly realise that large capacity batteries are required for an electric bicycle.

ENGLISH	DUTCH	GERMAN	SPANISH
abscissa	abscis	Abszisse	abscisa
AC controller	wisselstroominsteller	Wechselstromsteller	Controlador AC
acceleration torque	versnellingskoppel	Beschleunigungsmoment	torque de aceleración
AC-controller	AC controller	AC-Kontroller	Controlador AC
active component	actieve componente	aktive Komponente	componente activo
active power	aktief vermogen	Wirkleistung	potencia activa
aerodynamic	aërodynamisch	aerodynamisch	aerodinámico
air gap	luchtspleet	Luftspalte	entrehierro
algorithm	algoritme	Algorithmus	algoritmo
alternating current (AC)	wisselstroom	Wechselstrom	corriente alterna
ambient temperature	omgevingstemperatuur	Umgebungstemperatur	temperatura ambiente
amplifier	versterker	Verstärker	amplificador
analogons	analogons	analoges Modell	modelo analógico
analogue computer	analoge computer	Analogrechner	ordenador analógico
analogue	analoog	analog	análogo
angular displacement sensor	hoekstandopnemer	Winkelencoder	sensor de posición angular
angular position	hoekpositie	Winkelposition	posición angular
anode	anode	Anode	ánodo
anodizing	geanodiseerd	eloxiert	anodizado
antistatic packaging	antistatische verpakking	antistatische Verpackung	Empaque antiestatico
apparent power	schijnbaar vermogen	Scheinleistung	potencia aparente
appliance motor	huishoudelijke motor	Haushaltmotor	motor doméstico
applications	toepassingen	Anwendungen	aplicaciones
arc furnace	boogoven	Lichtbogenofen	horno de arco
armature reaction	ankerreactie	Ankerrückwirkung	reacción de armadura
armature	anker	Anker	armadura
asynchronous motor	asynchrone motor	Asynchronmaschine	motor asincrono
auxiliary pole	hulppool	Hilfspol	polo auxiliar
auxiliary thyristor	doofthyristor	Löschthyristor	tiristor auxiliar
auxiliary winding	hulpwikkeling	Hilfswicklung	arrollamiento auxiliar
avalanche diode	zenerdiode	Zenerdiode	diodo zener
average resistivity	soortelijke weerstand	spezifischer Widerstand	resistividad media
average value	gemiddelde waarde	(arithmetischer) Mittelwert	valor medio
back-to-back (SCRs)	antiparallel (SCRs)	antiparallele Ventilen	SCRs en antiparalelos
ballast	ballast	Vorschaltgerät	balast
band gap (energy gap)	bandafstand	Energielücke (Bandabstand)	banda prohibida
bandwidth	bandbreedte	Bandbreite	anchura de banda
bank of accumulators	accubatterij	Akkumulatorenbatterie	bateria de acumuladores
barrier layer	sperlaag	Sperrschicht	capa de barrera
base voltage	basisspanning	Basisspannung	tensión de base
base	basis	Basis	base
battery charger	batterijlader	Batterieladegerät	cargador de batería
biasing	voormagnetisatie	Vormagnetisierung	polarización magnética
bidirectional	bidirectioneel	bidirektional	bidireccional
bipolar transistor	bipolaire junctietransistor	bipolarer Transistor	transistor bipolar
bistable	bistabiel	bistabil	biestable
block diagram	blokschema	Blockschaltbild	diagrama duncional
block orientated	blokgeoriënteerd	blockorientiert	bloque orientación

blocking state SCR	sperren SCR	löschen Thyristor	estado de bloque directo
blocking voltage	blokkeerspanning	Blockierspannung	tensión de bloquear
blocking	blokkering	Sperrung	bloqueo
blow up	opblazen	instabilisieren	desbordamiento
brake (chopper)	rem (chopper)	Bremse (Zehrhacker)	freno (recortador)
breakaway torque	lostrekkoppel	Losbrechdrehmoment	par de despegue
break-over voltage	doorslagspanning	Kippspannung	tensión de voltear
breakover	doorslag	Durchbruch	(tensión de) disparo
break-over	kippen	kippen	voltear
bridge	brug	Brücke	puente
brushless motor	borstelloze motor	Bürstenlose Motor	motor sin escobillas
burst firing	periodesturing	Schwingungspaketsteuerung	Control de periodo
cadmiumsulfide	cadmiumsulfide	Kadmiumsulfid	sulfuro de cadmio
capacitor (motor)	condensator (motor)	Kondensator (motor)	(motor con)condensador
capacity	capaciteit	Kapazität	capacidad
case (package)	behuizing	Behausung	caja
cathode	kathode	Kathode	cátodo
cell density	celdichtheid	Zelledichte	Densidad de celula
centrifugal switch	centrifugaal schakelaar	zentrifugaler Schalter	contacto centrifugada
channel	kanaal	Kanal	canal
characteristics	karakteristieken	Kennlinien	características
charge (to)	opladen	aufladen	carga
charging current	laadstroom	Ladestrom	corriente de carga
choke	smoorspoel	Glättungsdrossel	bobina de choque
chopper	hakker	Zehrhacker /Gleichstromsteller	recortador
circuit	kring	Kreis	circuito
circuit	schakeling	Stromkreis	circuito
circular frequency	pulsatie (hoekfrequentie)	Kreisfrequenz	pulsacion
circulating current-free	kringstroomvrij	kreisstromfrei	sin corriente circular
clamp diode	klemdiode	Klemmdiode	diodo de libre reccorido
code disc	codeerschijf	Kodierungscheibe	disco de codigo
coercitivy	coërcitieve kracht	Koerzitivkraft	coercitividad
coil	spoel	Drossel / Spule	bobina de inductancia
collector	collector	Kollektor	colector
colour code	kleurencode	Farbekode	código de colores
commissioning rules	instelregels	Optimierungsvorschriften	reglas de ajuste
commutation	commutatie	Kommutierung	conmutación
commutatormachine	commutator machine	Kommutatormaschine	máquina de comutación
comparator	comparator	Vergleicher (Komparator)	comparador
compatibility	uitwisselbaarheid	Kompatibilität	compatibilidad
compensation coil	compensatiewikkeling	Kompensationswicklung	arrollamiento de compensación
compliance	compliantie	Nadelnachgiebigkeit	cumplimiento
compressor	compressor	Kompressor	compresor
conducting pattern	geleidingspatroon	Leitungsvorlage	modelo de conducción
conduction angle	geleidingshoek (vloeihoek)	Flusswinkel	ángulo de conducción
conductor	geleider	Leiter	conductor
connection wire	aansluitdraad	Anschlussdraht	conductor del conexión
connection	aansluiting	Anschluss	conexión
connection	schakeling	Schaltung	conexión

constant current (source)	constante (stroombron)	Konstantstromquelle	fuente de alimantación de corriente constante
construction (structure)	opbouw	Aufbau	estructura
consumer	verbruiker	Verbraucher	consumidor
contactor	contactor	Magnetschalter	contactor
continuous (duty)	continu (bedrijf)	Dauer (betrieb)	servicio continuo
continuous current	continue stroom	nichtlückender Betrieb	corriente permanente
continuous service	leemtevrij bedrijf	nichtlückender Betrieb	corriente continua
control (thyristor)	controle (thyristor)	Steuersätzer (Stromrichter)	control de tiristor
control characteristic	stuurkarakteristiek	Steuerkennlinie	caracteristic de control
control circuit	stuurcircuit	Steuerungskreis	circuito de excitación
control circuitry	triggerketen	Steuersatz	circuito de control
control IC	stuur-IC	integrierte Zündimpulsstufe	IC de control
control loop	regelkring	Regelkreis	bucle de control
control signal	controlesignaal	Steuersignal	control señal
control	besturing	Steuerung	control
controller	regelaar	Regler	regulador
convergence	convergentie	Konvergenz	convergencia
converter	convertor	Stromrichter	convertidor
cooking plate	kookplaat	Kochplatte	hornillo
cooling fin	koelvin	Kühlrippe	aleta de refrigeración
cooling	koeling	Kühlung	refrigeración
coordinate system	assenkruis	Systemkoordinatensystem	coordenadas
copper losses	koperverliezen	Kupferverluste	pérdidas de cobre
core	kern	Kern	núcleo
counter emf	tegen-emk	gegen emk	fuerza contra electromotriz
criterion (equal surface)	criterium (gelijke opp.)	Kriterium	criterio
current	stroom	Strom	corriente
current (rate of rise)	stroom(steilheid)	Strom (Anstiegsteilheit)	corriente (pendiente)
current circuit	stroomkring	Stromkreis	circuito de corriente
current control	stroomregeling	Stromregulierung	control de corriente
current density	stroomdichtheid	Stromdichte	densidad de corriente
current direction	stroomrichting	Stromrichtung	sentido de corriente
current flow	stroomsterkte	Stromstärke	ensidad de corriente
current form	stroomvorm	Ausbildung der Ströme	forma de la corriente
current gain factor	stroomversterkingsfactor	Stromverstärkung	factor del amplificación de la corriente
current limit	stroombegrenzing	Stromgrenzwert	límite de corriente
current propagation	stroomverloop	Stromverlauf	propagación de corriente
current source (controlled)	gestuurde stroombron	gesteuerte Stromquelle	fuente de corriente controlado
current source inverter	stroombroninvertor	Umrichter mit Zwischenkreis	invertor de fuente de corriente
current surge	stroomstoot	Stromstoss	golpe de corriente
curve	curve	Kennlinie	curva
cut off	gesperd	Sperrzustand	estado de bloqueo
cycloconvertor	cycloconverter	Direktumrichter	convertidor ciclón
dam (model)	dam-mode	Damm-Modell	dama (modelo)
damping	demping	Dämpfung	amortiguamiento
data (sheet)	gegevens (-blad)	Daten (-blatt)	datos (- página)

DC motor	gelijkstroommotor	Gleichstrommotor	motor corriente continua
DC voltage link	spanningstussenkring	Spannungszwischenkreis	circuito intermedio de tensión
DC-controller	mutator	Stromrichter	mutador
DC-link converter	tussenkringomzetter	Zwischenkreisumrichter	regulador de frecuencia con circuito intermedio
dead time	dode tijd	Totzeit	tiempo muerto
delay angle	ontsteekhoek	Zündwinkel	el ángulo de encendido
delay dissipation time	ontsteekuitbreidingstijd	Zündausbreitungszeit	(delay dissipation time)
delay of firing	ontsteekvertraging	Zündverzögerung	duración de precebado
delay time	looptijd	Laufzeit	retardo
delay time	vertragingstijd	Einschaltverzugszeit	tiempo de retardo
delay	vertraging	Verzögerung	retardo
depletion region	verarmingslaag	Übergangsgebiet	región de transición
derivative	afgeleide	Differentialquotient	derivada
detection	detectie	Demodulation	detección
detent torque	restkoppel	Rastmoment	par de detencion
diac	diac	Zweiwegschaltdiode	diac
diagram	diagram	Diagramm	diagrama
diagram	schema	Schaltbild	diagrama
dielectric	diëlectricum	Dielektrikum	dieléctrico
differential amplifier	verschilversterker	Differenzverstärker	amplificador diferencial
differential equation	differentiaalvergelijking	Differentialgleichung	ecuación diferencial
digital computer	digitale computer	digitale Rechenanlage	digital computador
digital	digitaal	digital	digital
direct current (DC)	gelijkstroom	Gleichstrom	corriente continua
discharge current	ontladingsstroom	Entladestrom	corriente de descarga
discharge lamp	ontladingslamp	Entladunglampe	lámpara de descarga
discontinuous current	stroomleemte	Lückgrenze	corriente discontinua
displacement factor	verschuivingsfactor	Phasenverschiebungswinkel	factor de desviación de fase
displacement rule	verschuivingsregel	Verschiebungsvorschrift	(displacement rule)
distortion factor	distorsiefactor	Verzerrungsfaktor	distorsión factor
distortion	vervorming	Verzerrung	distorsión
disturbance	storing	Störung	perturbación
doping	doteren	dotieren	dopado
double pulse	dubbelpuls	Doppelimpuls	impulso doble
double star connection	dubbele sterschakeling	Doppelsternschaltung	conexión en doble estrella
drain (current)	drain (stroom)	Drain (strom)	(corriente de) drenaje
drain	afvoerelektrode	Abfluss (drain)	drenador
drift velocity	driftsnelheid	Driftgeschwindigkeit	velocidad de deriva
drive system	aandrijfsystemen	Antriebe	sistema de propulsión
drive system	aandrijvingen	Antriebe	accionamiento
driver stage	stuurtrap	Treiberstufe	excitador
driving the base	uitsturen basis	steuern Basis	control de base
driving voltage	drijvende spanning	treibende Spannung	(driving voltage)
duty cycle	werkverhouding	Arbeitszyklus	ciclo de trabajo
duty	bedrijf	Betrieb	servicio
dynamic behavior	dynamisch (gedrag)	dynamisches (Verhalten)	conducta dinámica

dynamo	dynamo	Gleichstromgenerator	dinamo
eddy (currents)	wervelstroom (verliezen)	Wirbelstrom (verluste)	corriente Foucault (pérdidas)
efficiency	rendement	Wirkungsgrad	rendimiento
elasticity	elasticiteit	Elastizität	elasticidad
electric charge	lading	Ladung	carga
electric equivalent	elektrisch equivalent	Ersatzschaltbild	eléctrica equivalente
electric field	elektrisch veld	elektrisches Feld	campo eléctrico
electric machine	elektrische machine	elektrische Maschine	máquina eléctrica
electric motor	motor	Elektromotor	motor électro
electricity	elektriciteitsleer	Elektrizität	electricidad
electrochemical	elektro-chemisch	elektrochemisch	electroquímico
electrolysis	elektrolyse	Elektrolyse	electrólisis
electromagnetic	elektromagnetische	elektromagnetisch	elecromagnética
electromechanical	elektromechanisch	elektromechanisch	electromecánico
electronic motor control	motorcontrole	elektronische Motorkontrolle	control electrónica de motores
electrostatic (shield)	elektrostatisch (scherm)	elektrostatische Abschirmung	(pantalla) electrostático
elektrode	electrode	Elektrode	electrodo
elementary charge (electron)	elementaire lading	Elementarladung	carga elementaria
emergency supply	noodvoeding	Notspeisung	grid de reserva
emf (counter-)	emk (tegenwerkende-)	emk (Gegen-)	f.e.m. (contra)
emitter	emitter	Emitter	emisor
enamelled winding wire	wikkeldraad (geëmailleerd)	Lackdraht	hilo esmaltado para bobina
encoder	encoder	Kodierer	codificador
encremental	incrementeel	inkremental	incremental
energy (renewable)	energie (hernieuwbaar)	Energie (erneute)	energía (renovada)
energy converter	energie-omzetter	Umrichter elektrischer Energie	energía inversor
enhancement (FET)	verrijkingstype (FET)	Anreicherungstyp (FET)	(FET) incrementado
enhancement type	verrijkingstype	Anreicherungstyp	tipo de enriquecimiento
equal surface area	gelijke oppervlakte	gleichbleibende Oberfläche	superficie igual
equation	uitdrukking (vergelijking)	Beziehung (Relation/Gleichung)	ecuación
equivalent (circuit)	vervangings(schema)	Ersatz(schaltbild)	(diagrama) equivalente
equivalent resistance	vervangingsweerstand	Ersatzwiderstand	resistencia equivalente
excerpts from databook	uittreksels uit databoek	Auszug aus Datenbuch	extractos de datos
excit. from coupled exciter	afzonderlijk bekrachtigd	Eigenerregung	excitación separada
excitation	excitatie	Erregung	excitación
excite	bekrachtigen	erregen	excitación
external P-layer	uitwendige P-laag	äussere P-Schicht	(external P-layer)
extinction angle	doofhoek	Löschwinkel	ángulo de extinción
fall time	daaltijd (afvaltijd)	Abfallzeit	tiempo de caída
fast recovery diode	snelle hersteldiode	Schnelle Sperrdiode	(fast recovery diode)
fault current	foutstroom	Teilfehlerstrom	corriente de falta
feedback	terugkoppeling	Rückkopplung	realimentación
feeder clamp	voedingsklem	Stromklemme	grifa de alimentación
feeding cable	voedingsleiding	Speiseleitung	arteria alimentadora
ferrite core	ferrietkern	Ferritkern	núcleo de ferrita
ferromagnetic	ferromagnetisch	ferromagnetisch	ferromagnético
field control	veldregeling	Feldregelung	control de campo
field coordinates	veldcoördinaten	Feldkoordinate	coordenadas de campo

field intensity	veldsterkte	Feldstärke	intensidad de campo
field line	veldlijn	Feldlinie	linea de campo
field weakening	veldverzwakking	Feldschwächung	disminución de campo
filter	filter	Filter	filtro
final value theorem	eindwaardetheorema	Endwerttheorem	teorema del valor final
finit	eindig	endlich	finito
firing	ontsteken	zünden	encendido
first order system	eerste-orde systeem	Verzögerungsgliede 1.Ordnung	sistema del primer orden
flashing light	flikkerlicht	Funkelfeuer	luz centelleante
floating net	zwevend net	erdfreies Netz	red flotante
fluorescent lamp	fluorescentielamp	Fluoreszenzlampe	lámpara fluorescente
flux	flux	Fluss	flujo
forced commutation	gedwongen commutatie	gezwungene Kommutierung	conmutación forzado
form factor	vormfactor	Formfaktor	factor de forma
forward (current)	doorlaatrichting (stroom)	Durchlass(strom)	dirección de paso (corriente)
forward characteristic	doorlaatkarakteristiek	Durchlasskennlinie	característica directa
forward	voorwaarts	Durchlass(Vorwärtsrichtung)	dirección de paso
Fourier components	Fouriercomponenten	Fourier-Komponenten	Fourier componentes
Fourier series	Fourierreeks	Fourier-Reihe	Fourier serie
Fourier series	reeks van Fourier	Fourier-Reihe	serie de Fourier
freewheel diode	vrijloopdiode	Freilaufdiode (Löschdiode)	diodo de libre reccorido
frequency (converter)	frequentie(regelaar)	Frequenz(umrichter)	(regulador de) frecuencia
frequency spectrum	frequentiespectrum	Frequenzspektrum	espectro de frecuencias
friction (coefficient)	wrijving (coëfficiënt)	Friktion (koeffizient)	fricción (coeficiente)
full controlled	volgestuurd	vollgesteuert	(full controlled)
full wave (rectifier)	dubbelzijdig (gelijkrichter)	Zweiweg (-gleichrichter)	(rectificador de) onda completa
function	functie	Funktion	función
fundamental wave	grondgolf	Grundschwingung	onda fundamental
fuse	smeltveiligheid	Sicherung	fisible
gallium arsenide	galliumarsenide	Galliumarsenid	arseniuro de galio
gate (terminal)	poort(aansluiting)	Gate(-Anschluss)	(borne de) puerta
gate current	poortstroom	Gatestrom	corriente de puerta
gate power	poortvermogen	Gateleistung	potencia de puerta
gate turn off thyristor	uitschakelbare thyristor	Abschaltthyristor (GTO)	tiristor bloqueable
gate voltage	poortspanning	Steuerspannung	tensión de puerta
gate	poort	Gate / Gatter	puerta
generator mode	generatorbedrijf	Generatorbetrieb	funcionando como generador
gondola	gondel	Gondel	góndola
grid / mains / supply	net	Netz	red
ground wave	grondgolf	Grundschwingung	onda fundamental
half wave	halve netperiode	Netzhalbwelle	red medio ciclo
half-wave, three-phase	enkelzijdig, driefasen	Einweg, Dreiphasen	media onda, trifásica
Hall-effect	Hall effect	Hall-Effekt	efecto Hall
halogenlamp	halogeen lamp	Jodlamp	lámpara halógena
harmonics (order)	harmonischen (orde)	harmonische (Ordnungszahl)	armonicas (orden)
heat dissipation	warmte-afvoer	Wärmeableitung	disipación de calor

heatsink	koellichaam/koelplaat	Kühlkörper	disipador
heavy saturation	harde verzadiging	harte Sättigung	saturación fuerte
high voltage	hoogspanning	Hochspannung	alta tensión
hoisting apparatus	hefwerktuig	Hebemaschine	utensilio de leventados
holding current	houdstroom	Haltestrom	corriente de cierre
holding torque	houdkoppel	Haltemoment	torque de mantenimiento
holes and electrons	gaten en elektronen	Löcher und Elektronen	huecos y electrones
hot spots	warme punten	heisse Punkte	(hot spots)
human eye curve	oogbalcurve	spektrale Hellempfindlichkeit	sensibilidad espectro del ojo
hydraulical (labo)	hydraulisch laboratorium	hydraulisches Laboratorium	(labo) hydraulicamente
hyperbole	hyperbool	Hyperbel	hipérbola
hysteresis loss	hysteresisverliezen	Hysteresisverlust	pérdidas de histéresis
I^2-t integral	lastintegraal	Lastintegral	I^2-t integral
impedance	impedantie	Impedanz	impedancia
Impulse	impuls	Impuls	impulso
impurity	verontreiniging	Verunreinigung	impureza
induced emf	inductiespanning	Induktionsspannung	f.e.m. de inducción
inductance	inductantie	Selbstinduktion	inductancia
induction motor	inductiemotor	Induktionsmotor	motor de inducción
inductive heating	inductieve verwarming	Induktionsheizung	calentamiento inductivo
inductive load	inductieve belasting	induktiver Last	carga inductiva
industry standard	industriestandaard	Industriestandard	estándar industrial
inertia force	inertiekracht	Trägheitskraft	fuerza inercia
inertia resistance	traagheidskracht	Trägheitswiderstand	Fuerza de inercia
infra-red	infrarood	ultrarot	infrarrojo
initial value theorem	beginwaardetheorema	Anfangwerttheorem	teorema del valor inicial
input	ingang	Eingang	entrada
instant	tijdstip	Zeitpunkt	instante
instantaneous value	momentele waarde	Momentanwert	valor instantáneo
instantaneous value	ogenblikkelijke waarde	Momentanwert	inmediatamente
instrument transformer	meettransformator	Messwandler	transformador de medida
insulator	isolator	Isolator	aislador
integrated circuit	geïntegreerde schakeling	integrierte Schaltung	microcircuito
integration	integreren	Integration	integrar
interference suppression	ontstoring	Entstörung	supresión de interferencias
interference	interferentie	Interferenz	interferencia
intersection	snijpunt	Schnittpunkt	punto de interseccion
inversion layer	inversielaag	Inversionsschicht	capa de inversión
inverter	invertor	Wechselrichter	inversor
inverter	wisselrichter	Umkehrstromrichter	ondulador
iron losses	ijzerverliezen	Eisenverluste	pérdidas en el hierro
iteration	iteratie	Iteration	iteración
Joule losses	Joule-verliezen	Joule-Verluste	perdidas de Joule
Joule-effect	Joule effect	Joule-Effekt	efecto de Joule
junction	junctie	Sperrschicht (PN-Übergänge)	Unión PN
junction diode	junctiediode	Flächendiode	diodo de Union PN
kinetic (energy)	kinetische (energie)	kinetische (Energie)	cinétic (energia)

knee voltage	knik of knie	Kniespannung	tensión de rodillo
laminated core	blikpakket	Blechpaket	nucleo de chapas
Laplacetransform	Laplace-transformatie	Laplace-Transformation	transformada de Laplace
latching current	vergrendelstroom	Einraststrom	corriente de retención
leading screw transmission	schroefspiloverbrenging	Leitspindelübertragung	transmissión por pivote de tornillo
leadsulphite	loodsulfide	Bleisulfid	sulfito de plomo
leakage current	lekstroom	Leckstrom	corriente de fuga
leakage inductance	spreidingsinductantie	Streuinduktivität	inductancia de fuga
light (guide)	licht (geleider)	Licht (führung)	(guía de) luz
light pulse	lichtpuls	Lichtimpuls	impulso de luz
lightbeam	lichtbundel	Strahlenbündel	haz luminoso
lighting	verlichting	Beleuchtung	iluminación
limitation resistance	begrenzingsweerstand	Begrenzungswiderstand	Resistencia de limitación
line(current)	lijn(stroom)	Leiter(strom)	(corriente de) linea
linear	lineair	linear	lineal
load impedance	belastingsimpedantie	Lastimpedanz	impedancia de carga
load line	belastingslijn	Arbeitskennlinie	linea de carga
load voltage	klemspanning	Klemmenspannung	tensión en carga
load	belasting	Last	carga
low voltage	laagspanning	Niederspannung	baja tensión
luminous intensity	lichtsterkte	Lichtstärke	intensidad de luz
machine	machine	Maschine	máquina
magnetic (flux)	magnetische (flux)	Magnet (fluss)	(flujo) magnético
magnetic charging	magnetisch opladen	magnetisches Aufladen	cargar magnetico amente
magnetic inductance	magnetiseringsinductantie	magnetische Selbstinduktion	inductancia de magnetización
magnetomotive force	magnetomotorische kracht	magnetomotorische Kraft	Fuerza magnetomotriz
main pole	hoofdpool	Hauptpol	polo principal
majority- carriers	meerderheidsladingsdragers	Majoritätsträger	portadores mayoritarios
mass inertia	massatraagheidsmoment	Massenträgheitsmoment	momento de inercia de masa
matrix	matrix	Matrix	matriz
maximum current	maximale stroom	Spitzenstrom	corriente máxima
maximum reverse current	maximale sperstroom	überhöhter Sperrstrom	corriente inversa máxima
maximum value	amplitude	Scheitelwert	amplitud
measurement equipment	meetapparatuur	Messgerät	equipo de medida
measurement equipment	meetopstelling	Messaufbau	Equipo de medición
mechatronics	mechatronica	Mechatronik	mecatrónica
melting energy	smeltenergie	Schmelzeenergie	energia de fundición
memory scope	geheugenscope	Speicheroszilloskop	memoria osciloscopio
metallization	metallisatie	Metallizierung	metalización
microcontroller	microcontroller	Mikrokontroller	microcontrolador
microprocessor	microprocessor	Mikroprozessor	microprocesador
minority carriers	minderheidsladingsdragers	Minoritätsträger	portadores minoritarios
mixing	mengen	mischen	mezcla
modulation depth	modulatiediepte	Modulationsgrad	modulación medio
moment of torque	werkkoppel	Drehmoment	torque de trabajo
monopolar	enkelpolig	einpolig	unipolar

Mos field effect transistor	Mos-veldeffecttransistor	Mos-Feldeffekttransistor	transistor de efecto de campo Mos
multimeter	multimeter	Vielfachmessgerät	Multimetro
multi-pulse circuits	meerpulsige schakelingen	mehrpulsige Schaltungen	Circuito multi impulsos
natural commutation	natuurlijke commutatie	natürliche Kommutierung	conmutación natural
negative feedback	tegenkoppeling	Gegenkopplung	enlace contrario
network transformer	transformator (Net)	Netztransformator	transformador de red
network	netwerk	Netzwerk	red
neutral point (transformer)	sterpunt	Sternpunkt	punto neutro
N-material	N-materiaal	N-leitendes Material	N-material
noise (electrical)	ruis (elektrisch)	Rauschen (elektrisches)	ruido (eléctrico)
no load (current)	nullast(stroom)	Leerlauf(strom)	(corriente) en vacio
no load speed	leegloop toerental	Leerlaufdrehzahl	velocidad de marcha en vacio
nominal current	nominale stroom	Nennstrom	corriente nominal
nomograph	nomogram	Nomogramm	nomograma
notes	notaties	Notationen	apuntes
number indicator	cijferindicator	Zifferindikator	indicado de número
numeric example	cijfervoorbeeld	Rechenbeispiel	ejemplo cifrado
numerical analysis	numerieke technieken	numerische Mathematik	analísis numérico
ohmic load	ohms belast	mit Ohmischer Last	carga óhmica
ON	AAN	EIN-Zustand	(ON)
on-characteristic	aan-karakteristiek	Durchlasskennlinie	(on-characteristic)
opamp	opamp	Operationsverstärker	amplificador operacional
open circuit test	nullastproef	Leerlaufversuch	ensayo en circuito abierto
open loop control	sturing	Steuerung	control de buclo abierto
operation	werking	Arbeitsweise	operación
optimum amount	bedragoptimum	Betragsoptimum	óptimo de comportamiento
optocoupler	optokoppeling	Optoelektronisches	(optocoupler)
ordinate	ordinaat	Ordinate	ordenada
oscillation	oscillatie	Schwingung	oscilación
OUT	UIT	AUS-Zustand	(OUT)
output line	uitgangslijn	Ausgangsleitung	linea de salida
output stage	eindtrap	Endstufe	etapa de salida
output	uitgang	Ausgang	salida
oven (furnace)	oven	Ofen	horno
overlap angle	overlappingshoek	Überlappungswinkel	ángulo de solape
overshoot	doorschot	Überschwingweite	sobre exceso
overvoltage	overspanning	Überspannung	sobretensión
partial fraction	partieelbreuk	partieller Bruch	(partial fraction)
peak value	piekwaarde	Spitzenwert	valor de cresta
peak value	topwaarde	Scheitelwert	valor de cresta
period	periode	Periode	periodo
permanent magnet	permanente magneet	Dauermagnet	imán permanente
permeability	permeabiliteit	Permeabilität	permeabilidad
phase conductor	fasegeleider	Phasenleiter	conductor de fase
phase control	fase-aansnijding (controle)	Phasenanschnitt(steuerung)	control de fase
phase displacement	faseverschuiving	Phasenverschiebung	desfase
phase drop off	fase-afsnijding	Abschnittsteuerung	(phase drop off)

phase	fase	Phase	fase
phase-sequence	fasevolgorde	Phasenfolge	orden de fases
photomask	fotomasker	Photomaske	fotomáscara
photosensitive	fotogevoelig	photoempfindlich	fotosensible
photovoltaic cell	zonnecel	Photovoltaikzelle	célula fotovoltaica
photovoltaic	fotovoltaïsch	photovoltaik	fotovoltaico
pilot signal lamp	signaallamp	Signallampe	lámpara de señal
PNPN diode	vierlagendiode	Vierschichtdiode	diodo PNPN
polarized	gepolariseerd	polarisiert	polarizado
pole pairs (number of)	poolparen (aantal)	Polpaare (Zahl)	pares de polos (número de)
pole	pool	Pol	polo
position (measurement)	positie(meting)	Position(-smessung)	(medidade la) posición
potential energy	potentiële energie	potentielle Energie	energia potencial
potentiometer	potentiometer	Potentiometer	potenciómetro
power (dirty)	netvervuiling	Fehlerstrom	contaminación del red ?
power circuit	vermogenkring	Hauptkreis	circuito de potencia
power factor	arbeidsfactor	Leistungsfaktor	factor de potencia
power losses	vermogenverlies	Leistungsverluste	perdidas de potencia
power mosfet	vermogenmosfet	Leistungsmosfet	MOSFET de potencia
power supply	stroomvoorziening	Stromversorgung	sistema de alimentación
power supply	voeding	Netzteil	alimentación
primary	primaire zijde	Primärseite	circuito primario
principle configuration	principe-opstelling	Prinzipschaltbild	conexión de principio
pulley	riemschijf	Riemscheibe	polea
pulse train	impulstrein	Pulszug / Impulsfolge	tren de pulsos
pulse-shaping circuit	pulsvormer	Impulsformer	modelador de impulsos
pump	pomp	Pumpe	bomba
push-button	duwknop	Druckknopf	botón pulsador
PWM wave	PBM-golf	PBM-Wellenform	onda con modulación de anchura por impulso
quadrant	kwadrant	Quadrant	cuadrante
quelle (Source)	source	Quelle	fuente
radian	radiaal	radial	tradián
radiator	radiator	Radiator(Heizkörper)	radiador
radio frequency	Hoogfrequent	Hochfrequenz	alta frecuencia
radio receiver	radio-ontvanger	Rundfunkgerät	receptor de radio
rate of rise	steilheid	Steilheit	pendiente
ratings	grenswaarden	Grenzdaten	especificaciones máximas
reactive component	reactieve componente	Blindkomponente	componente reactivo
reactive moment	tegenwerkend koppel	Widerstandsmoment	torque de carga
reactive power	blind vermogen (wattloos)	Blindleistung	energía reactiva
receiver	ontvanger	Empfänger	receptor
recovery time	herstel (tijd)	Erholungszeit	tiempo de restitución
rectangular wave	rechthoeksgolf	Rechteckschwingung	onda rectangular
rectifier (B_2) (B_6)	mutator (B_2) (B_6)	Brückenschaltung (B_2) (B_6)	rectificador regulado
rectifier	gelijkrichter	Gleichrichter	rectificador
redundancy	redundantie	Redundanz	redundancia
regenerative (braking)	recuperatie (remmen)	rekuperatives Bremsen	frenado regenerativo
regulated process	proces(geregeld)	Regelstrecke	campo de regulación
regulation	regeling	Regelung	regulación
reluctance motor	reluctantiemotor	Reluktanzmotor	motor de reluctancia

residual voltage	restspanning	Restspannung	tensión residual
resistance heated furnace	weerstandsoven	Widerstandsofen	horno de resistencia
resistance	weerstand	Widerstand	resistencia
resistive	resistief	Ohms	resistivo
resonance	resonantie	Resonanz	resonancia
response	responsie	Ansprechverhalten	respuesta
retardation test	uitloopproef	Auslaufversuch	ensayo de desaceleración
retarding torque	vertragingskoppel	Verzögerungsmoment	torque de desaceleracion
reverse (current)	inverse (sperstroom)	negativer Sperrstrom	corriente inversa
reverse	invers	Rückwärtsrichtung	inverso
ring core	ringkern	ringförmiger Kern	núcleo anular
ripple	rimpel	Welligkeit	ondulación
rise time	stijgtijd	Anstiegzeit	tiempo de subida
rise time	doorschakeltijd	Durchschaltzeit	tiempo de ascenso
RMS	kwadratisch gemiddelde	Effektivwert	valor eficaz
root-mean-square value	effectieve waarde	Effektivwert	valor eficaz
rotating field	draaiveld	Drehfeld	campo giratorio
rotation	rotatie	Rotation	rotacional
rotor (frequency)	rotor (frequentie)	Läufer (frequenz)	(frecuencia del) rotor
safety facto	veiligheidsfactor	Sicherheitsfaktor	factor de seguridad
safety transformer	veiligheidstransformator	Sicherheitstransformator	transformador de seguridad
salvo	salvo	Salve	salva
saturation (hard)	verzadiging (hard)	Sättigung (hart)	saturación (duro)
saturation (quasi)	verzadiging (quasi)	Sättigung (quasi)	saturación (casi)
saw tooth (voltage)	zaagtand (spanning)	Sägezahn (spannung)	dente de sierra (tensión)
scale model	schaalmodel	Modell wirkliche Grösse	escala modelo
Scott transformer	Scott-transformator	Transfo in Scott-Schaltung	transformador de Scott
SCR	eénrichtingsthyristor	Einwegthyristor (SCR)	tiristor (SCR)
SCR(controlled rectifier)	gestuurde gelijkrichter	netzgeführter Stromrichter	rectificador controlado
screen grid vacuum tube	schermroosterbuis	Schirmgitterröhre	pentodo
screw tap	draadtap	Schraubenbohrer	(screw tap)
secondary	secondaire	Sekundärseite	secundario (transfo)
selector switch	omschakelaar	Umschalter	conmutador selector
self-excitation	zelfbekrachtiging	Selbsterregung	auto-excitación
self-inductance	zelfinductie	Selbstinduktivität	inductancia propia
semiconductor	halfgeleider	Halbleiter	semiconductor
sensor	opnemer	Messfühler	captador
sensor	voeler	Fühler, Sensor	captor, sensor
series motor	seriemotor	Reihenschlussmotor	motor serie
series resistor	voorschakelweerstand	Vorschaltwiderstand	resistencia adicional
Servomechanism	servomechanisme	Stellantrieb	servomecanismo
set value	ingestelde waarde	Sollwertführungsgrösse	valor especificado
settling time	inslingertijd	Ausregelzeit	tiempo de corrección
shaded pole motor	spleetpoolmotor	Spaltpolmotor	motor de anillos de desfase
short circuit to earth	aardsluiting	Erdschluss	cortocircuito a tierra
short circuit voltage	kortsluitspanning	Kurzschluss-Spannung	tensión de cortocircuito
signal line	signaallijn	Signalleitungen	linea de señal
simulation	simulatie	Simulation	simulación
sine	sinus	Sinus	seno
sine wave	sinusgolf	sinusförmige Welle	onda sinusoidal

single-phase, half-wave	éénfase, enkelzijdig	Einphase, Einweg	monofásico, media onda
six pulse configuration	zespulsig	sechspulsige Schaltung	circuito con seis impulsos
skin effect	skineffect	Skineffekt	efecto pelicular
slew rate	flanksteilheid	Anstiegsgeschwindigkeit	ritmo limitado
slip ring (armature)	sleepring (anker)	Schleifring (läufer)	anillo de fricción (rotor)
slotted armature	trommelanker	genuteter Anker	inducido dentado
smoothed	afgevlakt	geglättet	nivelada
smoothing capacitor	afvlakcondensator	Glättungskondensator	capaidad de filtrado
smoothing coil	afvlakspoel	Glättungsdrossel	bobina de filtrado
snubber (RCD)	RCD-snubber	RCD-snubber	(snubber)
snubber circuit	snubber	Snubber Netzwerk	circuito (snubber)
solar cell array	zonnepaneel	Solarzellenanordnung	campo fotovoltaico
soldering iron	soldeerbout	Lötkolben	soldador
solid state relay	solid state relay	Halbleiterrelais (statisches Relais)	(solid state relay)
source	bron	Quelle	fuente
space charge	ruimtelading	Raumladung	carga de espacio
space vector	ruimtevector	Raumvektor	vector espacial
specification	specificatie	Anforderung (Spezifikation)	especificaciones
spectral response	spectrale gevoeligheid	Spektralempfindlichkeit	sensibilidad espectral
speed profile	snelheidsprofiel	Geschwindigkeitsprofil	perfil de velocidad
speed (rated)	toerental (nominaal)	(Nenn)drehzahl	velocidad (nominal)
speed control	snelheidsregeling	Drehzahlregelung	control de velocidad
square wave	blokgolf	Rechteckwelle	onda cuadrada
squirrel cage induct. Motor	kooiankermotor	Käfigläufer-Induktionsmotor	motor de inducción de jaula
star (connection)	sterschakeling	Stern(schaltung)	conectado en estrella
start	aanlopen	anfahren / anlaufen	arranque
starting (point in) time	aanlooptijdstip	Anfahraugenblick	momento de arranque
starting torque	startkoppel	Durchdrehmoment	par de arranque
start-stop	start-stop	Start-Stopp	(start-stop)
static	statisch	statisch	estático
stator (number of poles)	stator (aantal poolparen)	Ständer (Polpaarzahl)	estator (pares de polos)
steam turbine	stoomturbine	Dampfturbine	turbina de vapor
steaty state (current)	regime (stroom)	Ausgleich (strom)	(corriente) de régimen
step accurateness	stapnauwkeurigheid	Schrittpräzision	escalón exactitud
step function	sprongfunctie	Sprungfunktion	función escalón
step function	stapfunctie	Sprungfunktion	función escalón
step response	stapresponsie	Sprungantwort	repuesta en escalón
stepper motor	stappenmotor	Schrittmotor	motor paso a paso
storage time	opslagtijd	Speicherzeit	tiempo de almacenamiento
strong inductive load	sterk inductief belast	stark induktiv belastet	carga inductiva fuerte
submarine cable	onderzeekabel	Unterseekabel	cable submarino
subscripts	voetletter	Indexzeichen	tipificación
substrate	grondlaag	Grundkörper	sustrato
substrate	substraat (grondlaag)	Substrat (Grundkörper)	sustrato
summing amplifier	sommeerversterker	Summierverstärker	amplificador sumador
supply network	distributienet	Versorgungsnetz	red de distribución
supply voltage	voedingsspanning	Speisespannung	tensión de alimentación
surface	oppervlak	Oberfläche	superficie
surge current	éénmalige piekstroom	Einschaltstromspitze	(surge current)

sustained short-circuit test	kortsluitproef	Dauerkurzschlussversuch	ensayo en cortocircuito
switch (to)	schakelen	schalten	conmutación
switch mode supply	geschakelde voeding	Schaltnetzteil	alimentación conmutado
switched regulator	geschakelde regelaar	geschalteter Regler	regulador controlado
switching (cycle)	schakel (cyclus)	Schalt (zyklus)	(ciclo de) conmutación
symbol	symbool	Symbol (Schaltzeichen)	simbolo
symmetrical optimum	symmetrisch optimum	symmetrisches Optimum	óptimo simetrico
synchron motor	synchrone motor	Drehstrom Synchronmotor	motor sincrón
synchron velocity	synchroon toerental	Synchrondrehzahl	velocidad sincrona
system	systeem	System	sistema
tail current	staartstroom	Schwanzstrom	(tail current)
teletransmission	telecommunicatie	Nachrichtentechnik	teletransmisión
television receiver	TV-ontvanger	Fernsehempfänger	receptor de televisión
temperature (control)	temperatuur(regeling)	Temperatur(regelung)	termorregulado
terminal board	klemmenbord	Klemmbrett	tablero de bornes
test	test	Test	ensayo
theorem	theorema	Theorem	teorema
thermal resistance	thermische weerstand	Wärmewiderstand	resistencia térmica
thermal resistance	warmteweerstand	thermischer Widerstand	resistividad térmica
thermistor	thermistor	Thermistor (Heissleiter)	termistor
three layer diode	drielagendiode	Dreischichtdiode	(three layer diode)
three term controller	PID-regelaar	PID-Regler	regulador de acción propor., integr. y deriv.
three-phase (bridge)	driefasen (brug)	Drehstrom (brücke)	(puente) trifásico
three-phase current	draaistroom	Drehstrom	corriente trifásica
three-phase grid	driefasennet	Drehstromnetz	red trifásico
threshold voltage	drempelspanning	Schwellenspannung	tensión umbral
thyristor (auxiliary)	hulpthyristor	Löschthyristor	auxiliar tiristor
thyristor (disc-)	thyristor (schijf-)	Thyristor(tablette)	Tiristor (disco-)
time interval	tijdsinterval	Zeitlücke	intervalo de tiempo
time ratio control	tijdsverhouding (sturen)	Steuerung mit stellbarem Zeitverhältnis	(time ratio control)
time-constant	tijdconstante	Zeitkonstante	constante de tiempo
to degauss	ontmagnetiseren	entmagnetisieren	desmagnetizar
TO3-package	TO3-omhulling	TO3-Gehäuse	TO3-encapsulado
tooth gear case	tandwielkast	Zahnradkasten	caja de engranaje
topology	topologie	Topologie	topologia
torque control	koppelregeling	Momentregelung	regulación del torque
torque	koppel	(Dreh-)Moment	par
traction motor	tractiemotor	Fahrmotor	motor de tracción
transconductance	transconductantie	Steilheit	transconductancia
transducer	meet(waarde)-omvormer	Messwertumformer	convertidor de medida
transfer function	transferfunctie	Übertragungsfunktion	función de transferencia
transformer (isolating)	scheidingstransformator	Trenntransformator	transformador de separación
transformer (core)	transformator(kern)	Transformator(kern)	núcleo magnético
transformer (electronic)	elektronische transfo	elektronischer Transformator	transf. electrónico
transformer (one coil)	spaartransformator	Autotransformator	transformador de ahorro
transformer (supply)	(voedings)transformator	(Versorgungs-)Transformator	transformador (de distribución)

transformer coupling	transformatorkoppeling	Transformatorkopplung	acoplamiento por transformador
transformer ratio	transformatieverhouding	Übersetzungsverhältnis	relación de transformación
transient(term)	overgangs(term)	Ausgleich(term)	(término de) transición
transmitter	zender	Sender	transmisor
trapezium converter	trapeziumconvertor	Trapezumrichter	trapecio convertidor
triac	triac	Zweiwegthyristor	tiristor triac
triangular wave	driehoeksgolf	Dreieckspannung	onda triángula
trigger current (SCR, triac)	triggerstroom	Zündstrom	corriente de control
turn-off losses	uitschakelverliezen	Abschaltverluste	perdidas de cortar
turn-off behaviour	uitschakelgedrag	Ausschaltverhalten	conducta de desconexión
turn-off characteristics	turn-off eigenschappen	Abschalteigenschaften	(turn-off characteristics)
turn-off circuit	uitschakelkring	Abschaltkreis	circuito de desconexión
turn-off gain	uitschakelversterking	Abschaltverstärkung	(turn-off gain)
turn-off time	uitschakeltijd	Verzögerungszeit	tiempo de cortar
turn-off	uitschakelen	abschalten	desconexión
turn-on losses	aanschakelverliezen	Anschaltverluste	perdidas de ncendido
turn-on time	aanschakeltijd	Anspreichzeit	tiempo de encendido
turn-on time	inschakeltijd	Einschaltzeit	tiempo de encendido
turns ratio	windingsverhouding	Windungszahlenverhältnis	relación de las vueltas de las bobinas
type indication	type-aanduiding	Typkennzeichen	indice de tipificación
uninterruptable (UPS)	onderbrekingsvrij (UPS)	unterbrechungslos(e) (UPS)	(UPS)
unipolar transistor	unipolaire transistor	Unipolartransistor	transistor unipolar
universal motor	universele motor	Universalmotor	motor universal
useable emf	meewerkende emk	mithelfende Emk	emf de cooperación
vacuum tube	elektronenbuis	Elektronenröhre	tubo electronico
valence band	valentieband	Valenzband	banda de valencia
vector control	vectorregeling	Vektorregelung	control vectorial
very inductive	zeer inductief	sehr induktiv	fuerte inductivo
voltage drop	spanningsval	Spannungsabfall	caida de tensión
voltage drop	spanningsverlies	Spannungsabfall	bajada de tensión
voltage peak (line transient)	spanningspiek	Spannungsspitze	cresta de tensión
voltage	spanning	Spannung	tensión
water heater (boiler)	waterverwarmer	Heisswasserspeicher (Boiler)	calentador
wave guide	golfgeleider	Hohlleiter	guiaondas
wave	golf	Welle	onda
waveform	golfvorm	Wellenform	forma de onda
welding transformer	lastransformator	Schweissung (transformator)	transformador de soldadura
wind park	windpark	Windpark	parque de viento
wind turbine	windturbine	Windturbine	turbina eólica
winding	wikkeling	Wicklung	devanado
work coil	werkspoel	Heizinduktor	inductor de calentamiento
working point	instelpunt	Arbeitspunkt	punto defuncionamiento
working quadrant	werkingskwadrant	Arbeitsquadrant	cuadrante de trabajo
Zener knee voltage	Zenerspanning	Zenerspannung	tension de Zener
zero crossing	nulpunt (nuldoorgang)	Nulldurchgang	cruce de cero
zerocross reference	referentie-nulpunt	Referenz-Nulldurchgang	(zerocross reference)
zigzag connection	zigzagschakeling	Zickzackschaltung	conexión en zigzag

INDEX

D

E

F

G

H

I

J

L

M

U

V

W

Z

Kolibri 360 GmbH
Ferdinand-Jühlke-Straße 7
99095 Erfurt, Deutschland
produktsicherheit@kolibri360.de